CAMBRIDGE MONOGRAPHS ON APPLIED AND COMPUTATIONAL MATHEMATICS

17 Scattered Data Approximation

The *Cambridge Monographs on Applied and Computational Mathematics* reflects the crucial role of mathematical and computational techniques in contemporary science. The series publishes expositions on all aspects of applicable and numerical mathematics, with an emphasis on new developments in this fast-moving area of research.

State-of-the-art methods and algorithms as well as modern mathematical descriptions of physical and mechanical ideas are presented in a manner suited to graduate research students and professionals alike. Sound pedagogical presentation is a prerequisite. It is intended that books in the series will serve to inform a new generation of researchers.

Also in this series:

1. A Practical Guide to Pseudospectral Methods, *Bengt Fornberg*
2. Dynamical Systems and Numerical Analysis, *A. M. Stuart and A. R. Humphries*
3. Level Set Methods and Fast Marching Methods, *J. A. Sethain*
4. The Numerical Solution of Integral Equations of the Second Kind, *Kendall E. Atkinson*
5. Orthogonal Rational Functions, *Adhemar Bultheel, Pablo González-Vera, Erik Hendiksen, and Olav Njastad*
6. The Theory of Composites, *Graeme W. Milton*
7. Geometry and Topology for Mesh Generation *Herbert Edelsfrunner*
8. Schwarz-Christoffel Mapping *Tofin A. Dirscoll and Lloyd N. Trefethen*
9. High-Order Methods for Incompressible Fluid Flow, *M. O. Deville, P. F. Fischer and E. H. Mund*
10. Practical Extrapolation Methods, *Avram Sidi*
11. Generalized Riemann Problems in Computational Fluid Dynamics, *Matania Ben-Artzi and Joseph Falcovitz*
12. Radial Basis Functions: Theory and Implementations, *Martin D. Buhmann*
13. Iterative Krylov Methods for Large Linear Systems, *Henk A. van der Vorst*
14. Simulating Hamiltonian Dynamics, *Ben Leimkuhler & Sebastian Reich*
15. Collocation Methods for Volterra Integral and Related Functional Equations, *Hermann Brunner*
16. Topology for Computing, *Afra J. Zomorodian*

Scattered Data Approximation

HOLGER WENDLAND
Institut für Numerische und Angewandte Mathematik
Universität Göttingen

CAMBRIDGE UNIVERSITY PRESS
Cambridge, New York, Melbourne, Madrid, Cape Town, Singapore,
São Paulo, Delhi, Dubai, Tokyo

Cambridge University Press
The Edinburgh Building, Cambridge CB2 8RU, UK

Published in the United States of America by Cambridge University Press, New York

www.cambridge.org
Information on this title: www.cambridge.org/9780521131018

First published 2005
This digitally printed version 2010

A catalogue record for this publication is available from the British Library

Library of Congress Cataloguing in Publication data

Wendland, Holger, 1968–
Scattered data approximation / Holger Wendland.
p. cm. – (Cambridge monographs on applied and computational mathematics; 17)
Includes bibliographical references and index.
ISBN 0 521 84335 9
1. Approximation theory. 2. Multivariate analysis. I. Title. II. Series.
QA221.W46 2004
511′.42–dc22 2004049270

ISBN 978-0-521-84335-5 Hardback
ISBN 978-0-521-13101-8 Paperback

Contents

Preface

Scattered data approximation is a recent, fast growing research area. It deals with the problem of reconstructing an unknown function from given scattered data. Naturally, it has many applications, such as terrain modeling, surface reconstruction, fluid–structure interaction, the numerical solution of partial differential equations, kernel learning, and parameter estimation, to name a few. Moreover, these applications come from such different fields as applied mathematics, computer science, geology, biology, engineering, and even business studies.

This book is designed to give a thorough, self-contained introduction to the field of multivariate scattered data approximation without neglecting the most recent results.

Having the above-mentioned applications in mind, it immediately follows that any competing method has to be capable of dealing with a very large number of data points in an arbitrary number of space dimensions, which might bear no regularity at all and which might even change position with time.

Hence, in my personal opinion a true scattered data method has to be *meshless*. This is an assumption that might be challenged but it will be the fundamental assumption throughout this book. Consequently, certain methods, that generally require a mesh, such as those using wavelets, multivariate splines, finite elements, box splines, etc. are immediately ruled out. This does not at all mean that such methods cannot sometimes be used successfully in the context of scattered data approximation; on the contrary, it just explains why these methods are not discussed in this book. The requirement of being truly meshless reduces the number of efficient multivariate methods dramatically. Amongst them, radial basis functions, or, more generally, approximation by (conditionally) positive definite kernels, the moving least squares approximation, and, to a certain extent, partition-of-unity methods, appear to be the most promising. Because of this, they will be given a thorough treatment.

A brief outline of the book is as follows. In Chapter 1 we discuss a few typical applications and then turn to natural cubic splines as a motivation for (conditionally) positive definite kernels. The following two chapters can be seen as an introduction to the problems of multivariate approximation theory. Chapter 4 is devoted to the moving least squares approximation. In Chapter 5 we collect certain auxiliary results necessary for the rest of the book, and the impatient or advanced reader might skip the details and come back to this chapter whenever necessary. The theory of radial basis functions starts with the discussion of positive definite and completely monotone functions in Chapters 6 and 7 and continues with conditionally positive definite and compactly supported functions. In Chapters 10 and 11 we deal with the error analysis of the approximation process. In the following chapter we start the numerical part of this book by discussing the stability of the process. After a short interplay on optimal recovery in Chapter 13, we continue the numerical treatment with chapters on data structures and efficient algorithms, where partition-of-unity methods are investigated also. In Chapter 16 we deal with generalized interpolation, which is important if, for example, partial differential equations are to be solved numerically using scattered data methods, and in Chapter 17 we consider applications to the sphere.

It is impossible to thank everyone who has helped me in the writing of this book. But it is my pleasure to point out at least a few of those persons without diminishing the respect and gratitude I owe to those not mentioned. First of all, I have to thank R. Schaback from the University of Göttingen and J. D. Ward and F. J. Narcowich from Texas A&M University. It was they who first attracted my attention to the field of scattered data approximation, and they have had a great influence on my point of view. Further help either by discussion or by proofreading parts of the text has been provided by A. Beckert, R. Brownlee, G. Fasshauer, P. Hähner, J. Miranda, and R. Opfer. Finally, I am more than grateful to Cambridge University Press, in particular to David Tranah and Ken Blake for their kind and efficient support.

1
Applications and motivations

In practical applications over a wide field of study one often faces the problem of reconstructing an unknown function f from a finite set of discrete data. These data consist of data sites $X = \{x_1, \ldots, x_N\}$ and data values $f_j = f(x_j)$, $1 \le j \le N$, and the reconstruction has to approximate the data values at the data sites. In other words, a function s is sought that either *interpolates* the data, i.e. that satisfies $s(x_j) = f_j$, $1 \le j \le N$, or at least *approximates* the data, $s(x_j) \approx f_j$. The latter case is in particular important if the data contain noise.

In many cases the data sites are scattered, i.e. they bear no regular structure at all, and there is a very large number of them, easily up to several million. In some applications, the data sites also exist in a space of very high dimensions. Hence, for a unifying approach methods have to be developed which are capable of meeting this situation. But before pursuing this any further let us have a closer look at some possible applications.

1.1 Surface reconstruction

Probably the most obvious application of scattered data interpolation and approximation is the reconstruction of a surface $\mathcal{S}$. Here, it is crucial to distinguish between explicit and implicit surfaces. *Explicit* surfaces play an important role in terrain modeling, for example. They can be represented as the graph of a function $f : \Omega \to \mathbb{R}$ defined on some region $\Omega \subseteq \mathbb{R}^d$, where d is in general given by $d = 2$. Staying with the terminology of terrain modeling, the data sites $X \subseteq \Omega$ depict certain points on a map, while a data value $f_j = f(x_j)$ describes the height at the point x_j. The data sites might form a regular grid, they might be situated on isolines (as in Figure 1.1), or they might have no structure at all. The region Ω itself might also carry some additional information; for example, it could represent the earth. Such additional information should be taken into account during the reconstruction process.

The reconstruction of an *implicit* surface, or more precisely of a compact, orientable manifold, is even more demanding. Such surfaces appear for example as sculptures, machine parts, and archaeological artifacts. They are often digitized using laser scanners, which easily produce huge *point clouds* $X = \{x_1, \ldots, x_N\} \subseteq \mathcal{S}$ consisting of several million points in $\mathbb{R}^3$. In this situation, the surface $\mathcal{S}$ can no longer be represented as the graph of a single

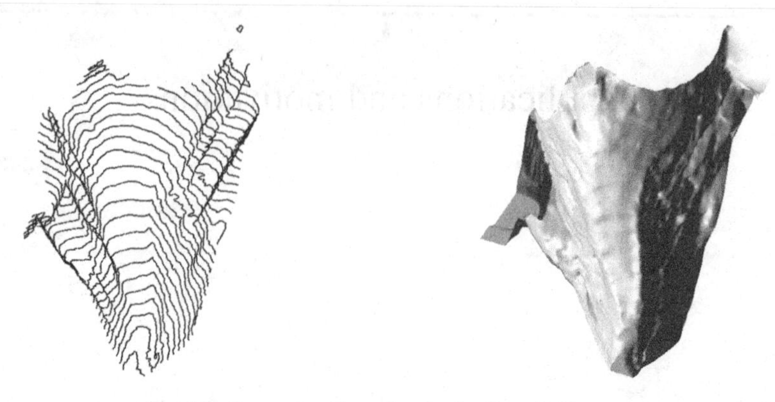

Fig. 1.1 Reconstruction of a glacier from isolines.

Fig. 1.2 Reconstruction (on the right) of the Stanford dragon (on the left).

function f. There are in the main two different approaches to building accurate models for implicit surfaces. In the first approach, one tries to find local *parameterizations* of the object that allow an efficient rendering. However, for complicated models (such as the dragon shown in Figure 1.2) this approach is limited. In the second approach, one tries to describe $\mathcal{S}$ as the zero-level set of a function F, i.e. $\mathcal{S} = \{x \in \Omega : F(x) = 0\}$. Such an implicit representation easily delivers function-based operations, for example shape blending or deformation or any other constructive solid geometry (CSG) operation such as the union, difference, or intersection of two or more objects.

The function F can be evaluated everywhere, which allows stepless zooming and smooth detail-extraction. Furthermore, it gives, to a certain extent, a measure of how far away a point $x \in \Omega$ is from the surface. Moreover, the surface normal is determined by the gradient of F whenever the representation is smooth enough.

The price we have to pay for such flexibility is that an implicit surface does not automatically lead to a fast visualization. An additional step is necessary, which is normally provided by either a ray-tracer or a polygonizer. But, for both, sufficiently good and appropriate solutions exist. Since our measured point cloud X is a subset of the surface $\mathcal{S}$ we are looking for an approximate solution s that satisfies $s(x_j) = 0$ for all $x_j \in X$. Obviously these interpolation conditions do not suffice to determine an accurate approximation to the surface, since, for example, the zero function satisfies them. The remedy for this problem is to add additional *off-surface* points. To make this approach work, we assume that our surface is the boundary of a compact set and that the function F can be chosen such that F is positive inside and negative outside that set. We also need surface normals to the unknown surface. If the data comes from a mesh or from a laser scanner that provides also normal information via its position during the scanning process these normals are immediately to hand. Otherwise, they have to be estimated from the point cloud itself, which can be done in two steps. In the first step, for each point $x_j \in X$ we search its $K \ll N$ nearest neighbors in X and try to determine a local tangent plane. This can be done by a *principal component analysis*. Let us assume that $\mathcal{N}(x_j)$ contains the indices of these neighbors. Then we compute the *center of gravity* of $\{x_k : k \in \mathcal{N}(x_j)\}$, i.e. $\hat{x}_j := K^{-1} \sum_{k \in \mathcal{N}(x_j)} x_k$, and the associated *covariance matrix*

$$\mathrm{Cov}\,(x_j) := \sum_{k \in \mathcal{N}(x_j)} (x_k - \hat{x}_j)(x_k - \hat{x}_j)^T \in \mathbb{R}^{3\times 3}.$$

The eigenvalues of this positive semi-definite matrix can be computed numerically or even analytically. They indicate how closely the neighborhood $\{x_k : k \in \mathcal{N}(x_j)\}$ of x_j determines a plane. To be more precise, if we have two eigenvalues that are close together and a third one, which is significantly smaller than the others, then the eigenvectors for the first two eigenvalues determine the plane, while the eigenvector for the smallest eigenvalue determines the normal to this plane. Hence, we have a tool for not only determining the normal but also deciding whether a normal can be fitted at all.

The second step deals with orienting consistently the normals just created. If two data points x_j and x_k are close then their associated normalized normals η_j and η_k must point in nearly the same direction, which means that $\eta_j^T \eta_k \approx 1$. This relation should hold for all points that are sufficiently close. To make this more precise, we use graph theory. First, we build a *Riemann graph*. This graph has a vertex for every normal η_j and an edge $e_{j,k}$ between the vertices of η_j and η_k if and only if $j \in \mathcal{N}(x_k)$ or $k \in \mathcal{N}(x_j)$. The *cost* or *weight* $w(e_{j,k})$ of such an edge measures the deviation of the normals η_j and η_k; for example, we could choose $w(e_{j,k}) = 1 - |\eta_j^T \eta_k|$. Hence, the normals are taken to be consistently oriented if we can find directions $b_j \in \{-1, 1\}$ such that $\sum_{e_{j,k}} b_j b_k w(e_{j,k})$ is minimized. Unfortunately, it is possible to show that this problem is NP-hard and hence that we can only find an approximate solution. The idea is simply to start with an arbitrary normal and then to propagate the orientation to neighboring normals. To this end, we compute the *minimal spanning tree* or *forest* for the Riemann graph. Since the number of edges in this graph is proportional to N, any reasonable algorithm for this problem, for example Kruskal's algorithm, will work fine in an acceptable amount of time. After that, we propagate the orientations by traversing the minimal spanning tree.

Once we have oriented the normals, this allows us to extend the given data sets by off-surface points. This can be done by for example adding one point along each normal on the outside and one on the inside of the surface. Special care is necessary to avoid the situation where an outside point belonging to one normal is actually an interior point in another part of the surface or that a supposedly interior point is so far away from its associated surface point that it is actually outside the surface at another place. The associated function values that s should attain are chosen to be proportional to the signed distance of the point from the surface.

Another possible way of adding off-surface points is based on the following fact. Suppose that x is a point which should be added. If x_j denotes its nearest neighbor in X and if X is a sufficiently dense sample of $\mathcal{S}$, then x_j comes close to the projection of x onto $\mathcal{S}$. Hence if x_j is approximately equal to x then the latter is a point of $\mathcal{S}$ itself. Otherwise, if the angle between the line through x_j and x on the one hand and the normal η_j (pointing outwards) on the other hand is less than 90 degrees then the point is outside the surface; if the angle is greater than 90 degrees then it is inside the surface.

After augmenting our initial data set by off-surface points, we are now back to a classical interpolation or approximation problem.

1.2 Fluid–structure interaction in aeroelasticity

Aereolasticity is the science that studies, among other things, the behavior of an elastic aircraft during flight. This behavior is influenced by the interaction between the deformations of the elastic structure caused by the fluid flow, and the impact that the aerodynamic forces would have on a rigid structural framework. To model these different aspects in a physically correct manner, different models have been developed, adapted to the specific problems.

The related aeroelastic problem can be described in a coupled-field formulation, where the interaction between the fluid and structural models is limited to the exchange of boundary conditions. This *loose* coupling has the advantage that each component of the coupled problem can be handled as an isolated entity. However, the challenging task is to reconcile the benefits of this isolated view with a realistic treatment of the new physical effects arising from the interaction.

Let us make this more precise. Suppose at first that we are interested only in computing the flow field around a given aircraft. This can be modeled mathematically by the Navier–Stokes or the Euler equations, which can be solved numerically using for example a finite-volume code. Such a solver requires a detailed model of the aircraft and its surroundings. In particular, the surface of the aircraft has to be rendered with a very high resolution, as indicated in the right-hand part of Figure 1.3. Let us suppose that our solver has computed a solution, which consists of a velocity field and a pressure distribution. For the time being, we are not interested in the problem of how such a solution can be computed. For us, it is crucial that the pressure distribution creates loads on the aircraft, which might and probably will lead to a deformation. So the next step is to compute the deformation from the loads or forces acting on the aircraft.

Obviously, though, a model having a fine resolution of the surface of the aircraft is not necessary for describing its structure; this might even impede the numerical stability. Hence,

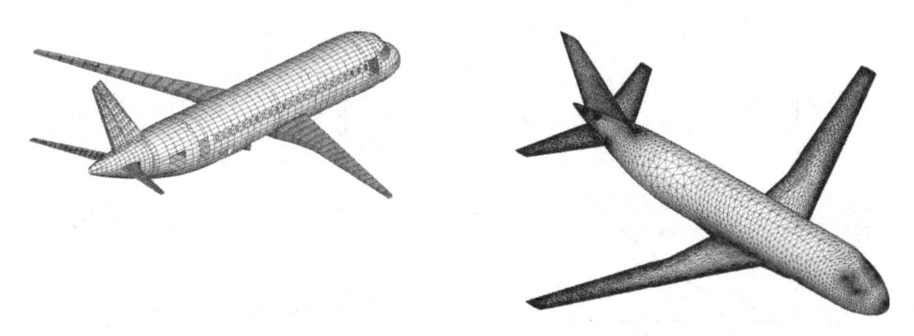

Fig. 1.3 The structural and aerodynamical model of a modern aircraft.

another model is required which is better suited to describing the structural deformation, for example the one shown in Figure 1.3 on the left. Again, along with the model comes a partial differential equation, this time from elasticity theory, which can again be solved, for example by finite elements. But before this can be done, the loads have to be transferred from one mesh to the other in a physically reasonable way. If this has been done and the deformation has been computed then we are confronted with another coupling problem. This time, the deformations have to be transferred from the structural to the aerodynamical model. If all these problems can be solved we can start to iterate the process until we find a steady state, which presumably exists.

Since we have the aerodynamical model, the structural model, and the coupling problem, one usually speaks in this context of a *three-field formulation*. As we said earlier, here we are interested only in the coupling process, which can be described as a scattered data approximation problem, as follows. Suppose that X denotes the nodes of the structural mesh and Y the nodes of the aerodynamical mesh (neither actually has to be a mesh). To transfer the deformations $u(x_j) \in \mathbb{R}^3$ from X to Y we need to find a vector-valued interpolant $s_{u,X}$ satisfying $s_{u,X}(x_j) = u(x_j)$. Then the deformations of Y are given simply by $s_{u,X}(y_j)$, $y_j \in Y$. Conversely, if $f(y_j) \in \mathbb{R}$ denotes the load at $y_j \in Y$ then we need another function $s_{f,Y}$ to interpolate f in Y. The loads on the mesh X are again simply given by evaluation at X. A few more things have to be said. First of all, if the loads are constant or if the displacements come from a linear transformation, this situation should be recovered exactly, which means that our interpolation process has to be exact for linear polynomials. Furthermore, certain physical entities such as energy and work should be conserved. This means at least that

$$\sum_{y \in Y} f(y) = \sum_{x \in X} s_{f,Y}(x)$$

and

$$\sum_{y \in Y} f(y) s_{u,X}(y) = \sum_{x \in X} s_{f,Y}(x) u(x),$$

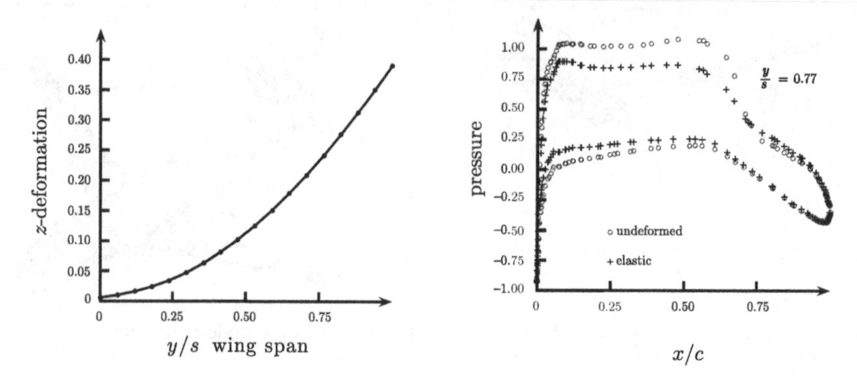

Fig. 1.4 Steady state of the deformed aircraft.

where the last equation is to be taken component-wise. If the models differ too much then both equations have to be understood in a more local sense. However, these equations make it obvious that in certain applications more has to be satisfied than just simple point evaluations. It is important to note that interpolation is crucial in this process since otherwise each coupling step would result in a loss of energy.

The advantage of this *scattered data* approach is that it allows us to couple any two models that have at least some node information. There is no additional information such as the elements or connectivity of the nodes involved. Moreover, the two models can be quite different. It often happens that the boundary of the aerodynamical aircraft has no joint node with the structural model. The latter might even degenerate into a two-dimensional object.

Figure 1.4 shows a typical result for the example from Figure 1.3 based on a speed $M = 0.8$, an angle of attack $\alpha = -0.087^\circ$, and an altitude $h = 10\,498$ meters. On the left the deformation of a wing is shown, while the right-hand graph gives the negative pressure distribution at 77% wing span, for a static and an elastic computation. The difference between the two pressure distributions indicates that elasticity causes a loss of buoyancy, which can become critical for highly flexible structures, as found for example in the case of a large civil aircraft.

It should, be clear that the coupling process described here is not limited to the field of aeroelasticity. It can be applied in any situation where a given problem is decomposed into several subproblems, provided that these subproblems exchange data over specified nodes.

1.3 Grid-free semi-Lagrangian advection

In this section we will discuss briefly how the scattered data approximation can be used to solve advection equations. For simplicity, we restrict ourselves here to the two-dimensional case and to the *transport equation*, which is given by

$$0 = \frac{\partial}{\partial t}u(x, y, t) + a_1(x, y)\frac{\partial}{\partial x}u(x, y, t) + a_2(x, y)\frac{\partial}{\partial y}u(x, y, t). \tag{1.1}$$

It describes, for example, the advection of a fluid with velocity field $a = (a_1, a_2)$ and it will serve us as a model problem straight away. Suppose that $(x(t), y(t))$ describes a curve for which the function $\widetilde{u}(t) := u(x(t), y(t), t)$ is constant, i.e. $\widetilde{u}(t) = \text{const}$. Such a curve is called a *characteristic curve* for (1.1). Differentiating $\widetilde{u}$ yields

$$0 = \frac{\partial u}{\partial t} + \dot{x}(t)\frac{\partial u}{\partial x} + \dot{y}(t)\frac{\partial u}{\partial y},$$

where $\dot{x} = dx/dt$. The similarity to (1.1) allows us to formulate the following approximation scheme for solving the transport equation (1.1) with initial data given by a known function u_0. Suppose that we know the distribution u at time t_n and at sites $X = \{(x_1, y_1), \ldots, (x_N, y_N)\}$ approximately, meaning that we have a vector $u^{(n)} \in \mathbb{R}^N$ with $u_j^{(n)} \approx u(x_j, y_j, t_n)$. To find the values of u at site (x_j, y_j) and time t_{n+1} we first have to find the *upstream point* (x_j^-, y_j^-) with $c := u(x_j^-, y_j^-, t_n) = u(x_j, y_j, t_{n+1})$ and then have to estimate the value c from the values of u at X and time t_n. Hence, in the first step we have to solve N ordinary differential equations

$$(\dot{\xi}_j, \dot{\eta}_j) = a(\xi_j, \eta_j), \qquad 1 \le j \le N,$$

with initial value $(\xi(t_{n+1}), \eta(t_{n+1})) = (x_j, y_j)$. The upstream point is the solution at t_n, i.e. $(x_j^-, y_j^-) = (\xi(t_n), \eta(t_n))$. Since this point will in general not be contained in X, the value $u(x_j^-, y_j^-, t_n)$ has to be estimated from $u^{(n)}$. This can be written as an interpolation problem. We need to find a function s_u that satisfies $s_u(x_j) = u_j^{(n)}$ for $1 \le j \le N$.

The method just described is called a *semi-Lagrangian method*. It is obviously not restricted to a two-dimensional setting. It also applies to advection equations other than the transport problem (even nonlinear ones), but then an interpolatory step might also be necessary when solving the ordinary differential equations.

Moreover, it is not necessary at all to use the same set of sites X in each time step. It is much more appropriate to adapt the set X as required.

Finally, if the concept of scattered data approximation is generalized to allow also functionals other than pure point-evaluation functionals, there are plenty of other possibilities for solving partial differential equations. We will discuss some of them later in this book.

1.4 Learning from splines

The previous sections should have given some insight into the application of scattered data interpolation and approximation in the multivariate case.

To derive some concepts, we will now have a closer look at the univariate setting. Hence we will suppose that the data sites are ordered as follows,

$$X : a < x_1 < x_2 < \ldots < x_N < b, \tag{1.2}$$

and that we have certain data values $f_1, \ldots, f_N$ to be interpolated at the data sites. In other words, we are interested in finding a continuous function $s : [a, b] \to \mathbb{R}$ with

$$s(x_j) = f_j, \qquad 1 \le j \le N.$$

At this point it is not necessary that the data values $\{f_j\}$ actually stem from a function f, but we will keep this possibility in mind for reasons that will become clear later.

In the univariate case it is well known that s can be chosen to be a polynomial p of degree at most $N - 1$, i.e. $p \in \pi_{N-1}(\mathbb{R})$. Or, more generally, if a *Haar space* $S \subseteq C(\mathbb{R})$ of dimension N is fixed then it is always possible to find a unique interpolant $s \in S$. In this context the space S has the remarkable property that it depends only on the number of points in X and not on any other information about the data sites, let alone about the data values. Thus it would be reasonable to look for such spaces also in higher dimensions. Unfortunately, a famous theorem of Mairhuber and Curtis (Mairhuber [115], see also Chapter 2) states that this is impossible. Thus if working in space dimension $d \ge 2$ it is impossible to fix an N-dimensional function space beforehand that is appropriate for *all* sets of N distinct data sites. However, probably no one with any experience in approximation theory would, even in the univariate case, try to interpolate a hundred thousand points with a polynomial.

The bottom line here is that for a successful interpolation scheme in $\mathbb{R}^d$ either conditions on the involved points have to be worked out, in such a way that a stable interpolation with polynomials is still possible, or the function space has to depend on the data sites. The last concept is also well known in the univariate case. It is a well-established fact that a large data set is better dealt with by splines than by polynomials. In contrast to polynomials, the accuracy of the interpolation process using splines is not based on the polynomial degree but on the spacing of the data sites. Let us review briefly properties of univariate splines in the special case of cubic splines. The set of cubic splines corresponding to a decomposition (1.2) is given by

$$S_3(X) = \{s \in C^2[a, b] : s|[x_i, x_{i+1}] \in \pi_3(\mathbb{R}), 0 \le i \le N\}, \tag{1.3}$$

where $x_0 := a$, $x_{N+1} := b$. It consists of all twice differentiable functions that coincide with cubic polynomials on the intervals given by X. The space $S_3(X)$ has dimension $\dim(S_3(X)) = N + 4$, so that the interpolation conditions $s(x_i) = f_i$, $1 \le i \le N$, do not suffice to guarantee a unique interpolant. Different strategies are possible to enforce uniqueness and one of these is given by the concept of natural cubic splines. The set of natural cubic splines

$$\mathcal{N}S_3(X) = \{s \in S_3(X) : s|[a, x_1], s|[x_N, b] \in \pi_1(\mathbb{R})\}$$

consists of all cubic splines that are linear polynomials on the outer intervals $[a, x_1]$ and $[x_N, b]$. It is easy to see that a cubic spline s is a natural cubic spline if and only if it satisfies $s''(x_1) = s^{(3)}(x_1) = 0$ and $s''(x_N) = s^{(3)}(x_N) = 0$. Since we have imposed four additional conditions it is natural to assume that the dimension of $\mathcal{N}S_3(X)$ is $\dim(\mathcal{N}S_3(X)) = N$, which is indeed true. Even more, it can be shown that the initial interpolation problem has a unique solution in $\mathcal{N}S_3(X)$. For this and all the other results on splines we refer the reader to Greville's article [75] or to the books by Schumaker [175] and de Boor [43].

This is not the end of the story, however; splines have several important properties and we state some of them for the cubic case.

(1) They are piecewise polynomials.
(2) An interpolating natural cubic spline satisfies a minimal norm property. This can be formulated as follows. Suppose f comes from the Sobolev space $H^2[a,b]$, i.e. $f \in C[a,b]$ has weak first- and second-order derivatives also in $L_2[a,b]$. (We will give a precise definition of this later on). Assume further that f satisfies $f(x_j) = f_j$, $1 \le j \le N$. If $s_{f,X}$ denotes the natural cubic spline interpolant then

$$(f'' - s''_{f,X}, s''_{f,X})_{L_2[a,b]} = 0.$$

This leads immediately to the Pythagorean equation

$$\|f'' - s''_{f,X}\|^2_{L_2[a,b]} + \|s''_{f,X}\|^2_{L_2[a,b]} = \|f''\|^2_{L_2[a,b]},$$

which means that the natural cubic spline interpolant is that function from $H^2[a,b]$ that minimizes the semi-norm $\|f''\|_{L_2[a,b]}$ under the conditions $f(x_j) = f_j$, $1 \le j \le N$.
(3) They possess a local basis (B-splines). These basis functions can be defined in various ways: by recursion, by divided differences, or by convolution.

Of course, this list gives only a few properties of splines. For more information, we refer the interested reader to the previously cited sources on splines.

The most dominant feature of splines, which has contributed most to their success, is that they are piecewise polynomials. This feature together with a local basis not only allows the efficient computation and evaluation of spline functions but also is the key ingredient for a simple error analysis. Hence, the natural way of extending splines to the multivariate setting is based on this property. To this end, a bounded region $\Omega \subseteq \mathbb{R}^d$ is partitioned into essentially disjoint subregions $\{\Omega_j\}_{j=1}^N$. Then the spline space consists simply of those functions s that are piecewise polynomials on each patch Ω_j and that have smooth connections on the boundaries of two adjacent patches. In two dimensions the most popular partition of a polygonal region is based on a triangulation. Even in this simplest case, however, the dimension of the spline space is in general unknown (see Schumaker [176]). Moreover, when coming to higher dimensions it is not at all clear what an appropriate replacement for the triangulation would be. Hence, even if substantial progress has been made in the two-dimensional setting, the method is not suited for general dimensions. Another possible generalization to the multivariate setting is based on the third property. In particular a construction based on convolution has led to the theory of Box splines (see de Boor *et al.* [44]). Again, even the two-dimensional setting is tough to handle, not to speak of higher-dimensional cases.

Hence, we want to take the second property as the motivation for a framework in higher dimensions. This approach leads to a remarkably beautiful theory, where all space dimensions can be handled in the same way. Since the resulting approximation spaces no longer consist of piecewise polynomials, we do not want to call the functions splines. The buzz phrase, which has become popular in this field, is *radial basis functions*.

To get an idea of radial basis functions let us stick a little longer with natural cubic splines. It is well known that the set $S_3(X)$ has the basis $(\cdot - x_j)_+^3$, $1 \le j \le N$, plus an arbitrary basis for $\pi_3(\mathbb{R})$. Here, x_+ takes the value of x for nonnegative x and zero in the other case. Hence, every $s \in \mathcal{N}S_3(X)$ has a representation of the form

$$s(x) = \sum_{j=1}^{N} \alpha_j (x - x_j)_+^3 + \sum_{j=0}^{3} \beta_j x^j, \qquad x \in [a, b]. \tag{1.4}$$

Because s is a natural spline we have the additional information that s is linear on the two outer intervals. On $[a, x_1]$ it has the representation $s(x) = \sum_{j=0}^{3} \beta_j x^j$ so that necessarily $\beta_2 = \beta_3 = 0$. Thus, (1.4) becomes

$$s(x) = \sum_{j=1}^{N} \alpha_j (x - x_j)_+^3 + \beta_0 + \beta_1 x, \qquad x \in [a, b]. \tag{1.5}$$

To derive the representation of s on $[x_N, b]$ we simply have to remove all subscripts $+$ on the functions $(\cdot - x_j)_+^3$ in (1.5). Expanding these cubics and rearranging the sums leads to

$$s(x) = \sum_{\ell=0}^{3} \binom{3}{\ell} (-1)^{3-\ell} \left(\sum_{j=1}^{N} \alpha_j x_j^{3-\ell} \right) x^\ell + \beta_0 + \beta_1 x, \qquad x \in [x_N, b].$$

Thus, for s to be a natural spline, the coefficients of s have to satisfy

$$\sum_{j=1}^{N} \alpha_j = \sum_{j=1}^{N} \alpha_j x_j = 0. \tag{1.6}$$

This is a first characterization of natural cubic splines. But we can do more. Using the identity $x_+^3 = (|x|^3 + x^3)/2$ leads, because of (1.6), to

$$\begin{aligned} s(x) &= \sum_{j=1}^{N} \frac{\alpha_j}{2} |x - x_j|^3 + \sum_{j=1}^{N} \frac{\alpha_j}{2} (x - x_j)^3 + \beta_0 + \beta_1 x \\ &= \sum_{j=1}^{N} \frac{\alpha_j}{2} |x - x_j|^3 + \sum_{\ell=0}^{3} \frac{1}{2} \binom{3}{\ell} (-1)^{3-\ell} \sum_{j=1}^{N} \alpha_j x_j^{3-\ell} x^\ell + \beta_0 + \beta_1 x \\ &= \sum_{j=1}^{N} \widetilde{\alpha}_j |x - x_j|^3 + \widetilde{\beta}_0 + \widetilde{\beta}_1 x, \end{aligned}$$

with $\widetilde{\alpha}_j = \frac{1}{2}\alpha_j$, $1 \le j \le N$, and $\widetilde{\beta}_0 = \beta_0 - \frac{1}{2} \sum \alpha_j x_j^3$, $\widetilde{\beta}_1 = \beta_1 + \frac{3}{2} \sum \alpha_j x_j^2$.

Proposition 1.1 *Every natural cubic spline s has a representation of the form*

$$s(x) = \sum_{j=1}^{N} \alpha_j \phi(|x - x_j|) + p(x), \qquad x \in \mathbb{R}, \tag{1.7}$$

where $\phi(r) = r^3$, $r \ge 0$, and $p \in \pi_1(\mathbb{R})$. The coefficients $\{\alpha_j\}$ have to satisfy (1.6). Vice versa, for every set $X = \{x_1, \dots, x_N\} \subseteq \mathbb{R}$ of pairwise distinct points and for every $f \in \mathbb{R}^N$

there exists a function s of the form (1.7) with (1.6) that interpolates the data, i.e. $s(x_j) = f_j$, $1 \le j \le N$.

This is our first result on radial basis functions. The resulting interpolant is up to a low-degree polynomial a linear combination of shifts of a radial function $\Phi = \phi(|\cdot|)$. The function is called radial because it is the composition of a univariate function with the Euclidean norm on $\mathbb{R}$. In this sense it generalizes to $\mathbb{R}^d$, where the name "radial" becomes even more apparent. A radial function is constant on spheres in $\mathbb{R}^d$. A straightforward generalization is to build interpolants of the form

$$s(x) = \sum_{j=1}^{N} \alpha_j \phi(\|x - x_j\|_2) + p(x), \qquad x \in \mathbb{R}^d, \tag{1.8}$$

where $\phi : [0, \infty) \to \mathbb{R}$ is a univariate fixed function and $p \in \pi_{m-1}(\mathbb{R}^d)$ is a low-degree d-variate polynomial. The additional conditions on the coefficients now become

$$\sum_{j=1}^{N} \alpha_j q(x_j) = 0 \qquad \text{for all } q \in \pi_{m-1}(\mathbb{R}^d). \tag{1.9}$$

In many cases it is possible to get along without the additional polynomial in (1.8), so that one does not need the additional conditions (1.9). In this particular case the interpolation problem boils down to the question whether the matrix $A_{\phi,X} = (\phi(\|x_k - x_j\|_2))_{1 \le k,j \le N}$ is nonsingular. To be more precise we could even ask:

Does there exist a function $\phi : [0, \infty) \to \mathbb{R}$ *such that for all* $d \in \mathbb{N}$, *for all* $N \in \mathbb{N}$, *and for all pairwise distinct* $x_1, \ldots, x_N \in \mathbb{R}^d$ *the matrix*

$$A_{\phi,X} := (\phi(\|x_i - x_j\|_2))_{1 \le i,j \le N}$$

is nonsingular?

Astonishingly, the answer is affirmative. Examples are the Gaussians $\phi(r) = e^{-\alpha r^2}$, $\alpha > 0$, the inverse multiquadric $\phi(r) = 1/\sqrt{c^2 + r^2}$, and the multiquadric $\phi(r) = \sqrt{c^2 + r^2}$, $c > 0$. In the first two cases it is even true that the matrix $A_{\phi,X}$ is always positive definite. We will give characterizations for ϕ to ensure this property.

Having the interpolant (1.8) with a huge N in mind, it would be even more useful to have a compactly supported basis function in order to reduce the number of necessary evaluations. Hence, almost impudently we reformulate the previous question as:

Does there exist a function $\phi : [0, \infty) \to \mathbb{R}$ *which satisfies all properties mentioned in the last question and which has in addition a compact support?*

This time the answer is negative. But if we sacrifice the dimension, i.e. if the function does not have to work for all dimensions but only for a fixed one then there exist also compactly supported functions yielding a positive definite interpolation matrix $A_{\phi,X}$.

Much of the theory we have to develop to prove the statements just made is independent of the form of the basis function. Several results hold if we replace $\phi(\|x - x_j\|_2)$ in (1.8) by $\Phi(x - x_j)$ with $\Phi : \mathbb{R}^d \to \mathbb{R}$ or even by $\Phi(x, x_j)$ with $\Phi : \Omega \times \Omega \to \mathbb{R}$. Of course, the latter case can only work if $X \subseteq \Omega$.

Even though we started with natural cubic splines, we have not yet used the minimal norm property. Instead, we have used the idea of shifting or translating a single function, which is also motivated by splines.

So it remains to shed some light on how the minimal norm property can be carried over to a multivariate setting. We start by computing the $L_2[a, b]$ inner product of the second-order derivatives of two natural cubic splines. Obviously, the polynomial part does not play a role, because differentiating it twice annihilates it.

Proposition 1.2 *Let $\phi(r) = r^3$, $r \geq 0$. Suppose the functions $s_X = \sum_{j=1}^N \alpha_j \phi(|\cdot - x_j|) + p_1$ and $s_Y = \sum_{k=1}^M \beta_k \phi(|\cdot - y_k|) + p_2$ are two natural cubic splines on $[a, b]$. Then*

$$(s_X'', s_Y'')_{L_2[a,b]} = 12 \sum_{j=1}^N \sum_{k=1}^M \alpha_j \beta_k \phi(|x_j - x_k|).$$

Proof Using $|x| = 2x_+ - x$ and (1.6) leads to

$$s_X''(x) = 6 \sum_{j=1}^N \alpha_j |x - x_j| = 12 \sum_{j=1}^N \alpha_j (x_j - x)_+ \tag{1.10}$$

and a corresponding result for s_Y. Next, on the one hand we can employ Taylor's formula for a function $f \in C^2[a, b]$ in the form

$$f(x) = f(a) + f'(a)(x - a) + \int_a^b f''(t)(x - t)_+ dt.$$

Setting $f(x) = (y - x)_+^3$ with a fixed $y \in [a, b]$ yields

$$(y - x)_+^3 = (y - a)^3 - 3(y - a)^2(x - a) + 6 \int_a^b (y - t)_+(x - t)_+ dt.$$

On the other hand, we can use the representation $(y - x)_+^3 = (|y - x|^3 + (y - x)^3)/2$ to derive

$$|y - x|^3 = 2(y - a)^3 - 6(y - a)^2(x - a) - (y - x)^3 + 12 \int_a^b (y - t)_+(x - t)_+ dt,$$

which leads to

$$\begin{aligned}\sum_{j=1}^N \sum_{k=1}^M \alpha_j \beta_k \phi(|x_j - y_k|) &= 2 \sum_{j,k} \alpha_j \beta_k (y_k - a)^3 - \sum_{j,k} \alpha_j \beta_k (y_k - x_j)^3 \\ &\quad - 6 \sum_{j,k} \alpha_j \beta_k (x_j - a)(y_k - a)^2 \\ &\quad + 12 \int_a^b \sum_j \alpha_j (x_j - t)_+ \sum_k \beta_k (y_k - t)_+ dt.\end{aligned}$$

All sums on the right-hand side except those under the integral are zero because of the annihilation effect (1.6) of the coefficients. Hence, recalling (1.10) gives the stated result. □

This result has the interesting consequence that we can introduce the linear space

$$F_\phi[a,b] := \left\{ \sum_{j=1}^{N} \alpha_j \phi(|\cdot - x_j|) : N \in \mathbb{N}, \quad \alpha \in \mathbb{R}^N, X = \{x_j\} \subseteq [a,b], \right.$$
$$\left. \text{with} \sum_{j=1}^{N} \alpha_j p(x_j) = 0 \text{ for all } p \in \pi_1(\mathbb{R}) \right\}.$$

This space is linear if the sum of two different functions based on the point sets X and Y is defined on the refined point set $X \cup Y$. Moreover, $F_\phi[a,b]$ carries the inner product

$$\left(\sum_{j=1}^{N} \alpha_j \phi(|\cdot - x_j|), \sum_{k=1}^{M} \beta_k \phi(|\cdot - y_k|) \right)_\phi := \sum_{j=1}^{N} \sum_{k=1}^{M} \alpha_j \beta_k \phi(|x_j - y_k|). \tag{1.11}$$

This reason that this is an inner product is that $(s,s)_\phi = 0$ means that the linear spline s'' has to be zero, which is only the case if all coefficients are already zero. Of course, we assume as usual that the x_j are pairwise distinct. Finally, completing the space $F_\phi[a,b]$ with respect to $\|\cdot\|_\phi$ means completing the space of piecewise linear functions with respect to the classical $L_2[a,b]$ inner product. Hence, standard arguments, which we will discuss thoroughly in Chapter 10, give the following characterization of $H^2[a,b]$.

Corollary 1.3 *Let $\phi(r) = r^3$, $r \geq 0$. The Sobolev space $H^2[a,b]$ coincides with*

$$\operatorname{clos}_{\|\cdot\|_\phi} F_\phi[a,b] + \pi_1(\mathbb{R}).$$

Moreover, (1.11) defines a semi-inner product on $H^2[a,b]$.

Now it is clear how a possible generalization can be made. We start with a function $\Phi : \Omega \times \Omega \to \mathbb{R}$ for $\Omega \subseteq \mathbb{R}^d$ and form the space $F_\Phi(\Omega)$ in the same way as we formed $F_\phi[a,b]$. We only have to replace $\phi(|\cdot - x_j|)$ by $\Phi(\cdot, x_j)$ and the annihilation conditions as appropriate. Of course this can only work for Φ's that allow us to equip $F_\Phi(\Omega)$ with an inner product $(\cdot,\cdot)_\Phi$. But we can turn things upside down and use this property as a definition for Φ. In any case, if Φ gives rise to an inner product on $F_\Phi(\Omega)$ we can form the closure $\mathcal{F}_\Phi(\Omega)$ of $F_\Phi(\Omega)$ with respect to this inner product. It will turn out that the new space can be interpreted as a space of functions again and that the interpolants are minimal semi-norm interpolants again.

1.5 Approximation and approximation orders

So far, we have only dealt with the concept of interpolation. But especially if the number of points is large and the data values contain noise it might be reasonable not to interpolate the given values exactly but to find a smooth function that only nearly fits the values. The standard method in this case is to compute a least squares solution. To this end, one chooses a finite-dimensional space $S \subseteq C(\mathbb{R}^d)$ and determines the function $s \in S$ that minimizes

$$\sum_{j=1}^{N} [s(x_j) - f_j]^2.$$

This approach has the disadvantage of being global in the sense that every data point has some influence on the solution function in any point. It would be more interesting to allow only the nearest neighbors of the evaluation point x to influence the approximate value $s(x)$. A method that realizes this idea is the moving least squares method. It is a variation on the classical least squares technique. Here, the dimension of the space S is supposed to be small. After fixing a weight function $w : \mathbb{R}^d \times \mathbb{R}^d \to \mathbb{R}$ one determines for every evaluation point x the solution s^* of

$$\min \left\{ \sum_{j=1}^{N} [s(x_j) - f_j]^2 w(x, x_j) : s \in S \right\}.$$

The approximate value is then given by $s^*(x)$. If the weight function is of the form $w(x, x_j) = w_0(x - x_j)$ with a compactly supported function w_0 having a support centered around zero this method uses only the data sites in a neighborhood of x to compute the approximate value. Thus, in this sense it is a local method, though for every evaluation point x a small linear system has to be solved. Nonetheless, this method turns out to be highly efficient.

After having discussed the concepts involved in reconstructing a function from discrete values we have to investigate the benefits of the methods. In general, this is done by analyzing the approximation properties of the approximation process.

Suppose that $X \subseteq \Omega \subseteq \mathbb{R}^d$, where Ω is a bounded set. Suppose further that the data values $\{f_j\}$ come from an unknown function $f \in C(\Omega)$, i.e. $f(x_j) = f_j$, $1 \le j \le N$. Then we are interested in how well the approximant or interpolant $s_{f,X}$ approximates the function f on Ω when the sets of data sites X become denser in Ω. This obviously needs a measure of how dense X is in Ω. For example, in case of splines it is usual to define $h_X = \max(x_{j+1} - x_j)$ as such a measure.

In case of natural cubic splines it is well known that there exists a constant $c > 0$ such that for all functions $f \in H^2[a, b]$ the error can be bounded as follows:

$$\|f - s_{f,X}\|_{L_\infty([a,b])} \le c h_X^{3/2} \|f''\|_{L_2[a,b]}.$$

The d-variate generalization is given by

Definition 1.4 *The fill distance of a set of points $X = \{x_1, \ldots, x_N\} \subseteq \Omega$ for a bounded domain Ω is defined to be*

$$h_{X,\Omega} := \sup_{x \in \Omega} \min_{1 \le j \le N} \|x - x_j\|_2.$$

The fill distance can be interpreted in various geometrical ways. For example, for any point $x \in \Omega$ there exists a data site x_j within a distance at most $h_{X,\Omega}$. Another possible interpretation is the following. The fill distance $h_{X,\Omega}$ denotes the radius of the largest ball which is completely contained in Ω and which does not contain a data site. In this sense $h_{X,\Omega}$ describes the largest data-site-free hole in Ω.

Definition 1.5 *An approximation process* $(f, X) \mapsto s_{f,X}$ *has* L_∞ *convergence order k for the function space* $\mathcal{F}$ *if there exists a constant* $c > 0$ *such that for all* $f \in \mathcal{F}$,

$$\|f - s_{f,X}\|_{L_\infty(\Omega)} \leq c h_{X,\Omega}^k \|f\|_{\mathcal{F}}.$$

The process possesses a spectral convergence order if there exists a constant $c > 0$ *and a constant* $\lambda \in (0, 1)$ *such that, for all* $f \in \mathcal{F}$,

$$\|f - s_{f,X}\|_{L_\infty(\Omega)} \leq c \lambda^{1/h_{X,\Omega}} \|f\|_{\mathcal{F}}.$$

1.6 Notation

It is time to fix certain items of notations that will be used throughout this book, though in fact some of them have already been employed. In many cases the symbols will be introduced where needed but some are important throughout, so that we want to collect them here.

Our main goal is to work with real-valued functions but sometimes it is necessary to employ complex-valued ones also. In this sense, the following function spaces contain real-valued functions if it is not stated otherwise.

We continue to denote the space of d-variate polynomials of absolute degree at most m by $\pi_m(\mathbb{R}^d)$. The function space $C^k(\Omega)$ is the set of k times continuously differentiable functions on Ω, where we assume $\Omega \subseteq \mathbb{R}^d$ to be open if $k \geq 1$. The intersection of all these spaces is denoted by $C^\infty(\Omega)$. The Lebesgue spaces are as usual denoted by $L_p(\Omega)$, $1 \leq p \leq \infty$, where $\Omega \subseteq \mathbb{R}^d$ should be measurable, i.e. they consist of all measurable functions f having a finite L_p-norm. The L_p-norm $\|\cdot\|_{L_p(\Omega)}$ is given by $\|f\|_{L_p(\Omega)}^p := \int_\Omega |f(x)|^p dx$ for $1 \leq p < \infty$ and by $\|f\|_{L_\infty(\Omega)} := \text{ess sup}_{x\in\Omega}|f(x)|$. The latter means the following. Every function $f \in L_\infty(\Omega)$ is essentially bounded on Ω, i.e. there exists a constant $K > 0$ such that $|f(x)| \leq K$ almost everywhere on Ω. The greatest lower bound of such constants K gives the norm. The space $L_p^{\text{loc}}(\Omega)$ consists of all measurable functions f that are locally in L_p, meaning their restriction $f|K$ belongs to $L_p(K)$ for every compact set $K \subseteq \Omega$. If Ω is an interval $[a, b]$ we will write $L_p[a, b]$ and $C^k[a, b]$ for $L_p([a, b])$ and $C^k([a, b])$, respectively, and similarly for open and semi-open intervals.

If dealing with several points in $\mathbb{R}^d$ one runs automatically into problems with indices. Throughout this book we will use the following notation. For a point $x \in \mathbb{R}^d$ its components will be given as $\mathrm{x}_1, \ldots, \mathrm{x}_d$, whereas $x_1, \ldots, x_N$ will denote N points in $\mathbb{R}^d$. The components of x_j are thus denoted $\mathrm{x}_{j,k}$, $1 \leq k \leq d$. We will use similar notation for y and z. These are the main critical cases and we will comply strictly with this usage. In the case of other letters we might sometimes relax the notation when there is no possibility of misunderstanding.

As usual we denote by $\|x\|_p$ the discrete p-norm $\|x\|_p^p = \sum_{j=1}^d |x_j|^p$ for $1 \le p < \infty$ and $\|x\|_\infty = \max |x_j|$.

For a multi-index $\alpha \in \mathbb{N}_0^d$ we denote its components by as usual $\alpha = (\alpha_1, \dots, \alpha_d)^T$. The length of α is given by $|\alpha| = \alpha_1 + \cdots + \alpha_d = \|\alpha\|_1$ and the factorial $\alpha!$ by $\alpha_1! \cdots \alpha_d!$. For two multi-indices α, β, the inequality $\alpha \le \beta$ is meant component-wise and

$$\binom{\alpha}{\beta} = \frac{\alpha!}{\beta!(\alpha - \beta)!}$$

is again defined component-wise. If $|\alpha| \le k$, $x \in \mathbb{R}^d$, and $f \in C^k(\Omega)$ are given, we denote the αth derivative of f and the αth power of x by

$$D^\alpha f := \frac{\partial^{|\alpha|}}{\partial x_1^{\alpha_1} \cdots x_d^{\alpha_d}} f \qquad \text{and} \qquad x^\alpha := x_1^{\alpha_1} \cdots x_d^{\alpha_d}.$$

Finally, there is a concept that will accompany us throughout this book. We will often meet multivariate functions that are constant on spheres around zero, i.e. they are radial. Hence, such a function depends only on the norm of its argument. We will always use a capital letter Φ to denote the multivariate function and a small letter ϕ to denote the univariate function, i.e. $\Phi = \phi(\|\cdot\|_2)$.

1.7 Notes and comments

The examples shown in this chapter use data sets that have courteously been provided by the following persons and institutions. The glacier data set can be found at Richard Franke's homepage. The dragon comes from the homepage of Stanford University. The aircraft model and data is due to the European Aeronautic Defence and Space Company (EADS).

It seems that the implicit reconstruction of a surface from unorganized points goes hand in hand with radial basis function interpolation and approximation. The first contributions in this direction dealt with Gaussians and are known under the name *blobby surfaces* or *metaballs*. They were invented independently by Blinn [27] and Hishimura *et al.* [84] in the early eighties. Then Turk and O'Brien together with various coauthors used thin-plate splines for modeling implicit surfaces; see for example [184–186,206]. Owing to the global character of these basis functions, however, all these examples were restricted to rather small data sets. The remedy to this problem has been twofold. On the one hand, Morse *et al.* [137] used the compactly supported radial basis functions devised by the present author [190]. On the other hand, fast evaluation methods based on far-field expansions have been developed for most of the globally supported radial basis functions. Results based on these methods can be found in a paper by Carr *et al.* [37]. Other more recent contributions are by Ohtake *et al.* [151, 152]. The idea of computing and orienting surface normals goes back to Hoppe; see [86, 87]. In [153], Pasko and Savchenko describe smooth ways for several CSG operations.

The idea of treating a fluid–structure interaction problem in a coupled-field formulation, so that the interaction is restricted to the exchange of boundary conditions, has been employed by Farhat [52] and Kutler [100]. Among others, Farhat and coworkers [51,52], Beckert [19], and Cebral and Löhner [38] have used different interpolation strategies to exchange the boundary conditions in an iterative or staggered procedure. Hounjet and Meijer [88] were the first to use thin-plate splines in this context. Beckert and Wendland [20] extended their approach to general radial basis functions, in particular to compactly supported ones.

The theory of semi-Lagrangian methods for the numerical solution of advection schemes was described and analyzed in [50] by Falcone and Ferretti. Behrens and Iske [21,76] were the first to use radial basis functions in the spatial reconstruction process. Other methods based on (generalized) scattered data interpolation for hyperbolic conservation laws have been investigated by Sonar [179] and Lorentz *et al.* [109].

2
Haar spaces and multivariate polynomials

Every book on numerical analysis has a chapter on univariate polynomial interpolation. For given data sites $x_1 < x_2 < \cdots < x_N$ and function values $f_1, \ldots, f_N$ there exists exactly one polynomial $p_f \in \pi_{N-1}(\mathbb{R}^1)$ that interpolates the data at the data sites. One of the interesting aspects of polynomial interpolation is that the space $\pi_{N-1}(\mathbb{R}^1)$ depends neither on the data sites nor on the function values but only on the number of points. While the latter is intuitively clear, it is quite surprising that the space can also be chosen independently of the data sites. In approximation theory the concept whereby the space is independent of the data has been generalized and we will shortly review the necessary results. Unfortunately, it will turn out that the situation is not as favorable in the multivariate case.

2.1 The Mairhuber–Curtis theorem

Motivated by the univariate polynomials we give

Definition 2.1 *Suppose that $\Omega \subseteq \mathbb{R}^d$ contains at least N points. Let $V \subseteq C(\Omega)$ be an N-dimensional linear space. Then V is called a Haar space of dimension N on Ω if for arbitrary distinct points $x_1, \ldots, x_N \in \Omega$ and arbitrary $f_1, \ldots, f_N \in \mathbb{R}$ there exists exactly one function $s \in V$ with $s(x_i) = f_i$, $1 \le i \le N$.*

In the sense of this definition, $V = \pi_{N-1}(\mathbb{R}^1)$ is an N-dimensional Haar space for any subset Ω of $\mathbb{R}$ that contains at least N points. Haar spaces, which are sometimes also called Chebychev spaces, can be characterized in many ways. Let us collect some alternative characterizations straight away.

Theorem 2.2 *Under the conditions of Definition 2.1 the following statements are equivalent.*

(1) V is an N-dimensional Haar space.

(2) Every $u \in V \setminus \{0\}$ has at most $N - 1$ zeros.

(3) For any distinct points $x_1, \ldots, x_N \in \Omega$ and any basis $u_1, \ldots, u_N$ of V we have that

$$\det(u_j(x_i)) \neq 0.$$

Proof Suppose that V is an N-dimensional Haar space and $u \in V \setminus \{0\}$ has N zeros, say $x_1, \dots, x_N$. In this case u and the zero function both interpolate zero on these N points. From the uniqueness we can conclude that $u \equiv 0$ in contrast with our assumption.

Next, let us assume that the second property is satisfied. If $\det A = 0$ with $A = (u_j(x_i))$ then there exists a vector $\alpha \in \mathbb{R}^N \setminus \{0\}$ with $A\alpha = 0$, i.e. $\sum_{j=1}^N \alpha_j u_j(x_i) = 0$ for $1 \le i \le N$. This means that the function $u := \sum \alpha_j u_j$ has N zeros and must therefore be identically zero. This is impossible since $\alpha \ne 0$.

Finally, if the third property is satisfied then we can make the Ansatz $u = \sum \alpha_j u_j$ for the interpolant. Obviously, the interpolation conditions become

$$\sum_{j=1}^{N} \alpha_j u_j(x_i) = f_i, \qquad 1 \le i \le N.$$

Now the coefficient vector, and hence u, is uniquely determined because $A = (u_j(x_i))$ is nonsingular. □

After this characterization of Haar spaces we turn to the question whether Haar spaces exist in higher space dimensions. The next result shows that this is the case only in simple situations.

Theorem 2.3 *(Mairhuber–Curtis) Suppose that $\Omega \subseteq \mathbb{R}^d$, $d \ge 2$, contains an interior point. Then there exists no Haar space on Ω of dimension $N \ge 2$.*

Proof Suppose that $U = \operatorname{span}\{u_1, \dots, u_N\}$ is a Haar space on Ω. As Ω contains an interior point there exists a ball $B(x_0, \delta) \subseteq \Omega$ with radius $\delta > 0$ and we can fix pairwise distinct $x_3, \dots, x_N \in B(x_0, \delta)$. Next we choose two continuous curves $x_1(t)$, $x_2(t)$, $t \in [0, 1]$, such that $x_1(0) = x_2(1)$, $x_1(1) = x_2(0)$ and such that the curves neither have any other points of intersection nor have any common points with $\{x_3, \dots, x_N\}$. This is possible since $d \ge 2$. Then on the one hand, since U is assumed to be a Haar space on Ω, the function

$$D(t) := \det((u_j(x_k))_{1 \le j,k \le N})$$

is continuous and does not change sign. On the other hand $D(1) = -D(0)$ because only the first two rows of the involved matrices are exchanged. Thus D must change signs, which is a contradiction. □

2.2 Multivariate polynomials

So far we know that, in contrast with their univariate sibling, multivariate polynomials cannot form a Haar space. Hence, it is impossible to interpolate all kinds of data at any set of data sites $X = \{x_1, \dots, x_Q\}$, $Q = \dim \pi_m(\mathbb{R}^d)$, by polynomials from $\pi_m(\mathbb{R}^d)$. Nonetheless, polynomial interpolation plays an important role even in the multivariate setting. But we have to restrict the sets of data sites for which interpolation is possible. This together with some other elementary facts about multivariate polynomials is the subject of this section.

Lemma 2.4 *For $d = 1, 2, 3, \dots$ and $m = 0, 1, 2, \dots$ we have*

$$\sum_{k=0}^{m} \binom{k+d}{d} = \binom{m+1+d}{d+1}.$$

Proof We use induction on m. For $m = 0$ there is nothing to prove. Now suppose that we have proven already the assertion for m and want to conclude it for $m+1$. This is done by writing

$$\begin{aligned}
\sum_{k=0}^{m+1} \binom{k+d}{d} &= \sum_{k=0}^{m} \binom{k+d}{d} + \binom{m+1+d}{d} \\
&= \binom{m+1+d}{d+1} + \binom{m+1+d}{d} \\
&= \frac{(m+1+d)!}{m!(d+1)!} + \frac{(m+1+d)!}{d!(m+1)!} \\
&= \frac{(m+1+d)!\,(m+d+2)}{(d+1)!\,(m+1)!} \\
&= \binom{m+d+2}{d+1}.
\end{aligned}$$

□

This auxiliary result helps us to understand the structure of polynomials in d dimensions better. The first part of the following result seems to be obvious but requires nevertheless a proof.

Theorem 2.5

(1) The monomials $x \mapsto x^\alpha$, $x \in \mathbb{R}^d$, $\alpha \in \mathbb{N}_0^d$, are linearly independent.

(2) $\dim \pi_m(\mathbb{R}^d) = \binom{m+d}{d}$.

Proof Let us start with the first property. The proof is by induction on d. For $d = 1$ this is obvious, since for any polynomial $p(x) = \sum_{j=0}^{n} c_j x^j$ the coefficients are determined by $c_j = p^{(j)}(0)/j!$. Thus if p is identically zero then all coefficients must be zero. Now suppose everything has been proven for $d-1$. Suppose that $J \subseteq \mathbb{N}_0^d$ is finite and $\sum_{\alpha \in J} c_\alpha x^\alpha = 0$. Define $J_k := \{\alpha \in J : \alpha_1 = k\}$. Then there exists an $n \in \mathbb{N}_0$ such that $J = J_0 \cup \cdots \cup J_n$. This means that

$$0 = \sum_{\alpha \in J} c_\alpha x^\alpha = \sum_{k=0}^{n} \sum_{\alpha \in J_k} c_\alpha x_1^{\alpha_1} \cdots x_d^{\alpha_d} = \sum_{k=0}^{n} x_1^k \sum_{\alpha \in J_k} c_\alpha x_2^{\alpha_2} \cdots x_d^{\alpha_d}.$$

Thus we can conclude from the univariate case that

$$\sum_{\alpha \in J_k} c_\alpha x_2^{\alpha_2} \cdots x_d^{\alpha_d} = 0, \qquad 0 \le k \le n.$$

By the induction hypothesis we have finally $c_\alpha = 0$ for $\alpha \in J$.

Next we come to the space dimension. From the first part we know that it suffices to show that

$$\#\{\alpha \in \mathbb{N}_0^d : |\alpha| \le m\} = \binom{m+d}{d},$$

when $\#J$ denotes the number of elements of the set J. This is done by induction on d. For $d = 1$ we have

$$\#\{\alpha \in \mathbb{N}_0 : \alpha \le m\} = m + 1 = \binom{m+1}{1}.$$

Assuming the correctness of this statement for $d - 1$ gives us

$$\begin{aligned}
\#\{\alpha \in \mathbb{N}_0^d : |\alpha| \le m\} &= \#\bigcup_{k=0}^{m} \left\{\alpha \in \mathbb{N}_0^d : \alpha_d = k, \sum_{i=1}^{d-1} \alpha_i \le m - k\right\} \\
&= \sum_{k=0}^{m} \#\{\alpha \in \mathbb{N}_0^{d-1} : |\alpha| \le m - k\} \\
&= \sum_{k=0}^{m} \binom{m-k+d-1}{d-1} \\
&= \binom{m+d}{d},
\end{aligned}$$

where the last equality comes from Lemma 2.4. □

In Definition 2.1 we formulated certain conditions for a function space to guarantee unique interpolation. Now we are in a situation to interpret these conditions as conditions for the points.

Definition 2.6 *The points $X = \{x_1, \dots, x_N\} \subseteq \mathbb{R}^d$ with $N \ge Q = \dim \pi_m(\mathbb{R}^d)$ are called $\pi_m(\mathbb{R}^d)$-unisolvent if the zero polynomial is the only polynomial from $\pi_m(\mathbb{R}^d)$ that vanishes on all of them.*

For an example take the linear polynomials on $\mathbb{R}^2$. From Theorem 2.5 we know that $\dim \pi_1(\mathbb{R}^2) = 3$. Since every bivariate linear polynomial describes a plane in three-dimensional space this plane is uniquely determined by three points if and only if these three points are not collinear. Thus three points in $\mathbb{R}^2$ are $\pi_1(\mathbb{R}^2)$-unisolvent if and only if they are not collinear.

A generalization of this fact is:

Theorem 2.7 *Suppose that $\{L_0, \dots, L_m\}$ is a set of $m + 1$ distinct lines in $\mathbb{R}^2$ and that $X = \{x_1, \dots, x_Q\}$ is a set of $Q = (m + 1)(m + 2)/2$ distinct points such that the first point lies on L_0, the next two points lie on L_1 but not on L_0, and so on, so that the last $m + 1$ points lie on L_m but not on any of the previous lines $L_0, \dots, L_{m-1}$. Then X is $\pi_m(\mathbb{R}^2)$-unisolvent.*

Proof Use induction on m. For $m = 0$ the result is trivial. Now take $p \in \pi_m(\mathbb{R}^2)$ with

$$p(x_i) = 0, \qquad i = 1, \ldots, Q;$$

then we have to show that p is the zero polynomial.

For each $i = 0, \ldots, m$ let the equation of the line L_i be given by

$$\alpha_i \mathrm{x}_1 + \beta_i \mathrm{x}_2 = \gamma_i,$$

remembering that x is given by $(\mathrm{x}_1, \mathrm{x}_2)^T$. Suppose now that p interpolates zero data at all the points x_i as stated above. Since p reduces to a univariate polynomial of degree m on L_m that vanishes at $m + 1$ distinct points on L_m, it follows that p vanishes identically on L_m, and so

$$p(x) = (\alpha_m \mathrm{x}_1 + \beta_m \mathrm{x}_2 - \gamma_m) q(x),$$

where q is a polynomial of degree $m - 1$. But now q satisfies the hypothesis of the theorem with m replaced by $m - 1$ and X replaced by an $\widetilde{X}$ consisting of the first $(m + 1)m/2$ points of U. By induction, therefore, $q \equiv 0$, and thus $p \equiv 0$. □

This theorem can be generalized to $\mathbb{R}^d$ by using hyperplanes in $\mathbb{R}^d$ and induction on d. Instead of doing so, we restrict ourselves to the following special case. Let

$$X_m^d = \{\beta \in \mathbb{N}_0^d : |\beta| \le m\}.$$

This set contains by earlier considerations exactly $Q = \dim \pi_m(\mathbb{R}^d)$ points.

Lemma 2.8 *The set X_m^d is $\pi_m(\mathbb{R}^d)$-unisolvent. Thus polynomial interpolation of degree m is uniquely possible.*

Proof We have to show that the only polynomial $p \in \pi_m(\mathbb{R}^d)$ with $p(\alpha) = 0$ for every $\alpha \in \mathbb{N}_0^d$ with $|\alpha| \le m$ is $p \equiv 0$. This is done by induction on the space dimension d. For $d = 1$ we have a univariate polynomial of degree m with $m + 1$ zeros, which can only be the zero polynomial itself. For the induction step from $d - 1$ to d let us write the polynomial as

$$p(x) = p(\widetilde{x}, \mathrm{x}_d) = \sum_{j=0}^{m} p_j(\widetilde{x}) \mathrm{x}_d^j$$

with polynomials $p_j \in \pi_{m-j}(\mathbb{R}^{d-1})$. We are finished if we can show that all the p_j are zero. This is done by induction on $m - j$. To be more precise, we will show that $p_{m-j} \equiv 0$ for $0 \le j \le m$ and $p_{m-i} | X_j^{d-1} = 0$ for $j + 1 \le i \le m$. Let us start with $j = 0$. Setting $x = (0, k)$ for $0 \le k \le m$ we see that $(0, k) \in X_m^d$ and that the univariate polynomial $p(0, \cdot)$ has $m + 1$ zeros. This means that the coefficients $p_i(0)$ have to vanish for $0 \le i \le m$, showing in particular that the constant polynomial p_m is zero. Now assume that our assumption is true for j. Then, $p_m \equiv \cdots \equiv p_{m-j} \equiv 0$ shows that

$$p(\widetilde{x}, x_d) = \sum_{i=0}^{m-j-1} p_i(\widetilde{x}) x_d^i$$

and that $p_i | X_j^{d-1} = 0$ for $0 \le i \le m - j - 1$. Next fix an arbitrary $\beta \in X_{j+1}^{d-1} \setminus X_j^{d-1}$. Then $|\beta| = j + 1$ and the data points (β, k) are in X_m^d for $0 \le k \le m - j - 1$. Thus the univariate polynomial $p(\beta, \cdot)$, which is of degree $m - j - 1$, has $m - j$ zeros and is therefore identically zero. This means $p_i(\beta) = 0$ for all $\beta \in X_{j+1}^{d-1}$ and all $0 \le i \le m - j - 1$, giving in particular $p_{m-j-1} \equiv 0$ by our induction assumption on $d - 1$. This finishes our induction on j and hence also the induction on d. □

This last proof gives an idea how complicated the situation in more than one dimension can be. If the data sites are part of a mildly disturbed grid, however, interpolation with polynomials is always possible.

3

Local polynomial reproduction

One reason for studying polynomials in approximation theory is that they allow a simple error analysis. For example, if the univariate function f possesses k continuous derivatives around a point $x_0 \in \mathbb{R}$ then the Taylor polynomial

$$p(x) = \sum_{j=0}^{k-1} \frac{f^{(j)}(x_0)}{j!}(x - x_0)^j$$

enjoys for $|x - x_0| \leq h$ the local approximation error

$$|f(x) - p(x)| = \frac{|f^{(k)}(\xi)|}{k!}|x - x_0|^k \leq Ch^k$$

with ξ between x and x_0. This local approximation order is inherited by every approximation process that recovers polynomials at least locally.

In this chapter we will work out these ideas more clearly by introducing the concept of a local polynomial reproduction and discussing its existence. In the next chapter we will give a concrete example of a local polynomial reproduction. But even the theoretical results provided here are useful, since they often allow error estimates for other approximation methods.

3.1 Definition and basic properties

Given a set $X = \{x_1, \ldots, x_N\}$ of pairwise distinct points in $\Omega \subseteq \mathbb{R}^d$ and function values $f(x_1), \ldots, f(x_N)$, we are concerned with finding an approximant s to the unknown function f. One way to form such an approximant is to separate the influence of the data sites and the data values by choosing functions $u_j : \Omega \to \mathbb{R}$, $1 \leq j \leq N$, which depend only on X and forming

$$s(x) = \sum_{j=1}^{N} f(x_j)u_j(x).$$

If the functions u_j are cardinal with respect to X, that is $u_j(x_k) = \delta_{j,k}$ for $1 \leq j, k \leq N$, we obviously have an interpolant. The reader might think of Lagrange interpolation with univariate polynomials as an example. If the functions u_j are not cardinal then s is sometimes called a *quasi-interpolant*.

In what follows the fill distance, which has already been introduced in Chapter 1, will play an important role as it will throughout this book. For convenience we restate its definition here. For a set of points $X = \{x_1, \ldots, x_N\}$ in a bounded domain $\Omega \subseteq \mathbb{R}^d$ the fill distance is defined to be

$$h_{X,\Omega} = \sup_{x \in \Omega} \min_{1 \le j \le N} \|x - x_j\|_2.$$

Now it is time to come to the basic definition of what we call a local polynomial reproduction.

Definition 3.1 *A process that defines for every set $X = \{x_1, \ldots, x_N\} \subseteq \Omega$ a family of functions $u_j = u_j^X : \Omega \to \mathbb{R}$, $1 \le j \le N$, provides a local polynomial reproduction of degree ℓ on Ω if there exist constants $h_0, C_1, C_2 > 0$ such that*

(1) $\sum_{j=1}^{N} p(x_j) u_j = p$ *for all* $p \in \pi_\ell(\mathbb{R}^d)|\Omega$,

(2) $\sum_{j=1}^{N} |u_j(x)| \le C_1$ *for all* $x \in \Omega$,

(3) $u_j(x) = 0$ *if* $\|x - x_j\|_2 > C_2 h_{X,\Omega}$ *and* $x \in \Omega$

is satisfied for all X with $h_{X,\Omega} \le h_0$.

Sometimes we will also say that $\{u_j\}$ forms a local polynomial reproduction or in a more sloppy way that the quasi-interpolant $s = \sum f(x_j) u_j$ provides a local polynomial reproduction.

The crucial point in the definition is that the constants involved are independent of the data sites. Moreover, so far we have not talked about the data values at all.

The first and the third condition justify the name *local polynomial reproduction*. The second condition is important for the approximation property of the associated quasi-interpolant. So far, we have not specified the functions u_j any further. Obviously they will depend on the set X, but they might not even be continuous. In the next chapter we will show how to construct such functions having an arbitrary smoothness.

The reason for investigating local polynomial reproductions is that they allow a very simple error analysis.

Theorem 3.2 *Suppose that $\Omega \subseteq \mathbb{R}^d$ is bounded. Define Ω^* to be the closure of $\cup_{x \in \Omega} B(x, C_2 h_0)$. Define $s_{f,X} = \sum_{j=1}^{N} f(x_j) u_j$, where $\{u_j\}$ is a local polynomial reproduction of order m on Ω. If $f \in C^{m+1}(\Omega^*)$ then there exists a constant $C > 0$ depending only on the constants from Definition 3.1 such that*

$$|f(x) - s_{f,X}(x)| \le C h_{X,\Omega}^{m+1} |f|_{C^{m+1}(\Omega^*)}$$

for all X with $h_{X,\Omega} \le h_0$. The semi-norm on the right-hand side is defined by $|f|_{C^{m+1}(\Omega^*)} := \max_{|\alpha|=m+1} \|D^\alpha f\|_{L_\infty(\Omega^*)}$.

Proof Let p be an arbitrary polynomial from $\pi_m(\mathbb{R}^d)$. Then, using the properties of a local polynomial reproduction gives

$$\begin{aligned}
|f(x) - s_{f,X}(x)| &\le |f(x) - p(x)| + \left| p(x) - \sum_{j=1}^{N} f(x_j)u_j(x) \right| \\
&\le |f(x) - p(x)| + \sum_{j=1}^{N} |p(x_j) - f(x_j)||u_j(x)| \\
&\le \|f - p\|_{L_\infty(B(x,C_2h_{X,\Omega}))} \left(1 + \sum_{j=1}^{N} |u_j(x)| \right) \\
&\le (1 + C_1)\|f - p\|_{L_\infty(B(x,C_2h_{X,\Omega}))}.
\end{aligned}$$

Here, $B(x, r)$ denotes the closed ball around x with radius $r > 0$. Now choose p to be the Taylor polynomial of f around x. This gives for $y \in \Omega$ a $\xi \in \Omega^*$ such that

$$f(y) = \sum_{|\alpha| \le m} \frac{D^\alpha f(x)}{\alpha!}(y - x)^\alpha + \sum_{|\alpha| = m+1} \frac{D^\alpha f(\xi)}{\alpha!}(y - x)^\alpha.$$

Hence, we can conclude that

$$\begin{aligned}
\left| f(x) - \sum_{j=1}^{N} u_j(x) f(x_j) \right| &\le (1 + C_1) \sum_{|\alpha| = m+1} \frac{1}{\alpha!} \|D^\alpha f\|_{L_\infty(\Omega^*)} (C_2 h)^{m+1} \\
&\le C h^{m+1} |f|_{C^{m+1}(\Omega^*)},
\end{aligned}$$

with $h = h_{X,\Omega}$. □

Let us point out that the result is indeed local. It always gives the approximation order that can be achieved by polynomial reproduction. Hence, if f is less smooth in a subregion of Ω, say it possesses only $\ell \le m$ continuous derivatives there, then the approximant still gives order $\ell - 1$ in that region and this is the best we can hope for.

3.2 Norming sets

After having learnt about the benefits of a local polynomial reproduction, we now want to prove its existence. The case $m = 0$ of constant polynomials is easy to handle. If one chooses for $x \in \Omega$ an index j such that $\|x - x_j\|_2$ is minimal, the functions can be defined pointwise by $u_j(x) = 1$ and $u_k(x) = 0$ for $k \ne j$. These functions obviously satisfy the conditions of Definition 3.1 with arbitrary h_0 and $C_1 = C_2 = 1$.

It remains to prove existence in case of polynomials of degree $m \ge 1$. This will be done in a very elegant way, which allows us to state all the involved constants explicitly. The drawback of this approach is that it is nonconstructive: it will give us no hint how the functions u_j might look.

Let V be a finite-dimensional vector space with norm $\|\cdot\|_V$ and let $Z \subseteq V^*$ be a finite set consisting of N functionals. Here, V^* denotes the dual space of V consisting of all linear and continuous functionals defined on V.

Definition 3.3 *We will say that Z is a norming set for V if the mapping $T : V \to T(V) \subseteq \mathbb{R}^N$ defined by $T(v) = (z(v))_{z \in Z}$ is injective. We will call T the sampling operator.*

If Z is a norming set for V then the inverse of the sampling operator exists on its range, $T^{-1} : T(V) \to V$. Let $\mathbb{R}^N$ have the norm $\|\cdot\|_{\mathbb{R}^N}$, and let $\|\cdot\|_{\mathbb{R}^{N*}}$ be the dual norm on $\mathbb{R}^{N*} = \mathbb{R}^N$. Equip $T(V)$ with the induced norm. Finally, let $\|T^{-1}\|$ be the norm defined as

$$\|T^{-1}\| = \sup_{x \in T(V)\setminus\{0\}} \frac{\|T^{-1}x\|_V}{\|x\|_{\mathbb{R}^N}} = \sup_{v \in V\setminus\{0\}} \frac{\|v\|_V}{\|Tv\|_{\mathbb{R}^N}}.$$

This norm will be called the *norming constant* of the norming set.

We are mainly interested in using the ℓ_∞-norm on $\mathbb{R}^N$, so that the dual norm is given by the ℓ_1-norm.

The term *norming set* comes from the fact that Z allows us to equip V with an equivalent norm via the operator T. To see this, first note that obviously $\|Tv\|_{\mathbb{R}^N} \le \|T\|\|v\|_V$. Conversely, since T is injective we have $\|v\|_V = \|T^{-1}(Tv)\|_V \le \|T^{-1}\|\|Tv\|_{\mathbb{R}^N}$. Hence, $\|\cdot\|_V$ and $\|T(\cdot)\|_{\mathbb{R}^N}$ are equivalent norms on V.

It is clear that we need at least $N \ge \dim V$ functionals to make the operator T injective. It is also obvious that we can get along with exactly $N = \dim V$ functionals. But what we have in mind is that a certain family of functionals is given, for example point-evaluation functionals. Then the natural question is how many of these functionals are necessary not only to make T injective but also to control the norm of T and its inverse. In case of point-evaluation functionals we might also ask for the locus of the points. These questions will be addressed in the next section. We now come to the main result on norming sets.

Theorem 3.4 *Suppose V is a finite-dimensional normed linear space and $Z = \{z_1, \ldots, z_N\}$ is a norming set for V, T being the corresponding sampling operator. For every $\psi \in V^*$ there exists a vector $u \in \mathbb{R}^N$ depending only on ψ such that, for every $v \in V$,*

$$\psi(v) = \sum_{j=1}^{N} u_j z_j(v)$$

and

$$\|u\|_{\mathbb{R}^{N*}} \le \|\psi\|_{V^*}\|T^{-1}\|.$$

Proof We define the linear functional $\widetilde{\psi}$ on $T(V)$ by $\widetilde{\psi}(x) = \psi(T^{-1}x)$. It has a norm that is bounded by $\|\widetilde{\psi}\| \le \|\psi\|_{V^*}\|T^{-1}\|$. By the Hahn–Banach theorem, $\widetilde{\psi}$ has a norm-preserving extension $\widetilde{\psi}_{\text{ext}}$ to $\mathbb{R}^N$. On $\mathbb{R}^N$ all linear functionals can be represented by the inner product with a fixed vector. Hence, there exists $u \in \mathbb{R}^N$ with

$$\widetilde{\psi}_{\mathrm{ext}}(x) = \sum_{j=1}^{N} u_j x_j$$

and $\|u\|_{\mathbb{R}^{N*}} \le \|\psi\|_{V^*}\|T^{-1}\|$. Finally, we find for an arbitrary $v \in V$, by setting $x = Tv$,

$$\psi(v) = \psi(T^{-1}x) = \widetilde{\psi}(x) = \widetilde{\psi}_{\mathrm{ext}}(x) = \sum_{j=1}^{N} u_j x_j = \sum_{j=1}^{N} u_j z_j(v),$$

which finishes the proof. □

It is important to notice that there are two different ingredients in this result. On the one hand we have the functionals in Z, and they determine the norming set. On the other hand there is the functional ψ. But the existence of the vector u does not depend on ψ, it only depends on Z. Of course, the actual vector u will depend on the specific ψ but not the general result on its existence. Hence, if we know that Z is a norming set, we can represent *every* functional ψ in this way.

3.3 Existence for regions with cone condition

To use norming sets in the context of local polynomial reproduction, we choose $V = \pi_m(\mathbb{R}^d)|\Omega$ and $Z = \{\delta_{x_1}, \dots, \delta_{x_N}\}$. Here δ_x denotes the point-evaluation functional defined by $\delta_x(f) = f(x)$. A first consequence of this choice is

Proposition 3.5 *The functionals $Z = \{\delta_{x_1}, \dots, \delta_{x_N}\}$ form a norming set for $\pi_m(\mathbb{R}^d)$ if and only if X is $\pi_m(\mathbb{R}^d)$-unisolvent.*

As mentioned earlier, we equip $\mathbb{R}^N$ with the ℓ_∞-norm. If we additionally choose ψ to be δ_x, Theorem 3.4 gives us a vector $u(x) \in \mathbb{R}^N$ that recovers polynomials in the sense of Definition 3.1 and has a bounded ℓ_1-norm. Hence, two of the three properties of a local polynomial reproduction are satisfied, provided that $Z = \{\delta_{x_1}, \dots, \delta_{x_N}\}$ is a norming set for the polynomial space. The third property will follow by a local argument. But before that we give conditions on the set of data sites X such that Z forms a norming set. This is hopeless in the case of a general domain Ω. Hence, we restrict ourselves to the case of domains satisfying an interior cone condition.

Definition 3.6 *A set $\Omega \subseteq \mathbb{R}^d$ is said to satisfy an interior cone condition if there exists an angle $\theta \in (0, \pi/2)$ and a radius $r > 0$ such that for every $x \in \Omega$ a unit vector $\xi(x)$ exists such that the cone*

$$C(x, \xi(x), \theta, r) := \{x + \lambda y \;:\; y \in \mathbb{R}^d, \|y\|_2 = 1, y^T\xi(x) \ge \cos\theta, \lambda \in [0, r]\} \tag{3.1}$$

is contained in Ω.

We will often make use of the following elementary geometric fact (see Figure 3.1).

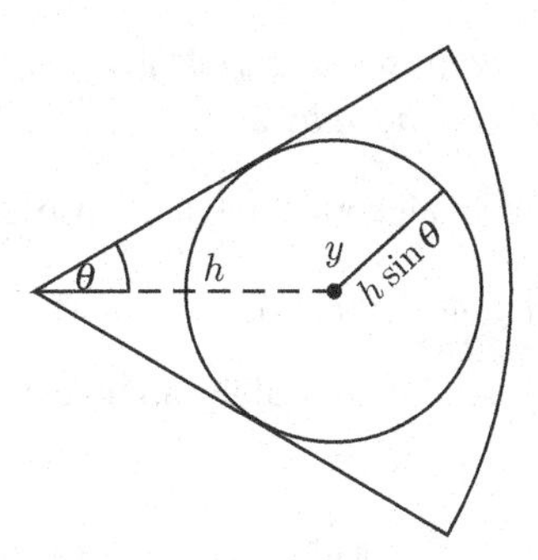

Fig. 3.1 Ball in a cone.

Lemma 3.7 *Suppose that $C = C(x, \xi, \theta, r)$ is a cone defined as in (3.1). Then for every $h \le r/(1 + \sin\theta)$ the closed ball $B = B(y, h \sin\theta)$ with center $y = x + h\xi$ and radius $h \sin\theta$ is contained in $C(x, \xi, \theta, r)$. In particular, if z is a point from this ball then the whole line segment $x + t(z - x)/\|z - x\|_2$, $t \in [0, r]$, is contained in the cone.*

Proof Without restriction we can assume $x = 0$. If $z \in B$ then $\|z\|_2 \le \|z - y\|_2 + \|y\|_2 \le h \sin\theta + h \le r$. Thus the ball B is contained in the larger ball around x with radius r and it remains to show that it is contained in the correct segment. Suppose that this is not the case. Then we can find a $z \in B$, $z \notin C$, i.e. $z^T\xi < \|z\|_2 \cos\theta$. This means that

$$\begin{aligned} h^2 \sin^2\theta &\ge \|z - y\|_2^2 = \|z - h\xi\|_2^2 \\ &= \|z\|_2^2 + h^2 - 2hz^T\xi > \|z\|^2 + h^2 - 2h\|z\|_2 \cos\theta, \end{aligned}$$

which leads to the contradiction

$$\begin{aligned} 0 &> \|z\|_2^2 + h^2(1 - \sin^2\theta) - 2h\|z\|_2 \cos\theta \\ &= \|z\|_2^2 + h^2 \cos^2\theta - 2h\|z\|_2 \cos\theta \\ &= (\|z\|_2 - h\cos\theta)^2 \ge 0. \end{aligned}$$

□

To prove the norming-set property we will use the fact that a multivariate polynomial reduces to a univariate polynomial when restricted to a line. Then we want to relate the Chebychev norm of the univariate polynomial on a line segment to the Chebychev norm of the multivariate polynomial on Ω. To do this we have to ensure that the line segment is completely contained in Ω.

Theorem 3.8 *Suppose $\Omega \subseteq \mathbb{R}^d$ is compact and satisfies an interior cone condition with radius $r > 0$ and angle $\theta \in (0, \pi/2)$. Let $m \in \mathbb{N}$ be fixed. Suppose $h > 0$ and the set $X = \{x_1, \dots, x_N\} \subseteq \Omega$ satisfy*

(1) $h \le \dfrac{r \sin\theta}{4(1 + \sin\theta)m^2}$,

(2) for every $B(x, h) \subseteq \Omega$ there is a center $x_j \in X \cap B(x, h)$;

then $Z = \{\delta_x : x \in X\}$ is a norming set for $\pi_m(\mathbb{R}^d)|\Omega$ and the norm of the inverse of the associated sampling operator is bounded by 2.

Proof Markov's inequality for an algebraic polynomial $p \in \pi_m(\mathbb{R}^1)$ is given by

$$|p'(t)| \le m^2 \|p\|_{L_\infty[-1,1]}, \qquad t \in [-1, 1];$$

see for example Cheney [41]. A simple scaling argument shows for $r > 0$ and all $p \in \pi_m(\mathbb{R}^1)$ that

$$|p'(t)| \le \frac{2}{r} m^2 \|p\|_{L_\infty[0,r]}, \qquad t \in [0, r].$$

Choose an arbitrary $p \in \pi_m(\mathbb{R}^d)$ with $\|p\|_{L_\infty(\Omega)} = 1$. Since Ω is compact there exists an $x \in \Omega$ with $|p(x)| = \|p\|_{L_\infty(\Omega)} = 1$. As Ω satisfies an interior cone condition we can find a $\xi \in \mathbb{R}^d$ with $\|\xi\|_2 = 1$ such that the cone $C(x) := C(x, \xi, \theta, r)$ is completely contained in Ω. Because $h/\sin\theta \le r/(1+\sin\theta)$ we can use Lemma 3.7 with h replaced by $h/\sin\theta$ to see that $B(y, h) \subseteq C(x)$ with $y = x + (h/\sin\theta)\xi$. For this y we find an $x_j \in X$ with $\|y - x_j\|_2 \le h$, i.e. $x_j \in B(y, h) \subseteq C(x)$. Thus the whole line segment

$$x + t\frac{x_j - x}{\|x_j - x\|_2}, \qquad t \in [0, r],$$

lies in $C(x) \subseteq \Omega$. If we finally apply Markov's inequality to

$$\widetilde{p}(t) := p\left(x + t\frac{x_j - x}{\|x_j - x\|_2}\right), \qquad t \in [0, r],$$

we see that

$$\begin{aligned} |p(x) - p(x_j)| &\le \int_0^{\|x - x_j\|_2} |\widetilde{p}\,'(t)| dt \\ &\le \|x - x_j\|_2 \frac{2}{r} m^2 \|\widetilde{p}\|_{L_\infty[0,r]} \\ &\le h\frac{2(1+\sin\theta)}{r\sin\theta} m^2 \|p\|_{L_\infty(\Omega)} \\ &\le \frac{1}{2} \end{aligned}$$

by using $\|x - x_j\|_2 \le \|x - y\|_2 + \|y - x_j\|_2 \le h + h/\sin\theta$. This shows that $|p(x_j)| \ge 1/2$ and proves the theorem. □

Note that in the case $h = h_{X,\Omega}$ condition (2) is automatically satisfied. However, setting $h = h_{X,\Omega}$ is somewhat too strong for what we have in mind. An immediate consequence of Theorems 3.4 and 3.8 is

Corollary 3.9 *If $X = \{x_1, \dots, x_N\} \subseteq \Omega$ and $h > 0$ satisfy the conditions of Theorem 3.8 then there exist for every $x \in \Omega$ real numbers $u_j(x)$ such that $\sum |u_j(x)| \le 2$ and $\sum u_j(x)p(x_j) = p(x)$ for all $p \in \pi_m(\mathbb{R}^d)$.*

Our first example of a region satisfying an interior cone condition is given by a ball. It is an example where the constants are independent of the space dimension.

Lemma 3.10 *Every ball with radius $\delta > 0$ satisfies an interior cone condition with radius $\delta > 0$ and angle $\theta = \pi/3$.*

Proof Without restriction we can assume that the ball is centered at zero. For every point x in the ball we have to find a cone with prescribed radius and angle. For the center $x = 0$ we can choose any direction to see that such a cone is indeed contained in the ball. For $x \neq 0$ we choose the direction $\xi = -x/\|x\|_2$. A typical point on the cone is given by $x + \lambda y$ with $\|y\|_2 = 1$, $y^T\xi \ge \cos\pi/3 = 1/2$ and $0 \le \lambda \le \delta$. For this point we find

$$\|x + \lambda y\|_2^2 = \|x\|_2^2 + \lambda^2 - 2\lambda\|x\|_2\xi^T y \le \|x\|_2^2 + \lambda^2 - \lambda\|x\|_2.$$

The last expression equals $\|x\|_2(\|x\|_2 - \lambda) + \lambda^2$, which can be bounded by $\lambda^2 \le \delta^2$ in the case $\|x\|_2 \le \lambda$. If $\lambda \le \|x\|_2$ then we can transform the last expression to $\lambda(\lambda - \|x\|_2) + \|x\|_2^2$, which can be bounded by $\|x\|_2^2 \le \delta^2$. Thus $x + \lambda y$ is contained in the ball. □

Corollary 3.11 *Define for $m \in \mathbb{N}$*

$$c_m = \frac{\sqrt{3}}{4(2 + \sqrt{3})m^2}. \tag{3.2}$$

If $Y = \{y_1, \ldots, y_M\} \subseteq B = B(x_0, \delta)$ satisfies $h_{Y,B} \le c_m\delta$ then $\mathrm{span}\{\delta_x : x \in Y\}$ *is a norming set for $\pi_m(\mathbb{R}^d)|B$ with norming constant $c = 2$. In particular, for every $x \in \Omega$ there exist real numbers $u_j(x)$ such that $\sum |u_j(x)| \le 2$ and $\sum u_j(x)p(y_j) = p(x)$ for all $p \in \pi_m(\mathbb{R}^d)$.*

While so far everything is still global, which means that we do not have any information on whether the u_j vanish, we now proceed to a local version. The main idea can be described as follows. If Ω satisfies an interior cone condition then we can find for every $x \in \Omega$ a cone with vertex x that is completely contained in Ω. Then we apply the global version given in Corollary 3.9 to this cone, using the fact that a cone itself satisfies an interior cone condition. The latter statement seems intuitively clear but its proof is rather technical. We will restrict ourselves here to cones with angle $\theta \in (0, \pi/5]$, which is actually no restriction at all.

Lemma 3.12 *Let $C = C(x_0, \xi, \theta, r)$ be a cone with angle $\theta \in (0, \pi/5]$ and radius $r > 0$. Define*

$$z = x_0 + \frac{r}{1 + \sin\theta}\xi.$$

Then we have, for every $x \in C$,

$$\|x - z\|_2 \le \frac{r}{1 + \sin\theta}.$$

Proof Without restriction we can assume that $x_0 = 0$. Because $0 < \theta \le \pi/5$ we have $2\cos\theta \ge 1 + \sin\theta$ or $2\cos\theta/(1+\sin\theta) \ge 1$. This means that

$$\begin{aligned}
\|x - z\|_2^2 &= \|x\|_2^2 + \|z\|_2^2 - 2x^T z \\
&= \|x\|_2^2 + \frac{r^2}{(1+\sin\theta)^2} - 2\frac{r}{1+\sin\theta}\xi^T x \\
&\le \|x\|_2^2 + \frac{r^2}{(1+\sin\theta)^2} - 2\frac{r\cos\theta}{1+\sin\theta}\|x\|_2 \\
&\le \|x\|^2 - r\|x\|_2 + \frac{r^2}{(1+\sin\theta)^2} \\
&= \|x\|_2(\|x\|_2 - r) + \frac{r^2}{(1+\sin\theta)^2} \\
&\le \frac{r^2}{(1+\sin\theta)^2}
\end{aligned}$$

for every $x \in C$. □

Proposition 3.13 *Suppose that $C = C(x_0, \xi, \theta, r)$ is a cone with angle $\theta \in (0, \pi/5]$ and radius $r > 0$. Then C satisfies a cone condition with angle $\widetilde{\theta} = \theta$ and radius*

$$\widetilde{r} = \frac{3\sin\theta}{4(1+\sin\theta)} r.$$

Proof Without restriction we set again $x_0 = 0$. Let

$$r_0 := \frac{\sin\theta}{1+\sin\theta} r,$$

and define

$$z := (r - r_0)\xi = \frac{r}{1+\sin\theta}\xi.$$

From Lemma 3.7 we see that the ball $B(z, r_0)$ is contained in the cone C. From Lemma 3.10 we know that we can find for every point $x \in B(z, r_0)$ a cone with angle $\pi/3 > \theta$ and radius $r_0 > (3/4)r_0 = \widetilde{r}$. This means that we can find a cone for every point inside this ball, and it remains to show the cone property for points inside the global cone but outside this ball.

Thus, we fix a point $x \in C$ with $\|x - z\|_2 \ge r_0$. Then we define the direction of the small cone to be $\zeta = (z - x)/\|z - x\|_2$ and we have to show that every point $y = x + \lambda_0 \eta$ with $\lambda_0 \in [0, \widetilde{r}]$, $\|\eta\|_2 = 1$, and $\eta^T \zeta \ge \cos\theta$ lies in C. Define

$$\lambda := \|z - x\|_2 \cos\theta + \left[r_0^2 + \|z - x\|_2^2(\cos^2\theta - 1)\right]^{1/2}.$$

Then λ is well defined, because Lemma 3.12 gives $\|z - x\|_2 \le r_0/\sin\theta$ and this means that

$$\begin{aligned}
r_0^2 + \|z - x\|_2^2(\cos^2\theta - 1) &= r_0^2 - \|z - x\|_2^2 \sin^2\theta \\
&\ge r_0^2 - \sin^2\theta \frac{r_0^2}{\sin^2\theta} = 0.
\end{aligned}$$

Furthermore, restricting θ to $(0, \pi/5]$ gives $\lambda \geq \|z - x\|_2 \cos\theta \geq r_0 \cos\theta \geq 3r_0/4 = \widetilde{r}$. This means that if the point $x + \lambda\eta$ is contained in C then the convexity of C shows that the point $x + \lambda_0\eta$ is also in C, which gives the cone property. This is done by an easy manipulation:

$$\begin{aligned}
\|z - (x + \lambda\eta)\|_2^2 &= \|z - x\|_2^2 + \lambda^2 - 2\lambda\eta^T(z - x) \\
&= \|z - x\|_2^2 + \lambda^2 - 2\lambda\|z - x\|_2\eta^T\zeta \\
&\leq \|z - x\|_2^2 + \lambda^2 - 2\lambda\|z - x\|_2 \cos\theta \\
&= (\lambda - \cos\theta\|z - x\|_2)^2 - \|z - x\|^2\cos^2\theta + \|z - x\|^2 \\
&= r_0^2.
\end{aligned}$$

Hence $x + \lambda\eta \in B(z, r_0) \subseteq C$. □

Now we are able to formulate and prove our local version of Corollary 3.9. Again let us point out that all the constants are in explicit form.

Theorem 3.14 *Suppose that $\Omega \subseteq \mathbb{R}^d$ is compact and satisfies an interior cone condition with angle $\theta \in (0, \pi/2)$ and radius $r > 0$. Fix $m \in \mathbb{N}$. Then there exist constants $h_0, C_1, C_2 > 0$ depending only on m, θ, r such that for every $X = \{x_1, \ldots, x_N\} \subseteq \Omega$ with $h_{X,\Omega} \leq h_0$ and every $x \in \Omega$ we can find real numbers $\widetilde{u}_j(x)$, $1 \leq j \leq N$, with*

(1) $\sum_{j=1}^N \widetilde{u}_j(x)p(x_j) = p(x)$ for all $p \in \pi_m(\mathbb{R}^d)$,
(2) $\sum_{j=1}^N |\widetilde{u}_j(x)| \leq C_1$,
(3) $\widetilde{u}_j(x) = 0$ provided that $\|x - x_j\|_2 > C_2 h_{X,\Omega}$.

Proof Without restriction we may assume $\theta \leq \pi/5$. We define the constants as

$$C_1 = 2, \qquad C_2 = \frac{16(1 + \sin\theta)^2 m^2}{3\sin^2\theta}, \qquad h_0 = \frac{r}{C_2}.$$

Let $h = h_{X,\Omega}$. Since $C_2 h \leq r$ the region Ω also satisfies an interior cone condition with angle θ and radius $C_2 h$. Hence, for every $x \in \Omega$ we can find a cone $C(x) := C(x, \xi, \theta, C_2 h)$, that is completely contained in Ω. By Proposition 3.13, the cone $C(x)$ itself satisfies a cone condition with angle θ and radius

$$\widetilde{r} = \frac{3\sin\theta}{4(1 + \sin\theta)} C_2 h.$$

Moreover, by the definition of C_2,

$$h = \widetilde{r}\frac{\sin\theta}{4(1 + \sin\theta)m^2},$$

so that we can apply Corollary 3.9 to the cone $C(x)$ and $Y = X \cap C(x)$ with $h = h_{X,\Omega}$. Hence, we find numbers $\widetilde{u}_j(x)$, for every j with $x_j \in Y$, such that

$$\sum_{x_j \in Y} \widetilde{u}_j(x)p(x_j) = p(x)$$

for all $p \in \pi_m(\mathbb{R}^d)$ and

$$\sum_{x_j \in Y} |\widetilde{u}_j(x)| \le 2.$$

By setting $\widetilde{u}_j(x) = 0$ for $x_j \in X \setminus Y$, we have (1) and (2). For (3), we realize that $\|x - x_j\|_2 > C_2 h$ means that x_j is not in the cone $C(x)$, and thus $\widetilde{u}_j(x) = 0$ by construction. □

Note that all constants can be improved if better estimates on the cone condition of a cone are used. Moreover, if the region Ω is a ball then it is better to use the results on balls, i.e. instead of using small cones to construct the functions u_j one should use small balls to derive better constants. This philosophy holds for other regions as well. The more information is known about a region the better are the constants.

Finally, we want to point out that

$$\sum_{j=1}^{N} |\widetilde{u}_j(x)| \le 2$$

means that the Lebesgue functions of the associated quasi-interpolants are uniformly bounded by 2. The reader should compare this with the known results on polynomial interpolation. The price we have to pay for this uniform bound is oversampling: we use significantly more points than $\dim(\pi_m(\mathbb{R}^d))$.

3.4 Notes and comments

The idea of using local polynomial reproductions in approximation theory is quite an old one. One could say that even splines are based on this background.

Norming sets have been introduced in the context of interpolation by positive definite functions on the sphere; see Jetter *et al.* [89]. With this approach it was for the first time possible to control the constants involved in the process of bounding certain Lebesgue functions. Since then, norming sets have been used for example to derive results on quadrature rules on the sphere (Mhaskar *et al.* [131, 132]), approximation with interpolatory constraints (Mhaskar *et al.* [130]), the moving least squares approximation (Wendland [196]), and the determination of explicit constants (Narcowich *et al.* [148]). The presentation given here is mainly motivated by Mhaskar *et al.* [132]. In the next chapter we will have a closer look at an application in the context of the moving least squares approximation.

4
Moving least squares

The crucial point in local polynomial reproduction is the compact support of the basis functions u_j. To be more precise, all supports have to be of the same size. The local support of the u_j means that data points far away from the current point of interest x have no influence on the function value at x. This is often a reasonable assumption.

The last chapter did not answer the question how to construct families with local polynomial reproductions efficiently. The moving least squares method provided in this chapter forms an example of this.

4.1 Definition and characterization

Suppose again that discrete values of a function f are given at certain data sites $X = \{x_1, \ldots, x_N\} \subseteq \Omega \subseteq \mathbb{R}^d$. Throughout this chapter Ω is supposed to satisfy an interior cone condition with angle θ and radius r.

The idea of the moving least squares approximation is to solve for every point x a locally weighted least squares problem. This appears to be quite expensive at first sight, but it will turn out to be a very efficient method. Moreover, in many applications one is only interested in a few evaluations. For such applications the moving least squares approximation is even more attractive, because it is not necessary to set up and solve a large system.

The influence of the data points is governed by a weight function $w : \Omega \times \Omega \to \mathbb{R}$, which becomes smaller the further away its arguments are from each other. Ideally, w vanishes for arguments $x, y \in \Omega$ with $\|x - y\|_2$ greater than a certain threshold. Such a behavior can be modeled by using a translation-invariant weight function. This means that w is of the form $w(x, y) = \Phi_\delta(x - y)$, $\Phi_\delta = \Phi(\cdot/\delta)$ being the scaled version of a compactly supported function $\Phi : \mathbb{R}^d \to \mathbb{R}$.

Definition 4.1 *For $x \in \Omega$, the value $s_{f,X}(x)$ of the moving least squares approximant is given by $s_{f,X}(x) = p^*(x)$ where p^* is the solution of*

$$\min \left\{ \sum_{i=1}^{N} [f(x_i) - p(x_i)]^2 w(x, x_i) \; : \; p \in \pi_m(\mathbb{R}^d) \right\}. \tag{4.1}$$

In what follows we will assume that $w(x, y) = \Phi_\delta(x - y)$ uses a nonnegative function Φ, with support in the unit ball $B(0, 1)$, which is positive on the ball $B(0, 1/2)$ with radius $1/2$. The latter will be important for our error analysis. In many applications it might be convenient to assume Φ to be radial as well, meaning that $\Phi(x) = \phi(\|x\|_2)$, $x \in \mathbb{R}^d$, with a univariate and nonnegative function $\phi : [0, \infty) \to \mathbb{R}$ that is positive on $[0, 1/2]$ and supported in $[0, 1]$. But this is not essential for what we have in mind.

In any case we can reformulate (4.1) more precisely as

$$\min\left\{\sum_{i\in I(x)} [f(x_i) - p(x_i)]^2 \Phi_\delta(x - x_i) \;:\; p \in \pi_m(\mathbb{R}^d)\right\}, \tag{4.2}$$

using the set of indices

$$I(x) \equiv I(x, \delta, X) := \{j \in \{1, \ldots, N\} : \|x - x_j\|_2 < \delta\}.$$

So far it is not clear at all why moving least squares provides local polynomial reproduction. It is quite instructive to have a look at the simplest case, namely $m = 0$. In this situation only constants are reproduced and the minimization problem has the explicit solution

$$s_{f,X}(x) = \sum_{j\in I(x)} f(x_j) \underbrace{\frac{\Phi_\delta(x - x_j)}{\sum_{i\in I(x)} \Phi_\delta(x - x_i)}}_{a_j(x):=}. \tag{4.3}$$

Such an approximant is sometimes also called the Shepard approximant (Shepard [177]). Obviously, it reproduces constants. The basis functions a_j have a support of radius δ, so that $\delta > C_2 h_{X,\Omega}$ is a good choice. Finally, one can immediately see that $\sum_j |a_j| = 1$. Hence we have indeed a local polynomial reproduction for polynomials of degree zero. Moreover, the smoothness of the approximant is ruled by the smoothness of Φ, provided that the denominator is always nonzero. The latter is always true for a sufficiently large $\delta > 0$.

Let us now come back to the case of a general m. We start with a reformulation of the initial minimization problem. To this end we need a result on quadratic optimization.

Lemma 4.2 *Let $a \in \mathbb{R}$, $b \in \mathbb{R}^n$, $A \in \mathbb{R}^{n\times n}$, and $P \in \mathbb{R}^{n\times m}$ be given. For $v \in \mathbb{R}^m$ define $M_v := \{x \in \mathbb{R}^n : P^T x = v\}$. Suppose that $A = A^T$ is positive definite on M_0. If M_v is not empty then the quadratic function*

$$f(x) := a + b^T x + x^T A x$$

has a unique minimum on M_v.

Proof Obviously, the set M_v is convex and closed. Hence, if f is strictly convex on M_v then we immediately have uniqueness. Since A is positive definite on the subspace M_0, there exists a $\lambda_{\min} > 0$ such that $x^T A x \ge \lambda_{\min}\|x\|_2^2$ for all $x \in M_0$. Then, an easy calculation gives, for $x, y \in M_v$ and $\lambda \in [0, 1]$,

$$\begin{aligned}(1 - \lambda) f(x) + \lambda f(y) - f((1 - \lambda)x + \lambda y) &= \lambda(1 - \lambda)(x - y)^T A(x - y)\\ &\ge \lambda(1 - \lambda)\lambda_{\min}\|x - y\|_2^2,\end{aligned}$$

showing that f is indeed strictly convex on M_v. To prove existence, we use a method standard in this context. Since M_v is not empty we can fix an $x_0 \in M_v$ and restrict the minimization problem to $\widetilde{M}_v := \{x \in M_v : f(x) \le f(x_0)\}$. Since f is continuous, $\widetilde{M}_v$ is also closed. Thus it remains to show that it is also bounded. Every $x \in \widetilde{M}_v$ can be represented by $x = x_0 + z$ with $z \in M_0$. Hence

$$\begin{aligned} 0 &\ge f(x) - f(x_0) \\ &= (b + 2Ax_0)^T(x - x_0) + (x - x_0)^T A(x - x_0) \\ &\ge \lambda_{\min}\|x - x_0\|_2^2 - \|b + 2Ax_0\|_2\|x - x_0\|_2 \end{aligned}$$

shows that $\|x - x_0\|_2$, and therefore also $\|x\|_2$, is uniformly bounded. □

We have formulated Lemma 4.2 more generally than is necessary for our purposes here, but we will need this more general version in a later chapter. Here, we consider only minimization problems where A is positive definite on all of $\mathbb{R}^d$. Note that setting $P = 0$ and $v = 0$ gives the global optimization problem on $M_v = \mathbb{R}^d$.

Theorem 4.3 *Suppose that for every $x \in \Omega$ the set $\{x_j : j \in I(x)\}$ is $\pi_m(\mathbb{R}^d)$-unisolvent. In this situation, problem (4.2) is uniquely solvable and the solution $s_{f,X}(x) = p^*(x)$ can be represented as*

$$s_{f,X}(x) = \sum_{i \in I(x)} a_i^*(x) f(x_i),$$

where the coefficients $a_i^*(x)$ are determined by minimizing the quadratic form

$$\sum_{i \in I(x)} a_i(x)^2 \frac{1}{\Phi_\delta(x - x_i)} \tag{4.4}$$

under the constraints

$$\sum_{j \in I(x)} a_j(x) p(x_j) = p(x), \qquad p \in \pi_m(\mathbb{R}^d). \tag{4.5}$$

Proof Denote a basis of $\pi_m(\mathbb{R}^d)$ by $p_1, \ldots, p_Q$. Suppose our polynomial has the form $p = \sum_{j=1}^{Q} b_j p_j$. This reduces the minimization problem (4.2) to finding the optimal coefficient vector b^*.

We use the following notation:

$$\begin{aligned} b &= (b_1, \ldots, b_Q)^T \in \mathbb{R}^Q, \\ \widetilde{f} &= (f(x_i) : i \in I(x))^T \in \mathbb{R}^{\#I(x)}, \\ P &= (p_j(x_i))_{\substack{i \in I(x) \\ 1 \le j \le Q}} \in \mathbb{R}^{\#I(x) \times Q}, \\ D = D(x) &= \operatorname{diag}(\Phi_\delta(x - x_i) : i \in I(x)) \in \mathbb{R}^{\#I(x) \times \#I(x)}, \\ R(x) &= (p_1(x), \ldots, p_Q(x))^T \in \mathbb{R}^Q. \end{aligned}$$

Then we have to minimize the function

$$\begin{aligned}
C(b) &= \sum_{i\in I(x)} \left[f(x_i) - \sum_{j=1}^{Q} b_j p_j(x_i) \right]^2 \Phi_\delta(x - x_i) \\
&= \sum_{i\in I(x)} [\widetilde{f}_i - (Pb)_i]^2 \Phi_\delta(x - x_i) \\
&= (\widetilde{f} - Pb)^T D(\widetilde{f} - Pb) \\
&= \widetilde{f}^T D\widetilde{f} - 2\widetilde{f}^T DPb + b^T P^T DPb
\end{aligned}$$

on $\mathbb{R}^Q$. Since $C(b)$ is a quadratic function in b we can apply Lemma 4.2. We get a unique solution if $P^T DP$ is positive definite. From

$$b^T P^T DPb = b^T P^T D^{1/2} D^{1/2} Pb = \|D^{1/2} Pb\|_2$$

it follows that $P^T DP$ is positive semi-definite. Moreover, $b^T P^T DPb = 0$ means that $Pb = 0$. Thus the polynomial $p = \sum b_j p_j$ vanishes on every $x_i, i \in I(x)$. Since this set is assumed to be $\pi_m(\mathbb{R}^d)$-unisolvent, p and hence b must be zero.

Now that we know the existence of a unique solution we can use the necessary condition $\nabla C(b^*) = 0$ to compute it. We find that

$$0 = \nabla C(b^*) = -2\widetilde{f}^T DP + 2(b^*)^T (P^T DP),$$

which gives $(b^*)^T = \widetilde{f}^T DP(P^T DP)^{-1}$, and we obtain the solution

$$p^*(x) = (b^*)^T R(x) = \widetilde{f}^T DP(P^T DP)^{-1} R(x).$$

In the final step we treat the problem of minimizing (4.4) under the constraints (4.5). This means that we have to minimize the function

$$C(a) := \sum_{i\in I(x)} a_i^2 \frac{1}{\Phi_\delta(x - x_i)} = a^T D^{-1} a$$

on the set

$$\begin{aligned}
M &:= \left\{ a \in \mathbb{R}^{\#I(x)} : \sum_{i\in I(x)} a_i p_j(x_i) = p_j(x), \quad 1 \le j \le Q \right\} \\
&= \{a \in \mathbb{R}^{\#I(x)} : P^T a = R(x)\}.
\end{aligned}$$

Since we have supposed $\{x_i : i \in I(x)\}$ to be $\pi_m(\mathbb{R}^d)$-unisolvent we can always find Q points that allow unique polynomial interpolation. Hence M is not empty. Moreover, D^{-1} is obviously positive definite. Thus Lemma 4.2 gives us a unique solution to this problem. To compute this solution we can use Lagrange multipliers. If $a^* \in M$ is a solution of the modified problem, there has to be a $\lambda \in \mathbb{R}^Q$ such that

$$\nabla C(a^*) = \lambda^T \frac{\partial}{\partial a} \left[P^T a - R(x) \right] \big|_{a=a^*}.$$

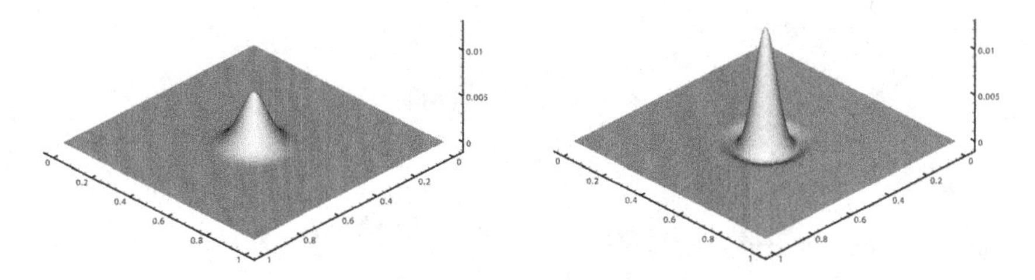

Fig. 4.1 Basis functions for $m = 0$ (on the left) and $m = 2$ (on the right).

This means that $2(a^*)^T D^{-1} = \lambda^T P^T$ or $a^* = DP\lambda/2$, showing in particular that a^* is the unique solution of the modified problem, which makes it also the solution of the initial problem. From $a^* \in M$ we can conclude that $R(x) = P^T a^* = P^T DP\lambda/2$, which gives the representation $\lambda = 2(P^T DP)^{-1} R(x)$. Finally, we find

$$\sum_{i \in I(x)} a_i^* f(x_i) = \widetilde{f}^T a^* = \widetilde{f}^T DP(P^T DP)^{-1} R(x) = p^*(x).$$

□

From the proof of the last theorem we can read off interesting properties of the basis functions a_j^*. The most important of these concern the form of a_j^* and the smoothness of the approximant.

Corollary 4.4 *The basis functions a_j^* are given by*

$$a_j^*(x) = \Phi_\delta(x - x_j) \sum_{k=1}^{Q} \lambda_k p_k(x_j), \tag{4.6}$$

where the λ_k are the unique solutions of

$$\sum_{k=1}^{Q} \lambda_k \sum_{j \in I(x)} \Phi_\delta(x - x_j) p_k(x_j) p_\ell(x_j) = p_\ell(x), \qquad 0 \le \ell \le Q. \tag{4.7}$$

Figure 4.1 shows two typical basis functions, for $m = 0$ and $m = 2$. The data sites form a regular 100×100 grid and the weight function is given by $\phi(r) = (1 - r)_+^4(4r + 1)$ (see Chapter 9). The support radius $\delta = 0.2$ was chosen to be large for illustration purposes; in practical applications such an ideal point placement would lead to a much smaller choice of δ, for example $\delta = 0.05$. Note that the basis function for $m = 0$ is nonnegative as it should be. The basis function for $m = 2$, however, becomes negative close to the boundary of its support. This will make the error analysis for higher-order moving least squares somewhat harder.

Corollary 4.5 *If Φ possesses k continuous derivatives then the approximant $s_{f,X}$ is also in C^k.*

Proof From the last corollary we know that the smoothness of $s_{f,X}$ is governed by the smoothness of Φ and the smoothness of the λ_j. Since Φ is supported in the unit ball, the

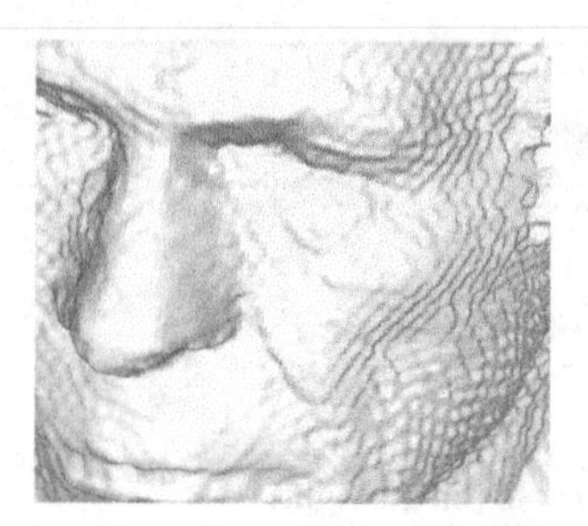  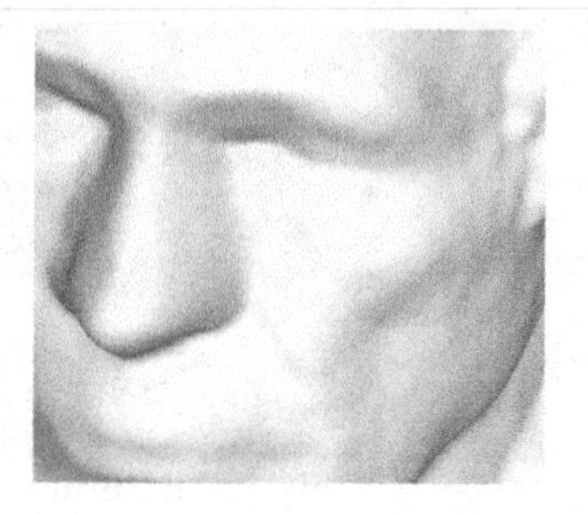

Fig. 4.2 Influence of the scaling parameter in the moving least squares approximation: from left to right, $\delta = 0.05, \delta = 0.07, \delta = 0.1$.

λ_j are also the solutions of

$$\sum_{k=1}^{Q} \lambda_k \sum_{j=1}^{N} \Phi_\delta(x - x_j) p_k(x_j) p_\ell(x_j) = p_\ell(x), \qquad 0 \le \ell \le Q,$$

and therefore as smooth as Φ. □

However, even if the smoothness of the approximant is governed by the smoothness of Φ, numerical examples show that the effect of the scaling parameter on the "visual" smoothness is more important. Too small a scaling parameter leads to a bumpy surface, while too large a parameter results in a smoothed-out representation; see Figure 4.2.

4.2 Local polynomial reproduction by moving least squares

In the case $m = 0$ it is easy to bound the approximation error. We know that $s_{f,X}$ has the representation (4.3), which shows immediately that the basis functions

$$a_i^*(x) = \frac{\Phi_\delta(x - x_i)}{\sum_{j \in I(x)} \Phi_\delta(x - x_j)}$$

satisfy $\sum |a_i^*(x)| = \sum a_i^*(x) = 1$. This gives the error bound

$$|f(x) - s_{f,X}(x)| \le 2\|f - p\|_{L_\infty(B(x,\delta))}.$$

For general $m \in \mathbb{N}$ we can draw conclusions on the approximation property by showing that the moving least squares approximation provides local polynomial reproduction. This needs a further assumption on the set of data sites. The general concept we are now going to introduce will be essential in the whole of this book.

So far, we have only measured how well a set of data sites covers the domain Ω. This is done by means of $h_{X,\Omega}$, which guarantees a data site x_j within a distance $h_{X,\Omega}$ for every $x \in \Omega$. Or, in other words, the largest ball in Ω that does not contain a data site has radius at most $h_{X,\Omega}$.

However, this covering should be achieved by as few data sites as possible.

Definition 4.6 *The separation distance of $X = \{x_1, \dots, x_N\}$ is defined by*

$$q_X := \frac{1}{2}\min_{i \neq j} \|x_i - x_j\|_2.$$

A set X of data sites is said to be quasi-uniform with respect to a constant $c_{\mathrm{qu}} > 0$ if

$$q_X \leq h_{X,\Omega} \leq c_{\mathrm{qu}} q_X. \tag{4.8}$$

The separation distance gives the largest possible radius for two balls centered at different data sites to be essentially disjoint. The definition of quasi-uniformity has to be seen in the context of more than one set of data sites. The idea is that a sequence of such sets is considered such that the region Ω is more and more filled out. Then it is important that condition (4.8) is satisfied by all sets in this sequence with the same constant c_{qu}. For example, if $\Omega = [0, 1]^d$ is the unit cube and $X_h = h\mathbb{Z}^d \cap \Omega$ then obviously the separation distance is given by $q_X = h/2$ while the fill distance is given by $h_{X_h,\Omega} = (\sqrt{d}/2)h$. Hence, we have quasi-uniformity with $c_{qu} = \sqrt{d}$. We will discuss quasi-uniform sets in more detail in Chapter 14.

Theorem 4.7 *Suppose that $\Omega \subseteq \mathbb{R}^d$ is compact and satisfies an interior cone condition with angle $\theta \in (0, \pi/2)$ and radius $r > 0$. Fix $m \in \mathbb{N}$. Let h_0, C_1, and C_2 denote the constants from Theorem 3.14. Suppose that $X = \{x_1, \dots, x_N\} \subseteq \Omega$ satisfies (4.8) and $h_{X,\Omega} \leq h_0$. Let $\delta = 2C_2 h_{X,\Omega}$. Then the basis functions $a_j^*(x)$ from Corollary 4.4 provide local polynomial reproduction, i.e.*

(1) $\sum_{j=1}^N a_j^*(x)p(x_j) = p(x)$ for all $p \in \pi_m(\mathbb{R}^d)$, $x \in \Omega$,
(2) $\sum_{j=1}^N |a_j^*(x)| \leq \widetilde{C}_1$,
(3) $a_j^*(x) = 0$ if $\|x - x_j\|_2 > \widetilde{C}_2 h_{X,\Omega}$,

with certain constants $\widetilde{C}_1, \widetilde{C}_2$ that can be derived explicitly.

Proof The first property follows from the construction process of moving least squares. The third property is a consequence of the compact support of Φ. By Corollary 4.4, a_j^* is supported in $B(x_j, \delta)$, and δ has been chosen as $2C_2 h_{X,\Omega}$. This shows that in particular $\widetilde{C}_2 = 2C_2$. For the second property we have to work harder. We start with

$$\sum_{i \in I(x)} |a_i^*(x)| \leq \left(\sum_{i \in I(x)} |a_i^*(x)|^2 \frac{1}{\Phi_\delta(x - x_i)}\right)^{1/2} \left(\sum_{i \in I(x)} \Phi_\delta(x - x_i)\right)^{1/2}$$

and bound each factor on the right-hand side separately, beginning with the first.

By Theorem 3.14 we know that there exists a set $\{\widetilde{u}_j\}$ providing local polynomial reproduction. The function $\widetilde{u}_j(x)$ vanishes if $\|x - x_j\|_2 > C_2 h_{X,\Omega} = \delta/2$. The minimal property of $\{a_j^*\}$ allows us the estimate

$$\begin{aligned}\sum_{i\in I(x)} |a_i^*(x)|^2 \frac{1}{\Phi_\delta(x-x_i)} &\le \sum_{i\in \widetilde{I}(x)} |\widetilde{u}_i(x)|^2 \frac{1}{\Phi_\delta(x-x_i)} \\ &\le \frac{1}{\min_{y\in B(0,1/2)}\Phi(y)} \sum_{i\in\widetilde{I}(x)} |\widetilde{u}_i(x)|^2 \\ &\le \frac{1}{\min_{y\in B(0,1/2)}\Phi(y)} \left(\sum_{i\in\widetilde{I}(x)} |\widetilde{u}_i(x)|\right)^2 \\ &\le \frac{C_1^2}{\min_{y\in B(0,1/2)}\Phi(y)}.\end{aligned}$$

Here, we have used the notation $\widetilde{I}(x) = \{j : x_j \in B(x, \delta/2)\}$. To bound the second factor we first note that obviously

$$\sum_{i\in I(x)} \Phi_\delta(x-x_i) \le \#I(x)\|\Phi\|_{L_\infty(\mathbb{R}^d)}.$$

To bound the number $\#I(x)$ of points in $I(x)$ we use a packing argument. Any ball with radius q_X centered at x_j is essentially disjoint to a ball centered at $x_k \ne x_j$ of the same radius, and all these balls with $j \in I(x)$ are contained in a ball with radius $q_X + \delta$. Comparing the volume of the union of the small balls with the volume of the large ball gives

$$\#I(x)\ \mathrm{vol}(B(0,1))\ q_X^d \le \mathrm{vol}(B(0,1))\ (\delta + q_X)^d.$$

Using quasi-uniformity finally leads to

$$\#I(x) \le (1+\delta/q_X)^d \le (1+2C_2c_{\mathrm{qu}})^d.$$

□

It is important to recognize that we have an explicit estimate on the constant C_2, because the latter determines the necessary scale factor δ. However, numerical tests show that the estimate we derived in Theorem 3.14 is rather pessimistic. One can often get away with a much smaller support radius.

Corollary 4.8 *In the situation of Theorem 4.7 define Ω^* to be the closure of $\cup_{x\in\Omega} B(x, 2C_2h_0)$. Then there exists a constant $c > 0$ that can be computed explicitly such that for all $f \in C^{m+1}(\Omega^*)$ and all quasi-uniform $X \subseteq \Omega$ with $h_{X,\Omega} \le h_0$ the approximation error is bounded as follows:*

$$\|f - s_{f,X}\|_{L_\infty(\Omega)} \le c h_{X,\Omega}^{m+1} |f|_{C^{m+1}(\Omega^*)}.$$

We end this section with an investigation of the computational complexity of the moving least squares method and with remarks on its practicability. We restrict our analysis to the situation of quasi-uniform data sites.

From Corollary 4.4 we can read off that the computational complexity of the approximant at a single point x is bounded by $\mathcal{O}(Q^3 + Q^2\#I(x) + Q\#I(x))$, if Q denotes the dimension of $\pi_m(\mathbb{R}^d)$ and if the time necessary to evaluate a polynomial is considered to be constant. We use the classical Landau symbol $\mathcal{O}$ to suppress unimportant constants. We know, however, that $\#I(x)$ is bounded by a constant independent of x. Thus the complexity for a single evaluation is constant if the set of indices $I(x)$ is known in advance. But this can be done in a preprocessing step that takes $\mathcal{O}(N)$ time and is based on a simple boxing strategy. To be more precise, it is not reasonable to collect the relevant data sites (i.e. $I(x)$) for each evaluation point separately. This would lead to an $\mathcal{O}(NM)$ complexity, if M denotes the number of evaluations. Instead, it is better to divide the domain Ω into boxes of side length h and collect for each box the data sites that it contains using a loop over all points. This can obviously be done in linear time. After that we can find for every $x \in \Omega$ the relevant boxes in constant time, so that the overall complexity for M evaluations of the approximant based on N points is $\mathcal{O}(N + M)$. A more thorough analysis of data structures for points in d-dimensional space is made in Chapter 14.

Finally, it is not reasonable to use a fixed polynomial basis such as a set of monomials for every evaluation point $x \in \Omega$. Instead, a local monomial basis centered at the evaluation point x and scaled by the support radius δ leads to a more stable method.

4.3 Generalizations

We now discuss two different matters. One deals with the assumption of quasi-uniformity; the other carries the initial problem over to a more general setting.

The assumption that the data sites are quasi-uniform is often violated, for example, if the data are clustered or if there are more points in regions where it is supposed that the unknown function has a difficult behavior. In such a situation our analysis fails. The reason for this is that we have chosen the same support radius δ everywhere. In our particular situation we have two possible ways to resolve this problem. On the one hand we could assign a different support radius to each basis function a_j^*. Even if in general this is a good idea, it would cause problems in the case of moving least squares. The better choice is to let δ depend on the current evaluation point. Hence, if x lies in a region with a high data density, $\delta(x)$ would be chosen small. If there are only few points around x, one has to choose $\delta(x)$ rather large. Of course δ, now considered as a function of x, should vary smoothly in x.

Sometimes one faces a more general approximation problem than the one discussed so far. For example, if partial differential equations have to be solved numerically, information is given not on the function itself but on its derivatives. In the case of moving least squares, most of what was done in the point-evaluation case holds also in a more general setting.

To describe the general problem, we will assume that we are working in a function space $\mathcal{F} \subseteq C(\Omega)$. The information we have is represented by a finite number of continuous and linear functionals $\lambda_1, \dots, \lambda_N \in \mathcal{F}^*$ and the information we seek is represented by a functional $\lambda \in \mathcal{F}^*$. Hence, we are given $\lambda_1(f), \dots, \lambda_N(f)$ and we want to use these data to get a good approximation $\widetilde{\lambda}(f)$ to $\lambda(f)$. The idea of moving least squares or, more generally of local polynomial reproductions requires the reproduction process to be exact on a finite-dimensional subspace $\mathcal{P}$ of $\mathcal{F}$. This has been a set of polynomials so far but we can also consider other function spaces.

In the spirit of moving least squares we choose a nonnegative weight function w, this time defined on $\mathcal{F}^* \times \mathcal{F}^*$, which measures the correlation of two functionals, and we define $\widetilde{\lambda}(f)$ to be $\lambda(p^*)$, where $p^* \in \mathcal{P}$ is the solution of the minimization problem

$$\min \left\{ \sum_{i=1}^{N} [\lambda_i(f) - \lambda_i(p)]^2 w(\lambda, \lambda_i) \ : \ p \in \mathcal{P} \right\}. \tag{4.9}$$

This new problem reduces to the old one by setting $\lambda_j = \delta_{x_j}$, $1 \le j \le N$, $\lambda = \delta_x$, and $w(\delta_{x_j}, \delta_x) = \Phi_\delta(x - x_j)$.

A closer look at what we did to prove existence and uniqueness shows that Theorem 4.3 remains true even in this more general situation. We only have to adapt the notion of unisolvent sets. But it should be clear that $\Lambda = \{\lambda_1, \ldots, \lambda_N\}$ is $\mathcal{P}$-unisolvent if and only if $\lambda_j(p) = 0$, $1 \le j \le N$, implies $p = 0$. Moreover, we have to replace the set of indices $I(x)$ by $I(\lambda) = \{j : w(\lambda_j, \lambda) > 0\}$.

Theorem 4.9 *Suppose that $\Lambda = \{\lambda_1, \ldots, \lambda_N\}$ is $\mathcal{P}$-unisolvent. In this situation, problem (4.9) is uniquely solvable and the solution $\widetilde{\lambda}(f) = \lambda(p^*)$ can be represented as*

$$\widetilde{\lambda}(f) = \sum_{i \in I(\lambda)}^{N} a_i^* \lambda_i(f),$$

where the coefficients a_i are determined by minimizing the quadratic form

$$\sum_{i \in I(\lambda)} a_i^2 \frac{1}{w(\lambda, \lambda_i)} \tag{4.10}$$

under the constraints

$$\sum_{i \in I(\lambda)} a_i \lambda_i(p) = \lambda(p), \qquad p \in \mathcal{P}. \tag{4.11}$$

It is crucial that $\Lambda = \{\lambda_1, \ldots, \lambda_N\}$ is $\mathcal{P}$-unisolvent, otherwise there will not be a unique solution. Hence, it is in general not possible to recover f from its derivatives in the interior of a domain Ω and its function values at the boundary of Ω as one would do in classical collocation methods for the numerical solution of partial differential equations. The information on the boundary gets lost in the interior because of the compact support of the weight function. But the compact support is necessary for an efficient evaluation.

4.4 Notes and comments

Approximation by moving least squares has its origin in the early paper [101] by Lancaster and Salkauskas from 1981 with special cases going back to McLain [120, 121] in 1974 and

1976 and to Shepard [177] in 1968. Other early papers are those by Farwig [53–55], who mainly concentrated on investigating the approximation order. It is interesting to see that Farwig remarked in [53] that the "least squares problem . . . varies in x" and hence "computing the global approximant . . . is generally very time-consuming". With the development of fast computers and efficient data structures for the data sites, this statement no longer holds, and the moving least squares approximation has in recent times attracted attention again, in the present context and also for the numerical solution of partial differential equations; see Belytschko *et al.* [23].

The equivalence of the two minimization problems given in Theorem 4.3 was first pointed out by Levin [104, 105]. The error estimates given here are based upon the present author's paper [196].

5

Auxiliary tools from analysis and measure theory

For our investigations on radial basis functions in the following chapters it is crucial to collect several results from different branches of mathematics. Hence, this chapter is a repository of such results. In particular, we are concerned with special functions such as Bessel functions and the Γ-function. We discuss the features of Fourier transforms and give an introduction to the aspects of measure theory that are relevant for our purposes. The reader who is not interested in these technical matters could skip this chapter and come back to it whenever necessary. Nonetheless, it is strongly recommended that one should at least have a look at the definition of Fourier transforms to become familiar with the notation we use. Because of the diversity of results presented here, we cannot give proofs in every case.

5.1 Bessel functions

Bessel functions will play an important role in what follows. Most of what we discuss here can be found in the fundamental book [187] by Watson.

The starting point for introducing Bessel functions is to remind the reader of the classical Γ-function and some of its features.

Definition 5.1 *The Γ-function is defined by*

$$\Gamma(z) := \lim_{n\to\infty} \frac{n!\,n^z}{z(z+1)\dots(z+n)}$$

for $z \in \mathbb{C}$.

It is a meromorphic function, well investigated in classical analysis. Some of its relevant properties are collected in the next proposition:

Proposition 5.2 *The Γ-function has the following properties:*

(1) $1/\Gamma(z)$ is an entire function;
(2) $\Gamma(1) = 1$, $\Gamma(1/2) = \sqrt{\pi}$;
(3) $\Gamma(z) = \int_0^\infty e^{-t}t^{z-1}dt$ for $\Re(z) > 0$ (Euler's representation);
(4) $\Gamma(z+1) = z\Gamma(z)$ (recurrence relation);

(5) $\Gamma(z)\Gamma(1-z) = \pi/\sin(\pi z)$ *(reflection formula);*

(6) $1 \le \dfrac{\Gamma(x+1)}{\sqrt{2\pi x}\, x^x e^{-x}} \le e^{1/(12x)}$, $x > 0$ *(Stirling's formula);*

(7) $\Gamma(2z) = \dfrac{2^{2z-1}}{\sqrt{\pi}}\Gamma(z)\Gamma(z+1/2)$ *(Legendre's duplication formula).*

The Γ-function and its properties are well known. Proofs of the formulas just stated can be found in any book on special functions. A particular choice would be the book [102] by Lebedev. Now, we continue by introducing Bessel functions.

Definition 5.3 *The Bessel function of the first kind of order $\nu \in \mathbb{C}$ is defined by*

$$J_\nu(z) := \sum_{m=0}^{\infty} \frac{(-1)^m (z/2)^{2m+\nu}}{m!\,\Gamma(\nu+m+1)}$$

for $z \in \mathbb{C} \setminus \{0\}$.

The power z^ν in this definition is defined by $\exp[\nu \log(z)]$, where log is the principal branch of the logarithm, i.e. $-\pi < \arg(z) \le \pi$.

The Bessel function can be seen as a function of z and also as a function of ν and the following remarks are easily verified. Obviously, $J_\nu(z)$ is holomorphic in $\mathbb{C} \setminus [0, \infty)$ as a function of z for every $\nu \in \mathbb{C}$. Moreover, the expansion converges pointwise also for $z < 0$. If $\nu \in \mathbb{N}$ then J_ν has an analytic extension to $\mathbb{C}$. If $\Re(\nu) \ge 0$ then we have a continuous extension of $J_\nu(z)$ to $z = 0$. Finally, if $z \in \mathbb{C} \setminus \{0\}$ is fixed then $J_\nu(z)$ is a holomorphic function in $\mathbb{C}$ as a function of ν. We state further elementary properties in the next proposition.

Proposition 5.4 *The Bessel function of the first kind has the following properties:*

(1) $J_{-n} = (-1)^n J_n$ *if* $n \in \mathbb{N}$;

(2) $\dfrac{d}{dz}\{z^\nu J_\nu(z)\} = z^\nu J_{\nu-1}(z)$;

(3) $\dfrac{d}{dz}\{z^{-\nu} J_\nu(z)\} = -z^{-\nu} J_{\nu-1}(z)$;

(4) $J_{1/2}(z) = \sqrt{\dfrac{2}{\pi z}} \sin z, \qquad J_{-1/2}(z) = \sqrt{\dfrac{2}{\pi z}} \cos z.$

Proof For the first property we simply use that $1/\Gamma(n) = 0$ for $n = 0, -1, -2, \ldots$. The second and the third property follow by differentiation under the sum using also the recurrence formula for the Γ-function. The latter together with $\Gamma(1/2) = \sqrt{\pi}$ demonstrates the last item. □

There exist many integral representations for Bessel functions. The one that matters here is given in the next proposition.

Proposition 5.5 *If we denote for $d \geq 2$ the unit sphere by $S^{d-1} = \{x \in \mathbb{R}^d : \|x\|_2 = 1\}$ then we have, for $x \in \mathbb{R}^d$,*

$$\int_{S^{d-1}} e^{ix^T y} dS(y) = (2\pi)^{d/2} \|x\|_2^{-(d-2)/2} J_{(d-2)/2}(\|x\|_2). \tag{5.1}$$

Proof Obviously, both sides of (5.1) are radially symmetric. Hence with spherical coordinates and $r = \|x\|_2$ we can derive

$$\int_{S^{d-1}} e^{ix^T y} dS(y) = \int_{S^{d-1}} e^{iry_1} dS(y) = \frac{2\pi^{(d-1)/2}}{\Gamma((d-1)/2)} \int_0^\pi e^{ir\cos\theta} \sin^{d-2}\theta \, d\theta$$

using that the surface area of S^{d-2} is given by

$$\omega_{d-2} = \frac{2\pi^{(d-1)/2}}{\Gamma((d-1)/2)}.$$

The initial and the last integral are obviously restrictions of entire functions to $r > 0$. Hence, we can calculate the last integral by expanding the exponent in the integral and integrating term by term, which gives

$$\int_0^\pi e^{ir\cos\theta} \sin^{d-2}\theta \, d\theta = \sum_{k=0}^\infty \frac{i^k r^k}{k!} a_k$$

with $a_k := \int_0^\pi \cos^k\theta \sin^{d-2}\theta \, d\theta$. By induction it is possible to show $a_{2k+1} = 0$ and

$$a_{2k} = \frac{(2k)!\,\Gamma((d-1)/2)\,\Gamma(1/2)}{2^{2k} k!\,\Gamma((k+d)/2)}.$$

Collecting everything together gives the stated representation. The exchange of integration and summation can easily be justified. Since we will give similar arguments in later proofs, this time we will leave the details to the reader. □

Our next result is concerned with the asymptotic behavior of the Bessel functions of the first kind.

Proposition 5.6 *The Bessel function has the following asymptotic behavior:*
(1) $J_\nu^{(\ell)}(r) = \mathcal{O}(1/\sqrt{r})$ for $r \to \infty$ and $\nu \in \mathbb{R}$, $\ell \in \mathbb{N}_0$;

(2) $J_\nu(r) = \sqrt{\dfrac{2}{\pi r}} \cos\left(r - \dfrac{\nu\pi}{2} - \dfrac{\pi}{4}\right) + \mathcal{O}(r^{-3/2})$

for $r \to \infty$ and $\nu \in \mathbb{R}$;

(3) $J_{d/2}^2(r) \leq \dfrac{2^{d+2}}{r\pi}$ for $r > 0$ and $d \in \mathbb{N}$;

(4) $\lim_{r\to 0} r^{-d} J_{d/2}^2(r) = \dfrac{1}{2^d \Gamma^2(d/2+1)}$ for $d \in \mathbb{N}$.

Proof The last property is an immediate consequence of the definition of the Bessel functions. The penultimate property is obviously true in the case $d = 1$, since in this case $J_{1/2}(r) = \sqrt{2/(\pi r)} \sin r$. The case $d \geq 2$ is more complicated. It is based mainly on Weber's "crude" inequality for Hankel functions (see Watson [187], p. 211). The complete proof needs too many details on Bessel functions to be presented here; it can be found in the paper [145] by Narcowich and Ward. To give the proof for the second property would go beyond the scope of this book; hence, we refer the reader to Watson [187], p. 199 or, alternatively, to Lebedev [102], p. 122. In the case $\ell = 0$, the first property is a weaker version of the second property. For higher derivatives we use repeatedly the recurrence relation $2J_\nu'(z) = J_{\nu-1}(z) - J_{\nu+1}(z)$, which is a consequence of the formulas given in Proposition 5.4, to derive the desired asymptotic behavior. □

We now turn to the Laplace transforms of some specific functions involving Bessel functions.

Lemma 5.7 *For $\nu > -1$ and every $r > 0$ it is true that*

$$\int_0^\infty J_\nu(t)t^{\nu+1}e^{-rt}dt = \frac{2^{\nu+1}\Gamma(\nu+3/2)r}{\sqrt{\pi}(r^2+1)^{\nu+3/2}}.$$

Proof Let us start by looking at the binomial series. For $0 \le r < 1$ and $\mu > 0$ we have

$$(1+r)^{-\mu} = \sum_{m=0}^\infty \frac{(-1)^m\Gamma(\mu+m)}{m!\,\Gamma(\mu)}r^m.$$

Hence, if we replace r by $1/r^2$ this gives

$$r^{2\mu}(1+r^2)^{-\mu} = \sum_{m=0}^\infty \frac{(-1)^m\Gamma(\mu+m)}{m!\,\Gamma(\mu)}r^{-2m}$$

for $r > 1$. Moreover, Legendre's duplication formula with $z = \nu + m + 1 > 0$ yields

$$\Gamma(\nu+m+3/2) = \frac{\sqrt{\pi}\,\Gamma(2\nu+2m+2)}{2^{2\nu+2m+1}\Gamma(\nu+m+1)}.$$

Thus, setting $\mu = \nu + 3/2 > 1/2$ shows that

$$\frac{2^{\nu+1}\Gamma(\nu+3/2)r}{\sqrt{\pi}(r^2+1)^{\nu+3/2}} = \sum_{m=0}^\infty \frac{(-1)^m\Gamma(2\nu+2m+2)}{2^{2m+\nu}m!\,\Gamma(\nu+m+1)}r^{-2m-2\nu-2}.$$

Now we will have a look at the integral. Using the definition of the Bessel function, and interchanging summation and integration, allows us to make the following derivation:

$$\begin{aligned}\int_0^\infty J_\nu(t)t^{\nu+1}e^{-rt}dt &= \sum_{m=0}^\infty \frac{(-1)^m}{2^{2m+\nu}m!\,\Gamma(\nu+m+1)}\int_0^\infty t^{2m+2\nu+1}e^{-rt}dt\\ &= \sum_{m=0}^\infty \frac{(-1)^m\Gamma(2m+2\nu+2)}{2^{2m+\nu}m!\,\Gamma(m+\nu+1)}r^{-2m-2\nu-2}.\end{aligned}$$

In the last step we used that the integral representation of the Γ-function can be expressed as

$$\int_0^\infty t^{2m+2\nu+1}e^{-rt}dt = r^{-2m-2\nu-2}\Gamma(2m+2\nu+2).$$

Moreover, the interchange of summation and integration can be justified as follows. First note that Stirling's formula allows the bound $\Gamma(\nu+m+1) \ge (1/c_\nu)m!$, where the constant

c_ν depends only on $\nu > -1$. Then

$$\sum_{m=0}^{\infty} \frac{(t/2)^{\nu+2m}}{m!\,\Gamma(\nu+m+1)} t^{\nu+1} e^{-rt} \leq c_\nu t^{2\nu+1} \sum_{m=0}^{\infty} \frac{t^{2m}}{2^{2m}(m!)^2} e^{-rt}$$
$$\leq c_\nu t^{2\nu+1} e^{-rt} \sum_{m=0}^{\infty} \frac{t^{2m}}{(2m)!}$$
$$\leq c_\nu t^{2\nu+1} e^{-rt} e^{t},$$

which is clearly in $L_1[0,\infty)$ provided that $r > 1$. Hence, Lebesgue's convergence theorem justifies the interchange.

Up to now we have shown that the stated equality holds for all $r > 1$. But, since both sides are analytic functions in $\Re(r) > 0$ and $|\Im(r)| < 1$, the equality extends to all $r > 0$ by analytic continuation. □

The following result is in the same spirit.

Lemma 5.8 *For $r > 0$ it is true that*

$$\int_0^\infty J_0(t) e^{-rt} dt = \frac{1}{(1+r^2)^{1/2}}.$$

Proof From the duplication formula of the Γ-function it follows that

$$\Gamma(m+1/2) = \frac{(2m)!\sqrt{\pi}}{2^{2m} m!}.$$

Hence, as in the proof of Lemma 5.7 we get the representation

$$(1+r^2)^{-1/2} = \sum_{m=0}^{\infty} \frac{(-1)^m \Gamma(m+1/2)}{m!\,\Gamma(1/2)} r^{-2m-1} = \sum_{m=0}^{\infty} \frac{(-1)^m (2m)!}{2^{2m}(m!)^2} r^{-2m-1}.$$

Thus, by interchanging summation and integration we derive

$$\int_0^\infty J_0(t) e^{-rt} dt = \sum_{m=0}^{\infty} \frac{(-1)^m}{2^{2m}(m!)^2} \int_0^\infty t^{2m} e^{-rt} dt$$
$$= \sum_{m=0}^{\infty} \frac{(-1)^m \Gamma(2m+1)}{2^{2m}(m!)^2} r^{-2m-1}$$
$$= (1+r^2)^{-1/2}$$

for $r > 1$. The interchange of summation and integration can be justified as in Lemma 5.7. The equality extends to $r > 0$ by analytic continuation as before. □

Our final result on Bessel functions of the first kind deals with J_0 again.

Lemma 5.9 *For the Bessel function J_0 the following two properties are satisfied:*

(1) $\int_0^r J_0(t) dt > 0$ *for all* $r > 0$;

(2) $\int_0^\infty J_0(t) dt = 1.$

The second integral is intended as an improper Riemann integral.

Proof The first result is known as Cooke's inequality. The easiest way to prove it is to use another representation of the Bessel function, namely

$$J_0(t) = \frac{2}{\pi}\int_1^\infty \frac{\sin(ut)}{(u^2-1)^{1/2}}\,du;$$

see Watson [187], p. 170. This shows that

$$\int_0^r J_0(t)dt = \frac{2}{\pi}\int_1^\infty \frac{1-\cos(ru)}{u(u^2-1)^{1/2}}\,du > 0$$

for all $r > 0$.

Finally, let us discuss the second result. We know by Lemma 5.8 that $\int_0^\infty J_0(t)e^{-rt}dt = (1+r^2)^{-1/2}$ for all $r > 0$. Hence we want to let $r \to 0$. Unfortunately, since J_0 is not in $L_1(\mathbb{R})$ we cannot use classical convergence arguments and so we have to be more precise. The idea is to use the triangle inequality twice to get the bound

$$\left|1 - \int_0^R J_0(t)dt\right| \le \left|1-(1+r^2)^{-1/2}\right| + \left|\int_0^R J_0(t)(e^{-rt}-1)dt\right| + \left|\int_R^\infty J_0(t)e^{-rt}dt\right|$$

for an arbitrary $r \in (0,1]$. Next suppose that for every $\epsilon > 0$ we can find an $R_0 > 0$ such that the last integral becomes uniformly less than $\epsilon/3$ for all $r \in (0,1]$ provided $R > R_0$. If this is true then we can fix such an $R > R_0$ and then choose $0 < r \le 1$ such that the first two terms also become small (using that J_0 is bounded), which establishes the second result. Hence it remains to prove this uniform bound on the last integral. To this end we use the asymptotic expansion of J_0 from Proposition 5.6, i.e.

$$J_0(t) = \sqrt{\frac{2}{\pi}}\frac{\cos(t-\pi/4)}{\sqrt{t}} + S(t),$$

with a remainder $S(t)$ satisfying $|S(t)| \le Ct^{-3/2}$ for $t \ge 1$. The remainder part of the integral in question can easily be bounded since $|\int_R^\infty S(t)e^{-rt}dt| \le C\int_R^\infty t^{-3/2}dt = CR^{-1/2}$ with a generic constant $C > 0$ that is independent of $r > 0$. The main part is bounded by integration by parts:

$$\int_R^\infty \frac{e^{-rt}}{\sqrt{t}}\cos\left(t-\frac{\pi}{4}\right)dt = -\frac{1}{\sqrt{2t}}e^{-rt}\frac{(1+r)\cos t + (r-1)\sin t}{1+r^2}\Bigg|_R^\infty - \int_R^\infty \frac{e^{-rt}}{\sqrt{8}\,t^{3/2}}\frac{(1+r)\cos t + (r-1)\sin t}{1+r^2}\,dt.$$

This expansion shows that the integral on the left-hand side can also be bounded uniformly by a constant times $R^{-1/2}$ for all $r \in (0,1]$. □

After discussing Bessel functions of the first kind we come to another family of Bessel functions, called *Bessel functions of imaginary argument* or *modified Bessel functions of the third kind*, sometimes also Mcdonald's functions.

Definition 5.10 *The modified Bessel function of the third kind of order $\nu \in \mathbb{C}$ is defined by*

$$K_\nu(z) := \int_0^\infty e^{-z\cosh t} \cosh(\nu t)\, dt$$

for $z \in \mathbb{C}$ with $|\arg(z)| < \pi/2$; $\cosh t = (e^t + e^{-t})/2$.

It follows immediately from the definition that $K_\nu(x) > 0$ for $x > 0$ and $\nu \in \mathbb{R}$. The modified Bessel functions satisfy recurrence relations similar to the Bessel functions of the first kind (see Watson [187], p. 79). The only one that matters here is

$$\frac{d}{dz}\left[z^\nu K_\nu(z)\right] = -z^\nu K_{\nu-1}(z). \tag{5.2}$$

Moreover, there exist several other representation formulas. The following one is of particular interest for us. Again, its proof goes beyond the scope of this book, so for this we refer the reader again to Watson [187], p. 206.

Proposition 5.11 *For $\nu \geq 0$ and $x > 0$ the modified Bessel function of the third kind has the representation*

$$K_\nu(x) = \left(\frac{\pi}{2x}\right)^{1/2} \frac{e^{-x}}{\Gamma(\nu + 1/2)} \int_0^\infty e^{-u} u^{\nu-1/2} \left(1 + \frac{u}{2x}\right)^{\nu-1/2} du.$$

This representation gives some insight into lower bounds on the decay of the modified Bessel functions.

Corollary 5.12 *For every $\nu \in \mathbb{R}$ the function $x \mapsto x^\nu K_{-\nu}(x)$ is nonincreasing on $(0, \infty)$. Moreover, it has the lower bound*

$$K_\nu(x) \geq \sqrt{\frac{\pi}{2}} \frac{e^{-x}}{\sqrt{x}}, \qquad x > 0,$$

if $|\nu| \geq 1/2$. In the case $|\nu| < 1/2$ the lower bound is given by

$$K_\nu(x) \geq \frac{\sqrt{\pi}\, 3^{|\nu|-1/2}}{2^{|\nu|+1}\Gamma(|\nu| + 1/2)} \frac{e^{-x}}{\sqrt{x}}, \qquad x \geq 1.$$

Proof The recurrence relation (5.2) together with $K_{-\nu} = K_\nu$ and the fact that $K_{\nu-1}$ is positive on $(0, \infty)$ gives the monotonic property of $x^\nu K_{-\nu}(x)$. To prove the lower bound we can restrict ourselves to $\nu \geq 0$ because of $K_\nu = K_{-\nu}$. If $\nu \geq 1/2$ then we get from Proposition 5.11

$$K_\nu(x) \geq \left(\frac{\pi}{2x}\right)^{1/2} \frac{e^{-x}}{\Gamma(\nu + 1/2)} \int_0^\infty e^{-u} u^{\nu-1/2} du = \left(\frac{\pi}{2x}\right)^{1/2} e^{-x}.$$

However, if $0 \leq \nu < 1/2$ then we have to be more careful. We use that for $u \in [0, 1]$ and

$x \ge 1$ it is obviously true that $u \le 1$ and $1 + u/(2x) \le 1 + 1/(2x) \le 3/2$, so that

$$\int_0^\infty e^{-u} u^{\nu-1/2} \left(1 + \frac{u}{2x}\right)^{\nu-1/2} du \ge \int_0^1 e^{-u} u^{\nu-1/2} \left(1 + \frac{u}{2x}\right)^{\nu-1/2} du$$
$$\ge \left(\frac{3}{2}\right)^{\nu-1/2} \int_0^1 e^{-u} du \ge \frac{1}{2}\left(\frac{3}{2}\right)^{\nu-1/2},$$

which finishes the proof. □

Finally, we derive upper bounds on K_ν. We will need these bounds for complex-valued ν. We start with their behavior if the argument tends to infinity.

Lemma 5.13 *The modified Bessel function K_ν, $\nu \in \mathbb{C}$, has the asymptotic behavior*

$$|K_\nu(r)| \le \sqrt{\frac{2\pi}{r}} e^{-r} e^{|\Re(\nu)|^2/(2r)}, \qquad r > 0. \tag{5.3}$$

Proof With $b = |\Re(\nu)|$ we have

$$|K_\nu(r)| \le \frac{1}{2} \int_0^\infty e^{-r\cosh t} |e^{\nu t} + e^{-\nu t}| dt$$
$$\le \frac{1}{2} \int_0^\infty e^{-r\cosh t} [e^{bt} + e^{-bt}] dt$$
$$= K_b(r).$$

Furthermore, from $e^t \ge \cosh t \ge 1 + t^2/2$ for $t \ge 0$ we can conclude that

$$K_b(r) \le \int_0^\infty e^{-r(1+t^2/2)} e^{bt} dt$$
$$= e^{-r} e^{b^2/(2r)} \frac{1}{\sqrt{r}} \int_{-b/\sqrt{r}}^\infty e^{-s^2/2} ds$$
$$\le \sqrt{2\pi} e^{-r} e^{b^2/(2r)} \sqrt{\frac{1}{r}}.$$

□

While the last lemma describes the asymptotic behavior of the modified Bessel functions for large arguments, the next lemma describes the behavior in a neighborhood of the origin. Nonetheless, in the case $\Re(\nu) \ne 0$ it holds for all $r > 0$.

Lemma 5.14 *For $\nu \in \mathbb{C}$ the modified Bessel functions satisfy, for $r > 0$,*

$$|K_\nu(r)| \le \begin{cases} 2^{|\Re(\nu)|-1}\Gamma(|\Re(\nu)|) r^{-|\Re(\nu)|}, & \Re(\nu) \ne 0, \\ \dfrac{1}{e} - \log \dfrac{r}{2}, & r < 2, \quad \Re(\nu) = 0. \end{cases} \tag{5.4}$$

Proof Let us first consider the case $\Re(\nu) \ne 0$. Again, we set $b = |\Re(\nu)|$ and know already that $|K_\nu(r)| \le K_b(r)$ from the proof of the last lemma.

From the definition of the modified Bessel function, however, we can conclude for every $a > 0$ that

$$
\begin{aligned}
K_b(r) &= \frac{1}{2}\int_{-\infty}^{\infty} e^{-r\cosh t} e^{bt}\, dt \\
&= \frac{1}{2}\int_{-\infty}^{\infty} e^{-r(e^t+e^{-t})/2} e^{bt}\, dt \\
&= a^{-b}\frac{1}{2}\int_{0}^{\infty} e^{(-r/2)(u/a+a/u)} u^{b-1}\, du
\end{aligned}
$$

by substituting $u = ae^t$. By setting $a = r/2$ we obtain

$$K_b(r) = 2^{b-1} r^{-b}\int_0^{\infty} e^{-u} e^{-r^2/(4u)} u^{b-1}\, du \le 2^{b-1}\Gamma(b) r^{-b}.$$

For $\Re(\nu) = 0$ we use $\cosh t \ge e^t/2$ to derive

$$
\begin{aligned}
K_0(r) &= \int_0^{\infty} e^{-r\cosh t}\, dt \le \int_0^{\infty} e^{-\frac{r}{2}e^t}\, dt \\
&= \int_{r/2}^{\infty} e^{-u}\frac{1}{u}\, du \le \int_1^{\infty} e^{-u}\, du + \int_{r/2}^{1}\frac{1}{u}\, du \\
&= \frac{1}{e} - \log\frac{r}{2}.
\end{aligned}
$$

□

5.2 Fourier transform and approximation by convolution

One of the most powerful tools in analysis is the Fourier transform. Not only will it help us to characterize positive definite functions, it will also be necessary in several other places. Hence, we will dwell on this subject maybe at first sight longer than necessary. We start with the classical L_1 theory.

Definition 5.15 *For $f \in L_1(\mathbb{R}^d)$ we define its Fourier transform by*

$$\widehat{f}(x) := (2\pi)^{-d/2}\int_{\mathbb{R}^d} f(\omega) e^{-ix^T\omega}\, d\omega$$

and its inverse Fourier transform by

$$f^{\vee}(x) := (2\pi)^{-d/2}\int_{\mathbb{R}^d} f(\omega) e^{ix^T\omega}\, d\omega.$$

We will always use this symmetric definition. But the reader should note that there are other definitions on the market, which differ from each other and this definition only by the way in which the 2π terms are distributed.

For a function $f \in L_1(\mathbb{R}^d)$ the Fourier transform is continuous. Moreover, the following rules are easily established. The overstrained reader might have a look at the book [180] by Stein and Weiss.

Theorem 5.16 *Suppose $f, g \in L_1(\mathbb{R}^d)$; then the following is true.*
(1) $\int_{\mathbb{R}^d} \widehat{f}(x)g(x)dx = \int_{\mathbb{R}^d} f(x)\widehat{g}(x)dx$.
(2) The Fourier transform of the convolution

$$f * g(x) := \int_{\mathbb{R}^d} f(y)g(x-y)\,dy$$

is given by $\widehat{f * g} = (2\pi)^{d/2}\widehat{f}\,\widehat{g}$.
(3) With $\widetilde{f}(x) := \overline{f(-x)}$ we find that $\widehat{f * \widetilde{f}} = (2\pi)^{d/2}|\widehat{f}|^2$.
(4) For $T_a f(x) := f(x-a)$, $a \in \mathbb{R}^d$, we have $\widehat{T_a f}(x) = e^{-ix^T a}\widehat{f}(x)$.
(5) For $S_\alpha f(x) := f(x/\alpha)$, $\alpha > 0$, we have $\widehat{S_\alpha f} = \alpha^d S_{1/\alpha}\widehat{f}$.
(6) If, in addition, $x_j f(x) \in L_1(\mathbb{R}^d)$ then $\widehat{f}$ is differentiable with respect to x_j and

$$\frac{\partial \widehat{f}}{\partial x_j}(x) = (-iy_j f(y))^\wedge(x).$$

If $\partial f/\partial x_j$ is also in $L_1(\mathbb{R}^d)$ then

$$\widehat{\frac{\partial f}{\partial x_j}}(x) = ix_j\widehat{f}(x).$$

Obviously the last item extends to higher-order derivatives in a natural way. The following space will turn out to be the natural playground for Fourier transforms.

Definition 5.17 *The Schwartz space $\mathcal{S}$ consists of all functions $\gamma \in C^\infty(\mathbb{R}^d)$ that satisfy*

$$|x^\alpha D^\beta \gamma(x)| \le C_{\alpha,\beta,\gamma}, \qquad x \in \mathbb{R}^d,$$

for all multi-indices $\alpha, \beta \in \mathbb{N}_0^d$ with a constant $C_{\alpha,\beta,\gamma}$ that is independent of $x \in \mathbb{R}^d$. The functions of $\mathcal{S}$ are called test functions or good functions.

In other words, a good function and all of its derivatives decay faster than any polynomial. Of course all functions from $C_0^\infty(\mathbb{R}^d)$ are contained in $\mathcal{S}$ but so also are the functions

$$\gamma(x) = e^{-\alpha\|x\|_2^2}, \qquad x \in \mathbb{R}^d,$$

for all $\alpha > 0$, and we are going to compute their Fourier transforms now. Because of the scaling property of the Fourier transform it suffices to pick one specific $\alpha > 0$.

Theorem 5.18 *The function $G(x) := e^{-\|x\|_2^2/2}$ satisfies $\widehat{G} = G$.*

Proof First note that, because

$$\widehat{G}(x) = (2\pi)^{-d/2}\int_{\mathbb{R}^d} e^{-\|y\|_2^2/2}e^{-ix^T y}dy$$
$$= \prod_{j=1}^d \left((2\pi)^{-1/2}\int_{-\infty}^{\infty} e^{-y_j^2/2}e^{-ix_j y_j}dy_j\right),$$

the Fourier transform of the d-variate Gaussian G is the product of the univariate Fourier transforms of the univariate Gaussian $g(t) = e^{-t^2/2}$, and it suffices to compute this univariate

Fourier transform. Cauchy's integral theorem yields

$$\begin{aligned}\widehat{g}(r) &= (2\pi)^{-1/2}\int_{-\infty}^{\infty} e^{-(t^2/2+irt)}dt\\ &= (2\pi)^{-1/2}e^{-r^2/2}\int_{-\infty}^{\infty} e^{-(t+ir)^2/2}dt\\ &= (2\pi)^{-1/2}e^{-r^2/2}\int_{-\infty}^{\infty} e^{-t^2/2}dt\\ &= e^{-r^2/2}.\end{aligned}$$

□

We need another class of functions. In a certain way they are the opposite of good functions. Actually, they define continuous linear functionals on $\mathcal{S}$, but this will not really matter for us.

Definition 5.19 *We say that a function f is slowly increasing if there exists a constant $m \in \mathbb{N}_0$ such that $f(x) = \mathcal{O}(\|x\|_2^m)$ for $\|x\|_2 \to \infty$.*

Our next result concerns the approximation of functions by convolution.

Theorem 5.20 *Define $g_m(x) = (m/\pi)^{d/2}e^{-m\|x\|_2^2}$, $m \in \mathbb{N}$, $x \in \mathbb{R}^d$. Then the following hold true:*

(1) $\int_{\mathbb{R}^d} g_m(x)dx = 1$;
(2) $\widehat{g}_m(x) = (2\pi)^{-d/2}e^{-\|x\|_2^2/(4m)}$,
(3) $\widehat{\widehat{g}}_m(x) = g_m(x)$,
(4) $\Phi(x) = \lim_{m\to\infty}\int_{\mathbb{R}^d}\Phi(\omega)g_m(\omega - x)d\omega$, provided that $\Phi \in C(\mathbb{R}^d)$ is slowly increasing.

Proof (1) follows from

$$\int_{\mathbb{R}^d} g_m(x)dx = (2\pi)^{-d/2}\int_{\mathbb{R}^d} e^{-\|y\|_2^2/2}dy = 1.$$

To prove (2) we remark that $g_m = (m/\pi)^{d/2}S_{1/\sqrt{2m}}G$. Thus Theorems 5.16 and 5.18 lead to

$$\widehat{g}_m = (m/\pi)^{d/2}(2m)^{-d/2}S_{\sqrt{2m}}\widehat{G} = (2\pi)^{-d/2}S_{\sqrt{2m}}G.$$

For (3) note that

$$\widehat{\widehat{g}}_m = (2\pi)^{-d/2}(S_{\sqrt{2m}}G)^\wedge = (2\pi)^{-d/2}(\sqrt{2m})^d S_{1/\sqrt{2m}}\widehat{G} = g_m.$$

For (4) we first restrict ourselves to the case $x = 0$. From (1) we see that

$$\int_{\mathbb{R}^d}\Phi(\omega)g_m(\omega)d\omega - \Phi(0) = \int_{\mathbb{R}^d}[\Phi(\omega) - \Phi(0)]g_m(\omega)d\omega.$$

Now choose an arbitrary $\epsilon > 0$. Then there exists a $\delta > 0$ such that $|\Phi(\omega) - \Phi(0)| < \epsilon/2$ for $\|\omega\|_2 \le \delta$. Furthermore, since Φ is slowly increasing, there exists an $\ell \in \mathbb{N}_0$ and an

$M > 0$ such that $|\Phi(\omega)| \le M(1 + \|\omega\|_2)^\ell$, $\omega \in \mathbb{R}^d$. This means that we can find a generic constant C_δ such that we can bound the integral as follows:

$$\begin{aligned}\left|\int_{\mathbb{R}^d} [\Phi(\omega) - \Phi(0)] g_m(\omega) d\omega\right| &\le \int_{\|\omega\|_2 \le \delta} |\Phi(\omega) - \Phi(0)| g_m(\omega) d\omega \\ &\quad + C_\delta \int_{\|\omega\|_2 > \delta} g_m(\omega) \|\omega\|_2^\ell d\omega \\ &\le \frac{\epsilon}{2} + C_\delta m^{-\ell/2} \int_{\|\omega\|_2/\sqrt{2m} > \delta} e^{-\|\omega\|_2^2/2} \|\omega\|_2^\ell d\omega \\ &\le \epsilon\end{aligned}$$

for sufficiently large m. The case $x \ne 0$ follows immediately by replacing Φ by $\Phi(\cdot + x)$ in the previous case. □

The approximation process described in Theorem 5.20, item (4), is a well-known method of approximating a function. It is sometimes also called *approximation by mollification or regularization*. Let us stay a little longer with this process. In particular we are now interested in replacing the Gaussians by an arbitrary compactly supported C^∞-function. Moreover, we are interested in weaker forms of convergence.

Lemma 5.21 *Suppose that* $f \in L_p(\mathbb{R}^d)$, $1 \le p < \infty$, *is given. Then we have* $\lim_{x \to 0} \|f - f(\cdot + x)\|_{L_p(\mathbb{R}^d)} = 0$.

Proof Let us denote $f(\cdot + x)$ by f_x. We start by showing the result for a continuous function g with compact support. Choose a compact set $K \subseteq \mathbb{R}^d$ such that the support of g_x is contained in K for all $x \in \mathbb{R}^d$ with $\|x\|_2 \le 1$. Since g is continuous, it is uniformly continuous on K. Hence for a given $\epsilon > 0$ we find a $\delta > 0$ such that $|g(y) - g(x + y)| < \epsilon$ for all $y \in K$ and all $\|x\|_2 < \delta$. This finishes the proof in this case because

$$\|g - g_x\|_{L_p(\mathbb{R}^d)} = \|g - g_x\|_{L_p(K)} \le \epsilon[\text{vol}(K)]^{1/p}.$$

Now, for an arbitrary $f \in L_p(\mathbb{R}^d)$ and $\epsilon > 0$ we choose a function $g \in C_0(\mathbb{R}^d)$ with $\|f - g\|_{L_p(\mathbb{R}^d)} < \epsilon/3$. By substitution this also means that $\|f_x - g_x\|_{L_p(\mathbb{R}^d)} < \epsilon/3$, giving

$$\begin{aligned}\|f - f_x\|_{L_p(\mathbb{R}^d)} &\le \|f - g\|_{L_p(\mathbb{R}^d)} + \|g - g_x\|_{L_p(\mathbb{R}^d)} + \|g_x - f_x\|_{L_p(\mathbb{R}^d)} \\ &< \frac{2\epsilon}{3} + \|g - g_x\|_{L_p(\mathbb{R}^d)},\end{aligned}$$

and this becomes smaller than ϵ for sufficiently small $\|x\|_2$. □

Note that the result is wrong in the case $p = \infty$. In fact, $\|f - f(\cdot - x)\|_{L_\infty(\mathbb{R}^d)} \to 0$ as $x \to 0$ implies that f is almost everywhere uniformly continuous.

The previous result allows us to establish the convergence of approximation by convolution, not only pointwise but also in L_p. The following results are formulated for $g \in C_0^\infty(\mathbb{R}^d)$, but the proofs show that for example the second and third items hold even if we have only $g \in L_1(\mathbb{R}^d)$.

Theorem 5.22 *Suppose an even and nonnegative $g \in C_0^\infty(\mathbb{R}^d)$ is given, normed by $\int g(x)dx = 1$. Define $g_m(x) = m^d g(mx)$. Then the following are true.*

(1) If $f \in L_1^{\mathrm{loc}}(\mathbb{R}^d)$ then $f * g \in C^\infty(\mathbb{R}^d)$ and $D^\alpha(f * g) = f * (D^\alpha g)$.
(2) If $f \in L_p(\mathbb{R}^d)$ with $1 \le p \le \infty$ then $f * g \in L_p(\mathbb{R}^d)$ and $\|f * g\|_{L_p(\mathbb{R}^d)} \le \|f\|_{L_p(\mathbb{R}^d)}\|g\|_{L_1(\mathbb{R}^d)}$.
(3) If $f \in L_p(\mathbb{R}^d)$ with $1 \le p < \infty$ then $\|f - f * g_n\|_{L_p(\mathbb{R}^d)} \to 0$ for $n \to \infty$.
(4) If $f \in C(\mathbb{R}^d)$ then $f * g_n \to f$ uniformly on every compact subset of $\mathbb{R}^d$.

Proof The first property is an immediate consequence of the theory of integrals depending on an additional parameter. The second property is obviously true for $p = 1, \infty$. Moreover, if $1 < p < \infty$ we obtain, with $q = p/(p-1)$,

$$\begin{aligned} |f * g(x)| &\le \int_{\mathbb{R}^d} |f(y)||g(x-y)|^{1/p}|g(x-y)|^{1/q}dy \\ &\le \left(\int_{\mathbb{R}^d} |f(y)|^p|g(x-y)|dy\right)^{1/p}\left(\int_{\mathbb{R}^d}|g(x-y)|dy\right)^{1/q} \end{aligned}$$

or in other words

$$\begin{aligned} \|f * g\|_{L_p(\mathbb{R}^d)}^p &\le \left(\int_{\mathbb{R}^d}|g(y)|dy\right)^{p/q}\int_{\mathbb{R}^d}\int_{\mathbb{R}^d}|f(y)|^p|g(x-y)|dydx \\ &= \|g\|_{L_1(\mathbb{R}^d)}^p\|f\|_{L_p(\mathbb{R}^d)}^p. \end{aligned}$$

Using Minkowski's inequality again, we can derive

$$\|f - f * g_n\|_{L_p(\mathbb{R}^d)}^p \le \int_{\mathbb{R}^d}\|f - f(\cdot + y/n)\|_{L_p(\mathbb{R}^d)}^p g(y)dy$$

in the same fashion as before. But since $\|f - f(\cdot + y/n)\|_{L_p(\mathbb{R}^d)} \to 0$ for $n \to \infty$ by Lemma 5.21 and since $\|f - f(\cdot + y/n)\|_{L_p(\mathbb{R}^d)} \le 2\|f\|_{L_p(\mathbb{R}^d)}$, Lebesgue's dominated convergence theorem yields the third property.

Finally, let f be a continuous function and $\epsilon > 0$ be given. If $K \subseteq \mathbb{R}^d$ is a compact set then we have $K \subseteq B(0, R)$ for $R > 0$. Choose $\delta > 0$ such that $|f(x) - f(y)| < \epsilon$ for all $x, y \in B(0, R+1)$ with $\|x - y\|_2 < \delta$. Without restriction we can assume that g is supported in $B(0, 1)$. This allows us to conclude, for $x \in K$, that

$$|f(x) - f * g_n(x)| \le \int_{B(x,1/n)} |f(x) - f(y)|g_n(x-y)dy \le \epsilon$$

whenever $1/n < \delta$, which establishes uniform convergence on K. □

We now come back to the initial form of approximation by convolution, as given in Theorem 5.20, to prove one of the fundamental theorems in Fourier analysis.

Theorem 5.23 *The Fourier transform defines an automorphism on $\mathcal{S}$. The inverse mapping is given by the inverse Fourier transformation. Furthermore, the $L_2(\mathbb{R}^d)$-norms of a function and its transform coincide: $\|\widehat{f}\|_{L_2(\mathbb{R}^d)} = \|f\|_{L_2(\mathbb{R}^d)}$.*

Proof From Theorem 5.16 we can conclude that Fourier transformation maps $\mathcal{S}$ back into $\mathcal{S}$ and that the same is valid for the inverse transformation. Using Theorem 5.16 and Theorem 5.20 we obtain, again using Lebesgue's convergence theorem,

$$\begin{aligned} f(x) &= \lim_{m\to\infty} \int_{\mathbb{R}^d} f(\omega) g_m(\omega - x) d\omega \\ &= \lim_{m\to\infty} \int_{\mathbb{R}^d} f(\omega + x) \widehat{\widehat{g}}_m(\omega) d\omega \\ &= \lim_{m\to\infty} \int_{\mathbb{R}^d} \widehat{f}(\omega) e^{ix^T\omega} \widehat{g}_m(\omega) d\omega \\ &= (2\pi)^{-d/2} \int_{\mathbb{R}^d} \widehat{f}(\omega) e^{ix^T\omega} d\omega. \end{aligned}$$

Finally, we now have for an arbitrary $g \in \mathcal{S}$,

$$\widehat{\widehat{\overline{g}}}(x) = (2\pi)^{-d/2} \int_{\mathbb{R}^d} \widehat{\overline{g}}(\omega) e^{-ix^T\omega} d\omega = (2\pi)^{-d/2} \overline{\int_{\mathbb{R}^d} \widehat{g}(\omega) e^{ix^T\omega} d\omega} = \overline{g}(x),$$

which, together with Theorem 5.16, leads to

$$\int_{\mathbb{R}^d} f(x)\overline{g}(x)dx = \int_{\mathbb{R}^d} f(x)\widehat{\widehat{\overline{g}}}(x)dx = \int_{\mathbb{R}^d} \widehat{f}(x)\widehat{\overline{g}}(x)dx,$$

so that not only are the norms equal but also the inner products. □

Obviously, the proof can be extended to the following situation.

Corollary 5.24 *If $f \in L_1(\mathbb{R}^d)$ is continuous and has a Fourier transform $\widehat{f} \in L_1(\mathbb{R}^d)$ then f can be recovered from its Fourier transform:*

$$f(x) = (2\pi)^{-d/2} \int_{\mathbb{R}^d} \widehat{f}(\omega) e^{ix^T\omega} d\omega, \qquad x \in \mathbb{R}^d.$$

Another consequence of Theorem 5.23 is that it allows us to extend the idea of the Fourier transform to $L_2(\mathbb{R}^d)$, even for functions that are not integrable and hence do not possess a classical Fourier transform. Theorem 5.23 asserts that Fourier transformation constitutes a bounded linear operator defined on the dense subset $\mathcal{S}$ of $L_2(\mathbb{R}^d)$. Therefore, there exists a unique bounded extension T of this operator to all $L_2(\mathbb{R}^d)$, which we will call Fourier transformation on $L_2(\mathbb{R}^d)$. We will also use the notation $\widehat{f} = T(f)$ for $f \in L_2(\mathbb{R}^d)$. In general, $\widehat{f}$ for $f \in L_2(\mathbb{R}^d)$ is given as the $L_2(\mathbb{R}^d)$-limit of $\{\widehat{f_n}\}$ if $f_n \in S$ converges to f in $L_2(\mathbb{R}^d)$.

Corollary 5.25 **(Plancherel)** *There exists an isomorphic mapping $T : L_2(\mathbb{R}^d) \to L_2(\mathbb{R}^d)$ such that:*

(1) $\|Tf\|_{L_2(\mathbb{R}^d)} = \|f\|_{L_2(\mathbb{R}^d)}$ for all $f \in L_2(\mathbb{R}^d)$;
(2) $Tf = \widehat{f}$ for all $f \in L_2(\mathbb{R}^d) \cap L_1(\mathbb{R}^d)$;
(3) $T^{-1}g = g^\vee$ for all $g \in L_2(\mathbb{R}^d) \cap L_1(\mathbb{R}^d)$.

The isomorphism is uniquely determined by these properties.

Proof The proof follows from the explanation given in the paragraph above this corollary and the fact that $\mathcal{S} \subseteq L_1(\mathbb{R}^d) \cap L_2(\mathbb{R}^d)$, so that $L_1(\mathbb{R}^d) \cap L_2(\mathbb{R}^d)$ is also dense in $L_2(\mathbb{R}^d)$ with respect to the $L_2(\mathbb{R}^d)$-norm. □

Finally, we will take a look at the Fourier transform of a radial function. Surprisingly, it turns out to be radial as well. This is of enormous importance in the theory to come.

Theorem 5.26 *Suppose $\Phi \in L_1(\mathbb{R}^d) \cap C(\mathbb{R}^d)$ is radial, i.e. $\Phi(x) = \phi(\|x\|_2)$, $x \in \mathbb{R}^d$. Then its Fourier transform $\widehat{\Phi}$ is also radial, i.e. $\widehat{\Phi}(\omega) = \mathcal{F}_d\phi(\|\omega\|_2)$ with*

$$\mathcal{F}_d\phi(r) = r^{-(d-2)/2} \int_0^\infty \phi(t)t^{d/2} J_{(d-2)/2}(rt)dt.$$

Proof The case $d = 1$ follows immediately from

$$J_{-1/2}(t) = \left(\frac{2}{\pi t}\right)^{1/2} \cos t.$$

In the case $d \geq 2$ we set $r = \|x\|_2$. Splitting the Fourier integral and using (5.1) yields

$$\begin{aligned}
\widehat{\Phi}(x) &= (2\pi)^{-d/2} \int_{\mathbb{R}^d} \Phi(\omega)e^{-ix^T\omega}d\omega \\
&= (2\pi)^{-d/2} \int_0^\infty t^{d-1} \int_{S^{d-1}} \Phi(t\|\omega\|_2)e^{-itx^T\omega}dS(\omega)dt \\
&= (2\pi)^{-d/2} \int_0^\infty \phi(t)t^{d-1} \int_{S^{d-1}} e^{-itx^T\omega}dS(\omega)dt \\
&= r^{-(d-2)/2} \int_0^\infty \phi(t)t^{d/2} J_{(d-2)/2}(rt)dt.
\end{aligned}$$

□

5.3 Measure theory

We assume the reader to have some familiarity with measure and integration theory. The convergence results of Fatou, Beppo Levi, and Lebesgue for integrals defined by general measures should be known. Other results like Riesz' representation theorem and Helly's theorem are perhaps not standard knowledge. Hence, we now review the material relevant to us. The reader should be aware of the fact that terms like Borel measure and Radon measure have different meanings throughout the literature. Hence, when using results from measure theory it is crucial to first have a look at the definitions. Here, we will mainly use the definitions and results of Bauer [9] since his definition of a measure is to a certain extent constructive. Another good source for the results here is Halmos [78]. We will not give proofs in this short section. Instead, we refer the reader to the books just mentioned. Moreover, Helly's theorem can be found in the book [46] by Donoghue.

Let Ω be an arbitrary set. We denote the set of all subsets of Ω by $\mathcal{P}(\Omega)$. We have to introduce several names and concepts now.

Definition 5.27 *A subset $\mathcal{R}$ of $\mathcal{P}(\Omega)$ is called a ring on Ω if*
(1) $\emptyset \in \mathcal{R}$,
(2) $A, B \in \mathcal{R}$ implies $A \setminus B \in \mathcal{R}$,
(3) $A, B \in \mathcal{R}$ implies $A \cup B \in \mathcal{R}$.

The name is motivated by the fact that a ring $\mathcal{R}$ is indeed a ring in the algebraical sense if one takes the intersection $\cap$ as multiplication and the symmetric difference Δ, defined by $A \Delta B := (A \setminus B) \cup (B \setminus A)$, as addition.

We are concerned with certain functions defined on a ring.

Definition 5.28 *Let $\mathcal{R}$ be a ring on a set Ω. A function $\mu : \mathcal{R} \to [0, \infty]$ is called a pre-measure if*
(1) $\mu(\emptyset) = 0$,
(2) for disjoint $A_j \in \mathcal{R}$, $j \in \mathbb{N}$, with $\cup A_j \in \mathcal{R}$ we have $\mu(\cup A_j) = \sum \mu(A_j)$.

The last property is called σ-additivity.

Note that for the second property we consider only those sequences $\{A_j\}$ of disjoint sets whose union is also contained in $\mathcal{R}$. This property is not automatically satisfied for a ring. It is different in the situation of a σ-algebra.

Definition 5.29 *A subset $\mathcal{A}$ of $\mathcal{P}(\Omega)$ is called a σ-algebra on Ω if*
(1) $\Omega \in \mathcal{A}$,
(2) $A \in \mathcal{A}$ implies $\Omega \setminus A \in \mathcal{A}$,
(3) $A_j \in \mathcal{A}$, $j \in \mathbb{N}$, implies $\cup_{j \in \mathbb{N}} A_j \in \mathcal{A}$.

Obviously, each σ-algebra is also a ring. Moreover, each ring $\mathcal{R}$, or more generally each subset $\mathcal{R}$ of $\mathcal{P}(\Omega)$, defines a smallest σ-algebra that contains $\mathcal{R}$. This σ-algebra is denoted by $\sigma(\mathcal{R})$ and is obviously given by

$$\sigma(\mathcal{R}) = \cap\{\mathcal{A} : \mathcal{A} \text{ is a } \sigma\text{-algebra and } \mathcal{R} \subseteq \mathcal{A}\}.$$

We also say that $\mathcal{R}$ generates the σ-algebra $\sigma(\mathcal{R})$.

Definition 5.30 *A pre-measure defined on a σ-algebra is called a measure. The sets in the σ-algebra are called measurable with respect to this measure.*

Obviously measurability depends actually more on the σ-algebra than on the actual measure.

Since any ring $\mathcal{R}$ is contained in $\sigma(\mathcal{R})$ it is natural to ask whether a pre-measure μ on $\mathcal{R}$ has an extension $\widetilde{\mu}$ to $\sigma(\mathcal{R})$, meaning that $\widetilde{\mu}(A) = \mu(A)$ for all $A \in \mathcal{R}$. The answer is affirmative.

Proposition 5.31 *Each pre-measure μ on a ring $\mathcal{R}$ on Ω has an extension $\widetilde{\mu}$ to $\sigma(\mathcal{R})$.*

The measures introduced so far will also be called *nonnegative measures* in contrast with *signed measures*. A signed measure is a function $\mu : \mathcal{A} \to \overline{\mathbb{R}} = \mathbb{R} \cup \{-\infty, \infty\}$ defined on a σ-algebra $\mathcal{A}$, which is σ-additive but not necessarily nonnegative.

A measure is called finite if $\mu(\Omega) < \infty$. The total mass of a nonnegative measure is given by $\|\mu\| := \mu(\Omega)$. A signed measure μ can be decomposed into $\mu = \mu_+ - \mu_-$ with two nonnegative measures μ_+, μ_-. In this case the total mass is defined by $\|\mu\| = \|\mu_+\| + \|\mu_-\|$. We will also use the notation $\|\mu\| = \int_\Omega |d\mu|$.

In the case where Ω is a topological space, further concepts are usually needed.

Definition 5.32 *Let Ω be a topological space and $\mathcal{O}$ denote its collection of open sets. The σ-algebra generated by $\mathcal{O}$ is called the Borel σ-algebra and denoted by $\mathcal{B}(\Omega)$. If Ω is a Hausdorff space then a measure μ defined on $\mathcal{B}(\Omega)$ that satisfies $\mu(K) < \infty$ for all compact sets $K \subseteq \Omega$ is called a Borel measure. The carrier of a Borel measure μ is the set* $\Omega \setminus \{U : U \text{ is open and } \mu(U) = 0\}$.

Note that a Borel measure is more than just a measure defined on Borel sets. The assumption that Ω is a Hausdorff space ensures that compact sets are closed and therefore measurable. A finite measure defined on Borel sets is automatically a Borel measure.

If Q is a subset of a Hausdorff space Ω then $\mathcal{B}(Q)$ is given by $\mathcal{B}(Q) = Q \cap \mathcal{B}(\Omega)$, using the induced topology on Q.

In case of $\mathbb{R}^d$, it is well known that $\mathcal{B}(\mathbb{R}^d)$ is also generated by the set of all semi-open cubes $[a, b) := \{x \in \mathbb{R}^d : a_j \le x_j < b_j\}$. To be more precise, it is known that the set $\mathcal{F}^d$, which contains all finite unions of such semi-open cubes, is a ring. Hence, any pre-measure defined on $\mathcal{F}^d$ has an extension to $\mathcal{B}(\mathbb{R}^d)$.

After introducing the notation for measures, the next step is to introduce measurable and integrable functions with respect to a certain measure. Since this is standard again, we omit the details here and proceed by stating those results that we will need later on.

Theorem 5.33 (Riesz) *Let Ω be a locally compact metric space. If λ is a linear and continuous functional on $C_0(\Omega)$, which is nonnegative, meaning that $\lambda(f) \ge 0$ for all $f \in C_0(\Omega)$ with $f \ge 0$, then there exists a nonnegative Borel measure μ on Ω such that*

$$\lambda(f) = \int_\Omega f(x) d\mu(x)$$

for all $f \in C_0(\Omega)$. If Ω possesses a countable basis then the measure μ is uniquely determined.

A metric space Ω possesses a countable basis if there exists a sequence $\{U_j\}_{j\in\mathbb{N}}$ of open sets such that each open set U is the union of some of these sets.

Theorem 5.34 (Helly) *Let $\{\nu_k\}$ be a sequence of (signed) Borel measures on the compact metric space Ω of uniformly bounded total mass. Then there exists a subsequence $\{\nu_{k_j}\}$ and a Borel measure ν on Ω such that*

$$\lim_{j\to\infty} \int_\Omega f(x) d\nu_{k_j}(x)$$

exists for all $f \in C(\Omega)$ and equals $\int_\Omega f(x) d\nu(x)$.

Finally, we need to know a result on the uniqueness of measures.

Theorem 5.35 (Uniqueness theorem) *Suppose that Ω is a metric space and μ and ν are two finite Borel measures on Ω. If*

$$\int f d\mu = \int f d\nu$$

for all continuous and bounded functions f, then $\mu \equiv \nu$.

In this book we are mainly confronted with situations where Ω is $\mathbb{R}^d$, $[0, 1]$, or $[0, \infty)$ endowed with the induced topology. These sets are obviously metric (hence Hausdorff), locally compact, and possess a countable basis. Moreover, they are complete with respect to their metric. Such spaces are sometimes called *Polish spaces* and have some remarkable properties. For example, every finite Borel measure is regular, meaning inner and outer regular. This will reassure readers who might have wondered about regularity.

Finally, whenever we work on $\mathbb{R}^d$ and do not specify a σ-algebra or a measure, we will assume tacitly that we are employing Borel sets and Lebesgue measure.

6
Positive definite functions

With moving least squares we have encountered an efficient method for approximating a multivariate function. The assumption that the weight function is continuous excludes interpolation as a possible approximation process. If the weight function Φ has a pole at the origin, however, interpolation is possible. Nonetheless, moving least squares is a method for approximation rather than for interpolation. The rest of this book is devoted to a very promising method that allows interpolation in an arbitrary number of space dimensions and for arbitrary choices of data sites.

6.1 Definition and basic properties

From the Mairhuber–Curtis theorem we know that there are no Haar spaces in the multivariate setting. If we still want to interpolate data values $f_1, \ldots, f_N$ at given data sites $X = \{x_1, \ldots, x_N\} \subseteq \mathbb{R}^d$ we have to take this into account. One simple way to do this is to choose a fixed function $\Phi : \mathbb{R}^d \to \mathbb{R}$ and to form the interpolant as

$$s_{f,X}(x) = \sum_{j=1}^{N} \alpha_j \Phi(x - x_j), \tag{6.1}$$

where the coefficients $\{\alpha_j\}$ are determined by the interpolation conditions

$$s_{f,X}(x_j) = f_j, \qquad 1 \le j \le N. \tag{6.2}$$

If we imagine for a moment that Φ is a bump function with center at the origin, the shifts $\Phi(\cdot - x_j)$ are functions that are centered at x_j. Motivated by this point of view, we will often call x_j a center and $X = \{x_1, \ldots, x_N\}$ a set of centers.

It would be nice if Φ could be chosen for all kinds of data sets, meaning for any number N and any possible combination $X = \{x_1, \ldots, x_N\}$. Obviously, the interpolation conditions (6.2) imposed on a function $s_{f,X}$ of the form (6.1) are equivalent to asking for an invertible interpolation matrix

$$A_{\Phi,X} := (\Phi(x_j - x_k))_{1 \le j,k \le N}.$$

From the numerical point of view it is desirable to have more conditions on the matrix $A_{\Phi,X}$, for example that it is positive definite. Later on, we will see that this requirement will turn out quite naturally.

Definition 6.1 *A continuous function $\Phi : \mathbb{R}^d \to \mathbb{C}$ is called positive semi-definite if, for all $N \in \mathbb{N}$, all sets of pairwise distinct centers $X = \{x_1, \dots, x_N\} \subseteq \mathbb{R}^d$, and all $\alpha \in \mathbb{C}^N$, the quadratic form*

$$\sum_{j=1}^{N} \sum_{k=1}^{N} \alpha_j \overline{\alpha_k} \Phi(x_j - x_k)$$

is nonnegative. The function Φ is called positive definite if the quadratic form is positive for all $\alpha \in \mathbb{C}^N \setminus \{0\}$.

Here, we have used a more general definition for complex-valued functions. The reason for this is that it allows us to use techniques such as Fourier transforms more naturally. However, we will see that for even, real-valued functions it suffices to investigate the quadratic form only for real vectors $\alpha \in \mathbb{R}^N$.

The reader should note that we call a function positive definite if the associated interpolation matrices are positive definite and positive semi-definite if the associated matrices are positive semi-definite. This seems to be natural. Unfortunately, for historical reasons there is an alternative terminology around in the literature: other authors call a function positive definite if the associated matrices are positive semi-definite and strictly positive definite if the matrices are positive definite. We do not follow this historical approach here, but the reader should always keep this in mind when looking at other texts.

From Definition 6.1 we can read off the elementary properties of a positive definite function.

Theorem 6.2 *Suppose Φ is a positive semi-definite function. Then the following properties are satisfied.*

(1) $\Phi(0) \ge 0$.
(2) $\Phi(-x) = \overline{\Phi(x)}$ *for all* $x \in \mathbb{R}^d$.
(3) *Φ is bounded, i.e.* $|\Phi(x)| \le \Phi(0)$ *for all* $x \in \mathbb{R}^d$.
(4) $\Phi(0) = 0$ *if and only if* $\Phi \equiv 0$.
(5) If $\Phi_1, \dots, \Phi_n$ are positive semi-definite and $c_j \ge 0$, $1 \le j \le n$, then $\Phi := \sum_{j=1}^{n} c_j \Phi_j$ is also positive semi-definite. If one of the Φ_j is positive definite and the corresponding c_j is positive then Φ is also positive definite.
(6) The product of two positive definite functions is positive definite.

Proof The first property follows by choosing $N = 1$ and $\alpha_1 = 1$ in the definition.

Next, setting $N = 2$, $\alpha_1 = 1$, $\alpha_2 = c$, $x_1 = 0$, and $x_2 = x$ gives

$$(1 + |c|^2)\Phi(0) + c\Phi(x) + \overline{c}\Phi(-x) \ge 0.$$

If we set $c = 1$ and $c = i$, respectively, this means that both $\Phi(x) + \Phi(-x)$ and $i[\Phi(x) - \Phi(-x)]$ must be real. This can only be satisfied if $\Phi(x) = \overline{\Phi(-x)}$, showing property (2).

To prove the third property we take $N = 2$, $\alpha_1 = |\Phi(x)|$, $\alpha_2 = -\overline{\Phi(x)}$, $x_1 = 0$ and $x_2 = x$. Then the condition in the definition and the fact that Φ satisfies $\Phi(-x) = \overline{\Phi(x)}$ leads to

$$2|\Phi(x)|^2\Phi(0) - 2|\Phi(x)|^3 \geq 0.$$

Property (4) follows immediately from the third. Property (5) is obvious. Property (6) is a consequence of a theorem of Schur; we include the proof here. Since the interpolation matrix $A_{\Phi_2,X}$ is positive definite there exists a unitary matrix $S \in \mathbb{C}^{N\times N}$ such that $A_{\Phi_2,X} = SD\overline{S}^T$, where $D = \text{diag}\{\lambda_1, \ldots, \lambda_N\}$ is the diagonal matrix with the eigenvalues $0 < \lambda_1 \leq \cdots \leq \lambda_N$ of $A_{\Phi_2,X}$ as diagonal entries. This means that

$$\Phi_2(x_\ell - x_j) = \sum_{k=1}^{N} s_{\ell k}\overline{s_{jk}}\lambda_k.$$

As Φ_1 is positive definite we have

$$\begin{aligned}
\alpha^T A_{\Phi_1\Phi_2,X}\overline{\alpha} &= \sum_{\ell=1}^{N}\sum_{j=1}^{N} \alpha_\ell\overline{\alpha_j}\Phi_1(x_\ell - x_j)\Phi_2(x_\ell - x_j) \\
&= \sum_{\ell=1}^{N}\sum_{j=1}^{N} \alpha_\ell\overline{\alpha_j}\Phi_1(x_\ell - x_j)\sum_{k=1}^{N} s_{\ell k}\overline{s_{jk}}\lambda_k \\
&= \sum_{k=1}^{N}\lambda_k\sum_{\ell=1}^{N}\sum_{j=1}^{N} \alpha_\ell s_{\ell k}\overline{\alpha_j s_{jk}}\Phi_1(x_\ell - x_j) \\
&\geq \lambda_1\sum_{\ell=1}^{N}\sum_{j=1}^{N} \alpha_\ell\overline{\alpha_j}\Phi_1(x_\ell - x_j)\sum_{k=1}^{N} s_{\ell k}\overline{s_{jk}} \\
&= \lambda_1\sum_{\ell=1}^{N} |\alpha_\ell|^2\Phi_1(0).
\end{aligned}$$

The last expression is nonnegative for all $\alpha \in \mathbb{C}^N$ and positive unless $\alpha = 0$. □

We now come back to the question of real-valued positive definite functions and their characterization. First of all, it is now clear that a positive semi-definite function is real-valued if and only if it is even. But we can also restrict ourselves to real coefficient vectors $\alpha \in \mathbb{R}^N$ in a quadratic form.

Theorem 6.3 *Suppose that $\Phi : \mathbb{R}^d \to \mathbb{R}$ is continuous. Then Φ is positive definite if and only if Φ is even and we have, for all $N \in \mathbb{N}$, for all $\alpha \in \mathbb{R}^N \setminus \{0\}$, and for all pairwise distinct $x_1, \ldots, x_N$,*

$$\sum_{j=1}^{N}\sum_{k=1}^{N} \alpha_j\alpha_k\Phi(x_j - x_k) > 0.$$

Proof If Φ is positive definite and real-valued then it is even by Theorem 6.2; it obviously satisfies the second condition.

If, however, Φ satisfies the given conditions then we have with $\alpha_j = a_j + ib_j$

$$\sum_{j,k=1}^{N} \alpha_j \overline{\alpha_k} \Phi(x_j - x_k) = \sum_{j,k=1}^{N} (a_j a_k + b_j b_k) \Phi(x_j - x_k) + i \sum_{j,k=1}^{N} a_k b_j [\Phi(x_j - x_k) - \Phi(x_k - x_j)].$$

As Φ is even the second sum on the right-hand side vanishes. The first sum is nonnegative because of the assumption and vanishes only if all the a_j and b_j vanish. □

6.2 Bochner's characterization

One of the most celebrated results on positive semi-definite functions is their characterization in terms of Fourier transforms, which was established by Bochner in 1932 for $d = 1$, [28], and 1933 for general d, [29]. This result is one of the reasons for introducing complex-valued positive definite functions. The idea behind it is easily described. Suppose $\Phi \in C(\mathbb{R}^d) \cap L_1(\mathbb{R}^d)$ has an integrable Fourier transform $\widehat{\Phi} \in L_1(\mathbb{R}^d)$. Then by the Fourier inversion formula we can recover Φ from its Fourier transform:

$$\Phi(x) = (2\pi)^{-d/2} \int_{\mathbb{R}^d} \widehat{\Phi}(\omega) e^{ix^T\omega} d\omega.$$

This means that a quadratic form involving Φ can be expressed as follows:

$$\sum_{j=1}^{N} \sum_{k=1}^{N} \alpha_j \overline{\alpha_k} \Phi(x_j - x_k) = (2\pi)^{-d/2} \sum_{j,k=1}^{N} \alpha_j \overline{\alpha_k} \int_{\mathbb{R}^d} \widehat{\Phi}(\omega) e^{i\omega^T(x_j - x_k)} d\omega$$
$$= (2\pi)^{-d/2} \int_{\mathbb{R}^d} \widehat{\Phi}(\omega) \left| \sum_{j=1}^{N} \alpha_j e^{ix_j^T\omega} \right|^2 d\omega.$$

Hence, if $\widehat{\Phi}$ is nonnegative then the function Φ is positive semi-definite. The analysis also shows that it is unimportant whether we recover Φ from its Fourier or its inverse Fourier transform.

Even though we will see that every positive definite and integrable function has an integrable Fourier transform, we cannot use this approach in the case of a nonintegrable function. For a complete characterization we have to replace the measure with Lebesgue density $\widehat{\Phi}$ by a more general Borel measure μ. We will see that this suffices to characterize every positive semi-definite function as the Fourier transform of such a measure.

To prove this characterization we need two auxiliary results. The first is a characterization of positive semi-definite functions as *integrally* positive semi-definite functions.

Proposition 6.4 *A continuous function $\Phi : \mathbb{R}^d \to \mathbb{C}$ is positive semi-definite if and only if Φ is bounded and satisfies*

$$\int_{\mathbb{R}^d} \int_{\mathbb{R}^d} \Phi(x - y) \gamma(x) \overline{\gamma(y)} dx dy \geq 0 \tag{6.3}$$

for all test functions γ from the Schwartz space $\mathcal{S}$.

Proof Suppose that Φ is a positive semi-definite function. Since Φ is bounded and $\gamma \in \mathcal{S}$ decays rapidly, the integral (6.3) is well defined. Moreover, for every $\epsilon > 0$ there exists a closed cube $W \subseteq \mathbb{R}^d$ such that

$$\left| \int_{\mathbb{R}^d} \int_{\mathbb{R}^d} \Phi(x-y)\gamma(x)\overline{\gamma(y)}dxdy - \int_W \int_W \Phi(x-y)\gamma(x)\overline{\gamma(y)}dxdy \right| < \frac{\epsilon}{2}.$$

But the double integral over the cubes is the limit of Riemannian sums. Hence, we can find $x_1, \ldots, x_N \in \mathbb{R}^d$ and weights $w_1, \ldots, w_N$ such that

$$\left| \int_W \int_W \Phi(x-y)\gamma(x)\overline{\gamma(y)}dxdy - \sum_{j,k=1}^N \Phi(x_j - x_k)\gamma(x_j)w_j\overline{\gamma(x_k)w_k} \right| < \frac{\epsilon}{2}.$$

This means that

$$\int_{\mathbb{R}^d} \int_{\mathbb{R}^d} \Phi(x-y)\gamma(x)\overline{\gamma(y)}dxdy > \sum_{j,k=1}^N \Phi(x_j - x_k)\gamma(x_j)w_j\overline{\gamma(x_k)w_k} - \epsilon.$$

Letting ϵ tend to zero and using that Φ is positive semi-definite shows indeed that (6.3) is true for all $\gamma \in \mathcal{S}$. As a positive semi-definite function Φ is also bounded.

Conversely, let us assume that Φ is bounded and satisfies (6.3). Because of the assumption imposed on Φ and γ we can rewrite the double integral as

$$\int_{\mathbb{R}^d} \int_{\mathbb{R}^d} \Phi(x-y)\gamma(x)\overline{\gamma(y)}dxdy = \int_{\mathbb{R}^d} \Phi(x)\gamma * \widetilde{\gamma}(x)dx$$

using the notation $\widetilde{\gamma}(x) = \overline{\gamma(-x)}$. Next, suppose that $x_1, \ldots, x_N \in \mathbb{R}^d$ and $\alpha_1, \ldots, \alpha_N \in \mathbb{C}$ are given. We choose

$$\gamma = \gamma_m = \sum_{j=1}^N \alpha_j g_{2m}(\cdot - x_j),$$

where $g_m(x) = (m/\pi)^{d/2} e^{-m\|\omega\|_2^2}$ is the function from Theorem 5.20. Then, using standard techniques to compute the Fourier transform, we see that

$$\widehat{\gamma}_m(\omega) = \sum_{j=1}^N \alpha_j e^{-i\omega^T x_j} \widehat{g}_{2m}(\omega) = (2\pi)^{-d/2} \sum_{j=1}^N \alpha_j e^{-i\omega^T x_j} e^{-\|x\|_2^2/(8m)},$$

which leads to

$$\begin{aligned}
(\gamma_m * \widetilde{\gamma}_m)^\wedge(\omega) &= (2\pi)^{d/2} \left|\widehat{\gamma}_m\right|^2(\omega) \\
&= (2\pi)^{-d/2} \left| \sum_{j=1}^N \alpha_j e^{-i\omega^T x_j} \right|^2 e^{-\|\omega\|_2^2/(4m)} \\
&= \sum_{j,k=1}^N \alpha_j \overline{\alpha_k} e^{-i\omega^T(x_j - x_k)} \widehat{g}_m(\omega) \\
&= \left(\sum_{j=1}^N \alpha_j \overline{\alpha_k} g_m(\cdot - (x_j - x_k)) \right)^\wedge (\omega).
\end{aligned}$$

This together with the uniqueness of the Fourier transform on $\mathcal{S}$ and Theorem 5.20 allows us conclude that

$$\sum_{j,k=1}^{N} \alpha_j \overline{\alpha_k} \Phi(x_j - x_k) = \sum_{j,k=1}^{N} \alpha_j \overline{\alpha_k} \lim_{m\to\infty} \int_{\mathbb{R}^d} \Phi(x) g_m(x - (x_j - x_k)) dx$$
$$= \lim_{m\to\infty} \int_{\mathbb{R}^d} \Phi(x) \gamma_m * \widetilde{\gamma}_m(x) dx,$$

and the last expression is nonnegative by our assumption. □

The second auxiliary result that we need on our way to proving Bochner's theorem is a generalization of Riesz' representation theorem, Theorem 5.33. For applying this theorem to $\mathbb{R}^d$, it is not necessary to have a linear functional that is bounded. It suffices that the functional is bounded only in a local sense. Moreover, the functional needs only to be defined on functions from $C_0^\infty(\mathbb{R}^d)$.

Proposition 6.5 *Suppose that $\lambda : C_0^\infty(\mathbb{R}^d) \to \mathbb{C}$ is a linear functional which is nonnegative, meaning that $\lambda(f) \geq 0$ for all $f \in C_0^\infty(\mathbb{R}^d)$ with $f \geq 0$. Then λ has an extension to $C_0(\mathbb{R}^d)$ and there exists a nonnegative Borel measure μ such that*

$$\lambda(f) = \int_{\mathbb{R}^d} f(x) d\mu(x)$$

for all $f \in C_0(\mathbb{R}^d)$.

Proof The proof is divided into three steps. In the first step we show that a nonnegative linear functional is locally bounded. Hence, let K be a compact subset of $\mathbb{R}^d$. Denote by $C_0^\infty(K)$ the subset of functions from $C_0^\infty(\mathbb{R}^d)$ having support in K. We use the notation $C_0(K)$ similarly. We want to show that $|\lambda(f)| \leq C_K \|f\|_{L_\infty(K)}$ for all $f \in C_0^\infty(K)$ with a constant C_K depending only on K. To this end we choose a function $\psi \in C_0^\infty(\mathbb{R}^d)$ with $\psi | K \equiv 1$ and $\psi \geq 0$. If $f \in C_0^\infty(K)$ is real-valued then

$$\|f\|_{L_\infty(K)} \psi \pm f \geq 0$$

obviously implies that $\lambda(f) \in \mathbb{R}$ and

$$|\lambda(f)| \leq \lambda(\psi) \|f\|_{L_\infty(K)}.$$

If f is complex-valued then we choose a $\theta \in \mathbb{R}$ such that $e^{i\theta}\lambda(f) \in \mathbb{R}$. Using $\lambda(\Re(f)) = \Re(\lambda(f))$, which we have just proven, a simple computation shows that $\lambda(\Re(e^{i\theta} f)) = e^{i\theta}\lambda(f)$. Applying the previous result to $\Re(e^{i\theta} f)$ gives therefore

$$|\lambda(f)| = |e^{i\theta}\lambda(f)| = |\lambda(\Re(e^{i\theta} f))| \leq \lambda(\psi) \|\Re(e^{i\theta} f)\|_{L_\infty(K)}$$
$$\leq \lambda(\psi) \|f\|_{L_\infty(K)}.$$

Hence we can define C_K to be $\lambda(\psi)$.

The second step deals with the local extension of λ. Being restricted to $C_0^\infty(K)$, the functional λ is continuous and thus has a unique extension to $C_0(K)$. The well-known construction starts for $f \in C_0(K)$ with a sequence $\{f_j\} \subseteq C_0^\infty(K)$ that converges uniformly to f and defines $\lambda(f)$ as the limit of $\lambda(f_j)$. Since $\{f_j\}$ can be chosen as the convolution of f with a family of nonnegative and compactly supported functions, the extension of λ to $C_0(K)$ is also nonnegative.

By Theorem 5.33 we can find a unique Borel measure μ_K defined on the Borel sets $\mathcal{B}(K)$ such that

$$\lambda(f) = \int_K f(x) d\mu_K(x) \qquad \text{for all } f \in C_0(K).$$

Since λ when restricted to $C_0(K)$ is continuous, the measure μ_K is finite.

In the final step we use this result to define a pre-measure on the ring $\mathcal{F}^d$, which is the collection of finite unions of semi-open cubes. See the start of Section 5.3 for the notation. To this end we let $K_j := \{x \in \mathbb{R}^d : \|x\|_2 \leq j\}$. The pre-measure $\widetilde{\mu}$ is now defined as follows. For $A \in \mathcal{F}^d$ we choose a $j \in \mathbb{N}$ sufficiently large that $A \subseteq K_j$ and define $\widetilde{\mu}(A) := \mu_{K_j}(A)$. The function $\widetilde{\mu}$ is well defined and indeed a pre-measure if its definition is independent of the choice of j. But this is the case since $\mu_{K_{j+1}}|\mathcal{B}(K_j) = \mu_{K_j}$ by uniqueness.

Hence, by Proposition 5.31 there exists a Borel measure μ on $\mathbb{R}^d$ such that

$$\lambda(f) = \int_{\mathbb{R}^d} f(x) d\mu(x)$$

for all $f \in C_0(\mathbb{R}^d)$. □

After these preparatory steps we are now in a position to formulate and prove Bochner's result.

Theorem 6.6 (Bochner) *A continuous function $\Phi : \mathbb{R}^d \to \mathbb{C}$ is positive semi-definite if and only if it is the Fourier transform of a finite nonnegative Borel measure μ on $\mathbb{R}^d$, i.e.*

$$\Phi(x) = \widehat{\mu}(x) = (2\pi)^{-d/2} \int_{\mathbb{R}^d} e^{-ix^T\omega} d\mu(\omega), \qquad x \in \mathbb{R}^d. \tag{6.4}$$

Proof We first assume that Φ is the Fourier transform of a finite nonnegative Borel measure and show that Φ is positive semi-definite. With the usual $x_1, \ldots, x_N \in \mathbb{R}^d$ and $\alpha \in \mathbb{C}^N$ we get, as in the introductory part of this section,

$$\begin{aligned} \sum_{j,k=1}^{N} \alpha_j \overline{\alpha_k} \Phi(x_j - x_k) &= (2\pi)^{-d/2} \sum_{j,k=1}^{N} \alpha_j \overline{\alpha_k} \int_{\mathbb{R}^d} e^{-i(x_j - x_k)^T\omega} d\mu(\omega) \\ &= (2\pi)^{-d/2} \int_{\mathbb{R}^d} \left| \sum_{j=1}^{N} \alpha_j e^{-ix_j^T\omega} \right|^2 d\mu(\omega) \geq 0. \end{aligned}$$

The last inequality holds because of the conditions imposed on the measure μ. As it is the Fourier transform of the finite measure μ, the function Φ must be continuous.

Next let us suppose that Φ is positive semi-definite. Then we define a functional λ on the Schwartz space $\mathcal{S}$ by

$$\lambda(\gamma) := \int_{\mathbb{R}^d} \Phi(x)\gamma^\vee(x)dx, \qquad \gamma \in \mathcal{S}.$$

If $\gamma \in S$ is of the form $\gamma = |\psi|^2$ with $\psi \in \mathcal{S}$ then we find with $f := \psi^\vee$ and $\widetilde{f}(x) = \overline{f(-x)}$ that

$$\begin{aligned}\lambda(\gamma) &= \int_{\mathbb{R}^d} \Phi(x)\left(|\widehat{f}|^2\right)^\vee(x)dx \\ &= (2\pi)^{-d/2}\int_{\mathbb{R}^d} \Phi(x) f * \widetilde{f}(x)dx \\ &= (2\pi)^{-d/2}\int_{\mathbb{R}^d}\int_{\mathbb{R}^d} \Phi(x-y)f(x)\overline{f(y)}dxdy \\ &\geq 0,\end{aligned}$$

where the last inequality follows from Proposition 6.4.

Hence λ is nonnegative on the set of all functions γ from the Schwartz space $\mathcal{S}$ of the form $\gamma = |\psi|^2$ with $\psi \in \mathcal{S}$.

We wish to extend this relation to all $\gamma \in C_0^\infty(\mathbb{R}^d)$ with $\gamma \geq 0$. To this end we form $\gamma + \varepsilon^2 G$, where G is the Gaussian $G(x) = e^{-\|x\|_2^2/2}$ and $\varepsilon > 0$. This function is in $\mathcal{S}$ and positive on $\mathbb{R}^d$. Thus it possesses a square root, say $\gamma_1 := \sqrt{\gamma + \varepsilon^2 G}$, which is clearly in C^∞. It is also in $\mathcal{S}$ because it coincides with $\varepsilon G(\cdot/2)$ for sufficiently large x. Altogether, this leads us to

$$0 \leq \lambda(|\gamma_1|^2) = \lambda(\gamma) + \varepsilon^2\lambda(G),$$

which establishes the desired result with $\varepsilon \to 0$. But we know that λ is a nonnegative linear functional defined on $C_0^\infty(\mathbb{R}^d)$ and this allows us to use Proposition 6.5 to obtain a nonnegative Borel measure μ on $\mathbb{R}^d$ such that

$$\lambda(\gamma) = \int_{\mathbb{R}^d} \gamma(x)d\mu(x), \qquad \gamma \in C_0(\mathbb{R}^d). \tag{6.5}$$

Next we apply approximation by convolution, as provided in Theorem 5.22, to show that μ is finite and that λ can be used to represent Φ. Approximation by convolution uses a family $\{g_m\}$ with $g_m = m^d g(m\cdot)$ where $g \in C(\mathbb{R}^d) \cap L_1(\mathbb{R}^d)$ has L_1-norm one and is nonnegative. Here, we need in addition that $\widehat{g}$ also is nonnegative and in $C_0(\mathbb{R}^d)$. This can be achieved by choosing a nonnegative function $g_0 \in C_0^\infty(\mathbb{R}^d)$ and defining g via $g = (g_0 * \widetilde{g_0})^\vee$ with possible renormalization. Note that this construction even leads to $g \in \mathcal{S}$ and $\widehat{g} \in C_0^\infty(\mathbb{R}^d)$.

Let us show that the measure μ has a finite total mass. First of all we have

$$\int_{\mathbb{R}^d} \Phi(x)g_m(x)dx = \int_{\mathbb{R}^d} \widehat{g}_m(x)d\mu(x) = \int_{\mathbb{R}^d} \widehat{g}(x/m)d\mu(x).$$

Since $\widehat{g}(x/m)$ tends to $(2\pi)^{-d/2}$ and is nonnegative we can apply Fatou's lemma to derive

$$
\begin{aligned}
(2\pi)^{-d/2}\int_{\mathbb{R}^d} d\mu(x) &= \int_{\mathbb{R}^d} \lim_{m\to\infty} \widehat{g}(x/m)d\mu(x) \\
&\le \lim_{m\to\infty}\int_{\mathbb{R}^d} \widehat{g}(x/m)d\mu(x) \\
&= \lim_{m\to\infty}\int_{\mathbb{R}^d} \Phi(x)g_m(x)dx \\
&= \Phi(0),
\end{aligned}
$$

since Φ is bounded. This shows that the total mass of μ is indeed bounded by $(2\pi)^{d/2}\Phi(0)$. Finally, Theorem 5.22 gives

$$
\begin{aligned}
\Phi(x) &= \lim_{m\to\infty}\int_{\mathbb{R}^d} \Phi(\omega)g_m(x-\omega)d\omega \\
&= \lim_{m\to\infty}\int_{\mathbb{R}^d} e^{-i\omega^T x}\widehat{g}_m(-\omega)d\mu(\omega) \\
&= (2\pi)^{-d/2}\int_{\mathbb{R}^d} e^{-ix^T\omega}d\mu(\omega),
\end{aligned}
$$

since μ is finite. This is what we intended to show. □

The advanced reader will certainly have noticed that the functional λ in the proof of Theorem 6.6 is nothing other than the distributional Fourier transform of Φ. Moreover, Proposition 6.5 just states that every nonnegative distribution can be interpreted as a Borel measure.

With the interpolation problem in mind, we are more interested in positive definite than in positive semi-definite functions. Unfortunately, a complete characterization of positive definite functions leads to a very technical discussion of the Borel measures involved. The interested reader might have a look at the paper [39] by Chang.

But all the important functions Φ that we will encounter in this book have either a discrete measure or a measure with Lebesgue density. In such a situation it is far easier to decide whether a positive semi-definite function is also positive definite. In fact, it will turn out that all positive semi-definite functions having a measure with a continuous Lebesgue density that is not identically zero are positive definite. We will now derive sufficient conditions on the measure μ to ensure that Φ is positive definite.

Lemma 6.7 *Suppose that $U \subseteq \mathbb{R}^d$ is open. Suppose, further, that $x_1, \dots, x_N \in \mathbb{R}^d$ are pairwise distinct and that $c \in \mathbb{C}^N$. If $\sum_{j=1}^N c_j e^{-ix_j^T\omega} = 0$ for all $\omega \in U$ then $c \equiv 0$.*

Proof By successive analytic continuation in each coordinate (or by the identity theorem, to be more precise) we can derive that the assumption actually means that $\sum_{j=1}^N c_j e^{-ix_j^T\omega} = 0$

for all $\omega \in \mathbb{R}^d$. Now take a test function $f \in S$. Then

$$0 = \sum_{j=1}^{N} c_j e^{-ix_j^T \omega} \widehat{f}(\omega) = \left(\sum_{j=1}^{N} c_j f(\cdot - x_j) \right)^{\wedge} (\omega)$$

for all $\omega \in \mathbb{R}^d$ implies that

$$\sum_{j=1}^{N} c_j f(x - x_j) = 0, \qquad x \in \mathbb{R}^d.$$

Now take f to be compactly supported with $f(0) = 1$ and support contained in the ball around zero with radius $\epsilon < \min_{j \neq k} \|x_j - x_k\|_2$. This gives

$$0 = \sum_{j=1}^{N} c_j f(x_k - x_j) = c_k f(0) = c_k$$

for $1 \leq k \leq N$. □

Knowing when the exponentials are linearly independent, we can give a sufficient condition for a function to be positive definite.

Theorem 6.8 *A positive semi-definite function Φ is positive definite if the carrier of the measure μ in the representation (6.4) contains an open subset.*

Proof Denote the open subset by U. Then $\mu(U) \neq 0$ and thus we can conclude from the proof of Theorem 6.6 that for any pairwise distinct $x_1, \ldots, x_N \in \mathbb{R}^d$ and any $\alpha \in \mathbb{C}^N$ we must have

$$\sum_{j=1}^{N} \alpha_j e^{-ix_j^T x} = 0, \qquad x \in U.$$

Then Lemma 6.7 leads to $\alpha \equiv 0$. □

Now we are able to construct positive definite functions just by choosing the measure. This is even simpler if μ has a Lebesgue density.

Corollary 6.9 *If $f \in L_1(\mathbb{R}^d)$ is continuous, nonnegative, and nonvanishing then*

$$\Phi(x) := \int_{\mathbb{R}^d} f(\omega) e^{-ix^T \omega} d\omega, \qquad x \in \mathbb{R}^d,$$

is positive definite.

Proof We use the measure μ defined for any Borel set B by

$$\mu(B) = \int_B f(x) dx.$$

Then the carrier of μ is equal to the support of f. However, since f is nonnegative and

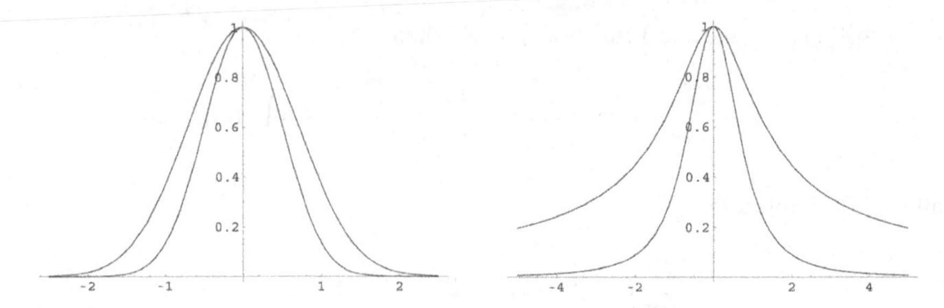

Fig. 6.1 The Gaussian for $\alpha = 1$ and $\alpha = 2$ (on the left) and the inverse multiquadrics with $c = 1$ for $\beta = 1/2$ and $\beta = 3/2$ (on the right).

nonvanishing, its support must contain an interior point, and hence the Fourier transform of f is positive definite by the preceding theorem. □

It is time for our first example. The reader can probably see that the Gaussian is a good candidate for a positive definite function. We formulate the result more generally using an additional scale parameter.

Theorem 6.10 *The Gaussian* $\Phi(x) = e^{-\alpha\|x\|_2^2}$, $\alpha > 0$, *is positive definite on every* $\mathbb{R}^d$.

Proof We reduce the proof to that for the function $G(x) := e^{-\|x\|_2^2/2}$, which satisfies $\widehat{G} = G$. If we introduce $G_\alpha = G(\cdot/\alpha)$ then we have $\Phi = G_{1/\sqrt{2\alpha}}$ and $\widehat{G_\alpha} = \alpha^d \widehat{G}(\alpha\cdot)$ by Theorem 5.16. Collecting these results we derive

$$\Phi(x) = 2^{-d}(\pi\alpha)^{-d/2} \int_{\mathbb{R}^d} e^{-\|\omega\|_2^2/(4\alpha)} e^{-ix^T\omega} d\omega,$$

which means that Φ is positive definite. □

The left-hand part of Figure 6.1 shows the univariate function $\phi(r) = e^{-\alpha r^2}$ for $\alpha = 1$ and $\alpha = 2$.

The Gaussian is positive definite for every scaling parameter $\alpha > 0$. However, the correct choice of α in a particular interpolation problem is crucial. On the one hand, if the parameter is too large then the Gaussian becomes a sharp peak, which immediately carries to the surface and leads to a rather poor representation; the interpolation matrix, however, is then almost diagonal and has a low condition number. On the other hand, a small value for α leads to a better surface reconstruction but corresponds to a very large condition number for the interpolation matrix. We shall come back to this relation between condition number and approximation property later on. Figure 6.2 demonstrates the effect of the scale parameter on the reconstruction of a smooth function from a 6×6 grid. The picture on the left comes closest to the original function.

Of course, Corollary 6.9 is useful when constructing positive definite functions. But if a function Φ is given it would be useful to have a tool to check whether it is positive definite. Motivated by the Gaussian we formulate:

Theorem 6.11 *Suppose that* $\Phi \in L_1(\mathbb{R}^d)$ *is a continuous function. Then* Φ *is positive definite if and only if* Φ *is bounded and its Fourier transform is nonnegative and nonvanishing.*

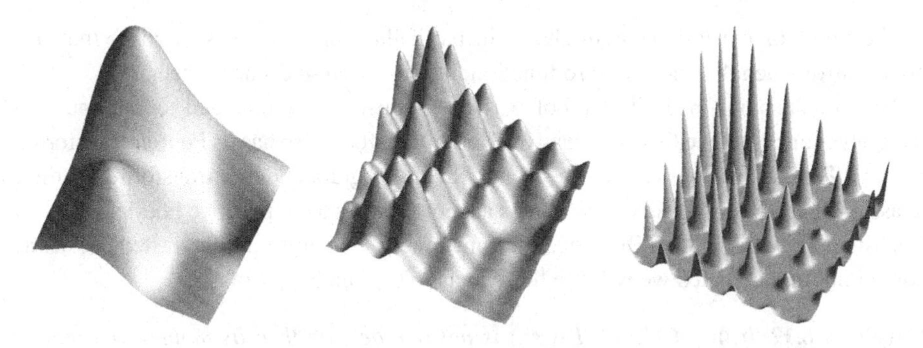

Fig. 6.2 The reconstruction of a smooth function using a Gaussian with scale parameter $\alpha = 10$, $\alpha = 100$, $\alpha = 1000$.

Proof Suppose that Φ is bounded and has a nonnegative and nonvanishing Fourier transform. It suffices to show that $\widehat{\Phi} \in L_1(\mathbb{R}^d)$ is satisfied. Then the Fourier inversion formula can be applied and Theorem 6.8 finishes the proof.

As in the proof of Bochner's theorem we choose g_m from Theorem 5.20 and get

$$\begin{aligned}(2\pi)^{d/2}\Phi(0) &= (2\pi)^{d/2} \lim_{m\to\infty} \int_{\mathbb{R}^d} \Phi(x) g_m(x) dx \\ &= (2\pi)^{d/2} \lim_{m\to\infty} \int_{\mathbb{R}^d} \widehat{\Phi}(\omega)\widehat{g}_m(\omega) d\omega \\ &= \int_{\mathbb{R}^d} \widehat{\Phi}(\omega) d\omega,\end{aligned}$$

since $\widehat{\Phi}$ is nonnegative. Thus $\widehat{\Phi}$ is in $L_1(\mathbb{R}^d)$.

If, conversely, Φ is positive definite then we know from Theorem 6.2 that Φ is bounded and from Bochner's theorem that Φ is the inverse Fourier transform of a nonnegative finite Borel measure μ on $\mathbb{R}^d$. Furthermore, using Theorem 5.16, Theorem 5.20 and Fubini's theorem we find that

$$\begin{aligned}\widehat{\Phi}(x) &= \lim_{m\to\infty} \int_{\mathbb{R}^d} \widehat{\Phi}(\omega) g_m(\omega - x) d\omega \\ &= \lim_{m\to\infty} \int_{\mathbb{R}^d} \Phi(\omega)\widehat{g}_m(\omega) e^{-ix^T\omega} d\omega \\ &= \lim_{m\to\infty} (2\pi)^{-d/2} \int_{\mathbb{R}^d}\int_{\mathbb{R}^d} e^{-i\omega^T\eta} d\mu(\eta) \widehat{g}_m(\omega) e^{-ix^T\omega} d\omega \\ &= \lim_{m\to\infty} (2\pi)^{-d/2} \int_{\mathbb{R}^d}\int_{\mathbb{R}^d} \widehat{g}_m(\omega) e^{-i\omega^T(\eta+x)} d\omega d\mu(\eta) \\ &= \lim_{m\to\infty} \int_{\mathbb{R}^d} g_m(-\eta - x) d\mu(\eta) \\ &\geq 0.\end{aligned}$$

Thus $\widehat{\Phi}$ is nonnegative. Now we can proceed as in the first part of the proof to show that $\widehat{\Phi} \in L_1(\mathbb{R}^d)$ and $\|\widehat{\Phi}\|_{L_1(\mathbb{R}^d)} = (2\pi)^{d/2}\Phi(0)$. Hence $\widehat{\Phi}$ cannot vanish identically. □

The proof of the last theorem shows in particular that it suffices to show that $\widehat{\Phi}$ is nonnegative when Φ is not the zero function, because then $\widehat{\Phi}$ cannot vanish.

The crucial point in the last proof was to establish that a bounded, continuous, and integrable function Φ with a nonnegative Fourier transform also has its Fourier transform in $L_1(\mathbb{R}^d)$. The assumptions can be weakened. For an integrable Fourier transform, it suffices to assume that Φ is integrable, continuous at zero, and has a nonnegative Fourier transform (see Stein & Weiss [180]). But since the functions in which we are interested are always continuous and bounded we will use these facts to shorten the proofs.

Corollary 6.12 *If $\Phi \in C(\mathbb{R}^d) \cap L_1(\mathbb{R}^d)$ is positive definite then its nonnegative Fourier transform is in $L_1(\mathbb{R}^d)$.*

Let us provide an additional example and apply Theorem 6.11 to prove the positive definiteness of another famous class of functions, namely inverse multiquadrics. Their Fourier transforms involve the modified Bessel functions K_ν.

Theorem 6.13 *The function $\Phi(x) = (c^2 + \|x\|_2^2)^{-\beta}$, $x \in \mathbb{R}^d$, with $c > 0$ and $\beta > d/2$ is positive definite. It possesses the Fourier transform*

$$\widehat{\Phi}(\omega) = \frac{2^{1-\beta}}{\Gamma(\beta)} \left(\frac{\|\omega\|_2}{c} \right)^{\beta - d/2} K_{d/2-\beta}(c\|\omega\|_2). \tag{6.6}$$

Proof Since $\beta > d/2$ the function Φ is in $L_1(\mathbb{R}^d)$. From the representation of the Γ-function for $\beta > 0$ we see that

$$\begin{aligned}\Gamma(\beta) &= \int_0^\infty t^{\beta-1} e^{-t} dt \\ &= s^\beta \int_0^\infty u^{\beta-1} e^{-su} du\end{aligned}$$

by substituting $t = su$ with $s > 0$. If we set $s = c^2 + \|x\|_2^2$ then we can conclude that

$$\Phi(x) = \frac{1}{\Gamma(\beta)} \int_0^\infty u^{\beta-1} e^{-c^2 u} e^{-\|x\|_2^2 u} du.$$

Inserting this into the Fourier transform and changing the order of integration, which can easily be justified, leads to

$$\begin{aligned}\widehat{\Phi}(\omega) &= (2\pi)^{-d/2} \int_{\mathbb{R}^d} \Phi(x) e^{-ix^T\omega} dx \\ &= (2\pi)^{-d/2} \frac{1}{\Gamma(\beta)} \int_{\mathbb{R}^d} \int_0^\infty u^{\beta-1} e^{-c^2 u} e^{-\|x\|_2^2 u} e^{-ix^T\omega} du dx \\ &= (2\pi)^{-d/2} \frac{1}{\Gamma(\beta)} \int_0^\infty u^{\beta-1} e^{-c^2 u} \int_{\mathbb{R}^d} e^{-\|x\|_2^2 u} e^{-ix^T\omega} dx du\end{aligned}$$

$$= \frac{1}{\Gamma(\beta)} \int_0^\infty u^{\beta-1} e^{-c^2u} (2u)^{-d/2} e^{-\|\omega\|_2^2/(4u)} du$$

$$= \frac{1}{2^{d/2}\Gamma(\beta)} \int_0^\infty u^{\beta-d/2-1} e^{-c^2u} e^{-\|\omega\|_2^2/(4u)} du,$$

where we have used Theorems 5.16 and 5.18. Moreover, we know from the proof of Lemma 5.14 that

$$K_\nu(r) = a^{-\nu} \frac{1}{2} \int_0^\infty e^{(-r/2)(u/a+a/u)} u^{\nu-1} du$$

for every $a > 0$. If we now set $r = c\|\omega\|_2$, $a = \|\omega\|_2/(2c)$, and $\nu = \beta - d/2$ for $\omega \neq 0$, we derive

$$K_{\beta-d/2}(c\|\omega\|_2) = \frac{1}{2} \left(\frac{\|\omega\|_2}{2c} \right)^{d/2-\beta} \int_0^\infty e^{-uc^2} e^{(-\|\omega\|_2^2)/(4u)} u^{\beta-d/2-1} du$$

$$= 2^{\beta-1}\Gamma(\beta) \left(\frac{\|\omega\|_2}{c} \right)^{d/2-\beta} \widehat{\Phi}(\omega),$$

which leads to the stated Fourier transform for $\omega \neq 0$ using $K_{-\nu} = K_\nu$. We can use continuity to see that (6.6) also holds for $\omega = 0$. Since the modified Bessel function is nonnegative and nonvanishing, Φ is positive definite. □

Examples of inverse multiquadrics for $\beta = 1/2$ and $3/2$ and $c = 1$ are provided on the right-hand side of Figure 6.1.

We made the restriction $\beta > d/2$ to ensure that Φ is integrable. This restriction makes the function dependent in a certain way on the space dimension. Later we will see that this restriction is artificial and that any $\beta > 0$ leads to a positive definite function on any $\mathbb{R}^d$.

We want to close this section by considering a remarkable property of positive definite functions concerning their smoothness.

Theorem 6.14 *Suppose that Φ is a positive definite function that belongs to C^{2k} in some neighborhood of the origin. Then Φ is in C^{2k} everywhere.*

Proof Since Φ is positive definite there exists a finite nonnegative Borel measure μ on $\mathbb{R}^d$ such that

$$\Phi(x) = (2\pi)^{-d/2} \int_{\mathbb{R}^d} e^{-ix^T\omega} d\mu(\omega). \tag{6.7}$$

This means that, for every test function $\gamma \in C_0^\infty(\mathbb{R}^d)$,

$$\int_{\mathbb{R}^d} \Phi(x)\gamma(x)dx = \int_{\mathbb{R}^d} \widehat{\gamma}(\omega)d\mu(\omega). \tag{6.8}$$

Again, we choose a regularization $g_\ell(x) = \ell^d g(\ell x)$ with a nonnegative function $g \in C_0^\infty(\mathbb{R}^d)$ with $\|g\|_{L_1(\mathbb{R}^d)} = 1$ and support $\{x \in \mathbb{R}^d : \|x\|_2 \leq 1\}$. Let Δ denote the usual Laplace operator, i.e. $\Delta = \sum_{j=1}^d \partial^2/\partial x_j^2$, and Δ^k the iterated Laplacian. By inserting

$(1-\Delta)^k g_\ell$ into (6.8) we find

$$\int_{\mathbb{R}^d} \widehat{g_\ell}(\omega)(1+\|\omega\|_2^2)^k d\mu(\omega) = \int_{\mathbb{R}^d} \Phi(x)(1-\Delta)^k g_\ell(x)dx$$
$$= \int_{\mathbb{R}^d} (1-\Delta)^k \Phi(x) g_\ell(x)dx,$$

since Φ possesses $2k$ continuous derivatives around zero and the integrals are actually only integrals over $\{x : \|x\|_2 \le 1/\ell\}$. The last integral converges for $\ell \to \infty$ to $(1-\Delta)^k\Phi(0)$. Hence, by Fatou's lemma and $\widehat{g_\ell}(\omega) \to (2\pi)^{-d/2}$ we get

$$\int_{\mathbb{R}^d} (1+\|\omega\|_2^2)^k d\mu(\omega) \le (2\pi)^{d/2}(1-\Delta)^k\Phi(0).$$

Thus we can differentiate up to $2k$ times under the integral in (6.7), which means that Φ is in C^{2k} everywhere. □

6.3 Radial functions

Even though we have used it already in several places we will now recall the definition of a radial function.

Definition 6.15 *A function $\Phi : \mathbb{R}^d \to \mathbb{R}$ is said to be radial if there exists a function $\phi : [0,\infty) \to \mathbb{R}$ such that $\Phi(x) = \phi(\|x\|_2)$ for all $x \in \mathbb{R}^d$.*

In the Gaussians and the multiquadrics we have already found examples of radial and positive definite functions without using the fact that they are radial. Now we want to exploit radiality in more detail.

A radial function has the advantage of a very simple structure. This is motivation enough to investigate whether such a univariate function is positive definite in the following sense.

Definition 6.16 *We will call a univariate function $\phi : [0,\infty) \to \mathbb{R}$ positive definite on $\mathbb{R}^d$ if the corresponding multivariate function $\Phi(x) := \phi(\|x\|_2)$, $x \in \mathbb{R}^d$, is positive definite.*

The smoothness of the multivariate function Φ is determined by the smoothness of the even extension of the univariate function ϕ. This is the reason why we always assume ϕ to be an even function defined on all of $\mathbb{R}$ by even extension.

Any radial function is obviously even. From Theorem 6.3 we know that we can restrict ourselves to real coefficients in the quadratic form.

Lemma 6.17 *Suppose that a univariate function ϕ is positive definite on $\mathbb{R}^d$; then it is also positive definite on $\mathbb{R}^k$ with $k \le d$.*

Proof The proof of this lemma is obvious because $\mathbb{R}^k$ is a subspace of $\mathbb{R}^d$. □

Theorem 5.26 tells us how to compute the Fourier transform of a radial function. The Fourier transform is again radial and thus can be expressed as a univariate function. Hence,

Bochner's characterization can be reduced to a univariate setting. For example, the radial version of Theorem 6.11 is given by

Theorem 6.18 *Suppose that $\phi \in C[0,\infty)$ satisfies $r \mapsto r^{d-1}\phi(r) \in L_1[0,\infty)$. Then ϕ is positive definite on $\mathbb{R}^d$ if and only if ϕ is bounded and*

$$\mathcal{F}_d\phi(r) := r^{-(d-2)/2} \int_0^\infty \phi(t) t^{d/2} J_{(d-2)/2}(rt)dt$$

is nonnegative and nonvanishing.

Proof From $r \mapsto r^{d-1}\phi(r) \in L_1[0,\infty)$ we know that $\Phi := \phi(\|\cdot\|_2)$ is in $L_1(\mathbb{R}^d)$. Furthermore, we have $\widehat{\Phi}(x) = \mathcal{F}_d\phi(\|x\|_2)$. □

Note that the operator $\mathcal{F}_d$ defined by $\mathcal{F}_d\phi = \widehat{\Phi}$ acts on univariate functions. Hence, if working with radial functions we are in a situation where we can do most of the analysis in a univariate setting, which often makes things easier.

We will demonstrate this concept by giving another example of a positive definite function. It suffices to show that the univariate function $\mathcal{F}_d\phi$ is nonnegative in contrast with having to show that the multivariate function $\widehat{\Phi}$ is nonnegative.

The function that we are now going to investigate differs from all positive definite functions we have encountered so far in two ways. First, it has a compact support. Second, it is not positive definite on every $\mathbb{R}^d$ as it was the case for Gaussians and inverse multiquadrics (see the remarks after the proof of Theorem 6.13). We will see that these two features do not appear together accidentally. On the contrary, the first implies the second.

Lemma 6.19 *Define the functions $f_0(r) = 1 - \cos r$ and*

$$f_n(r) = \int_0^r f_0(t) f_{n-1}(r-t)dt$$

for $n \geq 1$. Let

$$B_n = \frac{2^{n+1/2} n!(n+1)!}{\sqrt{\pi}}.$$

Then f_n has the representation

$$\int_0^r (r-t)^{n+1} t^{n+1/2} J_{n-1/2}(t)dt = B_n f_n(r). \tag{6.9}$$

Proof Define the integral on the left-hand side of (6.9) to be $g(r)$. Thus g is the convolution $g(r) = \int_0^r g_1(r-s)g_2(s)ds$ of the functions $g_1(s) := s^{n+1}$ and $g_2(s) := s^{n+1/2}J_{n-1/2}(s)$. This form of convolution differs slightly from the kind about which we have learnt in the context of Fourier transforms. This new form of convolution is compatible with the Laplace transform $\mathcal{L}$ in the sense that $\mathcal{L}g(r) = \int_0^\infty g(t)e^{-rt}dt$ is the product of the Laplace transforms of g_1 and g_2. These transforms can be computed for $r > 0$ in the following manner. The first function g_1 can be handled by using the integral representation for the

Γ-function:

$$\mathcal{L}g_1(r) = \int_0^\infty s^{n+1}e^{-rs}ds = r^{-n-2}\int_0^\infty t^{n+1}e^{-t}dt = \frac{\Gamma(n+2)}{r^{n+2}}$$
$$= \frac{(n+1)!}{r^{n+2}}.$$

The Laplace transform of g_2 was computed in Lemma 5.7. If we set $\nu = n - 1/2 > -1$, Lemma 5.7 yields

$$\mathcal{L}g_2(r) = \int_0^\infty s^{n+1/2}J_{n-1/2}(s)e^{-rs}ds = \frac{n!\,2^{n+1/2}r}{\sqrt{\pi}\,(1+r^2)^{n+1}}$$

for $r > 0$. Together these two expressions give

$$\mathcal{L}g(r) = \frac{2^{n+1/2}n!(n+1)!}{\sqrt{\pi}}\frac{1}{r^{n+1}(1+r^2)^{n+1}}.$$

It is easily shown, however, that the function $f_0(r) = 1 - \cos r$ has Laplace transform $1/[r(1+r^2)]$. Thus, since f_n is the n-fold convolution of this function with itself, we get

$$\mathcal{L}f_n(r) = \frac{1}{r^{n+1}(1+r^2)^{n+1}}.$$

By the uniqueness of the Laplace transform this leads to

$$g(r) = \frac{2^{n+1/2}n!(n+1)!}{\sqrt{\pi}}f_n(r)$$

as stated. □

The reason for proving this lemma is provided by the integral in (6.9). Obviously it represents the Fourier transform of a radial function. And we now know that this Fourier transform is nonnegative and nonvanishing. To give the associated function ϕ itself, let us introduce the cutoff function $(\cdot)_+$, which is defined by

$$(x)_+ = \begin{cases} x & \text{for } x \geq 0, \\ 0 & \text{for } x < 0, \end{cases}$$

and the notation $\lfloor x \rfloor$, which denotes the largest integer less than or equal to x.

Theorem 6.20 *The truncated power function*

$$\phi_\ell(r) = (1-r)_+^\ell$$

is positive definite on $\mathbb{R}^d$ *provided that* $\ell \in \mathbb{N}$ *satisfies* $\ell \geq \lfloor d/2 \rfloor + 1$.

Proof Let us start with the case of an odd space dimension $d = 2n+1$ and $\ell = \lfloor d/2 \rfloor + 1 = n+1$. We have to check whether the function $\mathcal{F}_{2n+1}\phi_{n+1}$ is nonnegative and nonvanishing. In this special situation it takes the form

$$r^{3n+2}\mathcal{F}_{2n+1}\phi_{n+1}(r) = \int_0^r (r-s)^{n+1}s^{n+1/2}J_{n-1/2}(s)ds.$$

Thus by (6.9) we see that

$$r^{3n+2}\mathcal{F}_{2n+1}\phi_{n+1}(r) = B_n f_n(r),$$

which is clearly nonnegative and nonvanishing. For an even space dimension $d = 2n$ and the same $\ell = n + 1$ we only need to remark that $\phi_{\lfloor 2n/2\rfloor+1} = \phi_{\lfloor (2n+1)/2\rfloor+1}$. Hence $\phi_{\lfloor 2n/2\rfloor+1}$ is positive definite on $\mathbb{R}^{2n+1}$ and therefore also on $\mathbb{R}^{2n}$. The same argument proves the positive definiteness of $\phi_\ell(\|\cdot\|_2)$ for $\ell > \lfloor d/2\rfloor + 1$. □

The restriction $\ell \in \mathbb{N}$ is actually not necessary but simplifies the proof. It is also possible to allow real values for ℓ.

A general way of constructing positive definite functions is to integrate a fixed positive definite function against a nonnegative measure. That is exactly the way the Fourier transform acts in Bochner's characterization.

Theorem 6.21 *Suppose that the continuous function $\phi : [0,\infty) \to \mathbb{R}$ is given by*

$$\phi(r) := \int_0^\infty (1 - rt)_+^{k-1} f(t)dt, \tag{6.10}$$

where $f \in C[0,\infty)$ is nonnegative and nonvanishing. Then ϕ is positive definite on $\mathbb{R}^d$ if $k \geq \lfloor d/2\rfloor + 2$.

Proof We find, for arbitrary coefficients and centers,

$$\sum_{j=1}^N \sum_{\ell=1}^N \alpha_j \alpha_\ell \phi(\|x_j - x_\ell\|_2) = \int_0^\infty \sum_{j,\ell=1}^N \alpha_j \alpha_\ell \phi_{k-1}(t\|x_j - x_\ell\|_2) f(t)dt \geq 0$$

because ϕ_{k-1} is positive definite on $\mathbb{R}^d$ under the given assumptions, by Theorem 6.20. Furthermore, since f is continuous, nonnegative, and nonvanishing the latter integral can vanish only if $\alpha \equiv 0$. □

A function of the form (6.10) obviously belongs to C^{k-2}, and the derivatives satisfy $(-1)^\ell \phi^{(\ell)}(r) \geq 0$ for $0 \leq \ell \leq k-2$. Such functions are called multiply monotone functions.

Definition 6.22 *Suppose that $k \in \mathbb{N}$ satisfies $k \geq 2$. A function $\phi : (0,\infty) \to \mathbb{R}$ is called k-times monotone if $(-1)^\ell \phi^{(\ell)}$ is nonnegative, nonincreasing, and convex. If $k = 1$ we require ϕ to be nonnegative and nonincreasing.*

As in the case of Bochner's theorem, the representation in (6.10) is not general enough to characterize all k-times monotone functions. Again, the solution is to allow a more general measure. Since this result will not play an important role for us, we just cite it for the interested reader. The proof can be found in Williamson [201].

Theorem 6.23 *A necessary and sufficient condition that $\phi : (0,\infty) \to \mathbb{R}$ is k-times monotone is that ϕ is of the form*

$$\phi(r) = \int_0^\infty (1 - rt)_+^{k-1} d\gamma(t),$$

where γ is a nonnegative Borel measure.

6.4 Functions, kernels, and other norms

Our investigation of positive definite functions was motivated by the interpolation problem in (6.2), using an interpolant of the form (6.1). This particular choice was very convenient for our analysis, because we could restrict everything to the investigation of a single function Φ. But of course (6.1) is not the only possible approach to the interpolation problem. More generally, one would start with a function $\Phi : \mathbb{R}^d \times \mathbb{R}^d \to \mathbb{C}$ and try to form the interpolant as

$$s_{f,X} = \sum_{j=1}^{N} \alpha_j \Phi(\cdot, x_j).$$

If we are interested only in sets of centers $X = \{x_1, \dots, x_N\}$ that are contained in a certain subset $\Omega \subseteq \mathbb{R}^d$, then it even suffices to have a $\Phi : \Omega \times \Omega \to \mathbb{C}$. To make the difference from the previous approach clearer, we will call such a Φ a *kernel* rather than a function.

Definition 6.24 *A continuous kernel $\Phi : \Omega \times \Omega \to \mathbb{C}$ is called positive definite on $\Omega \subseteq \mathbb{R}^d$ if for all $N \in \mathbb{N}$, all pairwise distinct $X = \{x_1, \dots, x_N\} \subseteq \Omega$, and all $\alpha \in \mathbb{C}^N \setminus \{0\}$ we have*

$$\sum_{j=1}^{N} \sum_{k=1}^{N} \alpha_j \overline{\alpha_k} \Phi(x_j, x_k) > 0.$$

The definition is in a certain way not precise. Since we have not specified the set Ω it might as well be a finite set. In that situation it would be impossible to find for all $N \in \mathbb{N}$ pairwise distinct points X. It should be clear that, in such a situation, only those $N \in \mathbb{N}$ would have to be considered that allow the choice of N pairwise distinct data sites.

Radial basis functions fit into this more general setting by defining $\Phi(x, y) = \phi(\|x - y\|_2)$ which leads to real-valued kernels. Most of the kernels we will discuss are radial, but we can also use tensor products to create multivariate positive definite functions from univariate ones.

Proposition 6.25 *Suppose that $\phi_1, \dots, \phi_d$ are positive definite and integrable functions on $\mathbb{R}$. Then*

$$\Phi(x) := \phi_1(x_1) \cdots \phi_d(x_d), \qquad x = (x_1, \dots, x_d)^T \in \mathbb{R}^d,$$

is a positive definite function on $\mathbb{R}^d$.

Proof Since the univariate functions $\{\phi_j\}$ are integrable, so also is the multivariate function Φ. Moreover, its d-variate Fourier transform $\widehat{\Phi}$ is the product of the univariate Fourier transforms:

$$\widehat{\Phi}(x) = \widehat{\phi_1}(x_1) \cdots \widehat{\phi_d}(x_d).$$

If we apply Theorem 6.11 to the univariate functions we see that their Fourier transforms are nonnegative and nonvanishing. This means that the multivariate Fourier transform $\widehat{\Phi}$

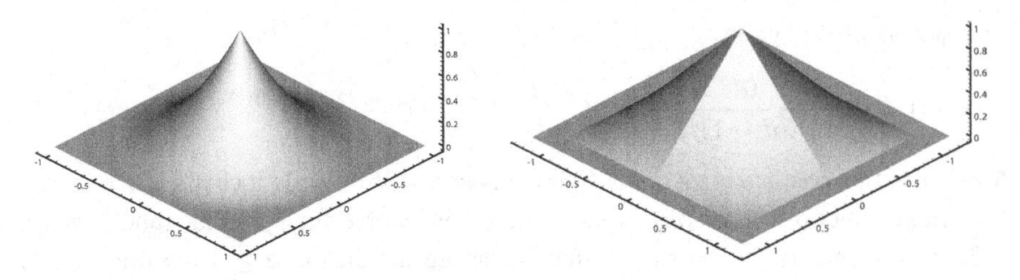

Fig. 6.3 The bivariate functions $\Phi(x) = (1 - \|x\|_2)_+^2$ (on the left) and $\Phi(x) = (1 - |x_1|)_+(1 - |x_2|)_+$ (on the right).

also possesses this property. Hence, a final application of Theorem 6.11 shows that Φ is positive definite. □

In Figure 6.3 two compactly supported functions are shown. The function on the left hand side is the radial function $\Phi(x) = (1 - \|x\|_2)_+^2$, which is positive definite on $\mathbb{R}^2$ by Theorem 6.20. The function on the right is the function $\Phi(x) = (1 - |x_1|)_+(1 - |x_2|)_+$, which is also positive definite on $\mathbb{R}^2$ by Theorem 6.20 and Proposition 6.25. We have chosen the smallest possible exponent in both cases.

If functions $\Phi : \Omega \times \Omega \to \mathbb{C}$ are considered then one might naturally arrive at the question whether there exists an extension of Φ defined on a bigger subset of $\mathbb{R}^d$. Such a question was discussed by Rudin in [159, 160] in the situation where the kernel is actually a function; to be more precise, where $\Phi(x, y) = \Phi_0(x - y)$ and Φ_0 is a function that is defined on $\Omega - \Omega = \{x - y : x, y \in \Omega\}$. The results are diverse. On the one hand, it was shown by Rudin [159] that if Ω is a closed cube in $\mathbb{R}^d$ with $d \geq 2$ then there always exists a positive semi-definite kernel of this form that does not have an extension to all of $\mathbb{R}^d$. On the other hand, Rudin [160] proved that every positive semi-definite kernel of this form, defined on a ball and in addition radial, has an extension to $\mathbb{R}^d$. In the case of a univariate setting, balls and cubes are the same and hence an extension for functions that are positive semi-definite on intervals to the whole real line exists. Questions about uniqueness in this context were considered in Akutowicz' article [1].

One reason for looking at radial functions is that they allow easier computation. Hence, one might think of investigating also ℓ_p-radial functions, i.e. functions $\Phi : \mathbb{R}^d \to \mathbb{R}$ of the form $\Phi(x) = \phi(\|x\|_p)$, $x \in \mathbb{R}^d$. So far, such functions have not played a role in the theory of radial basis function interpolation; hence we do not want to discuss them in much detail. Nonetheless, there might be some applications that would benefit from basis functions of this particular form. Thus we will give at least a certain amount of information on this challenging topic. The first thing is that the situation is similar to the ℓ_2 case if the ℓ_1-norm is used. In this situation it is also possible to characterize all positive semi-definite functions.

Theorem 6.26 *A function $\Phi : \mathbb{R}^d \to \mathbb{R}$ defined by $\Phi(x) = \phi(\|x\|_1)$, $x \in \mathbb{R}^d$, is positive semi-definite if and only if there exists a finite Borel measure α on $[0, \infty)$ such that*

$$\phi(r) = \int_0^\infty \phi_0(rt)d\alpha(t),$$

where $\phi_0(r)$ is given by

$$\phi_0(r) = \frac{2^{d/2}\Gamma^2(d/2)}{\sqrt{\pi}\,\Gamma((d-1)/2)} r^{-(d-2)/2} \int_1^\infty (t^2-1)^{(d-3)/2} t^{-(3d-4)/2} J_{(d-2)/2}(rt)dt.$$

A proof of this result was given by Cambanis *et al.* [36]. While it characterizes positive semi-definite functions that are ℓ_1-radial, things are worse for ℓ_p-radial functions when $p > 2$. In this case it can be shown that for space dimension $d \geq 3$ the only function $\Phi(x) = \phi(\|x\|_p)$ that is positive semi-definite is the trivial function $\Phi \equiv 0$. Nonetheless, there is some hope for space dimensions $d \leq 2$ and for $p \in [1,2]$. Details on the negative result can be found in Zastavnyi [209].

6.5 Notes and comments

Positive (semi-)definite functions play an important role not only in approximation theory but also in other mathematical areas, for example in probability theory. There, a positive (semi-)definite function is nothing other than the characteristic function of a probability distribution. Hence, the basic properties listed at the beginning of this chapter are folklore nowadays and the reader might find more information in the review article by Stewart [181] or the book by Lukacs [110]. But in contrast with probability theory, where semi-definite functions work as well as definite ones, approximation theory has to stick with positive definite functions, because being positive definite is crucial for interpolation.

The most important result of this chapter is without a question Bochner's theorem (Theorem 6.6). But even though nowadays all tribute goes to Bochner for this result, Mathias [117] had proved already, in 1923, that a univariate positive (semi-)definite function has a nonnegative Fourier transform. The proof of Bochner's theorem, given here in the modern language of measure theory, was motivated by Donoghue's presentation [46].

The truncated power function in Theorem 6.20 has been investigated by several authors, for example Askey [6] and Chanysheva [40]. It will play an important role in what follows.

7

Completely monotone functions

At the end of the last chapter we discussed radial positive definite functions. We investigated whether a univariate function $\phi : [0, \infty) \to \mathbb{R}$ defines a positive definite function $\Phi(x) = \phi(\|x\|_2)$ on a fixed $\mathbb{R}^d$. But we already had examples where the univariate function gives rise to a positive definite function on *every* $\mathbb{R}^d$. This is of course a very pleasant feature. The reader should reflect on this property for a moment. It means that we can use the same univariate function ϕ to interpolate any number of scattered data in any space dimension. Hence, we now want to discuss such functions in greater detail.

In the last chapter, we also encountered k-times monotone functions and noticed their connection with positive definite functions on $\mathbb{R}^d$ where k and d were related in a certain way.

It will turn out that there is a similar connection between completely monotone functions, which are the generalization of multiply monotone functions, and radial functions that are positive definite on every $\mathbb{R}^d$. To be more precise, suppose $\phi : [0, \infty) \to \mathbb{R}$ is given by

$$\phi(r) = \int_0^\infty e^{-tr^2} d\nu(t), \tag{7.1}$$

with a nonnegative and finite Borel measure ν defined on the Borel sets $\mathcal{B}([0, \infty)) = [0, \infty) \cap \mathcal{B}(\mathbb{R})$. Then, for arbitrary $x_1, \ldots, x_N$ and an arbitrary $\alpha \in \mathbb{R}^N$, we have

$$\sum_{j,k=1}^N \alpha_j \alpha_k \phi(\|x_j - x_k\|_2) = \int_0^\infty \sum_{j,k=1}^N \alpha_j \alpha_k e^{-t\|x_j - x_k\|_2^2} d\nu(t) \geq 0, \tag{7.2}$$

because the Gaussians involved are positive definite and the measure ν is nonnegative. Moreover, ϕ is continuous because ν is finite. This means that ϕ is positive semi-definite on every $\mathbb{R}^d$. Furthermore, obviously $f(r) := \phi(\sqrt{r})$ satisfies

$$(-1)^n \frac{d^n}{dr^n} f(r) = (-1)^n \frac{d^n}{dr^n} \int_0^\infty e^{-rt} d\nu(t) = \int_0^\infty t^n e^{-rt} d\nu(t) \geq 0.$$

Differentiation under the integral sign is justified because the measure ν is finite. Hence, if ϕ is a function of the form (7.1) then it is positive semi-definite on every $\mathbb{R}^d$ and the associated function $f = \phi(\sqrt{\cdot})$ satisfies $(-1)^n f^{(n)} \geq 0$. The goal of this chapter is to prove that all three properties are actually equivalent.

7.1 Definition and first characterization

We start by defining completely monotone functions.

Definition 7.1 *A function ϕ is called completely monotone on $(0, \infty)$ if it satisfies $\phi \in C^\infty(0, \infty)$ and*

$$(-1)^\ell \phi^\ell(r) \geq 0$$

for all $\ell \in \mathbb{N}_0$ and all $r > 0$. The function ϕ is called completely monotone on $[0, \infty)$ if it is in addition in $C[0, \infty)$.

The first equivalent characterization of completely monotone functions can be derived by using iterated forward differences.

Definition 7.2 *Let $k \in \mathbb{N}_0$. Suppose that $\{f_j\}_{j\in\mathbb{N}_0}$ is a sequence of real numbers. The kth-order iterated forward difference is*

$$\Delta^k\{f_j\}(\ell) \equiv \Delta^k f_\ell := \sum_{j=0}^{k} (-1)^{k-j} \binom{k}{j} f_{\ell+j}, \qquad \ell \in \mathbb{N}_0. \tag{7.3}$$

For a function $\phi : [0, \infty) \to \mathbb{R}$ we define the kth-order difference by

$$\Delta_h^k \phi(r) := \sum_{j=0}^{k} (-1)^{k-j} \binom{k}{j} \phi(r + jh), \tag{7.4}$$

for any $r \geq 0$ and $h > 0$. If ϕ is defined only on $(0, \infty)$ then we restrict r in (7.4) to $r > 0$.

Obviously, for a fixed $r \geq 0$ and a fixed $h > 0$, we have $\Delta_h^k \phi(r) = \Delta^k\{f_j\}(0)$ with the sequence $\{f_j\}$ given by $f_j := \phi(r + jh)$.

Lemma 7.3 *Suppose that $\phi : (0, \infty) \to \mathbb{R}$ satisfies $(-1)^n \Delta_h^n \phi(r) \geq 0$ for all $r, h > 0$ and $n = 0, 1, 2$. Then ϕ is nonnegative, nonincreasing, continuous, and convex on $(0, \infty)$.*

Proof If we set $n = 0$ and $n = 1$ in the assumptions of the lemma, we see immediately that ϕ is nonnegative and nonincreasing. As a nonincreasing function, ϕ possesses limiting values on the right and on the left for every $r \in (0, \infty)$, which satisfy $\phi(r+) \leq \phi(r) \leq \phi(r-)$. If we set $n = 2$, we see that ϕ is midpoint convex, meaning that $\phi(r + h) \leq [\phi(r) + \phi(r + 2h)]/2$ for all $r, h > 0$. This gives in particular $2\phi(r - h) \leq \phi(r) + \phi(r - 2h)$ and $2\phi(r) \leq \phi(r + h) + \phi(r - h)$ and leads to $\phi(r-) \leq \phi(r)$ and $2\phi(r) \leq \phi(r+) + \phi(r-)$. But this shows that ϕ is continuous at $r > 0$. Finally, a continuous and midpoint convex function is convex. □

After this preparatory step we are able to prove that the condition stated in the lemma is equivalent to the requirement that ϕ is completely monotone, if it holds for all $n \in \mathbb{N}_0$. This seems to be quite natural, since Δ^n represents the discretization of the nth-order derivative.

Theorem 7.4 *For $\phi : (0, \infty) \to \mathbb{R}$ the following statements are equivalent:*
(1) ϕ is completely monotone on $(0, \infty)$;
(2) ϕ satisfies $(-1)^n \Delta_h^n \phi(r) \geq 0$ for all $r, h > 0$ and $n \in \mathbb{N}_0$.

Proof If ϕ is completely monotone on $(0, \infty)$ then we have

$$\Delta_h^n \phi(r) = \Delta_h^{n-1}\{\phi(r+h) - \phi(r)\} = h\Delta_h^{n-1}\phi'(\xi_1)$$

with $\xi_1 \in (r, r+h)$. Iterating this statement gives for every $r, h > 0$ and every $n \in \mathbb{N}_0$ a $\xi_n \in (r, r+nh)$ such that $\Delta_h^n \phi(r) = h^n \phi^{(n)}(\xi_n)$. Hence, (1) implies that $(-1)^n \Delta_h^n \phi(r) \geq 0$.

Conversely, if (2) is satisfied we have to show that $\phi \in C^\infty(0, \infty)$ and that the alternation condition is satisfied.

From Lemma 7.3 we know that ϕ is nonnegative and continuous on $(0, \infty)$. This means that ϕ satisfies $(-1)^n \phi^{(n)} \geq 0$ for $n = 0$. We also know from Lemma 7.3 that ϕ is nonincreasing and convex. Hence, it possesses left- and right-hand derivatives satisfying

$$\phi'_-(r) \leq \phi'_+(r) \leq \phi'_-(y)$$

whenever $0 < r < y < \infty$. If we can show that $g := -\phi'_+$ satisfies the alternation condition in (2), it follows, again from Lemma 7.3, that g is continuous on $(0, \infty)$, meaning that ϕ is in $C^1(0, \infty)$ and that $(-1)^n \phi^{(n)} \geq 0$ for $n = 1$ is satisfied. Moreover, since g satisfies the alternation condition in (2), we can carry out everything for g instead of ϕ. Iterating this idea finally proves that ϕ is completely monotone on $(0, \infty)$.

To see that g satisfies the alternation condition, we start by showing that $(-1)^k \Delta_h^k \phi$ is a nonincreasing function. First of all we have

$$\begin{aligned}\sum_{i=0}^{n-1} \Delta_{h/n} \phi\left(r + \frac{i}{n}h\right) &= \sum_{i=0}^{n-1} \left[\phi\left(r + \frac{i+1}{n}h\right) - \phi\left(r + \frac{i}{n}h\right)\right] \\ &= \phi(r+h) - \phi(r) \\ &= \Delta_h^1 \phi(r).\end{aligned}$$

Thus we can expand $\Delta_h^k \phi(r)$ as follows:

$$\Delta_h^k \phi(r) = \sum_{i_1=0}^{n-1} \cdots \sum_{i_k=0}^{n-1} \Delta_{h/n}^k \phi\left(r + (i_1 + \cdots + i_k)\frac{h}{n}\right).$$

If we apply $(-1)^k \Delta_{h/n}^1$ to both sides of this equality we get

$$(-1)^k \Delta_{h/n}^1 \Delta_h^k \phi(r) = \sum_{i_1=0}^{n-1} \cdots \sum_{i_k=0}^{n-1} (-1)^k \Delta_{h/n}^{k+1} \phi\left(r + (i_1 + \cdots + i_k)\frac{h}{n}\right) \leq 0.$$

The last inequality holds because ϕ satisfies the alternation condition in (2). Hence we know that $(-1)^k \Delta_h^k \phi(r) \geq (-1)^k \Delta_h^k \phi(r + h/n)$ for any $n \in \mathbb{N}$. Iterating this process gives us

$$(-1)^k \Delta_h^k \phi(r) \geq (-1)^k \Delta_h^k \phi\left(r + \frac{h}{n}\right) \geq \cdots \geq (-1)^k \Delta_h^k \phi\left(r + \frac{m}{n}h\right)$$

for any $n, m \in \mathbb{N}$. As ϕ is continuous this shows that $(-1)^k \Delta_h^k \phi$ is nonincreasing. But this implies that

$$\frac{(-1)^k \Delta_h^k \phi(r) - (-1)^k \Delta_h^k \phi(r+\delta)}{-\delta} \leq 0$$

for any $\delta > 0$ and thus that $(-1)^k \Delta_h^k \phi'_+(r) \leq 0$, which finishes the proof. □

For later use we state the following obvious extension for completely monotone functions on $[0, \infty)$.

Corollary 7.5 *If ϕ is completely monotone on $[0, \infty)$ then we have*

$$(-1)^\ell \Delta_h^\ell \phi(r) \geq 0$$

for all $h > 0$, all $r \geq 0$, and all $\ell \in \mathbb{N}_0$.

7.2 The Bernstein–Hausdorff–Widder characterization

Bochner's characterization of positive semi-definite functions demonstrated how powerful an integral representation can be. Hence, it is now our goal to represent completely monotone functions in such a way. This can be achieved in different ways. Here, we choose an approach over completely monotone sequences. In the style of Theorem 7.4, a sequence $\{\mu_j\}_{j\in\mathbb{N}_0}$ is called completely monotone if $(-1)^k \Delta^k \mu_m \geq 0$ for all $k, m \in \mathbb{N}_0$.

The Bernstein polynomials will play an important role in this context.

Definition 7.6 *For $k \in \mathbb{N}_0$ we define the Bernstein polynomials by*

$$B_{k,m}(t) = \binom{k}{m} t^m (1-t)^{k-m}, \qquad 0 \leq m \leq k.$$

Associated with these Bernstein polynomials is the Bernstein operator, defined by

$$B_k(f)(t) = \sum_{m=0}^{k} f\left(\frac{m}{k}\right) B_{k,m}(t)$$

for any $f : [0, 1] \to \mathbb{R}$.

We will also need an operator that is defined by a sequence of numbers $\{\mu_j\}$ and acts on polynomials.

Definition 7.7 *For a sequence $\mu = \{\mu_j\}_{j\in\mathbb{N}_0}$ of real numbers, a linear operator $M_\mu : \pi(\mathbb{R}) \to \mathbb{R}$ is defined by*

$$M_\mu\left(\sum_{j=0}^{n} a_j t^j\right) = \sum_{j=0}^{n} a_j \mu_j.$$

The operator M_μ is defined on polynomials given in the monomial basis. We need to know how it acts on the Bernstein polynomials.

Lemma 7.8 *For $k \in \mathbb{N}_0$, $0 \le m \le k$, and a sequence $\{\mu_j\}_{j\in\mathbb{N}_0}$ of real numbers we set*

$$\lambda_{k,m} := \binom{k}{m}(-1)^{k-m}\Delta^{k-m}\mu_m.$$

Then the operator M_μ applied to the Bernstein polynomial $B_{k,m}$ gives the value $M_\mu(B_{k,m}) = \lambda_{k,m}$. Moreover, we have

$$\sum_{m=0}^{k}\lambda_{k,m} = \mu_0$$

for all $m \in \mathbb{N}_0$.

Proof Elementary calculus gives

$$B_{k,m}(t) = \binom{k}{m}\sum_{j=0}^{k-m}\binom{k-m}{j}(-1)^j t^{j+m}.$$

Hence, application of M_μ to $B_{k,m}$ leads to the representation

$$\begin{aligned} M_\mu(B_{k,m}) &= \binom{k}{m}\sum_{j=0}^{k-m}\binom{k-m}{j}(-1)^j\mu_{j+m} \\ &= \binom{k}{m}(-1)^{k-m}\Delta^{k-m}\mu_m \\ &= \lambda_{k,m}. \end{aligned}$$

Finally, since M_μ is a linear operator we have

$$\sum_{m=0}^{k}\lambda_{k,m} = \sum_{m=0}^{k}M_\mu(B_{k,m}) = M_\mu\left(\sum_{m=0}^{k}B_{k,m}\right) = M_\mu(1) = \mu_0.$$

□

Obviously, if $\{\mu_j\}$ is completely monotone then all the $\lambda_{k,m}$ are nonnegative. The next lemma shows how the coefficients $\{\mu_j\}$ can be recovered by the operator M_μ from the $\lambda_{k,m}$ if the latter are nonnegative.

Lemma 7.9 *Suppose that for $\mu = \{\mu_j\}_{j\in\mathbb{N}_0}$ we have $\lambda_{k,m} \ge 0$ for $k \in \mathbb{N}_0$ and $0 \le m \le k$. Then*

$$\mu_n = \lim_{k\to\infty} M_\mu(B_k(t^n)), \qquad n \in \mathbb{N}_0.$$

Proof First, we prove the statement for $n \ge 1$. Note that

$$\prod_{i=0}^{n-1}\frac{kt-i}{k-i}$$

converges uniformly on [0, 1] to t^n for $k \to \infty$ because every factor converges uniformly to t and the number of factors is finite. Furthermore, we find for $k \geq n$

$$\begin{aligned} t^n &= t^n[(1-t)+t]^{k-n} \\ &= \sum_{m=0}^{k-n} \binom{k-n}{m} t^{m+n}(1-t)^{k-n-m} \\ &= \sum_{m=n}^{k} \frac{m(m-1)\cdots(m-n+1)}{k(k-1)\cdots(k-n+1)} B_{k,m}(t). \end{aligned}$$

This means that

$$\mu_n = M_\mu(t^n) = \sum_{m=n}^{k} \frac{m(m-1)\cdots(m-n+1)}{k(k-1)\cdots(k-n+1)} \lambda_{k,m}.$$

However, we can conclude from the representation of $M_\mu(B_{k,m})$ that

$$M_\mu(B_k(t^n)) = \sum_{m=0}^{k} \left(\frac{m}{k}\right)^n \lambda_{k,m},$$

so that

$$\begin{aligned} \mu_n - M_\mu(B_k(t^n)) = &\sum_{m=n}^{k} \left[\frac{m(m-1)\cdots(m-n+1)}{k(k-1)\cdots(k-n+1)} - \left(\frac{m}{k}\right)^n\right] \lambda_{k,m} \\ &- \sum_{m=0}^{n-1} \left(\frac{m}{k}\right)^n \lambda_{k,m} \end{aligned}$$

From the remark at the beginning of the proof and from $\sum_{m=0}^{k} |\lambda_{k,m}| = \sum_{m=0}^{k} \lambda_{k,m} = \mu_0$ we know that we can bound the first summand by $\epsilon\mu_0$ for a given $\epsilon > 0$ if k is sufficiently large. Hence for $n \geq 1$ we have

$$|\mu_n - M_\mu(B_k(t^n))| \leq \epsilon\mu_0 + \left(\frac{n}{k}\right)^n \mu_0 < 2\epsilon\mu_0,$$

for sufficiently large k. This shows convergence in the case $n \in \mathbb{N}$. For $n = 0$ note that $M_\mu(B_k(1)) = M_\mu(1) = \mu_0$. □

The proof implies that the condition $\lambda_{k,m} \geq 0$ can be replaced by the condition $\sum_{m=0}^{k} |\lambda_{k,m}| \leq L$ for all $k \in \mathbb{N}_0$ with a uniform constant $L > 0$ without altering the result. The same is true for the next proposition if the nonnegative measure therein is replaced by a signed measure.

This next result shows that every sequence $\{\mu_j\}$ with $\lambda_{k,m} \geq 0$ is actually a moment sequence, meaning that it can be represented as a sequence of the moments of a certain measure. This is in particular the case if the sequence is completely monotone.

Proposition 7.10 *Suppose that the sequence $\mu = \{\mu_j\}_{j\in\mathbb{N}_0}$ satisfies $\lambda_{k,m} \geq 0$ for $k \in \mathbb{N}_0$ and $0 \leq m \leq k$. Then there exists a finite nonnegative Borel measure α on $[0, 1]$ such that*

$$\mu_n = \int_0^1 t^n d\alpha, \qquad n \in \mathbb{N}_0.$$

Proof Let us denote the Dirac measure centered at $t \in [0, 1]$ by ϵ_t. To be more precise, this measure is defined on the Borel sets $\mathcal{B}([0, 1])$ by

$$\epsilon_t(A) := \begin{cases} 1 & \text{if } t \in A, \\ 0 & \text{otherwise,} \end{cases}$$

for $A \in \mathcal{B}([0, 1])$. This allows us to define for $k \in \mathbb{N}$ the discrete nonnegative measure

$$\alpha_k = \sum_{m=0}^{k} \lambda_{k,m}\epsilon_{m/k},$$

which obviously satisfies

$$\int_0^1 t^n d\alpha_k = \sum_{m=0}^{k} \lambda_{k,m} \left(\frac{m}{k}\right)^n = M_\mu(B_k(t^n)).$$

Thus, by Lemma 7.9 we get

$$\mu_n = \lim_{k\to\infty} \int_0^1 t^n d\alpha_k.$$

Moreover, the α_k have finite total mass and this is also uniformly bounded:

$$\|\alpha_k\| = \int_0^1 d\alpha_k = \sum_{m=0}^{k} \lambda_{k,m} = \mu_0.$$

Hence, Helly's theorem (Theorem 5.34) now guarantees the existence of a finite nonnegative measure α with total mass $\|\alpha\| \leq \mu_0$ and a sequence α_{k_j}, so that

$$\lim_{j\to\infty} \int_0^1 f(t)d\alpha_{k_j} = \int_0^1 f(t)d\alpha$$

for all $f \in C[0, 1]$. Setting $f(t) = t^n$ finishes the proof. □

After these preparatory steps we are able to prove a characterization that was independently treated by Bernstein in 1914 and 1928, by Hausdorff in 1921, and by Widder in 1931.

Theorem 7.11 (Hausdorff–Bernstein–Widder) *A function $\phi : [0, \infty) \to \mathbb{R}$ is completely monotone on $[0, \infty)$ if and only if it is the Laplace transform of a nonnegative finite Borel measure ν, i.e. it is of the form*

$$\phi(r) = \mathcal{L}\nu(r) = \int_0^\infty e^{-rt} d\nu(t).$$

Proof In the introductory part of this chapter we have seen already that every ϕ that is the Laplace transform of a nonnegative and finite measure ν satisfies $(-1)^k\phi^{(k)}(r) \geq 0$ for $r > 0$. Moreover, because ν is finite, ϕ is also continuous at zero.

Conversely, let us assume that ϕ is completely monotone. For a fixed $N \in \mathbb{N}$ we consider the sequences $\mu_n = \phi(n/N)$, $n \in \mathbb{N}_0$. Since

$$\Delta^k \mu_m = \Delta^k_{1/N}\phi\left(\frac{m}{N}\right),$$

we find that

$$\lambda_{k,m} = \binom{k}{m}(-1)^{k-m}\Delta^{k-m}_{1/N}\phi\left(\frac{m}{N}\right) \geq 0$$

for $0 \leq m \leq k$ and $k \in \mathbb{N}_0$, by Corollary 7.5. Proposition 7.10 gives for every $N \in \mathbb{N}$ a finite nonnegative Borel measure α_N on $[0, 1]$ such that

$$\phi\left(\frac{n}{N}\right) = \int_0^1 t^n d\alpha_N, \qquad n \in \mathbb{N}_0.$$

If we define the measurable maps $T_N : [0, 1] \to [0, 1]$, $t \mapsto t^N$, and $S : [0, \infty) \to (0, 1]$, $t \mapsto e^{-t}$, we can conclude, on the one hand, that

$$\phi(n) = \int_0^1 t^{nN} d\alpha_N = \int_0^1 t^n dT_N(\alpha_N).$$

Since both measures $T_N(\alpha_N)$ and α_1 are finite, we can use the approximation theorem of Weierstrass to derive

$$\int_0^1 f(t)d\alpha_1 = \int_0^1 f(t)dT_N(\alpha_N)$$

for all $f \in C[0, 1]$, which gives by the uniqueness theorem $T(\alpha_N) = \alpha_1$. On the other hand, we have

$$\phi\left(\frac{n}{N}\right) = \int_0^1 t^n d\alpha_N = \int_{0+}^1 t^{n/N} d\alpha_1 = \int_0^\infty e^{-nt/N} dS^{-1}\alpha_1 =: \int_0^\infty e^{-nt/N} d\nu.$$

Since $\nu = S^{-1}\alpha_1$ inherits the properties of α_1, it is nonnegative and finite. Using that ϕ is continuous and Lebesgue's convergence theorem leads finally to

$$\phi(r) = \int_0^\infty e^{-rt} d\nu(t)$$

for all $r \geq 0$. □

For later reasons we now state the corresponding result for completely monotone functions on $(0, \infty)$. This will play an important role in the theory of conditionally positive definite functions. The difference from Theorem 7.11 is that the measure does not need to be finite. Actually, it is finite if and only if ϕ is continuous at zero.

Corollary 7.12 *A function ϕ is completely monotone on $(0, \infty)$ if and only if there exists a nonnegative Borel measure ν on $[0, \infty)$ such that*

$$\phi(r) = \int_0^\infty e^{-rt} d\nu(t)$$

for all $r > 0$.

Proof The proof of the sufficient part follows as before by successive differentiation, which can be justified because

$$t^n e^{-rt} = t^n e^{-rt/2} e^{-rt/2} \leq C_{n,r} e^{-rt/2}$$

for all $t \geq 0$ and $C_{n,r}$ is uniformly bounded for all r in a fixed compact subset of $(0, \infty)$.

For the necessary part, note that for each $\delta > 0$ the function $\phi(\cdot + \delta)$ is completely monotone on $[0, \infty)$. Thus by Theorem 7.11 we find a measure α_δ such that

$$\phi(r + \delta) = \int_0^\infty e^{-rt} d\alpha_\delta(t).$$

If we define the measure ν by $\nu(A) = \int_A e^{\delta t} d\alpha_\delta$ for $A \in \mathcal{B}([0, \infty))$ then we can derive

$$\phi(r) = \int_0^\infty e^{-rt} d\nu(t) \tag{7.5}$$

for $r > \delta$. But from the uniqueness property of the Laplace transform we can conclude that ν actually does not depend on $\delta > 0$, so that (7.5) remains valid for $r > 0$. □

7.3 Schoenberg's characterization

After having established that completely monotone functions are nothing other than Laplace transforms of nonnegative and finite Borel measures, we turn to the connection between positive semi-definite radial and completely monotone functions, which was first pointed out by Schoenberg in 1938.

Theorem 7.13 (Schoenberg) *A function ϕ is completely monotone on $[0, \infty)$ if and only if $\Phi := \phi(\|\cdot\|_2^2)$ is positive semi-definite on every $\mathbb{R}^d$.*

The univariate function ϕ acts here as a d-variate function via $\phi(\|\cdot\|_2^2)$, which differs from our definition of a radial function. We will reformulate the result from the point of view of a positive definite function after the proof.

Proof If ϕ is completely monotone on $[0, \infty)$ then it has a representation

$$\phi(r) = \int_0^\infty e^{-rt} d\nu(t)$$

for some nonnegative finite Borel measure ν, by Theorem 7.11. Thus the d-variate function Φ can be represented by

$$\Phi(x) = \phi(\|x\|_2^2) = \int_0^\infty e^{-\|x\|_2^2 t} d\nu(t),$$

and is hence positive semi-definite, as we have already seen in the introductory part of this chapter.

Next, let us suppose that $\phi(\|\cdot\|_2^2)$ is positive semi-definite on every $\mathbb{R}^d$. Since ϕ is obviously continuous in zero, we know from Theorem 7.4 that it suffices to show that $(-1)^k \Delta_h^k \phi(r) \geq 0$ for all $k \in \mathbb{N}_0$ and all $r, h > 0$. This can be done by induction on k. For $k = 0$ we have to show that $\phi(r) \geq 0$ for all $r \in (0, \infty)$. To this end we choose $x_j = \sqrt{r/2}\, e_j$, $1 \leq j \leq N$, where e_j denotes the jth unit coordinate vector in $\mathbb{R}^N$. Since $\phi(\|\cdot\|_2^2)$ is positive semi-definite on every $\mathbb{R}^N$ we get

$$0 \leq \sum_{j,\ell=1}^{N} \phi(\|x_j - x_\ell\|_2^2) = N\phi(0) + N(N-1)\phi(r),$$

because our special choice of data sites gives $\|x_j - x_\ell\|_2^2 = r$ for $j \neq \ell$. Dividing by $N(N-1)$ and letting N tend to infinity allows us to conclude that $\phi(r) \geq 0$.

For the induction step it obviously suffices to show that $-\Delta_h^1 \phi(\|\cdot\|_2^2)$ is also positive semi-definite on every $\mathbb{R}^d$, if $\phi(\|\cdot\|_2^2)$ is positive semi-definite on every $\mathbb{R}^d$. To do this, suppose that $x_1, \ldots, x_N \in \mathbb{R}^d$ and $\alpha \in \mathbb{R}^N$ are given. We take the x_j as elements of $\mathbb{R}^{d+1}$ and define

$$y_j := \begin{cases} x_j, & 1 \leq j \leq N, \\ x_{j-N} + \sqrt{h} e_{d+1}, & N < j \leq 2N, \end{cases}$$

and

$$\beta_j := \begin{cases} \alpha_j, & 1 \leq j \leq N, \\ -\alpha_{j-N}, & N < j \leq 2N. \end{cases}$$

Since $\phi(\|\cdot\|_2^2)$ is also positive semi-definite on $\mathbb{R}^{d+1}$ we have

$$\begin{aligned} 0 &\leq \sum_{j,k=1}^{2N} \beta_j \beta_k \phi(\|y_j - y_k\|_2^2) \\ &= \sum_{j,k=1}^{N} \alpha_j \alpha_k \phi(\|x_j - x_k\|_2^2) - \sum_{j=1}^{N} \sum_{k=N+1}^{2N} \alpha_j \alpha_{k-N} \phi(\|x_j - x_{k-N}\|_2^2 + h) \\ &\quad - \sum_{j=N+1}^{2N} \sum_{k=1}^{N} \alpha_{j-N} \alpha_k \phi(\|x_{j-N} - x_k\|_2^2 + h) \\ &\quad + \sum_{j,k=N+1}^{2N} \alpha_{j-N} \alpha_{k-N} \phi(\|x_{j-N} - x_{k-N}\|_2^2) \end{aligned}$$

$$= 2 \sum_{j,k=1}^{N} \alpha_j \alpha_k \left[\phi(\|x_j - x_k\|_2^2) - \phi(\|x_j - x_k\|_2^2 + h) \right]$$

$$= -2 \sum_{j,k=1}^{N} \alpha_j \alpha_k \Delta_h^1 \phi(\|x_j - x_k\|_2^2).$$

Thus $-\Delta_h^1 \phi(\| \cdot \|_2^2)$ is positive semi-definite. □

Again, we are more interested in positive definite functions than in positive semi-definite ones. This time, a complete characterization of radial functions ϕ as being positive definite on every $\mathbb{R}^d$ is simpler than in the case of general positive definite functions on a fixed $\mathbb{R}^d$.

Theorem 7.14 *For a function $\phi : [0, \infty) \to \mathbb{R}$ the following three properties are equivalent:*

(1) ϕ is positive definite on every $\mathbb{R}^d$;
(2) $\phi(\sqrt{\cdot})$ is completely monotone on $[0, \infty)$ and not constant;
(3) there exists a finite nonnegative Borel measure ν on $[0, \infty)$ that is not concentrated at zero, such that

$$\phi(r) = \int_0^\infty e^{-r^2 t} d\nu(t).$$

Proof From Theorems 7.13 and 7.11 we know already that ϕ is positive semi-definite on every $\mathbb{R}^d$ if and only if $\phi(\sqrt{\cdot})$ is completely monotone and if and only if it has the stated integral representation. Hence it only remains to discuss the additional properties. For pairwise distinct $x_1, \ldots, x_N \in \mathbb{R}^d$ and $\alpha \in \mathbb{R}^N \setminus \{0\}$ the quadratic form can be represented by (7.2). Since the Gaussian is positive definite we see that the function ϕ is not positive definite (and thus only positive semi-definite) if and only if the measure ν is up to a constant nonnegative factor the Dirac measure centered at zero. This shows the equivalence of the first and the third property. Finally, the second and the third property are obviously equivalent. □

We finish this chapter by providing a further example, returning to inverse multiquadrics. We mentioned earlier that the restriction $\beta > d/2$ in Theorem 6.13 is artificial. The last result allows us to get rid of this restriction. Moreover, the proof of positive definiteness becomes much simpler than before.

Theorem 7.15 *The inverse multiquadrics $\phi(r) := (c^2 + r^2)^{-\beta}$ are positive definite functions on every $\mathbb{R}^d$ provided that $\beta > 0$ and $c > 0$. Moreover, ϕ has the representation*

$$\phi(r) = \int_0^\infty e^{-r^2 t} d\nu(t),$$

with measure ν allowing the representation

$$d\nu(t) = \frac{1}{\Gamma(\beta)} t^{\beta - 1} e^{-c^2 t} dt.$$

Proof Set $f(r) = \phi(\sqrt{r})$. Then f is completely monotone since

$$(-1)^\ell f^{(\ell)}(r) = (-1)^{2\ell}\beta(\beta+1)\cdots(\beta+\ell-1)(r+c^2)^{-\beta-\ell} \geq 0.$$

Since f is not constant ϕ must be positive definite. For the representation, see the proof of Theorem 6.13. □

7.4 Notes and comments

Completely monotone functions are obviously closely related to absolutely monotone functions, which satisfy $f^{(\ell)} \geq 0$ for all ℓ. The latter were introduced by Bernstein [25] in 1914 and characterized as Laplace–Stieltjes integrals in 1928 [26]. Somewhat earlier than the latter date, namely in 1921 (see [81, 82]), Hausdorff considered completely monotone sequences. His work essentially contained the characterization by Bernstein. But Bernstein was obviously not aware of the work of Hausdorff and gave an independent proof. Later, in 1931, Widder gave another independent proof [199].

The first two sections of this chapter are based on Widder's work, as it is represented in his book [200] but employing a measure-theoretical approach rather than Laplace–Stieltjes integrals.

Schoenberg proved his characterization (Theorem 7.13) in 1938 [173] by considering Bochner's characterization via Fourier transforms in the case of radial functions. He investigated what happens if the space dimension tends to infinity.

The approach taken here seems to be more elementary and, from a certain point of view, also more elegant. It originates from work by Wells and Williams [188] and Kuelbs [99].

8

Conditionally positive definite functions

The interpolation problem (6.2) led us to the idea of using positive definite functions. But not all popular choices of radial basis functions that are used fit into this scheme. The thin-plate spline may serve us as an example. Suppose the basis function is given by $\Phi(x) = \|x\|_2^2 \log(\|x\|_2)$, $x \in \mathbb{R}^d$. Let $N = d + 1$ and let the centers be the vertices of a regular simplex whose edges are all of unit length. Then all entries $\Phi(x_j - x_k)$ of the interpolation matrix are zero.

In this chapter we generalize the notion of positive definite functions in a way that covers all the relevant possibilities for basis functions. We will derive characterizations that can be seen as the generalizations of Bochner's and Schoenberg's results.

8.1 Definition and basic properties

As in the case of positive definite functions we distinguish between real-valued and complex-valued functions, but we now have to be more careful.

Definition 8.1 *A continuous function $\Phi : \mathbb{R}^d \to \mathbb{C}$ is said to be conditionally positive semi-definite of order m (i.e. to have conditional positive definiteness of order m) if, for all $N \in \mathbb{N}$, all pairwise distinct centers $x_1, \ldots, x_N \in \mathbb{R}^d$, and all $\alpha \in \mathbb{C}^N$ satisfying*

$$\sum_{j=1}^{N} \alpha_j p(x_j) = 0 \tag{8.1}$$

for all complex-valued polynomials of degree less than m, the quadratic form

$$\sum_{j,k=1}^{N} \alpha_j \overline{\alpha_k} \Phi(x_j - x_k) \tag{8.2}$$

is nonnegative. Φ is said to be conditionally positive definite of order m if the quadratic form is positive, unless α is zero.

A first important fact on conditionally positive (semi-)definite functions concerns their order.

Proposition 8.2 *A function that is conditionally positive (semi-)definite of order m is also conditionally positive (semi-)definite of order $\ell \geq m$. A function that is conditionally positive (semi-)definite of order m on $\mathbb{R}^d$ is also conditionally positive (semi-)definite of order m on $\mathbb{R}^n$ with $n \leq d$.*

This means for example that every positive definite function has also conditional positive definiteness of any order.

Another consequence is that it is natural to look for the smallest possible order m. Hence, when speaking of a conditionally positive definite function of order m we give in general the minimal possible m.

As in the case of positive definite functions the definition reduces to real coefficients and polynomials if the basis function is real-valued and even. This is in particular the case whenever Φ is radial.

Theorem 8.3 *A continuous, even function $\Phi : \mathbb{R}^d \to \mathbb{R}$ is conditionally positive definite of order m if and only if, for all $N \in \mathbb{N}$, all pairwise distinct centers $x_1, \ldots, x_N \in \mathbb{R}^d$, and all $\alpha \in \mathbb{R}^N \setminus \{0\}$ satisfying*

$$\sum_{j=1}^{N} \alpha_j p(x_j) = 0$$

for all real-valued polynomials of degree less than m, the quadratic form

$$\sum_{j,k=1}^{N} \alpha_j \alpha_k \Phi(x_j - x_k)$$

is positive.

We cannot conclude from the definition of a conditionally positive semi-definite function Φ alone that $\Phi(x) = \overline{\Phi(-x)}$ is automatically satisfied, as was the case for positive semi-definite functions. This is a consequence of the following proposition.

Proposition 8.4 *Every polynomial q of degree less than $2m$ is conditionally positive semi-definite of order m. More precisely, for all sets $\{x_1, \ldots, x_N\} \subseteq \mathbb{R}^d$ and all $\alpha \in \mathbb{C}^N$ satisfying (8.1) for all $p \in \pi_{m-1}(\mathbb{R}^d)$, the quadratic form (8.2) for $\Phi = q$ is identically zero.*

Proof With multi-indices $\beta, \kappa \in \mathbb{N}_0^d$ and $q(x) = \sum_{|\beta|<2m} c_\beta x^\beta$ we can use the multinomial theorem to derive

$$\begin{aligned}\sum_{j,k=1}^{N} \alpha_j \overline{\alpha_k} q(x_j - x_k) &= \sum_{|\beta|<2m} \sum_{j,k=1}^{N} \alpha_j \overline{\alpha_k} c_\beta (x_j - x_k)^\beta \\ &= \sum_{|\beta|<2m} c_\beta \sum_{\kappa \leq \beta} (-1)^{|\kappa|} \binom{\beta}{\kappa} \sum_{j=1}^{N} \alpha_j x_j^{\beta-\kappa} \sum_{k=1}^{N} \overline{\alpha_k} x_k^\kappa .\end{aligned}$$

Since $|\beta| < 2m$ it is impossible that both $|\kappa| \geq m$ and $|\beta - \kappa| \geq m$. Thus either the sum $\sum_{j=1}^N \alpha_j x_j^{\beta-\kappa}$ or the sum $\sum_{k=1}^N \overline{\alpha_k} x_k^\kappa$ must vanish due to (8.1) for all pairs β, κ. □

It can be shown that every polynomial with a degree greater than $2m$ cannot be conditionally positive semi-definite of order m (cf. Sun [182]).

The conditional positive definiteness of order m of a function Φ can also be interpreted as the positive definiteness of the matrix $A_{\Phi,X} = (\Phi(x_j - x_k))$ on the space of vectors α such that

$$\sum_{j=1}^{N} \alpha_j p_\ell(x_j) = 0, \qquad 1 \leq \ell \leq Q = \dim \pi_{m-1}(\mathbb{R}^d).$$

Thus, in this sense, $A_{\Phi,X}$ is positive definite on the space of vectors α "perpendicular" to polynomials. Let us dwell on this subject a little more. Each pair consisting of a vector $\alpha \in \mathbb{C}^N$ and a set of pairwise distinct points $X = \{x_1, \ldots, x_N\}$ that together satisfy (8.1) for all polynomials of degree less than m define a linear functional

$$\lambda_{\alpha,X} := \sum_{j=1}^{N} \alpha_j \delta_{x_j},$$

where δ_x denotes point evaluation at x. Define $\pi_{m-1}(\mathbb{R}^d)^\perp$ to be the space of all such functionals. Then α is admissible in the definition of a conditionally positive semi-definite function if and only if $\lambda_{\alpha,X} \in \pi_{m-1}(\mathbb{R}^d)^\perp$.

The case $m = 1$, which also appears in the linear algebra literature, is usually dealt with using $\leq$ and is then referred to as (conditionally or almost) *negative* definite. In this case the constraint on the α_j is simply $\sum_{j=1}^{N} \alpha_j = 0$.

Since the matrix $A_{\Phi,X}$ is conditionally positive definite of order m, it is positive definite on a subspace of dimension $N - Q$, $Q = \dim \pi_{m-1}(\mathbb{R}^d)$. Thus it has the interesting property that at least $N - Q$ of its eigenvalues are positive. This follows immediately from the Courant–Fischer theorem. In the case $m = 1$ we can make an even stronger statement.

Theorem 8.5 *Suppose that Φ is conditionally positive definite of order 1 and that $\Phi(0) \leq 0$. Then the matrix $A_{\Phi,X} \in \mathbb{R}^{N\times N}$ has one negative and $N - 1$ positive eigenvalues and in particular it is invertible.*

Proof From the Courant–Fischer theorem we conclude that $A_{\Phi,X}$ has at least $N - 1$ positive eigenvalues. But since $0 \geq N\Phi(0) = \operatorname{tr}(A_{\Phi,X}) = \sum_{i=1}^{n} \lambda_i$, where the λ_i denote the eigenvalues of $A_{\Phi,X}$ and $\operatorname{tr}(A_{\Phi,X})$ its trace, $A_{\Phi,X}$ must also have at least one negative eigenvalue. □

As in the case of positive semi-definite functions, conditionally positive semi-definite functions can be characterized to be integrally conditionally positive semi-definite. See Proposition 6.4 for the corresponding result on positive semi-definite functions.

Proposition 8.6 *Let Φ be continuous. Then Φ is conditionally positive semi-definite of order m if and only if*

$$\int_{\mathbb{R}^d} \int_{\mathbb{R}^d} \Phi(x - y)\gamma(x)\overline{\gamma(y)}dx dy \geq 0 \tag{8.3}$$

for all $\gamma \in C_0^\infty(\mathbb{R}^d)$ *that satisfy*

$$\int_{\mathbb{R}^d} \gamma(x)p(x)dx = 0 \qquad \text{for all} \quad p \in \pi_{m-1}(\mathbb{R}^d). \tag{8.4}$$

Proof Suppose Φ possesses the stated property. If we choose a nonnegative even function $g \in C_0^\infty(\mathbb{R}^d)$ with $\|g\|_{L_1(\mathbb{R}^d)} = 1$ and set $g_\ell(x) = \ell^d g(\ell x)$ then we know that

$$f(x) = \lim_{\ell\to\infty} \int_{\mathbb{R}^d} f(y)g_\ell(x-y)dy$$

for every continuous f. If $x_1, \dots, x_N \in \mathbb{R}^d$ and $\alpha \in \mathbb{C}^N$ such that $\sum_{j=1}^N \alpha_j p(x_j) = 0$ for all $p \in \pi_{m-1}(\mathbb{R}^d)$ are given then the functions

$$\gamma_\ell(x) := \sum_{j=1}^N \overline{\alpha_j} g_\ell(x - x_j)$$

are in $C_0^\infty(\mathbb{R}^d)$ and satisfy

$$\int_{\mathbb{R}^d} \gamma_\ell(x)p(x)dx = \int_{\mathbb{R}^d} \sum_{j=1}^N \overline{\alpha_j} p(x+x_j)g_\ell(x)dx = 0$$

for all $p \in \pi_{m-1}(\mathbb{R}^d)$. Thus we find that

$$\begin{aligned} 0 &\le \int_{\mathbb{R}^d}\int_{\mathbb{R}^d} \Phi(x-y)\gamma_\ell(x)\overline{\gamma_\ell(y)}dxdy \\ &= \sum_{j,k=1}^N \alpha_j\overline{\alpha_k} \int_{\mathbb{R}^d}\int_{\mathbb{R}^d} \Phi(x-y-(x_j-x_k))g_\ell(x)g_\ell(y)dxdy, \end{aligned}$$

which converges for $\ell \to \infty$ to

$$\sum_{j,k=1}^N \alpha_j\overline{\alpha_k}\Phi(x_j - x_k).$$

Conversely, let us suppose that Φ is conditionally positive semi-definite of order m. We want to employ Riemann sums to show that (8.3) is satisfied for all $\gamma \in C_0(\mathbb{R}^d)$ with (8.4). Unfortunately, if $m > 0$ then the discretizations gained by Riemann sums will not satisfy (8.1) in general. Hence, we have to modify the approach. Fix a $\gamma \in C_0(\mathbb{R}^d)$ that satisfies (8.4). Let K be a compact cube that contains the support of γ. Then we have to discretize the integral

$$\int_K\int_K \gamma(x)\overline{\gamma(y)}\Phi(x-y)dxdy. \tag{8.5}$$

A scaling argument shows that without loss of generality we can assume K to be the unit cube $[0,1]^d$.

We divide K into N equally sized subcubes of volume $1/N$ and pick an $x_j^{(N)}$ from each of these subcubes. Let Q be the dimension of $\pi_{m-1}(\mathbb{R}^d)$ and $p_1, \dots, p_Q$ be a basis for $\pi_{m-1}(\mathbb{R}^d)$. Let $N \ge Q$. We choose the points $\{x_j^{(N)}\}$ and their ordering in such a way that the first Q points are always the same for every N, i.e. $x_1^{(N)} = x_1, \dots, x_Q^{(N)} = x_Q$ for all $N \ge Q$. Moreover, these points should be chosen to be $\pi_{m-1}(\mathbb{R}^d)$-unisolvent. Then we discretize the integrals (8.4) to define the numbers

$$A_k^{(N)} := \sum_{j=1}^{N} \frac{1}{N} \gamma(x_j^{(N)}) p_k(x_j^{(N)}), \qquad 1 \le k \le Q \quad \text{and} \quad N \ge Q.$$

From the conditions imposed on γ we know that $A_k^{(N)}$ tends to zero for $N \to \infty$ and this holds for all $1 \le k \le Q$. Since $\{x_1, \dots, x_Q\}$ is $\pi_{m-1}(\mathbb{R}^d)$-unisolvent, we find for every $N \ge Q$ a unique vector $\beta^{(N)} \in \mathbb{R}^Q$ with

$$\sum_{j=1}^{Q} \beta_j^{(N)} p_k(x_j) = A_k^{(N)}, \qquad 1 \le k \le Q.$$

This $\beta^{(N)}$ is obviously given by $\beta^{(N)} = P^{-1} A^{(N)}$ if $P = (p_k(x_j))$. Hence each coefficient $\beta_j^{(N)}$ tends to zero as $N \to \infty$. Finally let us define the corrected coefficients

$$\alpha_j := \begin{cases} \dfrac{1}{N} \gamma(x_j^{(N)}) - \beta_j^{(N)} & \text{for } 1 \le j \le Q, \\ \dfrac{1}{N} \gamma(x_j^{(N)}) & \text{for } Q+1 \le j \le N. \end{cases}$$

These coefficients satisfy condition (8.1) by construction:

$$\begin{aligned} \sum_{j=1}^{N} \alpha_j p_k(x_j^{(N)}) &= \sum_{j=1}^{N} \frac{1}{N} \gamma(x_j^{(N)}) p_k(x_j^{(N)}) - \sum_{j=1}^{Q} \beta_j^{(N)} p_k(x_j^{(N)}) \\ &= A_k^{(N)} - A_k^{(N)} = 0 \end{aligned}$$

for $1 \le k \le Q$. Hence, we can insert them into the quadratic form for Φ, obtaining

$$\begin{aligned} 0 &\le \sum_{j=1}^{N} \sum_{k=1}^{N} \alpha_j \overline{\alpha_k} \Phi(x_j^{(N)} - x_k^{(N)}) \\ &= \sum_{j=1}^{N} \sum_{k=1}^{N} \frac{1}{N^2} \gamma(x_j^{(N)}) \overline{\gamma(x_k^{(N)})} \Phi(x_j^{(N)} - x_k^{(N)}) \\ &\quad - \sum_{j=1}^{Q} \beta_j^{(N)} \sum_{k=1}^{N} \frac{1}{N} \overline{\gamma(x_k^{(N)})} \Phi(x_j - x_k^{(N)}) \\ &\quad - \sum_{j=1}^{N} \frac{1}{N} \gamma(x_j^{(N)}) \sum_{k=1}^{Q} \overline{\beta_k^{(N)}} \Phi(x_j^{(N)} - x_k) \\ &\quad + \sum_{j,k=1}^{Q} \beta_j^{(N)} \overline{\beta_k^{(N)}} \Phi(x_j - x_k). \end{aligned}$$

Let us analyze the behavior of these four double sums as $N \to \infty$. The first is a Riemann sum for $\int_K \int_K \gamma(x)\overline{\gamma(y)}\Phi(x-y)dxdy$ and hence converges to this integral. The second and third double sums can each be bounded by

$$\|\gamma\|_{L_\infty(K)}\|\Phi\|_{L_\infty([-1,1]^d)}\frac{1}{N}\sum_{j=1}^{Q}|\beta_j^{(N)}|,$$

and they tend to zero as $N \to \infty$. The last double sum can be bounded by

$$\|\Phi\|_{L_\infty([-1,1]^d)}\left(\sum_{j=1}^{Q}|\beta_j^{(N)}|\right)^2$$

and also tends to zero as $N \to \infty$. This proves that the integral in (8.5) is indeed nonnegative. □

We end this section with an interesting construction. A conditionally positive semi-definite function of order m can easily be used to construct a conditionally positive semi-definite function of order less than m.

Proposition 8.7 *Suppose that Φ is conditionally positive semi-definite of order $m > 0$ and that $\ell \le m$ is fixed. If $y_1, \dots, y_M \in \mathbb{R}^d$ and $\beta \in \mathbb{C}^M \setminus \{0\}$ satisfy $\sum_{j=1}^M \beta_j p(y_j) = 0$ for all $p \in \pi_{\ell-1}(\mathbb{R}^d)$ then the function*

$$\Psi(x) := \sum_{j,k=1}^{M} \beta_j \overline{\beta_k} \Phi(x - y_j + y_k)$$

is conditionally positive semi-definite of order $m - \ell$.

Proof We will use the monomials as a basis for the polynomial space. Suppose $x_1, \dots, x_N \in \mathbb{R}^d$ and $\alpha \in \mathbb{C}^N$ are given so that $\sum_{j=1}^N \alpha_j x_j^\nu = 0$ for all $\nu \in \mathbb{N}_0^d$ with $|\nu| < m - \ell$. Then

$$\begin{aligned}\sum_{\ell,n=1}^{N} \alpha_\ell \overline{\alpha_n} \Psi(x_\ell - x_n) &= \sum_{\ell,n=1}^{N}\sum_{j,k=1}^{M} \alpha_\ell \beta_j \overline{\alpha_n \beta_k} \Phi((x_\ell - y_j) - (x_n - y_k)) \\ &= \sum_I \sum_J C_I \overline{C_J} \Phi(z_I - z_J),\end{aligned}$$

where each of the last sums runs over MN terms, $C_I = \alpha_\ell \beta_j$, and $z_I = x_\ell - y_j$. The last expression is nonnegative by the assumptions imposed on Φ, if we can show that the new centers and new coefficients satisfy the side conditions for polynomials of degree less than m. But this is true because

$$\begin{aligned}\sum_I C_I z_I^\nu &= \sum_{\ell=1}^{N}\sum_{j=1}^{M} \alpha_\ell \beta_j (x_\ell - y_j)^\nu \\ &= \sum_{\mu \le \nu} (-1)^{|\mu|} \binom{\nu}{\mu} \left(\sum_{\ell=1}^{N} \alpha_\ell x_\ell^{\nu-\mu}\right)\left(\sum_{j=1}^{M} \beta_j y_j^\mu\right) \\ &= 0.\end{aligned}$$

The last equality holds for all $|\nu| < m$ because either $|\nu - \mu| < m - \ell$ or $|\mu| < \ell$ is satisfied. □

8.2 An analogue of Bochner's characterization

In the case of positive definite functions we found the integral characterization by Bochner to be very helpful. For all relevant basis functions, the version given in Theorem 6.11 for integrable functions was sufficient.

For a conditionally positive definite function also there exists a characterization comparable to Bochner's and we will state it at the end of this section. But we want to start with a version that is, rather, an analogue of the result in Theorem 6.11. Of course, for a conditionally positive definite function we cannot hope for integrability. But the crucial point in the positive definite case was actually not integrability but the existence of a classical Fourier transform. If we want to apply this idea here, we have to modify the notion of the Fourier transform for our purposes. To this end, a special subspace of the Schwartz space $\mathcal{S}$ will be of importance.

Definition 8.8 *For $m \in \mathbb{N}_0$ the set of all functions $\gamma \in \mathcal{S}$ that satisfy $\gamma(\omega) = \mathcal{O}(\|\omega\|_2^m)$ for $\|\omega\|_2 \to 0$ will be denoted by $\mathcal{S}_m$.*

In what follows, we restrict ourselves to slowly increasing basis functions. The reader should remember that a function is called slowly increasing if it grows at most like any particular fixed polynomial. To discuss only such functions is actually not a restriction, because one can show that every conditionally positive definite function of order m grows at most like a polynomial of degree $2m$ (see Madych and Nelson [113]).

Definition 8.9 *Suppose that $\Phi : \mathbb{R}^d \to \mathbb{C}$ is continuous and slowly increasing. A measurable function $\widehat{\Phi} \in L_2^{\mathrm{loc}}(\mathbb{R}^d \setminus \{0\})$ is called the generalized Fourier transform of Φ if there exists an integer $m \in \mathbb{N}_0$ such that*

$$\int_{\mathbb{R}^d} \Phi(x)\widehat{\gamma}(x)dx = \int_{\mathbb{R}^d} \widehat{\Phi}(\omega)\gamma(\omega)d\omega$$

is satisfied for all $\gamma \in \mathcal{S}_{2m}$. The integer m is called the order of $\widehat{\Phi}$.

Note that the order m of a generalized Fourier transform corresponds to the space $\mathcal{S}_{2m}$ and not $\mathcal{S}_m$. Furthermore, if $\widehat{\Phi}$ is a generalized Fourier transform of order m then it has also order $\ell \geq m$. Hence, in general we will refer to the smallest possible m when speaking of the order.

Several remarks are necessary. If the generalized Fourier transform exists in this way, it is uniquely determined up to Lebesgue-zero sets. If $\Phi \in L_1(\mathbb{R}^d)$ then its classical Fourier transform and its generalized Fourier transform coincide. The order is zero. The same is true for $\Phi \in L_2(\mathbb{R}^d)$. The generalized Fourier transform and the distributional Fourier transform coincide on the set $\mathcal{S}_{2m}$.

In this chapter, we are concerned only with generalized Fourier transforms $\widehat{\Phi}$ that are continuous on $\mathbb{R}^d \setminus \{0\}$ and have an algebraic singularity at the origin. The order of the singularity determines the minimal order m of the generalized Fourier transform. Later, we will need the more general form.

The next result not only gives an example of a generalized Fourier transform, it also shows in which way the function Φ is determined by its generalized Fourier transform.

Proposition 8.10 *Suppose $\Phi = p$ is a polynomial of degree less than m. Then for every test function $\gamma \in \mathcal{S}_m$ we have*

$$\int_{\mathbb{R}^d} \Phi(x)\widehat{\gamma}(x)dx = 0. \tag{8.6}$$

Hence the generalized Fourier transform of p is the zero function and has order $m/2$.

Conversely, if Φ is a continuous function satisfying (8.6) for all $\gamma \in \mathcal{S}_m$ then Φ is a polynomial of degree less than m.

Proof For the first part let us assume that Φ has the representation $\Phi(x) = \sum_{|\beta|<m} c_\beta x^\beta$. Then

$$\begin{aligned}\int_{\mathbb{R}^d} \Phi(x)\widehat{\gamma}(x)dx &= \sum_{|\beta|<m} c_\beta i^{-|\beta|} \int_{\mathbb{R}^d} (ix)^\beta \widehat{\gamma}(x)dx \\ &= \sum_{|\beta|<m} c_\beta i^{-|\beta|} \int_{\mathbb{R}^d} \widehat{D^\beta \gamma}(x)dx \\ &= (2\pi)^{d/2} \sum_{|\beta|<m} c_\beta i^{-|\beta|} D^\beta \gamma(0) \\ &= 0,\end{aligned}$$

since $\gamma \in \mathcal{S}_m$.

For the second part we choose a fixed test function $\chi \in C_0^\infty(\mathbb{R}^d)$, which is identically equal to one in a neighborhood of the origin. Then we define for an arbitrary $g \in \mathcal{S}$ the function

$$\gamma(x) = g(x) - \sum_{|\beta|<m} \frac{D^\beta g(0)}{\beta!} x^\beta \chi(x), \qquad x \in \mathbb{R}^d,$$

which is clearly in $\mathcal{S}_m$ and has Fourier transform

$$\widehat{\gamma}(\omega) = \widehat{g}(\omega) - \sum_{|\beta|<m} \frac{D^\beta g(0)}{\beta!} i^{|\beta|} D^\beta \widehat{\chi}(\omega).$$

Hence (8.6) yields

$$\begin{aligned}0 &= \int_{\mathbb{R}^d} \Phi(x)\widehat{\gamma}(x)dx \\ &= \int_{\mathbb{R}^d} \Phi(x)\widehat{g}(x)dx - \sum_{|\beta|<m} \frac{D^\beta g(0)}{\beta!} i^{|\beta|} \int_{\mathbb{R}^d} \Phi(x) D^\beta \widehat{\gamma}(x)dx \\ &= \int_{\mathbb{R}^d} \Phi(x)\widehat{g}(x)dx - \sum_{|\beta|<m} \frac{i^{|\beta|}}{\beta!} c_\beta D^\beta g(0),\end{aligned}$$

the constants c_β being defined by $c_\beta = \int \Phi(x) D^\beta \widehat{\chi}(x)dx$. If we finally use $D^\beta g(0) = (2\pi)^{-d/2} i^{|\beta|} \int \widehat{g}(x) x^\beta dx$ we derive

$$\int_{\mathbb{R}^d} \left(\Phi(x) - (2\pi)^{-d/2} \sum_{|\beta|<m} \frac{c_\beta(-1)^{|\beta|}}{\beta!} x^\beta \right) \widehat{g}(x)dx = 0$$

for all $\widehat{g} \in \mathcal{S}$. Approximation by convolution from Theorem 5.20 shows that Φ is indeed a polynomial of degree less than m. □

On our way to deriving a Bochner-type result we need to know how to construct functions from $\mathcal{S}_{2m}$. A simple trick is to employ centers and coefficients, which satisfy the side condition for a conditionally positive definite function.

Lemma 8.11 *Suppose that pairwise distinct $x_1, \ldots, x_N \in \mathbb{R}^d$ and $\alpha \in \mathbb{C}^N \setminus \{0\}$ are given such that (8.1) is satisfied for all $p \in \pi_{m-1}(\mathbb{R}^d)$. Then*

$$\sum_{j=1}^N \alpha_j e^{i x_j^T \omega} = \mathcal{O}(\|\omega\|_2^m)$$

holds for $\|\omega\|_2 \to 0$.

Proof The expansion of the exponential function leads to

$$\sum_{j=1}^N \alpha_j e^{i x_j^T \omega} = \sum_{k=0}^\infty \frac{i^k}{k!} \sum_{j=1}^N \alpha_j (x_j^T \omega)^k.$$

For fixed $\omega \in \mathbb{R}^d$ we have $p_k(x) := (x^T\omega)^k \in \pi_k(\mathbb{R}^d)$. Thus (8.1) ensures that the first m terms vanish:

$$\sum_{j=1}^N \alpha_j e^{i x_j^T \omega} = \sum_{k=m}^\infty \frac{i^k}{k!} \sum_{j=1}^N \alpha_j (x_j^T \omega)^k,$$

which gives the stated behavior. □

Now it is time to state and prove our main theorem. It states that the order of the generalized Fourier transform, which is nothing other than the order of the singularity of the Fourier transform at the origin, determines the minimal order of a conditionally positive definite function.

Theorem 8.12 *Suppose $\Phi : \mathbb{R}^d \to \mathbb{C}$ is continuous, slowly increasing, and possesses a generalized Fourier transform $\widehat{\Phi}$ of order m, which is continuous on $\mathbb{R}^d \setminus \{0\}$. Then Φ is conditionally positive definite of order m if and only if $\widehat{\Phi}$ is nonnegative and nonvanishing.*

Proof Suppose that $\widehat{\Phi}$ is nonnegative and nonvanishing. Suppose further that pairwise distinct $x_1, \ldots, x_N \in \mathbb{R}^d$ and $\alpha \in \mathbb{C}^N \setminus \{0\}$ satisfy (8.1) for all $p \in \pi_{m-1}(\mathbb{R}^d)$. Define

$$f(x) := \sum_{j,k=1}^N \alpha_j \overline{\alpha_k} \Phi(x + (x_j - x_k))$$

and

$$\gamma_\ell(x) := \left|\sum_{j=1}^N \alpha_j e^{ix^T x_j}\right|^2 \widehat{g_\ell}(x) = \sum_{j,k=1}^N \alpha_j \overline{\alpha_k} e^{ix^T(x_j-x_k)} \widehat{g_\ell}(x)$$

where $g_\ell(x) = (\ell/\pi)^{d/2} e^{-\ell\|x\|_2^2}$ is the test function from Theorem 5.20. On account of $\gamma_\ell \in \mathcal{S}$ and Lemma 8.11 we have $\gamma_\ell \in \mathcal{S}_{2m}$. Furthermore, we can compute the Fourier transform:

$$\begin{aligned}
\widehat{\gamma_\ell}(x) &= (2\pi)^{-d/2} \int_{\mathbb{R}^d} \sum_{j,k=1}^N \alpha_j \overline{\alpha_k} e^{i\omega^T(x_j-x_k)} \widehat{g_\ell}(\omega) e^{-ix^T\omega} d\omega \\
&= \sum_{j,k=1}^N \alpha_j \overline{\alpha_k} (2\pi)^{-d/2} \int_{\mathbb{R}^d} \widehat{g_\ell}(\omega) e^{-i\omega^T(x-(x_j-x_k))} d\omega \\
&= \sum_{j,k=1}^N \alpha_j \overline{\alpha_k} g_\ell(x-(x_j-x_k)),
\end{aligned}$$

because $\widehat{\widehat{g_\ell}} = g_\ell$. Collecting these facts gives, together with Definition 8.9,

$$\begin{aligned}
\int_{\mathbb{R}^d} f(x) g_\ell(x) dx &= \int_{\mathbb{R}^d} \Phi(x) \sum_{j,k=1}^N \alpha_j \overline{\alpha_k} g_\ell(x-(x_j-x_k)) dx \\
&= \int_{\mathbb{R}^d} \Phi(x) \widehat{\gamma_\ell}(x) dx \\
&= \int_{\mathbb{R}^d} \widehat{\Phi}(\omega) \gamma_\ell(\omega) d\omega \\
&= \int_{\mathbb{R}^d} \left|\sum_{j=1}^N \alpha_j e^{i\omega^T x_j}\right|^2 \widehat{g_\ell}(\omega) \widehat{\Phi}(\omega) d\omega \\
&\geq 0.
\end{aligned}$$

Thus we have by Theorem 5.20,

$$\sum_{j,k=1}^N \alpha_j \overline{\alpha_k} \Phi(x_j - x_k) = \lim_{\ell\to\infty} \int_{\mathbb{R}^d} f(x) g_\ell(x) dx \geq 0.$$

Moreover, since $\left|\sum_{j=1}^N \alpha_j e^{i\omega^T x_j}\right|^2 \widehat{g_\ell}(\omega)\widehat{\Phi}(\omega)$ is nondecreasing in $\ell \in \mathbb{N}$, the Beppo–Levi convergence theorem guarantees the integrability of the limit function $(2\pi)^{-d/2} \left|\sum_{j=1}^N \alpha_j e^{i\omega^T x_j}\right|^2 \widehat{\Phi}(\omega)$ and also the identity

$$\sum_{j,k=1}^N \alpha_j \overline{\alpha_k} \Phi(x_j - x_k) = (2\pi)^{-d/2} \int_{\mathbb{R}^d} \left|\sum_{j=1}^N \alpha_j e^{i\omega^T x_j}\right|^2 \widehat{\Phi}(\omega) d\omega.$$

The same arguments as in the proof of Theorem 6.11 show that the quadratic form cannot vanish if $\widehat{\Phi}$ is nonvanishing.

Now suppose that Φ is conditionally positive definite of order m. Because of Proposition 8.6, the function Φ satisfies (8.3) for all $\gamma \in C_0^\infty(\mathbb{R}^d)$ with (8.4).

Next choose a nonnegative function $k \in C_0^\infty(\mathbb{R}^d)$ having support $B(0,1) := \{x \in \mathbb{R}^d : \|x\|_2 \le 1\}$ with $\|k\|^2_{L_2(\mathbb{R}^d)} = (2\pi)^{-d/2}$. If we set $k_\ell(x) := \ell^{d/2}k(\ell x)$, $x \in \mathbb{R}^d$, and

$$\gamma_\ell(x) := k_\ell(\cdot - y)^\wedge(x) = e^{-ix^Ty}\widehat{k_\ell}(x)$$

for a fixed $y \neq 0$, we find by application of Theorem 5.16, for every multi-index $\alpha \in \mathbb{N}_0^d$,

$$\begin{aligned}\int_{\mathbb{R}^d} \gamma_\ell(x)x^\alpha dx &= \int_{\mathbb{R}^d} x^\alpha \widehat{k_\ell}(x)e^{-ix^Ty}dx\\ &= (2\pi)^{d/2} i^{-|\alpha|}(D^\alpha k_\ell)(-y)\\ &= 0,\end{aligned}$$

provided that $\ell > 1/\|y\|_2$. Thus on the one hand γ_ℓ satisfies (8.4) for these ℓ-values and can be inserted into (8.3), which gives

$$\int_{\mathbb{R}^d} \Phi(x)\gamma_\ell * \widetilde{\gamma}_\ell(x)dx \ge 0$$

with $\widetilde{\gamma}(x) := \overline{\gamma(-x)}$. On the other hand we have

$$(\gamma_\ell * \widetilde{\gamma}_\ell)^\vee(0) = (2\pi)^{d/2}|k_\ell(-y)|^2 = 0$$

if $\ell > 1/\|y\|_2$. Thus we can conclude that $(\gamma_\ell * \widetilde{\gamma}_\ell)^\vee$ lies in $\mathcal{S}_{2m}$ and can be inserted into the definition of the generalized Fourier transform. Using

$$(\gamma_\ell * \widetilde{\gamma}_\ell)^\vee(x) = (2\pi)^{d/2}|(\gamma_\ell)^\vee(x)|^2 = (2\pi)^{d/2}|k_\ell(x-y)|^2$$

leads to

$$\begin{aligned}0 &\le \int_{\mathbb{R}^d}\int_{\mathbb{R}^d} \Phi(x-z)\gamma_\ell(x)\overline{\gamma_\ell(z)}dxdz\\ &= \int_{\mathbb{R}^d} \Phi(x)(\gamma_\ell * \widetilde{\gamma}_\ell)(x)dx\\ &= \int_{\mathbb{R}^d} \widehat{\Phi}(\omega)(\gamma_\ell * \widetilde{\gamma}_\ell)^\vee(\omega)d\omega\\ &= (2\pi)^{d/2}\int_{\mathbb{R}^d} \widehat{\Phi}(\omega)|k_\ell(\omega-y)|^2 d\omega,\end{aligned}$$

which converges for $\ell \to \infty$ to $\widehat{\Phi}(y)$. □

For reasons that will become clear later, we will restate the representation for the quadratic form we derived in the last proof.

Corollary 8.13 *Suppose that $\Phi : \mathbb{R}^d \to \mathbb{C}$ is continuous and slowly increasing. Suppose further that Φ possesses a nonnegative, nonvanishing, generalized Fourier transform $\widehat{\Phi}$ of order m that is continuous on $\mathbb{R}^d \setminus \{0\}$. Then for all pairwise distinct $x_1, \dots, x_N \in \mathbb{R}^d$ and all $\alpha \in \mathbb{C}^N$ with (8.1) for all $p \in \pi_{m-1}(\mathbb{R}^d)$, we have*

$$\sum_{j,k=1}^{N} \alpha_j \overline{\alpha_k} \Phi(x_j - x_k) = (2\pi)^{-d/2} \int_{\mathbb{R}^d} \left| \sum_{j=1}^{N} \alpha_j e^{i\omega^T x_j} \right|^2 \widehat{\Phi}(\omega) d\omega.$$

Theorem 8.12 will be sufficient for all our goals. Nonetheless, one might be interested in finding a complete characterization of all conditionally positive semi-definite functions of a given order m. For completeness, we will state the result here and prove the sufficient part. For the proof of the necessary part, we refer the interested reader to Sun's paper [182].

Theorem 8.14 *Let $\Phi \in C(\mathbb{R}^d)$. In order for Φ to be conditionally positive semi-definite of order m it is necessary and sufficient that Φ has the following integral representation:*

$$\Phi(x) = \int_{\mathbb{R}^d \setminus \{0\}} \left(e^{-ix^T\omega} - \kappa(\omega) \sum_{|\beta| < 2m} \frac{(-ix)^\beta}{\beta!} \right) d\mu(\omega) + \sum_{|\beta| \le 2m} a_\beta \frac{(-ix)^\beta}{\beta!}.$$

Here μ is a positive Borel measure on $\mathbb{R}^d \setminus \{0\}$ satisfying

$$\int_{0<\|\omega\|_2\le 1} \|\omega\|_2^{2m} d\mu(\omega) < \infty \quad \text{and} \quad \int_{\|\omega\|_2 \ge 1} d\mu(\omega) < \infty.$$

The function κ is an analytic function in S such that $\kappa(\omega) - 1$ has a zero of order $2m+1$ at the origin. The numbers a_β, $|\beta| = 2m$, satisfy $\sum_{|\beta|=m,|\gamma|=m} \alpha_\beta \overline{\alpha_\gamma} a_{\beta+\gamma} \ge 0$ for all $\alpha_\beta \in \mathbb{C}$.

Proof As stated previously we want to prove only the sufficient part of this theorem. Suppose $x_1, \dots, x_N \in \mathbb{R}^d$ and $\alpha \in \mathbb{C}^N$ with (8.1) are given. From Proposition 8.4 we know that

$$\begin{aligned}
&\sum_{j,k=1}^{N} \alpha_j \overline{\alpha_k} \sum_{|\beta| \le 2m} \frac{a_\beta}{\beta!} [-i(x_j - x_k)]^\beta \\
&= (-1)^m \sum_{|\beta|=2m} \frac{a_\beta}{\beta!} \sum_{j,k=1}^{N} \alpha_j \overline{\alpha_k} (x_j - x_k)^\beta \\
&= (-1)^m \sum_{|\beta|=2m} a_\beta \sum_{\gamma+\nu=\beta} (-1)^{|\nu|} \sum_{j=1}^{N} \frac{\alpha_j x_j^\gamma}{\gamma!} \sum_{k=1}^{N} \frac{\overline{\alpha_k} x_k^\nu}{\nu!} \\
&= \sum_{|\gamma|=m} \sum_{|\nu|=m} a_{\gamma+\nu} A_\gamma \overline{A_\nu} \\
&\ge 0,
\end{aligned}$$

with $A_\gamma = \sum_{j=1}^N \alpha_j x_j^\gamma/\gamma!$, which vanishes for $|\gamma| < m$. Thus we have by Proposition 8.4 and Lemma 8.11

$$\begin{aligned}\sum_{j,k=1}^N \alpha_j\overline{\alpha_k}\Phi(x_j - x_k) &= \int_{\mathbb{R}^d\setminus\{0\}} \left|\sum_{j=1}^N \alpha_j e^{-ix_j^T\omega}\right|^2 d\mu(\omega)\\ &\quad + \sum_{j,k=1}^N \alpha_j\overline{\alpha_k} \sum_{|\beta|\le 2m} \frac{a_\beta}{\beta!}[-i(x_j - x_k)]^\beta\\ &\ge 0.\end{aligned}$$

□

8.3 Examples of generalized Fourier transforms

In this section we will compute the generalized Fourier transforms of the most popular basis functions. They can be used to show that the basis functions are conditionally positive definite. Even if the latter follows in most cases more easily from a characterization given in the next section, the knowledge of the generalized Fourier transforms is of great importance for error estimates and for estimates on the stability of the interpolation process to be derived in the chapters that follow.

Our first example concerns the generalized Fourier transform of the multiquadrics. The basic idea of the proof is to start with the classical Fourier transform of the inverse multiquadrics given in Theorem 6.13 and then to use analytic continuation. We will use the notation $\lceil t\rceil$ for the smallest integer greater than or equal to $t \in \mathbb{R}$.

Theorem 8.15 *The function* $\Phi(x) = (c^2 + \|x\|_2^2)^\beta$, $x \in \mathbb{R}^d$, *with* $c > 0$ *and* $\beta \in \mathbb{R}\setminus\mathbb{N}_0$ *possesses the (generalized) Fourier transform*

$$\widehat{\Phi}(\omega) = \frac{2^{1+\beta}}{\Gamma(-\beta)}\left(\frac{\|\omega\|_2}{c}\right)^{-\beta-d/2} K_{d/2+\beta}(c\|\omega\|_2), \qquad \omega \ne 0, \tag{8.7}$$

of order $m = \max(0, \lceil\beta\rceil)$.

Proof Define $G = \{\lambda \in \mathbb{C} : \Re(\lambda) < m\}$ and denote the right-hand side of (8.7) by $\varphi_\beta(\omega)$. We are going to show by analytic continuation that

$$\int_{\mathbb{R}^d} \Phi_\lambda(\omega)\widehat{\gamma}(\omega)d\omega = \int_{\mathbb{R}^d} \varphi_\lambda(\omega)\gamma(\omega)d\omega, \qquad \gamma \in \mathcal{S}_{2m} \tag{8.8}$$

is valid for all $\lambda \in G$ where $\Phi_\lambda(\omega) = (c^2 + \|\omega\|_2^2)^\lambda$. First of all, note that (8.8) is valid for $\lambda \in G$ with $\lambda < -d/2$ by Theorem 6.13 and in the case $m > 0$ also for $\lambda = 0, 1, \dots, m-1$ by Proposition 8.10 and the fact that $1/\Gamma(-\lambda)$ is zero in these cases. Analytic continuation will lead us to our stated result when we can show that both sides of (8.8) exist and are analytic functions in λ. We will do this only for the right-hand side, since it is obvious for the left-hand side. Thus let us define

$$f(\lambda) = \int_{\mathbb{R}^d} \varphi_\lambda(\omega)\gamma(\omega)d\omega.$$

Suppose $\mathcal{C}$ is a closed curve in G. Since φ_λ is an analytic function in $\lambda \in G$ it has the representation

$$\varphi_\lambda(\omega) = \frac{1}{2\pi i}\int_{\mathcal{C}} \frac{\varphi_z(\omega)}{z-\lambda}dz$$

for λ in the interior $\operatorname{Int}\mathcal{C}$ of the curve $\mathcal{C}$. Now suppose that we have already shown that the integrand in the definition of $f(\lambda)$ can be bounded uniformly on $\mathcal{C}$ by an integrable function. This ensures that $f(\lambda)$ is well defined in G and by Fubini's theorem we can conclude that

$$\begin{aligned} f(\lambda) &= \int_{\mathbb{R}^d} \varphi_\lambda(\omega)\gamma(\omega)d\omega \\ &= \frac{1}{2\pi i}\int_{\mathbb{R}^d}\int_{\mathcal{C}} \frac{\varphi_z(\omega)}{z-\lambda}dz\gamma(\omega)d\omega \\ &= \frac{1}{2\pi i}\int_{\mathcal{C}} \frac{1}{z-\lambda}\int_{\mathbb{R}^d} \varphi_z(\omega)\gamma(\omega)d\omega dz \\ &= \frac{1}{2\pi i}\int_{\mathcal{C}} \frac{f(z)}{z-\lambda}dz \end{aligned}$$

for $\lambda \in \operatorname{Int}\mathcal{C}$, which means that f is analytic in G. Thus it remains to bound the integrand uniformly. Let us first consider the asymptotic behavior in a neighborhood of the origin, say for $\|\omega\|_2 < \min\{1/c, 1\}$. If we set $b = \Re(\lambda)$ we can use Lemma 5.14 to get, in the case $b \neq -d/2$,

$$|\varphi_\lambda(\omega)\gamma(\omega)| \le C_\gamma \frac{2^{b+|b+d/2|}\Gamma(|b+d/2|)}{|\Gamma(-\lambda)|} c^{b+d/2-|b+d/2|}\, \|\omega\|_2^{-b-d/2-|b+d/2|+2m},$$

and, in the case $b = -d/2$,

$$|\varphi_\lambda(\omega)\gamma(\omega)| \le C_\gamma \frac{2^{1-d/2}}{|\Gamma(-\lambda)|}\left(\frac{1}{e} - \log\frac{c\|\omega\|_2}{2}\right)\|\omega\|_2^{2m}.$$

Since $\mathcal{C}$ is compact and $1/\Gamma$ is analytic this gives for all $\lambda \in \mathcal{C}$ and $\|\omega\|_2 < \min\{1/c, 1\}$

$$|\varphi_\lambda(\omega)\gamma(\omega)| \le C_{\gamma,m,c,\mathcal{C}}\left(1 + \|\omega\|_2^{-d+2\epsilon} - \log\frac{c\|\omega\|_2}{2}\right),$$

with $\epsilon = m - b > 0$. For large arguments the integrand in the definition of $f(\lambda)$ can be estimated by Lemma 5.13:

$$|\varphi_\lambda(\omega)\gamma(\omega)| \le C_\gamma \frac{2^{1+b}\sqrt{2\pi}}{|\Gamma(-\lambda)|} c^{b+(d-1)/2}\|\omega\|_2^{-b-(d+1)/2} e^{-c\|\omega\|_2} e^{|b+d/2|^2/(2c\|\omega\|_2)},$$

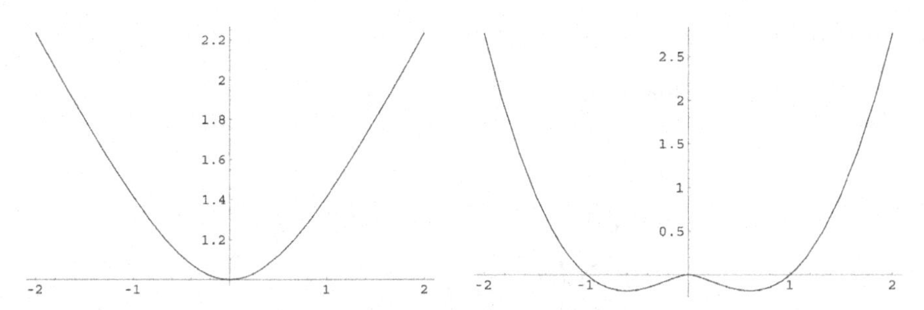

Fig. 8.1 The multiquadric $\phi(r) = \sqrt{1+r^2}$ (on the left) and the thin-plate spline $\phi(r) = |r|^2 \log(|r|)$ (on the right).

using the fact that $\gamma \in \mathcal{S}$ is certainly bounded. Since $\mathcal{C}$ is compact, this can be bounded independently of $\lambda \in \mathcal{C}$ by

$$|\varphi_\lambda(\omega)\gamma(\omega)| \le C_{\gamma,\mathcal{C},m,c} e^{-c\|\omega\|_2}.$$

This completes the proof. □

The left-hand half of Figure 8.1 shows the function $\phi(r) = \sqrt{1+r^2}$, for which the name multiquadric has been coined.

Theorem 8.16 *The function $\Phi(x) = \|x\|_2^\beta$, $x \in \mathbb{R}^d$, with $\beta > 0$, $\beta \notin 2\mathbb{N}$, has the generalized Fourier transform*

$$\widehat{\Phi}(\omega) = \frac{2^{\beta+d/2}\Gamma((d+\beta)/2)}{\Gamma(-\beta/2)} \|\omega\|_2^{-\beta-d}, \qquad \omega \ne 0,$$

of order $m = \lceil \beta/2 \rceil$.

Proof Let us start with the function $\Phi_c(x) = (c^2 + \|x\|_2^2)^{\beta/2}$, $c > 0$. This function possesses a generalized Fourier transform of order $m = \lceil \beta/2 \rceil$ given by

$$\widehat{\Phi}_c(\omega) = \varphi_c(\omega) = \frac{2^{1+\beta/2}}{\Gamma(-\beta/2)} \|\omega\|_2^{-\beta-d} (c\|\omega\|_2)^{(\beta+d)/2} K_{(\beta+d)/2}(c\|\omega\|_2),$$

owing to Theorem 8.15. Here we use the subscript c instead of β, since β is fixed and we want to let c go to zero. Moreover, we can conclude from the proof of Theorem 8.15 that for $\gamma \in \mathcal{S}_{2m}$ the product can be bounded by

$$|\varphi_c(\omega)\gamma(\omega)| \le C_\gamma \frac{2^{\beta+d/2}\Gamma((\beta+d)/2)}{|\Gamma(-\beta/2)|} \|\omega\|_2^{2m-\beta-d}$$

for $\|\omega\|_2 \to 0$ and by

$$|\varphi_c(\omega)\gamma(\omega)| \le C_\gamma \frac{2^{\beta+d/2}\Gamma((\beta+d)/2)}{|\Gamma(-\beta/2)|} \|\omega\|_2^{-\beta-d}$$

for $\|\omega\|_2 \to \infty$, independently of $c > 0$. Since $|\Phi_c(\omega)\widehat{\gamma}(\omega)|$ can also be bounded independently of c by an integrable function, we can use the convergence theorem of Lebesgue twice to derive

$$\begin{aligned}\int_{\mathbb{R}^d} \|x\|_2^\beta \widehat{\gamma}(x)dx &= \lim_{c\to 0}\int_{\mathbb{R}^d} \Phi_c(x)\widehat{\gamma}(x)dx \\ &= \lim_{c\to 0}\int_{\mathbb{R}^d} \varphi_c(\omega)\gamma(\omega)dx \\ &= \frac{2^{1+\beta/2}}{\Gamma(-\beta/2)}\int_{\mathbb{R}^d}\frac{\gamma(\omega)}{\|\omega\|_2^{\beta+d}}\lim_{c\to 0}(c\|\omega\|_2)^{(\beta+d)/2}K_{(\beta+d)/2}(c\|\omega\|_2)d\omega \\ &= \frac{2^{\beta+d/2}\Gamma((d+\beta)/2)}{\Gamma(-\beta/2)}\int_{\mathbb{R}^d}\|\omega\|_2^{-\beta-d}\gamma(\omega)d\omega\end{aligned}$$

for $\gamma \in \mathcal{S}_{2m}$. The last equality follows from

$$\lim_{r\to 0} r^\nu K_\nu(r) = \lim_{r\to 0} 2^{\nu-1}\int_0^\infty e^{-t}e^{-r^2/(4t)}t^{\nu-1}dt = 2^{\nu-1}\Gamma(\nu);$$

see also the proof of Lemma 5.14. □

Our final example deals with the thin-plate or surface splines. The right-hand half of Figure 8.1 shows the most popular representative of this class.

Theorem 8.17 *The function $\Phi(x) = \|x\|_2^{2k}\log\|x\|_2$, $x \in \mathbb{R}^d$, $k \in \mathbb{N}$, possesses the generalized Fourier transform*

$$\widehat{\Phi}(\omega) = (-1)^{k+1}2^{2k-1+d/2}\Gamma(k+d/2)k!\,\|\omega\|_2^{-d-2k}$$

of order $m = k+1$.

Proof For $r > 0$ fixed and $\beta \in (2k, 2k+1)$ we expand the function $\beta \mapsto r^\beta$ in a Taylor series, obtaining

$$r^\beta = r^{2k} + (\beta - 2k)r^{2k}\log r + \int_{2k}^{\beta}(\beta - t)r^t\log^2(r)\,dt. \tag{8.9}$$

From Theorem 8.16 we know the generalized Fourier transform of the function $x \mapsto \|x\|_2^\beta$ of order $m = \lceil\beta/2\rceil = k+1$. From Proposition 8.10 we see that the generalized Fourier transform of order m of the function $x \mapsto \|x\|_2^{2k}$ equals zero. Thus we can conclude from (8.9) for any test function $\gamma \in \mathcal{S}_{2m}$ that

$$\begin{aligned}&\int_{\mathbb{R}^d}\widehat{\gamma}(x)\|x\|_2^{2k}\log\|x\|_2\,dx \\ &= \frac{1}{\beta-2k}\int_{\mathbb{R}^d}\widehat{\gamma}(x)\left(\|x\|_2^\beta - \|x\|_2^{2k}\right)dx \\ &\quad -\frac{1}{\beta-2k}\int_{\mathbb{R}^d}\int_{2k}^{\beta}(\beta-t)\widehat{\gamma}(x)\|x\|_2^t\log^2\|x\|_2\,dtdx \\ &= \frac{2^{\beta+d/2}\Gamma((d+\beta)/2)}{(\beta-2k)\Gamma(-\beta/2)}\int_{\mathbb{R}^d}\|\omega\|_2^{-\beta-d}\gamma(\omega)d\omega + \mathcal{O}(\beta-2k),\end{aligned}$$

for $\beta \to 2k$. Furthermore, we know from Proposition 5.2 that

$$\frac{1}{\Gamma(-\beta/2)(\beta - 2k)} = -\frac{\sin(\pi\beta/2)\Gamma(1+\beta/2)}{\pi(\beta - 2k)}.$$

Because

$$\lim_{\beta\to 2k} \frac{\sin(\pi\beta/2)}{\beta - 2k} = \lim_{\beta\to 2k} \frac{\pi}{2}\cos(\pi\beta/2) = \frac{\pi}{2}(-1)^k$$

we see that

$$\lim_{\beta\to 2k} \frac{1}{\Gamma(-\beta/2)(\beta - 2k)} = (-1)^{k+1}\frac{k!}{2}.$$

Now we can apply the dominated convergence theorem to derive

$$\int_{\mathbb{R}^d} \|x\|_2^{2k} \log \|x\|_2 \, \widehat{\gamma}(x) dx = 2^{2k+d/2}\Gamma(k+d/2)(-1)^{k+1}\frac{k!}{2}\int_{\mathbb{R}^d} \frac{\gamma(\omega)}{\|\omega\|_2^{d+2k}} d\omega$$

for all $\gamma \in \mathcal{S}_{2m}$, which gives the stated Fourier transform. □

Now it is easy to decide whether the functions just investigated are conditionally positive definite. As mentioned before we state the minimal m. The case of inverse multiquadrics was treated in Theorem 7.15.

Corollary 8.18 *The following functions $\Phi : \mathbb{R}^d \to \mathbb{R}$ are conditionally positive definite of order m:*

(1) $\Phi(x) = (-1)^{\lceil\beta\rceil}(c^2 + \|x\|_2^2)^\beta, \quad \beta > 0, \quad \beta \notin \mathbb{N}, \quad m = \lceil\beta\rceil,$
(2) $\Phi(x) = (-1)^{\lceil\beta/2\rceil}\|x\|_2^\beta, \quad \beta > 0, \quad \beta \notin 2\mathbb{N}, \quad m = \lceil\beta/2\rceil,$
(3) $\Phi(x) = (-1)^{k+1}\|x\|_2^{2k}\log\|x\|_2, \quad k \in \mathbb{N}, \quad m = k+1.$

8.4 Radial conditionally positive definite functions

As in case of (unconditionally) positive semi-definite functions it is possible to derive a characterization of conditionally positive semi-definite and radial functions from Theorem 8.14. We omit the details here and refer the interested reader to the article [77] by Guo *et al.* Instead, we turn to univariate functions that are conditionally positive (semi-)definite on *every* $\mathbb{R}^d$ and derive a result along the lines of the characterization of Schoenberg. In particular, the univariate function ϕ acts again as a multivariate function Φ via $\phi(\|\cdot\|_2^2)$.

Theorem 8.19 **(Micchelli)** *Suppose that $\phi \in C[0,\infty) \cap C^\infty(0,\infty)$ is given. Then the function $\Phi = \phi(\|\cdot\|_2^2)$ is conditionally positive semi-definite of order $m \in \mathbb{N}_0$ on every $\mathbb{R}^d$ if and only if $(-1)^m\phi^{(m)}$ is completely monotone on $(0,\infty)$.*

Proof Suppose that $(-1)^m\phi^{(m)}$ is completely monotone on $(0,\infty)$. We know from Theorem 7.11 that it can be represented by

$$(-1)^m\phi^{(m)}(r) = \int_0^\infty e^{-rt}\,d\mu(t)$$

with a nonnegative Borel measure μ on $[0,\infty)$. As we do not want to assume $\phi^{(m)}$ to be continuous in zero, the measure μ will not be finite. Hence, we define $\phi_\epsilon(r) = \phi(r+\epsilon)$ for $\epsilon > 0$. Using Taylor's formula gives us

$$\begin{aligned}\phi_\epsilon(r) &= \sum_{\ell=0}^{m-1}\frac{\phi_\epsilon^{(\ell)}(0)}{\ell!}r^\ell + \frac{1}{(m-1)!}\int_0^r (r-t)^{m-1}\phi_\epsilon^{(m)}(t)dt\\ &= \sum_{\ell=0}^{m-1}\frac{\phi_\epsilon^{(\ell)}(0)}{\ell!}r^\ell + \frac{(-1)^m}{(m-1)!}\int_0^r\int_0^\infty (r-t)^{m-1}e^{-(t+\epsilon)s}d\mu(s)dt\\ &= \sum_{\ell=0}^{m-1}\frac{\phi_\epsilon^{(\ell)}(0)}{\ell!}r^\ell + \frac{(-1)^m}{(m-1)!}\int_0^\infty e^{-\epsilon s}\int_0^r (r-t)^{m-1}e^{-ts}dtd\mu(s),\end{aligned}$$

where we have applied Fubini's theorem in the last step. A further application of Taylor's formula to the function $r \mapsto e^{-rs}$ shows that

$$e^{-rs} = \sum_{j=0}^{m-1}\frac{(-1)^j}{j!}(rs)^j + \frac{(-s)^m}{(m-1)!}\int_0^r (r-t)^{m-1}e^{-st}dt.$$

Inserting this representation for the inner integral into the representation of ϕ_ϵ leads finally to

$$\phi_\epsilon(r) = \sum_{\ell=0}^{m-1}\frac{\phi_\epsilon^{(\ell)}(0)}{\ell!}r^\ell + \int_0^\infty\left(e^{-rs} - \sum_{j=0}^{m-1}\frac{(-1)^j}{j!}r^js^j\right)e^{-\epsilon s}\frac{d\mu(s)}{s^m}.$$

Now suppose pairwise distinct $x_1,\dots,x_N \in \mathbb{R}^d$ and $\alpha \in \mathbb{R}^N$ satisfying (8.1) for all $p \in \pi_{m-1}(\mathbb{R}^d)$ are given. Then we can conclude from Proposition 8.4 that

$$\sum_{j,k=1}^N \alpha_j\alpha_k\phi_\epsilon(\|x_j-x_k\|_2^2) = \int_0^\infty \sum_{j,k=1}^N \alpha_j\alpha_k e^{-s\|x_j-x_k\|_2^2}e^{-\epsilon s}\frac{d\mu(s)}{s^m} \geq 0$$

for every $\epsilon > 0$, since the Gaussian is positive definite. Note that the last integral is well defined by an argument similar to that in the proof of Lemma 8.11. But as the function on the left-hand side is a continuous function in ϵ we can let ϵ tend to zero, showing that the quadratic form is nonnegative for $\epsilon = 0$ also.

For the necessary part let us assume that $\phi(\|\cdot\|_2^2)$ is conditionally positive definite of order m. We want to show by induction on m that in this case $(-1)^m\phi^{(m)}$ is completely monotone. For $m = 0$ this is Schoenberg's result given in Theorem 7.13. For the induction step we assume that the result is true for m and we want to conclude it for $m+1$. Suppose $\phi(\|\cdot\|_2^2)$ is conditionally positive semi-definite of order $m+1$ on every $\mathbb{R}^d$. Fix a dimension d. For

$h > 0$ we consider the $(d+1)$-variate function

$$\Psi_h(x) = 2\phi(\|x\|_2^2) - \phi(\|x + \sqrt{h}e_{d+1}\|_2^2) - \phi(\|x - \sqrt{h}e_{d+1}\|_2^2), \qquad x \in \mathbb{R}^{d+1},$$

where e_{d+1} denotes the $(d+1)$-th unit vector. By Proposition 8.7 we know that Ψ_h is conditionally positive semi-definite of order m on $\mathbb{R}^{d+1}$ and hence on $\mathbb{R}^d$. But the restriction to $\mathbb{R}^d$ is given by

$$\Psi_h(x) = 2\left[\phi(\|x\|_2^2) - \phi(\|x\|_2^2 + h)\right] =: 2\psi_h(\|x\|_2^2), \qquad x \in \mathbb{R}^d.$$

Thus, by the induction hypothesis, $(-1)^m \psi_h^{(m)}$ is completely monotone on $(0, \infty)$ for every $h > 0$, i.e.

$$(-1)^{m+\ell}\psi_h^{(m+\ell)} = (-1)^{m+\ell}\left[\phi^{(m+\ell)}(r) - \phi^{(m+\ell)}(r+h)\right] \geq 0$$

for $r > 0$, $\ell \in \mathbb{N}_0$. But this means in particular that

$$(-1)^{m+\ell+1}\frac{\phi^{(m+\ell)}(r+h) - \phi^{(m+\ell)}(r)}{h} \geq 0$$

for all $r > 0$, $\ell \in \mathbb{N}_0$, and $h > 0$. Letting h tend to zero results in

$$(-1)^{m+1+\ell}\phi^{(m+1+\ell)}(r) \geq 0$$

for all $r > 0$, $\ell \in \mathbb{N}_0$, which finishes the induction proof. □

An argument similar to that in the proof of Theorem 7.14 yields:

Corollary 8.20 *Suppose that the function ϕ of Theorem 8.19 is not a polynomial of degree at most m; then $\phi(\|\cdot\|_2^2)$ is conditionally positive definite of order m on every $\mathbb{R}^d$.*

Micchelli's result (Theorem 8.19) gives a very powerful tool for deciding whether a given radial function is conditionally positive definite on every $\mathbb{R}^d$. To demonstrate its usefulness we investigate inverse multiquadrics, power functions, and thin-plate splines again.

Example The multiquadrics $\phi(r) = (-1)^{\lceil\beta\rceil}(c^2 + r^2)^\beta$, $c, \beta > 0$, $\beta \notin \mathbb{N}$, are conditionally positive definite of order $m = \lceil\beta\rceil$ on every $\mathbb{R}^d$.

Proof If we define $f_\beta(r) = (-1)^{\lceil\beta\rceil}(c^2 + r)^\beta$, we see that

$$f_\beta^{(k)}(r) = (-1)^{\lceil\beta\rceil}\beta(\beta-1)\cdots(\beta-k+1)(c^2+r)^{\beta-k}.$$

Thus $(-1)^{\lceil\beta\rceil} f_\beta^{(\lceil\beta\rceil)}(r) = \beta(\beta-1)\cdots(\beta-\lceil\beta\rceil+1)(c^2+r)^{\beta-\lceil\beta\rceil}$ is completely monotone and $m = \lceil\beta\rceil$ is the smallest possible choice that makes $(-1)^m f_\beta^{(m)}$ completely monotone. □

Example The functions $\phi(r) = (-1)^{\lceil\beta/2\rceil} r^\beta$, $\beta > 0$, $\beta \notin 2\mathbb{N}$, are conditionally positive definite of order $m = \lceil\beta/2\rceil$ on every $\mathbb{R}^d$.

Proof Define $f_\beta(r) = (-1)^{\lceil\beta/2\rceil} r^{\beta/2}$ to see that

$$f_\beta^{(k)}(r) = (-1)^{\lceil\beta/2\rceil} \frac{\beta}{2}\left(\frac{\beta}{2}-1\right)\cdots\left(\frac{\beta}{2}-k+1\right) r^{\beta/2-k}.$$

Thus again $(-1)^{\lceil\beta/2\rceil} f_\beta^{(\lceil\beta/2\rceil)}(r)$ is completely monotone and $m = \lceil\beta/2\rceil$ is the smallest possible choice. □

Example The thin-plate or surface splines $\phi(r) = (-1)^{k+1} r^{2k} \log(r)$ are conditionally positive definite of order $m = k+1$ on every $\mathbb{R}^d$.

Proof Since $2\phi(r) = (-1)^{k+1} r^2 \log(r^2)$ we set $f_k(r) = (-1)^{k+1} r^k \log(r)$. Then it is easy to see that

$$f_k^{(\ell)}(r) = (-1)^{k+1} k(k-1)\cdots(k-\ell+1) r^{k-\ell} \log(r) + p_\ell(r), \qquad 1 \le \ell \le k,$$

where p_ℓ is a polynomial of degree $k-\ell$. This means in particular that $f_k^{(k)}(r) = (-1)^{k+1} k! \log(r) + c$ and finally that $(-1)^{k+1} f_k^{(k+1)}(r) = k!\, r^{-1}$, which is obviously completely monotone on $(0, \infty)$. □

8.5 Interpolation by conditionally positive definite functions

The investigation of positive definite functions was motivated by the interpolation problem (6.2) and the Ansatz (6.1) for the interpolating function. The example mentioned at the beginning of the present chapter showed that this definition of the interpolating function does not work in the case of conditionally positive definite functions. But a slight change in the definition of the interpolation function ensures solvability of the interpolation matrix. Instead of (6.1) we now define the interpolant to a function f at the centers $X = \{x_1, \dots, x_N\}$ as

$$s_{f,X}(x) = \sum_{j=1}^{N} \alpha_j \Phi(x - x_j) + \sum_{k=1}^{Q} \beta_k p_k(x).$$

Here, Q denotes again the dimension of the polynomial space $\pi_{m-1}(\mathbb{R}^d)$ and $p_1, \dots, p_Q$ denote a basis of $\pi_{m-1}(\mathbb{R}^d)$. To cope with the additional degrees of freedom, the interpolation conditions

$$s_{f,X}(x_j) = f(x_j), \qquad 1 \le j \le N,$$

are completed by the additional conditions

$$\sum_{j=1}^{N} \alpha_j p_k(x_j) = 0, \qquad 1 \le k \le Q.$$

Solvability of this system is therefore equivalent to solvability of the system

$$\begin{pmatrix} A_{\Phi,X} & P \\ P^T & 0 \end{pmatrix} \begin{pmatrix} \alpha \\ \beta \end{pmatrix} = \begin{pmatrix} f|X \\ 0 \end{pmatrix} \tag{8.10}$$

where $A_{\Phi,X} = (\Phi(x_j - x_k)) \in \mathbb{R}^{N \times N}$ and $P = (p_k(x_j)) \in \mathbb{R}^{N \times Q}$. This last system is obviously solvable if the matrix on the left-hand side, which we will denote by $\widetilde{A}_{\Phi,X}$, is invertible.

Theorem 8.21 *Suppose that Φ is conditionally positive definite of order m and X is a $\pi_{m-1}(\mathbb{R}^d)$-unisolvent set of centers. Then the system (8.10) is uniquely solvable.*

Proof Suppose that $(\alpha, \beta)^T$ lies in the null space of the matrix $\widetilde{A}_{\Phi,X}$. Then we have

$$A_{\Phi,X}\alpha + P\beta = 0,$$
$$P^T\alpha = 0.$$

The second equation means that α satisfies condition (8.1) for all $p \in \pi_{m-1}(\mathbb{R}^d)$. Multiplying the first equation by α^T gives $0 = \alpha^T A_{\Phi,X}\alpha + (P^T\alpha)^T\beta = \alpha^T A_{\Phi,X}\alpha$. Since Φ is conditionally positive definite of order m we can conclude that $\alpha = 0$ and thus $P\beta = 0$. Finally, since X is $\pi_{m-1}(\mathbb{R}^d)$-unisolvent, this means that $\beta = 0$. □

For a solution of the system (8.10) it is not necessary to require X to be $\pi_{m-1}(\mathbb{R}^d)$-unisolvent. This is only needed for uniqueness. To see the solvability in the general case let us set $V := P(\mathbb{R}^Q) \subseteq \mathbb{R}^N$ and $A := A_{\Phi,X}$. Then the orthogonal complement $V^\perp$ of V is the null space of P^T. The system (8.10) is solvable for every $f|X$ if $\mathbb{R}^N = AV^\perp + V$. But this sum is a direct sum, because $x \in AV^\perp \cap V$ means that $x = A\alpha = P\beta$ with a certain $\alpha \in V^\perp$ and a $\beta \in \mathbb{R}^Q$; this implies that $\alpha^T A\alpha = (P^T\alpha)^T\beta = 0$ and hence $\alpha = 0$ and $A\alpha = 0$. Knowing that the intersection of $AV^\perp$ and V contains only the zero vector gives

$$N \geq \dim(AV^\perp + V) = \dim AV^\perp + \dim V = \dim V^\perp + \dim V = N,$$

because $A|V^\perp : V^\perp \to AV^\perp$ is bijective. But this means solvability.

It is obvious that the addition of polynomial terms of total degree at most $m - 1$ to the expansion guarantees polynomial reproduction, i.e. if the data come from a polynomial of total degree less than m then they are fitted by that polynomial.

The method described in this section can be generalized in a straightforward way by using arbitrary linearly independent functions $p_1, \ldots, p_Q$ on $\mathbb{R}^d$ instead of polynomials. Moreover, the conditionally positive definite function can be replaced by a conditionally positive definite kernel $\Phi : \Omega \times \Omega \to \mathbb{C}$.

8.6 Notes and comments

One might say that the whole radial basis function theory started with the practical work of Hardy in 1971 on multiquadrics (see [79] and also the review article [80]) and with the theoretical work of Duchon on thin-plate splines (see [47–49]) in the late 1970s. Shortly thereafter, Meinguet [122–124] popularized thin-plate splines as a practical numerical method for multivariate interpolation. In the mid 1980s, Micchelli [133] put Hardy's multiquadrics on a firm mathematical basis by solving a conjecture, drawn from computational experience, of Franke by connecting it with the classical results of Bochner and Schoenberg on positive definite functions and thereby releasing it from the limitations of the variational perspectives of thin-plate splines and the specialized form of multiquadrics.

The proof of Micchelli's theorem, which consists in its original form of only the sufficient part, was completed in [77] by Guo *et al.* seven years later in an even stronger version. As for positive definite and radial functions, it is not necessary to assume ϕ to be in $C^\infty(0, \infty)$. This can be concluded from the fact that $\phi(\|\cdot\|_2^2)$ is conditionally positive definite on every $\mathbb{R}^d$. The simple proof given here was based upon Sun's paper [183].

Several results on conditionally positive definite functions employ distribution theory, pseudo-functions, or both; see for example Madych and Nelson [113] and Gel'fand and Vilenkin [69]. The approach here tries to avoid the use of such tools even if it sometimes recovers the ideas behind them. In any case, to introduce straightforwardly the concept that the generalized Fourier transform of a function with a possible singularity at the origin is a function itself seems to be a major step in simplifying this theory. It came first up in the unpublished preprint [111] by Madych and Nelson. Some of the examples of generalized Fourier transforms can be found in the books by Gel'fand and Vilenkin [69] and by Jones [95].

9
Compactly supported functions

In numerical analysis, the concept of locally supported basis functions is of general importance. Several function spaces used for approximation possess locally supported basis functions. The most prominent examples in the one-dimensional case are the well-known B-splines. The general advantages of compactly supported basis functions are a sparse interpolation matrix on the one hand, and the possibility of a fast evaluation of the interpolant on the other.

Thus, it seems to be natural to look for locally supported functions also in the context of radial basis function interpolation and we will give an introduction to this field in this chapter.

At the outset, though, we want to point out one crucial difference from classical spline theory. While the support radius of the B-splines can be chosen proportional to the maximal distance between two neighboring centers, something similar will not lead to a convergent scheme in the theory of radial basis functions. The correct choice of the support radius is a very delicate question, which we will address in a later chapter on numerical methods.

9.1 General remarks

Gaussians, (inverse) multiquadrics, powers, and thin-plate splines share two joint features. They are all globally supported and are positive definite on *every* $\mathbb{R}^d$. The truncated powers from Theorem 6.20, however, are compactly supported but are also restricted to a finite number of space dimensions.

We will see that the two features are connected. But let us first comment on conditionally positive definite functions.

Theorem 9.1 *Assume that the function* $\Phi : \mathbb{R}^d \to \mathbb{C}$ *is continuous and compactly supported. If* Φ *is conditionally positive definite of minimal order* $m \in \mathbb{N}_0$ *then* m *is necessarily zero, i.e.* Φ *is must be positive definite.*

Proof Since Φ is integrable, it possesses a classical Fourier transform $\widehat{\Phi}$ that is continuous. In this situation the generalized Fourier transform coincides with the classical one. Hence, by Theorem 8.12 the Fourier transform is nonnegative in $\mathbb{R}^d \setminus \{0\}$ and not identically zero. Since it is continuous we also have $\widehat{\Phi}(0) \geq 0$, and Theorem 6.11 ensures that Φ is positive definite. □

Thus we can concentrate on positive definite radial functions with compact support and use the classical Fourier transform instead of the generalized Fourier transform to handle them.

The next theorem shows that the Fourier transform is indeed the right tool to handle such functions, not the Laplace transform, which we have seen to be important in the context of completely monotone functions.

Theorem 9.2 *Suppose the continuous and nonvanishing function $\phi : [0, \infty) \to \mathbb{R}$ is positive definite on every $\mathbb{R}^d$. Then $\phi(r) \neq 0$ for all $r \in [0, \infty)$.*

Proof Since ϕ is positive definite on every $\mathbb{R}^d$ there exists a finite nonnegative Borel measure μ on $[0, \infty)$ such that

$$\phi(r) = \int_0^\infty e^{-r^2 u} d\mu(u).$$

If ϕ had a zero r_0, this would mean that

$$0 = \int_0^\infty e^{-r_0^2 u} d\mu(u).$$

As $e^{-r_0^2 u} > 0$ for all $u \geq 0$, we must have $\mu([0, \infty)) = 0$ and hence $\phi \equiv 0$, which contradicts the fact that ϕ is nonvanishing. □

An immediate consequence of the preceding theorem is that the dimensions d, on which a compactly supported ϕ is positive definite, are restricted to a finite number. If ϕ is not positive definite on a fixed $\mathbb{R}^{d_0}$ then it cannot be positive definite on any higher-dimensional space.

Corollary 9.3 *A continuous, univariate, and compactly supported function ϕ cannot be positive definite on every $\mathbb{R}^d$.*

9.2 Dimension walk

From the results of the last section we know that if we want to construct locally supported, radial, and positive definite functions we have to work with a fixed space dimension d. In this case the Fourier transform is the right tool. Following Bochner, we know that a positive definite function on $\mathbb{R}^d$ is characterized by a nonnegative d-variate Fourier transform. In the case of a radial function $\Phi = \phi(\|\cdot\|_2) \in L_1(\mathbb{R}^d)$ this is a radial function $\widehat{\Phi} = \mathcal{F}_d\phi(\|\cdot\|_2)$ again (see Theorem 5.26), where

$$\mathcal{F}_d\phi(r) = r^{-(d-2)/2} \int_0^\infty \phi(t) t^{d/2} J_{(d-2)/2}(rt)\, dt.$$

This operator $\mathcal{F}_d$, which acts on univariate functions, can be manipulated by operators that we now want to introduce.

Definition 9.4

(1) Let ϕ be given such that $t \mapsto \phi(t)t$ is in $L_1[0,\infty)$; then we define for $r \geq 0$

$$(\mathcal{I}\phi)(r) = \int_r^\infty t\phi(t)dt.$$

(2) For even $\phi \in C^2(\mathbb{R})$ we define for $r \geq 0$

$$(\mathcal{D}\phi)(r) = -\frac{1}{r}\phi'(r).$$

In both cases the resulting functions should be seen as even functions by even extension.

Thus $\mathcal{I}$ and $\mathcal{D}$ map even univariate functions to even univariate functions by even extension. Both operators respect a compact support.

Note that the function $\mathcal{D}\phi$ is continuous at zero. Since $\phi \in C^2(\mathbb{R})$ is even we have $\phi'(t) = -\phi'(-t)$ and in particular $\phi'(0) = 0$. This means that $\phi'(t) = \mathcal{O}(t)$ for $t \to 0$ and hence $\mathcal{D}\phi(t) = \mathcal{O}(1)$ for $t \to 0$. Moreover, the operators $\mathcal{I}$ and $\mathcal{D}$ are inverse in the following sense.

Lemma 9.5 *If ϕ is continuous and satisfies $t \mapsto t\phi(t) \in L_1[0,\infty)$ then $\mathcal{D}\mathcal{I}\phi = \phi$. Conversely, if $\phi \in C^2(\mathbb{R})$ is even and satisfies $\phi' \in L_1[0,\infty)$ then $\mathcal{I}\mathcal{D}\phi = \phi$.*

The interaction between the operators $\mathcal{I}$, $\mathcal{D}$ and $\mathcal{F}_d$ is given by the next theorem. Remember that $t \mapsto \phi(t)t^{d-1} \in L_1[0,\infty)$ implies in particular that $\Phi = \phi(\|\cdot\|_2) \in L_1(\mathbb{R}^d)$.

Theorem 9.6 *Suppose that ϕ is continuous.*

(1) If $t \mapsto \phi(t)t^{d-1} \in L_1[0,\infty)$ and $d \geq 3$ then $\mathcal{F}_d(\phi) = \mathcal{F}_{d-2}(\mathcal{I}\phi)$.

(2) If $\phi \in C^2(\mathbb{R})$ is even and $t \mapsto \phi'(t)t^d \in L_1[0,\infty)$ then $\mathcal{F}_d(\phi) = \mathcal{F}_{d+2}(\mathcal{D}\phi)$.

Proof To prove the first statement, we start by showing that the function $r \mapsto \mathcal{I}\phi(r)r^{d-3} \in L_1[0,\infty)$ and hence that $\mathcal{I}\phi(\|\cdot\|_2) \in L_1(\mathbb{R}^{d-2})$. Since $d \geq 3$ the function $\mathcal{I}\phi$ is well defined and continuous. Moreover, for $R > 0$ we find that

$$\begin{aligned}
\int_0^R |\mathcal{I}\phi(r)|r^{d-3}dr &\leq \int_0^R \int_r^\infty |\phi(t)|tr^{d-3}dtdr \\
&= \int_0^R \int_r^R |\phi(t)|tr^{d-3}dtdr + \int_0^R \int_R^\infty |\phi(t)|tr^{d-3}dtdr
\end{aligned}$$

and we have to bound each of the last two integrals uniformly in R. For the first of these we exchange the order of integration to get

$$\begin{aligned}
\int_0^R \int_r^R |\phi(t)|tr^{d-3}dtdr &= \int_0^R \int_0^t |\phi(t)|tr^{d-3}drdt \\
&= \frac{1}{d-2}\int_0^R t^{d-1}|\phi(t)|dt,
\end{aligned}$$

which is obviously bounded by $\|t^{d-1}\phi(t)\|_{L_1[0,\infty)}/(d-2)$. The second integral is actually a product of two univariate integrals and allows the bound

$$\begin{aligned}\int_0^R \int_R^\infty |\phi(t)| t r^{d-3} dt dr &= \frac{R^{d-2}}{d-2} \int_R^\infty t^{d-1} |\phi(t)| t^{-d+2} dt \\ &\le \frac{1}{d-2} \int_R^\infty t^{d-1} |\phi(t)| dt \\ &\le \frac{1}{d-2} \int_0^\infty t^{d-1} |\phi(t)| dt.\end{aligned}$$

Now that we know about integrability, we can apply $\mathcal{F}_{d-2}$ to $\mathcal{I}\phi$. Using integration by parts and $(d/dz)[z^\nu J_\nu(z)] = z^\nu J_{\nu-1}(z)$ (see Proposition 5.4) leads to

$$\begin{aligned}\mathcal{F}_{d-2}(\mathcal{I}\phi)(r) &= r^{-(d-4)/2} \int_0^\infty (\mathcal{I}\phi)(t) t^{(d-2)/2} J_{(d-4)/2}(rt) dt \\ &= r^{-(d-2)/2} \left[(\mathcal{I}\phi)(t) t^{(d-2)/2} J_{(d-2)/2}(rt) \Big|_0^\infty + \int_0^\infty \phi(t) t^{d/2} J_{(d-2)/2}(rt) dt \right] \\ &= \mathcal{F}_d \phi(r).\end{aligned}$$

The boundary terms vanish for the following reasons. Because of the integrability of $t \mapsto \mathcal{I}\phi(t) t^{d-3}$ we have at least $\mathcal{I}\phi(t) = \mathcal{O}(t^{-d+2})$ for $t \to \infty$. The asymptotic behavior of the Bessel functions gives $J_\nu(t) = \mathcal{O}(1/\sqrt{t})$ (see Proposition 5.6). Hence, $(\mathcal{I}\phi)(t) t^{(d-2)/2} J_{(d-2)/2}(rt) = \mathcal{O}(t^{-(d-1)/2})$ for $t \to \infty$ and vanishes at infinity. For the lower bound we use the asymptotic behavior of the Bessel functions $J_\nu(t) = \mathcal{O}(t^\nu)$ for $\nu \ge 0$ and $t \to 0$ together with the boundedness of $\mathcal{I}\phi$ to derive $(\mathcal{I}\phi)(t) t^{(d-2)/2} J_{(d-2)/2}(rt) = \mathcal{O}(t^{d-2})$ for $t \to 0$, so that this function also vanishes at zero. This finishes the proof of the first part.

For the second part, define $\psi := \mathcal{D}\phi$. Then ψ is well defined, continuous, and satisfies $t \mapsto \psi(t) t^{d+1} \in L_1[0,\infty)$. This means in particular that $\mathcal{I}\psi = \mathcal{I}\mathcal{D}\phi = \phi$. Finally, we can apply the first part to ψ instead of ϕ and $d+1$ instead of d to derive

$$\mathcal{F}_{d+2}(\mathcal{D}\phi) = \mathcal{F}_{d+2}(\psi) = \mathcal{F}_d(\mathcal{I}\psi) = \mathcal{F}_d(\phi),$$

and this finishes the proof. □

This interaction between these operators allows us to express the higher-dimensional Fourier transforms of radial functions by lower-dimensional ones and vice versa. Since positive definite integrable functions are characterized by a nonnegative and nonvanishing Fourier transform we can draw the following conclusion.

Corollary 9.7 *Suppose that ϕ is continuous. If on the one hand $t \mapsto \phi(t) t^{d-1} \in L_1[0,\infty)$ and $d \ge 3$ then ϕ is positive definite on $\mathbb{R}^d$ if and only if $\mathcal{I}\phi$ is positive definite on $\mathbb{R}^{d-2}$. On the other hand, if $\phi \in C^2(\mathbb{R})$ is even and $t \mapsto \phi'(t) t^d \in L_1[0,\infty)$ then ϕ is positive definite on $\mathbb{R}^d$ if and only if $\mathcal{D}\phi$ is positive definite on $\mathbb{R}^{d+2}$.*

Proof This follows immediately from the preceding theorem and Bochner's characterization for radial and integrable functions given in Theorem 6.18. □

The operators $\mathcal{I}$ and $\mathcal{D}$ that we have introduced vary the space dimension in steps of width 2, which means that one deals with a sequence of either odd-dimensional spaces or even-dimensional spaces. A generalization of the operators $\mathcal{I}$ and $\mathcal{D}$ to a whole family of operators $\mathcal{I}_\nu$ with $\nu \in \mathbb{R}$ where $\mathcal{I}_1 = \mathcal{I}$ and $\mathcal{I}_{-1} = \mathcal{D}$ was made by Schaback and Wu in [172]. This family allows us to walk through the space dimensions in the Fourier domain not only in steps of width 2 but also in steps of width 1 and even, in a generalized way, in steps of arbitrary width. Unfortunately, these operators no longer have a simple form and thus are difficult to apply.

9.3 Piecewise polynomial functions with local support

A local support of the basis function is only one step on the way to an efficient numerical approximation scheme. The next step is to ensure that the basis function is easily evaluated. This is why from now on we will concentrate on functions of the form

$$\phi(r) = \begin{cases} p(r), & 0 \le r \le 1, \\ 0, & r > 1, \end{cases} \tag{9.1}$$

where p denotes a univariate polynomial. Of course, these functions are extended to the whole real line, again by even extension. We can restrict ourselves to functions with support in $[0, 1]$ or $[-1, 1]$, respectively. Other intervals can be obtained by scaling, because this does not change a function from being positive definite. The d-variate Fourier transform of $\phi(\cdot/\delta)$, $\delta > 0$, is $\delta^d(\mathcal{F}_d\phi)(\delta\cdot)$, which is nonnegative if and only if the Fourier transform of ϕ is nonnegative.

From Theorem 6.20 we already know a positive definite function of the form (9.1): the function

$$\phi_\ell(r) = (1 - r)_+^\ell \tag{9.2}$$

is positive definite on $\mathbb{R}^d$ provided that $\ell \ge \lfloor d/2 \rfloor + 1$.

These functions, when seen as even functions, are only continuous, even for large ℓ. Since the basis function determines the smoothness of the approximant, it is necessary to have smoother functions of the form (9.1) as well. Numerical considerations, however, ask for a polynomial of the lowest possible degree. Hence it is quite natural to look for a function of the form (9.1) with a polynomial of minimal degree, if its smoothness and space dimension are prescribed. We will answer this question completely in the next section. But beforehand we will give certain general results concerning functions of the form (9.1).

It is obvious that every even function ϕ of the form (9.1) possesses an even number of continuous derivatives around zero and that this number is determined by the first odd coefficient of the polynomial p that does not vanish. Furthermore ϕ is obviously in C^∞ at $(0, 1)$ and $(1, \infty)$, so that the only critical point is 1.

The proof of the following lemma, which describes the influence of the operators $\mathcal{I}$ and $\mathcal{D}$ on the smoothness, is straightforward if one takes the special form of ϕ into account. One only has to integrate and differentiate polynomials.

Lemma 9.8 *Suppose that ϕ is an even function of the form (9.1) and that it possesses $2k$ continuous derivatives around 0 and ℓ continuous derivatives around 1. Then $\mathcal{I}\phi$ possesses $2k+2$ continuous derivatives around 0 and $\ell+1$ continuous derivatives around 1. If $k, \ell \geq 1$ then $\mathcal{D}\phi$ possesses $2k-2$ continuous derivatives around 0 and $\ell - 1$ continuous derivatives around 1.*

The results of Lemma 9.8 remain true for an arbitrary ϕ that is sufficiently smooth outside 0 and 1 and allows the application of $\mathcal{I}$ and $\mathcal{D}$. In such a situation, the only part that needs a closer look is the smoothness at zero. For example, if ϕ is continuous at zero then $\mathcal{I}\phi$ possesses the derivative $(\mathcal{I}\phi)'(t) = -t\phi(t)$, so that $(\mathcal{I}\phi)''(0) = -\phi(0)$. Higher orders are dealt with in the same way. Finally, we have already discussed the fact that an even $\phi \in C^2(\mathbb{R})$ leads to a continuous function $\mathcal{D}\phi$.

It is time for another example, which straight away proves the beauty of the whole concept. It comes from Wu [203]. Define $f_\ell(r) = (1-r^2)_+^\ell$ for $\ell \in \mathbb{N}$. Then $g_\ell := f_\ell * f_\ell(2\,\cdot)$ is positive definite on $\mathbb{R}$, because its Fourier transform is the squared Fourier transform of f_ℓ. Moreover, it is of the form (9.1) with a polynomial of degree $4\ell + 1$ and it is in $C^{2\ell}$. Thus $g_{k,\ell} := \mathcal{D}^k g_\ell$ is positive definite on $\mathbb{R}^{2k+1}$, is of the form (9.1) with a polynomial of degree $4\ell - 2k + 1$ and is in $C^{2\ell-2k}$ if $\ell \geq k$. Later on we will see that these functions do not have minimal degree for a given smoothness.

The first step in characterizing functions of the form (9.1) with minimal degree by their smoothness and positive definiteness (i.e. by the space dimension) is to show that none of them has an odd number of continuous derivatives. To do this we first take a closer look at the smoothness of ϕ in a neighborhood of 1.

Theorem 9.9 *Let ϕ be an even function of the form (9.1) that is positive definite on $\mathbb{R}^d$. Then ϕ satisfies $\phi \in C^{\lfloor d/2 \rfloor}(0, \infty)$.*

Proof On account of the form (9.1), the function $\Phi = \phi(\|\cdot\|_2)$ is in $L_1(\mathbb{R}^d)$. Since ϕ is also positive definite on $\mathbb{R}^d$ we know from Corollary 6.12 that the Fourier transform $\widehat{\Phi}$ of Φ also belongs to $L_1(\mathbb{R}^d)$. This means in other words that

$$\int_0^\infty |\mathcal{F}_d\phi(t)| t^{d-1} dt < \infty.$$

Since both Φ and $\widehat{\Phi}$ are in $L_1(\mathbb{R}^d)$ we can recover Φ from its Fourier transform $\widehat{\Phi}$. But as both functions are radial this can be written in the form

$$\phi(r) = r^{-(d-2)/2} \int_0^\infty \mathcal{F}_d\phi(t) t^{d/2} J_{(d-2)/2}(rt) dt. \tag{9.3}$$

Using Leibniz' formula and the fact that the Bessel functions and their derivatives possess the asymptotic behavior $J_\nu^{(m)}(r) = \mathcal{O}(r^{-1/2})$ for $r \to \infty$ (see Proposition 5.6), we can conclude for $r > 0$ that

$$\frac{d^\ell}{dr^\ell}\left(r^{-(d-2)/2}J_{(d-2)/2}(rt)\right) \le c(r)t^{\ell-1/2} \qquad \text{for } t \to \infty.$$

Thus we can differentiate ℓ times under the integral in (9.3) as long as $\ell \le \lfloor (d-1)/2 \rfloor$. In this case the integral in the representation of $\phi^{(\ell)}$ can be bounded by a constant factor times

$$\int_0^\infty \mathcal{F}_d\phi(t)t^{d/2+\ell-1/2}dt.$$

This means that we can form up to $\lfloor (d-1)/2 \rfloor$ derivatives of ϕ. This finishes the proof for odd d because then $\lfloor d/2 \rfloor = \lfloor (d-1)/2 \rfloor$.

For even space dimension $d = 2\ell$ it remains to show that the ℓth derivative of p vanishes at 1. Therefore, suppose that $p(r) = \sum_{j=0}^n c_j r^j$; then, for $r > 0$ the d-variate Fourier transform of ϕ is given by

$$\begin{aligned}
\mathcal{F}_d\phi(r) &= r^{-d}\int_0^r p\left(\frac{t}{r}\right)t^{d/2}J_{(d-2)/2}(t)dt \\
&= r^{-d}\sum_{j=0}^n c_j r^{-j}\int_0^r t^{j+\ell}J_{\ell-1}(t)dt \\
&=: r^{-d}I(r).
\end{aligned}$$

Since Φ is positive definite on $\mathbb{R}^d$ we have $I \ge 0, \not\equiv 0$. Now, making use of $(d/dt)[t^{-\nu}J_\nu(t)] = -t^{-\nu}J_{\nu+1}(t)$ we obtain via integration by parts

$$\begin{aligned}
I(r) &= \sum_{j=0}^n c_j r^{-j}\int_0^r t^{j+\ell}J_{\ell-1}(t)dt \\
&= \sum_{j=0}^n c_j r^{-j}\int_0^r t^{j+2\ell-2}t^{-\ell+2}J_{\ell-1}(t)dt \\
&= \sum_{j=0}^n c_j r^{-j}\left(t^{j+\ell}J_{\ell-2}(t)\Big|_r^0 + (j+2\ell-2)\int_0^r t^{j+\ell-1}J_{\ell-2}(t)dt\right) \\
&= -\sum_{j=0}^n c_j r^\ell J_{\ell-2}(r) + \sum_{j=0}^n c_j r^{-j}(j+2\ell-2)\int_0^r t^{j+\ell-1}J_{\ell-2}(t)dt \\
&= \sum_{j=0}^n c_j r^{-j}(j+2\ell-2)\int_0^r t^{j+\ell-1}J_{\ell-2}(t)dt
\end{aligned}$$

because $p(1) = \sum c_j = 0$. Knowing that $p(1) = p'(1) = \cdots = p^{(\ell-1)}(1) = 0$, we can iterate this process to derive finally

$$I(r) = -\sum_{j=1}^{n} c_j(j+2\ell-2)(j+2\ell-4)\dots jJ_{-2}(r)$$

$$+\sum_{j=1}^{n} c_j(j+2\ell-2)(j+2\ell-4)\dots j(j-2)r^{-j}\int_0^r t^{j-1}J_{-2}(t)dt.$$

The asymptotic behavior of the first summand is determined by

$$J_\nu(r) = \sqrt{\frac{2}{\pi r}}\cos\left(r - \frac{\nu\pi}{2} - \frac{\pi}{4}\right) + \mathcal{O}(r^{-3/2})$$

(see Proposition 5.6), which means that it decays as $\mathcal{O}(1/\sqrt{r})$ for $r \to \infty$, with varying sign unless the sum

$$\sum_{j=1}^{n} c_j(j+2\ell-2)(j+2\ell-4)\cdots j \tag{9.4}$$

is equal to zero. We are now going to prove that the second summand in the last representation of $I(r)$ decays as $\mathcal{O}(1/r)$ for $r \to \infty$. This means that the second summand cannot compensate the change in sign of the first summand for large r and thus the sum (9.4) must equal zero.

Using $J_n = (-1)^n J_{-n}, n \in \mathbb{N}$, and $(d/dt)[t^\nu J_\nu(t)] = t^\nu J_{\nu-1}(t)$ we obtain, via integration by parts,

$$\begin{aligned} r^{-j}\int_0^r t^{j-1}J_{-2}(t)dt &= r^{-j}\int_0^r t^{j-1}J_2(t)dt \\ &= r^{-j}\int_0^r t^{j-4}t^3 J_2(t)dt \\ &= r^{-j}\, t^{j-1}J_3(t)\Big|_0^r - (j-4)r^{-j}\int_0^r t^{j-2}J_3(t)dt \\ &= r^{-1}J_3(r) - (j-4)r^{-j}\int_0^r t^{j-2}J_3(t)dt. \end{aligned}$$

Since $J_\nu(r)$ decays as $\mathcal{O}(1/\sqrt{r})$ we can bound the integral term in the last expression:

$$\begin{aligned} \left| r^{-j}\int_0^r t^{j-2}J_3(t)dt \right| &\le c_1 r^{-j} + c_2 r^{-j}\int_{r_0}^r t^{j-2-1/2}dt \\ &\le c_3 r^{-j} + c_4 r^{-3/2} \le c r^{-1} \end{aligned}$$

for every $j \ge 1$. Knowing that the sum (9.4) is zero leads us finally to

$$0 = \sum_{j=1}^{n} c_j(j+2\ell-2)\cdots j = (-1)^\ell D^\ell[r^{2\ell-2}p(r)](1) = \sum_{j=0}^{\ell} \gamma_j p^{(j)}(1)$$

with certain constants γ_j, $0 \le j \le \ell$, $\gamma_\ell = 1$. But as the first $\ell - 1$ derivatives of p vanish at 1, we derive $p^{(\ell)}(1) = 0$. □

Obviously, the result of Theorem 9.9 remains true for odd space dimensions and general functions ϕ with $t \mapsto \phi(t)t^{d-1} \in L_1[0,\infty)$ and is simply a consequence of the radiality. For even space dimensions the result is in this generality wrong; see Gneiting [70]. Hence, for arbitrary space dimensions and functions the generalization becomes $\phi \in C^{\lfloor (d-1)/2 \rfloor}(0,\infty)$.

Theorem 9.10 *Suppose that ϕ is a continuous and even function of the form (9.1) that is positive definite on $\mathbb{R}^d$. Then there exist integers $k, \ell \in \mathbb{N}_0$ such that ϕ possesses $2k$ continuous derivatives around* 0 *and $2k + \ell + \lfloor d/2 \rfloor$ continuous derivatives around* 1.

Proof For brevity we will use the following abbreviations. If ϕ is positive definite on $\mathbb{R}^d$ we will denote this by $\phi \in \mathrm{PD}_d$. Furthermore $\phi \in C^\ell(x_0)$ will mean that there is an open neighborhood U of x_0 with $\phi \in C^\ell(U)$.

For $k = 0$ this is simply Theorem 9.9. The case where the space dimension $d = 1$ gives a special version of Theorem 6.14.

Hence, it remains to prove this theorem for $d \ge 2$. Assume that the function ϕ satisfies

$$\phi \in C^{2k}(0) \cap C^m(1) \cap \mathrm{PD}_d,$$

with k and m chosen maximal. Then we have $m \ge \lfloor d/2 \rfloor$ by Theorem 9.9. We will discuss the cases $m \ge k$ and $m < k$ separately. If $m \ge k$ is satisfied then we have

$$\psi := \mathcal{D}^k \phi \in C^0(0) \cap C^{m-k}(1) \cap \mathrm{PD}_{d+2k},$$

which implies by Theorem 9.9 that ψ possesses at least $\lfloor d/2 \rfloor + k$ continuous derivatives around 1, i.e.

$$\psi \in C^{\lfloor d/2 \rfloor + k}(1).$$

This means that $m - k \ge \lfloor d/2 \rfloor + k$, i.e. $m \ge \lfloor d/2 \rfloor + 2k$.

We are finished as soon as we have shown that $0 \le m < k$ is impossible. So let us take $m < k$; this implies that

$$\psi := \mathcal{D}^m \phi \in C^{2(k-m)}(0) \cap C^0(1) \cap \mathrm{PD}_{d+2m};$$

but this means that $0 \ge \lfloor d/2 \rfloor + m$ by the same arguments as before. The last inequality cannot be satisfied because we have $d \ge 2$. Hence $m < k$ is impossible and this concludes the proof. □

9.4 Compactly supported functions of minimal degree

Knowing that functions of the form (9.1) must necessarily have an even number of continuous derivatives, it is our aim in this section to find those functions that are of minimal degree with respect to a given space dimension d and a given smoothness $2k$. Of course, we mean by the degree of the function ϕ the degree of the polynomial p.

Definition 9.11 *With $\phi_\ell(r) = (1-r)_+^\ell$ we define*

$$\phi_{d,k} = \mathcal{I}^k \phi_{\lfloor d/2 \rfloor + k + 1}. \tag{9.5}$$

As the operator $\mathcal{I}$ respects a compact support and maps polynomials to polynomials, the functions $\phi_{d,k}$ are of the form (9.1). A possible iterative scheme to compute the polynomials is stated in the next theorem. A proof can be easily achieved by induction.

Theorem 9.12 *Within its support* $[0, 1]$ *the function $\phi_{d,k}$ has the representation*

$$p_{d,k}(r) = \sum_{j=0}^{\ell+2k} d_{j,k}^{(\ell)} r^j$$

with $\ell = \lfloor d/2 \rfloor + k + 1$. The coefficients can be computed recursively for $0 \le s \le k-1$:

$$d_{j,0}^{(\ell)} = (-1)^j \binom{\ell}{j}, \qquad 0 \le j \le \ell,$$

$$d_{0,s+1}^{(\ell)} = \sum_{j=0}^{\ell+2s} \frac{d_{j,s}^{(\ell)}}{j+2}, \qquad d_{1,s+1}^{(\ell)} = 0, \qquad s \ge 0,$$

$$d_{j,s+1}^{(\ell)} = -\frac{d_{j-2,s}^{(\ell)}}{j}, \qquad s \ge 0, \qquad 2 \le j \le \ell + 2s + 2.$$

Furthermore, precisely the first k odd coefficients $d_{j,k}^{(\ell)}$ vanish.

These functions are not only of the form (9.1) but also of minimal degree and hence the answer to our initial question.

Theorem 9.13 *The functions $\phi_{d,k}$ are positive definite on $\mathbb{R}^d$ and are of the form*

$$\phi_{d,k}(r) = \begin{cases} p_{d,k}(r), & 0 \le r \le 1, \\ 0, & r > 1, \end{cases}$$

with a univariate polynomial $p_{d,k}$ of degree $\lfloor d/2 \rfloor + 3k + 1$. They possess continuous derivatives up to order 2k. They are of minimal degree for given space dimension d and smoothness 2k and are up to a constant factor uniquely determined by this setting.

Proof We already know that these functions are of the specific form. They are positive definite on $\mathbb{R}^d$, because we have for the Fourier transform $\mathcal{F}_d \phi_{d,k} = \mathcal{F}_{d+2k} \phi_{\lfloor d/2 \rfloor + k + 1}$ by Theorem 9.6. Moreover, the function $\phi_{\lfloor d/2 \rfloor + k + 1}$ is positive definite on $\mathbb{R}^{d+2k}$ by Theorem 6.20.

From Lemma 9.8 we know that $\phi_{d,k}$ possesses $2k$ continuous derivatives, which is the stated smoothness. The degree is given by Theorem 9.12.

Finally, suppose that there is a function ψ which is positive definite on $\mathbb{R}^d$, is of the form (9.1), and possesses $2k$ continuous derivatives. Assume further that ψ is of minimal degree. Then we can form the function $\tilde{\psi} := \mathcal{D}^k \phi$, which is still of the form (9.1) and

Table 9.1 *Compactly supported functions of minimal degree*

Space dimension	Function	Smoothness
	$\phi_{1,0}(r) = (1-r)_+$	C^0
$d = 1$	$\phi_{1,1}(r) \doteq (1-r)_+^3(3r+1)$	C^2
	$\phi_{1,2}(r) \doteq (1-r)_+^5(8r^2+5r+1)$	C^4
	$\phi_{3,0}(r) = (1-r)_+^2$	C^0
$d \le 3$	$\phi_{3,1}(r) \doteq (1-r)_+^4(4r+1)$	C^2
	$\phi_{3,2}(r) \doteq (1-r)_+^6(35r^2+18r+3)$	C^4
	$\phi_{3,3}(r) \doteq (1-r)_+^8(32r^3+25r^2+8r+1)$	C^6
	$\phi_{5,0}(r) = (1-r)_+^3$	C^0
$d \le 5$	$\phi_{5,1}(r) \doteq (1-r)_+^5(5r+1)$	C^2
	$\phi_{5,2}(r) \doteq (1-r)_+^7(16r^2+7r+1)$	C^4

is at least continuous by Lemma 9.8. Furthermore, $\tilde{\psi}$ is positive definite on $\mathbb{R}^{d+2k}$. Thus $\tilde{\psi}$ must possess at least $\lfloor d/2 \rfloor + k$ continuous derivatives at 1, which means that $\tilde{\psi}(r) = (1-r)_+^{\lfloor d/2 \rfloor + k + 1} q(r)$ with a polynomial q. But, because of the minimal degree, q must be a constant. Thus $\psi = I^k \tilde{\psi}$ and $\phi_{d,k}$ can differ only by a constant factor. □

For convenience, we list the simplest cases in Table 9.1, where $\doteq$ denotes equality up to a positive constant factor. We use this notation also in the next corollary, in which we give the explicit form for the most important cases.

Corollary 9.14 *The functions $\phi_{d,k}$, $k = 0, 1, 2, 3$ have the following form:*

$$\begin{aligned}
\phi_{d,0}(r) &= (1-r)_+^{\lfloor d/2 \rfloor + 1}, \\
\phi_{d,1}(r) &\doteq (1-r)_+^{\ell+1}[(\ell+1)r + 1], \\
\phi_{d,2}(r) &\doteq (1-r)_+^{\ell+2}[(\ell^2 + 4\ell + 3)r^2 + (3\ell + 6)r + 3], \\
\phi_{d,3}(r) &\doteq (1-r)_+^{\ell+3}[(\ell^3 + 9\ell^2 + 23\ell + 15)r^3 \\
&\qquad + (6\ell^2 + 36\ell + 45)r^2 + (15\ell + 45)r + 15],
\end{aligned}$$

where we have used $\ell := \lfloor d/2 \rfloor + k + 1$.

Proof The form for $k = 0$ is obvious. For $k = 1$ we have to apply $\mathcal{I}$ once. A simple computation shows for $r \in [0, 1]$ that

$$\phi_{d,1}(r) = I\phi_\ell(r) = \int_r^1 t(1-t)^\ell dt = \frac{(1-r)^{\ell+1}}{(\ell+1)(\ell+2)}[(\ell+1)r + 1].$$

The other cases are dealt with in the same spirit. □

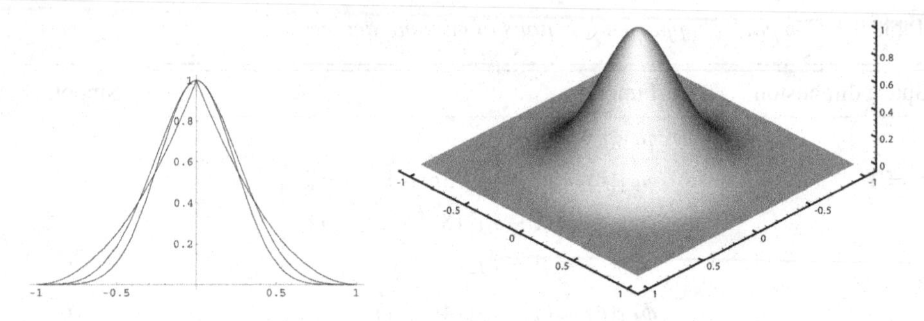

Fig. 9.1 The C^0-, C^2-, and C^4-function for $\mathbb{R}^3$ (on the left) and the C^2-, function in $\mathbb{R}^2$ (on the right).

The left-hand part of Figure 9.1 shows the univariate functions $\phi_{3,0}$, $\phi_{3,1}$, and $\phi_{3,2}$, which are positive definite on $\mathbb{R}^d$ for $d \leq 3$; the right-hand part shows the C^2-function $\phi_{3,1}(r) = (1-r)^4(4r+1)$ in $\mathbb{R}^2$.

9.5 Generalizations

In the example following Lemma 9.8 we have already seen another class of compactly supported radial basis functions. While these basis functions are still of the form (9.1), one can also consider basis functions that do not have a polynomial representation within their support. The reason for choosing univariate polynomials was to get a simple evaluation. But obviously a simple evaluation can be achieved by other (univariate) functions also.

While it took rather a long time for the first compactly supported radial basis functions to be found, it is now quite simple to construct a variety of them by different tools. Besides the operators $\mathcal{I}$ and $\mathcal{D}$ we want to discuss two further techniques here. Both use existing positive definite functions to construct new ones.

The first idea is to apply an operator T to the basis function Φ. This operator has to be nonnegative in the Fourier domain, i.e. it has to satisfy $\widehat{T\Phi} \geq 0$ if $\widehat{\Phi} \geq 0$. Then the resulting function is also positive definite. If we are interested in compactly supported functions then we have to make sure that T respects the compact support of a function. The most important example is the classical Laplace operator.

Lemma 9.15 *Suppose that $\Phi \in C^2(\mathbb{R}^d) \cap L_1(\mathbb{R}^d)$ is positive definite. Denote by Δ the Laplace operator $\Delta = \sum_{j=1}^d \partial^2/\partial x_j^2$. If $-\Delta\Phi$ is integrable then it is also positive definite. If Φ is radial, so is $\Delta\Phi$. If Φ has compact support, so has $\Delta\Phi$.*

Proof Since all involved functions are integrable, the Fourier transform of $\Delta\Phi$ is given by $\widehat{\Delta\Phi} = -\|\cdot\|_2^2\widehat{\Phi}$. This proves the positive definiteness of $-\Delta\Phi$. The Laplace operator applied to a radial function is a radial operator. If $\Phi = \phi(\|\cdot\|_2)$ then

$$\Delta\Phi(x) = \phi''(\|x\|_2) + \frac{d-1}{\|x\|_2}\phi'(\|x\|_2).$$

The statement about the support is obvious. □

The second possible construction technique we want to mention is a specialization of the first one and is crucial to Bochner's and Schoenberg's characterizations. It uses the fact that if a positive definite function is integrated against a nonnegative measure then the result is also positive definite.

Proposition 9.16 *Suppose that $f \in L_1[0, \infty)$ is nonnegative and positive on a set U of positive Lebesgue measure. Suppose further that a bounded kernel $K : [0, \infty) \times [0, \infty) \to \mathbb{R}$ is given satisfying*

(1) $K(\cdot, r)$ is measurable for all $r \geq 0$,
(2) $K(t, \cdot)$ is positive semi-definite on $\mathbb{R}^d$ for all $t > 0$ and positive definite on $\mathbb{R}^d$ for $t \in U$.

Then

$$\phi(r) := \int_0^\infty K(t, r) f(t) dt \tag{9.6}$$

is positive definite on $\mathbb{R}^d$.

Proof Since $K(t, \cdot)$ is positive semi-definite for every $t > 0$, it is in particular continuous. Standard arguments yield that ϕ is also continuous. Finally, if pairwise distinct $x_1, \ldots, x_N \in \mathbb{R}^d$ and $\alpha \in \mathbb{R}^N \setminus \{0\}$ are given, we can see at once that

$$\sum_{j,k=1}^N \alpha_j\alpha_k\phi(\|x_j - x_k\|_2) = \int_0^\infty \sum_{j,k=1}^N \alpha_j\alpha_k K(t, \|x_j - x_k\|_2) f(t) dt \geq 0.$$

Actually, the quadratic form must be positive because otherwise the set of points where the integrand does not vanish must have measure zero. This is impossible since it contains the set U. □

A typical choice for K is $K(t, r) = \psi(r/t)$ with a compactly supported function ψ. The compact support makes K well defined. For example, one could use the compactly supported functions $\phi_{d,k}$.

Proposition 9.16 does not work only for kernels K such that $K(t, \cdot)$ is positive definite. It is also possible to relax this condition if the choices for the function f are further restricted. For example, in [34, 35] Buhmann makes the choices $K(t, r) = (1 - r^2/t)_+^\lambda$ and $f(t) = t^\alpha(1 - t^\delta)_+^\rho$. He discusses thoroughly all choices of the parameters α, δ, λ, and ρ such that ϕ from (9.6), which now takes the form

$$\phi(r) = \int_0^\infty (1 - r^2/t)_+^\lambda t^\alpha (1 - t^\delta)_+^\rho dt, \tag{9.7}$$

becomes positive definite. The result is

Theorem 9.17 *Let $0 < \delta \le \frac{1}{2}$, $\rho \ge 1$ be real numbers, and suppose that $\lambda \neq 0$ and α are real quantities with*

$$\lambda \in \begin{cases} (-\frac{1}{2}, \infty), & -1 < \alpha \le \min\left\{\frac{1}{2}, \lambda - \frac{1}{2}\right\}, & \text{if } d = 1 \\ [1, \infty), & -\frac{1}{2} < \alpha \le \frac{1}{2}\lambda, & \text{if } d = 1 \\ (-\frac{1}{2}, \infty), & -1 < \alpha \le \min\left\{\frac{1}{2}(\lambda - \frac{1}{2}), \lambda - \frac{1}{2}\right\}, & \text{if } d = 2, \\ [0, \infty), & -1 < \alpha \le \frac{1}{2}(\lambda - 1), & \text{if } d = 3, \\ (\frac{1}{2}(d-5), \infty), & -1 < \alpha \le \frac{1}{2}\left[\lambda - \frac{1}{2}(d-1)\right], & \text{if } d > 3. \end{cases}$$

Then the radial basis function (9.7) gives rise to a positive definite function on $\mathbb{R}^d$.

9.6 Notes and comments

Astonishingly, it needed quite some time for compactly supported radial basis functions to be found. Everything started with the explicit construction of 'Euclid's hat' in the present author's thesis [189], see also Schaback [162]; this is nothing other than the d-variate convolution of the characteristic function of the unit ball with itself. A little earlier, Narcowich and Ward had used this function in [145] in a different context and without an explicit form.

The construction of the compactly supported functions of minimal degree was done initially by the present author in [191] and partially published afterwards in [190, 192]. Nowadays, they are often simply called "Wendland's functions". A basis for these results was given by the earlier publications by Chanysheva [40], Askey [6], and Gasper [68].

The operators $\mathcal{I}$ and $\mathcal{D}$ have become known to the radial basis function community through Wu's paper [203], but it seems that these operators have been known longer in the field of probability theory. In particular, Matheron [116] called them *montée* and *descentée*.

10
Native spaces

So far we have encountered positive definite functions in the context of a scattered data interpolation problem in $\mathbb{R}^d$. In this chapter we want to take another point of view, which also prepares us for the error analysis of the interpolation process. Our approach is motivated by the following example. The Sobolev spaces on $\mathbb{R}^d$ can be defined by

$$H^s(\mathbb{R}^d) = \{f \in L_2(\mathbb{R}^d) : \widehat{f}(\cdot)(1 + \|\cdot\|_2^2)^{s/2} \in L_2(\mathbb{R}^d)\}.$$

They can be equipped with an inner product

$$(f, g)_{H^s(\mathbb{R}^d)} := (2\pi)^{-d/2} \int_{\mathbb{R}^d} \widehat{f}(\omega)\overline{\widehat{g}(\omega)}(1 + \|\omega\|_2^2)^s d\omega.$$

By the Sobolev embedding theorem it is well known that for $s > d/2$ the inclusion $H^s(\mathbb{R}^d) \subseteq C(\mathbb{R}^d)$ holds, or, to be more precise, that every equivalence class in $H^s(\mathbb{R}^d)$ contains a continuous representer. We will always interpret $H^s(\mathbb{R}^d)$ as a set of continuous functions in this way. A closer look at the inner product shows that it contains a nonnegative weight function. In the case $s > d/2$ this weight function can be used to define a positive definite function Φ by $\widehat{\Phi}(\omega) = (1 + \|\omega\|_2^2)^{-s}$. Actually, we know by Theorem 6.13 that Φ is given by

$$\Phi(x) = \frac{2^{1-s}}{\Gamma(s)} \|x\|_2^{s-d/2} K_{d/2-s}(\|x\|_2).$$

Formal computations, which will be justified later on, give

$$(f, \Phi(\cdot - x))_{H^s(\mathbb{R}^d)} = (2\pi)^{-d/2} \int_{\mathbb{R}^d} \frac{\widehat{f}(\omega)\overline{\widehat{\Phi}(\omega)e^{-ix^T\omega}}}{\widehat{\Phi}(\omega)} d\omega = f(x).$$

This reproducing property emphasizes the role of the function Φ for the Sobolev space and we will focus on it in the next section.

10.1 Reproducing-kernel Hilbert spaces

We are interested in vector spaces $\mathcal{F}$ consisting of functions $f : \Omega \to \mathbb{R}$ defined on a region $\Omega \subseteq \mathbb{R}^d$. The region Ω can be quite arbitrary except that it should contain at least one point. We consider only real vector spaces of real-valued functions. Very soon we will see that on the one hand they correspond to real-valued positive semi-definite kernels and that on the other hand every real-valued positive definite kernel leads naturally to a real Hilbert space of real-valued functions. To include complex-valued kernels also, we would have to discuss complex-valued function spaces. There are three reasons for not doing so. First, both cases can be handled in a very similar way. The only difference is that in the complex case special care has to be taken with the complex conjugate sign. Second, this time there is no fundamental gain in a complex setting. While complex-valued functions were indeed useful to derive Bochner's and related results, the theory of reproducing-kernel Hilbert spaces does not benefit from them. The main reason, however, is that all relevant positive definite functions are real-valued (because they are radial) and hence their associated function spaces are also real spaces of real-valued functions. Nonetheless, we will comment on the complex situation when appropriate and the capable reader will have no problem in stating and proving the corresponding results.

Definition 10.1 *Let $\mathcal{F}$ be a real Hilbert space of functions $f : \Omega \to \mathbb{R}$. A function $\Phi : \Omega \times \Omega \to \mathbb{R}$ is called a reproducing kernel for $\mathcal{F}$ if*

(1) $\Phi(\cdot, y) \in \mathcal{F}$ for all $y \in \Omega$,
(2) $f(y) = (f, \Phi(\cdot, y))_{\mathcal{F}}$ for all $f \in \mathcal{F}$ and all $y \in \Omega$.

The reproducing kernel of a Hilbert space is uniquely determined. Suppose there are two reproducing kernels Φ_1 and Φ_2. Then property (2) gives $(f, \Phi_1(\cdot, y) - \Phi_2(\cdot, y))_{\mathcal{F}} = 0$ for all $f \in \mathcal{F}$ and all $y \in \Omega$. Setting $f = \Phi_1(\cdot, y) - \Phi_2(\cdot, y)$ for a fixed y shows the uniqueness.

Let us give a first characterization of a Hilbert function space with a reproducing kernel.

Theorem 10.2 *Suppose that $\mathcal{F}$ is a Hilbert space of functions $f : \Omega \to \mathbb{R}$. Then the following statements are equivalent:*

(1) the point evaluation functionals are continuous, i.e. $\delta_y \in \mathcal{F}^*$ for all $y \in \Omega$;
(2) $\mathcal{F}$ has a reproducing kernel.

Proof Suppose that the point evaluation functionals are continuous. By Riesz' representation theorem we find, for every $y \in \Omega$, a $\Phi_y \in \mathcal{F}$ such that $\delta_y(f) = (f, \Phi_y)_{\mathcal{F}}$ for all $f \in \mathcal{F}$. Thus, $\Phi(x, y) := \Phi_y(x)$ is the reproducing kernel of $\mathcal{F}$. Now suppose that $\mathcal{F}$ has a reproducing kernel Φ. This means that $\delta_y = (\cdot, \Phi(\cdot, y))_{\mathcal{F}}$ for $y \in \Omega$. Since the inner product is continuous, so is δ_y. □

A reproducing-kernel Hilbert space has many special features. We collect some of them now.

Theorem 10.3 *Suppose $\mathcal{F}$ is a Hilbert space of functions $f : \Omega \to \mathbb{R}$ with reproducing kernel Φ. Then we have*

(1) $\Phi(x, y) = (\Phi(\cdot, x), \Phi(\cdot, y))_{\mathcal{F}} = (\delta_x, \delta_y)_{\mathcal{F}^*}$ for $x, y \in \Omega$,
(2) $\Phi(x, y) = \Phi(y, x)$ for $x, y \in \Omega$,
(3) if $f, f_n \in \mathcal{F}$, $n \in \mathbb{N}$, are given such that f_n converges to f in the Hilbert space norm then f_n also converges pointwise to f.

Proof The Riesz' representation $F : \mathcal{F}^* \to \mathcal{F}$ reduces for point evaluations to $F(\delta_y) = \Phi(\cdot, y)$ because of the reproducing-kernel properties. This means that

$$(\delta_x, \delta_y)_{\mathcal{F}^*} = (F(\delta_x), F(\delta_y))_{\mathcal{F}} = (\Phi(\cdot, x), \Phi(\cdot, y))_{\mathcal{F}}.$$

Furthermore,

$$\Phi(x, y) = \delta_x(\Phi(\cdot, y)) = (\Phi(\cdot, y), \Phi(\cdot, x))_{\mathcal{F}} = (\Phi(\cdot, x), \Phi(\cdot, y))_{\mathcal{F}}.$$

Hence, property (1) is proven. Property (2) follows immediately from Property (1). Property (3) is a consequence of

$$|f_n(x) - f(x)| = |(f_n - f, \ \Phi(\cdot, x))_{\mathcal{F}}| \le \|f_n - f\|_{\mathcal{F}} \|\Phi(\cdot, x)\|_{\mathcal{F}}.$$

□

Our next result discloses the connection between reproducing-kernel Hilbert spaces and positive definite kernels.

Theorem 10.4 *Suppose that $\mathcal{F}$ is a reproducing-kernel Hilbert function space with reproducing kernel $\Phi : \Omega \times \Omega \to \mathbb{R}$. Then Φ is positive semi-definite. Moreover, Φ is positive definite if and only if the point evaluation functionals are linearly independent in $\mathcal{F}^*$.*

Proof Since the kernel Φ is real-valued and symmetric, we can restrict ourselves to real coefficients in the quadratic form. For pairwise distinct $x_1, \dots, x_N$ and $\alpha \in \mathbb{R}^N \setminus \{0\}$ we have

$$\sum_{j=1}^{N} \sum_{k=1}^{N} \alpha_j \alpha_k \Phi(x_j, x_k) = \left(\sum_{j=1}^{N} \alpha_j \delta_{x_j}, \sum_{k=1}^{N} \alpha_k \delta_{x_k} \right)_{\mathcal{F}^*} = \left\| \sum_{j=1}^{N} \alpha_j \delta_{x_j} \right\|_{\mathcal{F}^*}^2 \ge 0.$$

The last expression can and will only be zero if the point evaluation functionals are linearly dependent. □

Hence, the reproducing kernel of a function space $\mathcal{F}$ leads to a real-valued positive semi-definite kernel. Obviously, if the function space $\mathcal{F}$ is a complex vector space containing complex-valued functions then everything said so far remains true with mild modifications. In particular the reproducing-kernel is now a complex-valued positive semi-definite function.

From the first property of Definition 10.1 we know that $\mathcal{F}$ contains all functions of the form $f = \sum_{j=1}^{N} \alpha_j \Phi(\cdot, x_j)$ if $x_j \in \Omega$. Furthermore, we know that

$$\|f\|_{\mathcal{F}}^2 = \sum_{j=1}^{N} \sum_{k=1}^{N} \alpha_j \alpha_k (\Phi(\cdot, x_j), \Phi(\cdot, x_k))_{\mathcal{F}} = \sum_{j=1}^{N} \sum_{k=1}^{N} \alpha_j \alpha_k \Phi(x_j, x_k).$$

This feature will be used to construct a reproducing-kernel Hilbert space for a given positive definite kernel in the next section.

But before this we want to have a look at how invariance properties of the space influence the kernel.

Definition 10.5 *Let $\mathcal{T}$ be a group of transformations $T : \Omega \to \Omega$. We say $\mathcal{F}$ is invariant under the group $\mathcal{T}$ if*

(1) $f \circ T \in \mathcal{F}$ for all $f \in \mathcal{F}$ and $T \in \mathcal{T}$,
(2) $(f \circ T, g \circ T)_{\mathcal{F}} = (f, g)_{\mathcal{F}}$ for all $f, g \in \mathcal{F}$ and all $T \in \mathcal{T}$.

The invariance of the function space is inherited by the kernel.

Theorem 10.6 *Suppose that the reproducing-kernel Hilbert function space $\mathcal{F}$ is invariant under the transformations of $\mathcal{T}$; then the reproducing kernel Φ satisfies*

$$\Phi(Tx, Ty) = \Phi(x, y)$$

for all $x, y \in \Omega$ and all $T \in \mathcal{T}$.

Proof From the reproducing-kernel properties and the invariance properties we can read off

$$f(y) = f \circ T^{-1}(Ty) = (f \circ T^{-1}, \Phi(\cdot, Ty))_{\mathcal{F}} = (f, \Phi(T\cdot, Ty))_{\mathcal{F}}, \qquad y \in \Omega.$$

Of course, we also have $f(y) = (f, \Phi(\cdot, y))_{\mathcal{F}}$, and hence the uniqueness of the reproducing kernel gives $\Phi(\cdot, y) = \Phi(T\cdot, Ty)$ for all $y \in \Omega$. □

The following examples show that radial basis functions arise quite naturally within the concept of reproducing kernels. For our first example suppose that $\Omega = \mathbb{R}^d$. Let $\mathcal{T}$ be the group of translations on $\mathbb{R}^d$. If we choose the translation $T\xi = x - \xi$ for a fixed $x \in \mathbb{R}^d$ then Theorem 10.6 shows that

$$\Phi(x, y) = \Phi(0, x - y) =: \Phi_0(x - y),$$

i.e. the kernel is translation invariant.

For our second example suppose that Ω is still $\mathbb{R}^d$ but $\mathcal{T}$ consists now of the translations and the orthogonal transformations. Then we can choose an orthogonal transformation $A \in \mathbb{R}^{d\times d}$ such that $A\xi = \|\xi\|_2 e_1$, where e_1 is the first unit vector in $\mathbb{R}^d$ and $\xi = x - y$. Hence

$$\Phi(x, y) = \Phi(Ax, Ay) = \Phi_0(A(x - y)) = \Phi_0(\|x - y\|_2 e_1) =: \phi(\|x - y\|_2),$$

i.e. Φ is radial.

10.2 Native spaces for positive definite kernels

So far, we have seen that a positive definite function or kernel appears naturally as the reproducing kernel of a Hilbert function space. But since we normally do not start with a function space but with a positive definite kernel we are confronted by the problem of finding the associated function space that has this kernel as its reproducing kernel. In this section we want to solve this problem by constructing the corresponding Hilbert function space. Hence, assume that $\Phi : \Omega \times \Omega \to \mathbb{R}$ is a symmetric positive definite kernel.

Motivated by the second remark after the proof of Theorem 10.4 we define the $\mathbb{R}$-linear space

$$F_\Phi(\Omega) := \operatorname{span}\{\Phi(\cdot, y) \ : \ y \in \Omega\}$$

and equip this space with the bilinear form

$$\left(\sum_{j=1}^{N} \alpha_j \Phi(\cdot, x_j), \sum_{k=1}^{M} \beta_k \Phi(\cdot, y_k)\right)_\Phi := \sum_{j=1}^{N} \sum_{k=1}^{M} \alpha_j \beta_k \Phi(x_j, y_k).$$

Theorem 10.7 *If $\Phi : \Omega \times \Omega \to \mathbb{R}$ is a symmetric positive definite kernel then $(\cdot, \cdot)_\Phi$ defines an inner product on $F_\Phi(\Omega)$. Furthermore, $F_\Phi(\Omega)$ is a pre-Hilbert space with reproducing kernel Φ.*

Proof Obviously $(\cdot, \cdot)_\Phi$ is bilinear and symmetric because Φ is symmetric. Moreover, if we choose an arbitrary function $f = \sum_{j=1}^N \alpha_j \Phi(\cdot, x_j) \not\equiv 0$ from $F_\Phi(\Omega)$ we find that

$$(f, f)_\Phi = \sum_{j=1}^{N} \sum_{k=1}^{N} \alpha_j \alpha_k \Phi(x_j, x_k) > 0,$$

because Φ is positive definite. Finally, we obtain for this f

$$(f, \Phi(\cdot, y))_\Phi = \sum_{j=1}^{N} \alpha_j \Phi(x_j, y) = f(y),$$

which establishes also the reproducing kernel. □

The completion $\mathcal{F}_\Phi(\Omega)$ of this pre-Hilbert space with respect to the $\|\cdot\|_\Phi$-norm is the first candidate for a Hilbert function space with reproducing kernel Φ. But the elements of the completion are abstract elements and we have to interpret them as functions. Since the point-evaluation functionals are continuous on $F_\Phi(\Omega)$, their extensions to the completion remain continuous. This can be used to define function values for elements of the completion. To represent this connection we could use the sloppy notation

$$f(x) := (f, \Phi(\cdot, x))_\Phi$$

for every $f \in \mathcal{F}_\Phi(\Omega)$. But we want to repeat these arguments in the more technical situation of conditionally positive definite kernels and will therefore be more precise. Thus we define the linear mapping

$$R : \mathcal{F}_\Phi(\Omega) \to C(\Omega), \qquad R(f)(x) := (f, \Phi(\cdot, x))_\Phi.$$

The resulting functions are indeed continuous because

$$|Rf(x) - Rf(y)| = (f, \Phi(\cdot, x) - \Phi(\cdot, y))_\Phi \le \|f\|_\Phi \|\Phi(\cdot, x) - \Phi(\cdot, y)\|_\Phi$$

and

$$\|\Phi(\cdot, x) - \Phi(\cdot, y)\|_\Phi^2 = \Phi(x, x) + \Phi(y, y) - 2\Phi(x, y).$$

Furthermore, we have $Rf(x) = f(x)$ for all $x \in \Omega$ and all $f \in F_\Phi(\Omega)$.

Lemma 10.8 *The linear mapping $R : \mathcal{F}_\Phi(\Omega) \to C(\Omega)$ is injective.*

Proof $Rf = 0$ for an $f \in \mathcal{F}_\Phi(\Omega)$ would mean that $(f, \Phi(\cdot, x))_\Phi = 0$ for all $x \in \Omega$ or $f \perp F_\Phi(\Omega)$. But $\mathcal{F}_\Phi(\Omega)$ is the completion of $F_\Phi(\Omega)$. Hence, the only element from $\mathcal{F}_\Phi(\Omega)$ perpendicular to $F_\Phi(\Omega)$ is $f = 0$. □

After this technical interlude we are able to define the native Hilbert space of a positive definite kernel Φ.

Definition 10.9 *The native Hilbert function space corresponding to the symmetric positive definite kernel $\Phi : \Omega \times \Omega \to \mathbb{R}$ is defined by*

$$\mathcal{N}_\Phi(\Omega) := R(\mathcal{F}_\Phi(\Omega)).$$

It carries the inner product

$$(f, g)_{\mathcal{N}_\Phi(\Omega)} := (R^{-1} f, R^{-1} g)_\Phi.$$

Indeed, the space so defined is a Hilbert space of continuous functions on Ω with reproducing kernel Φ. Since $\Phi(\cdot, x)$ is an element of $F_\Phi(\Omega)$ for $x \in \Omega$ it is unchanged under R and hence

$$f(x) = (R^{-1} f, \Phi(\cdot, x))_\Phi = (f, \Phi(\cdot, x))_{\mathcal{N}_\Phi(\Omega)} \tag{10.1}$$

for all $f \in \mathcal{N}_\Phi(\Omega)$ and $x \in \Omega$.

Theorem 10.10 *Suppose that $\Phi : \Omega \times \Omega \to \mathbb{R}$ is a symmetric positive definite kernel. Then its associated native space $\mathcal{N}_\Phi(\Omega)$ is a Hilbert function space with reproducing kernel Φ.*

Hence, positive (semi-)definite kernels and reproducing kernels of Hilbert function spaces are the same thing. In the rest of this section we will discuss the uniqueness of the native space and will give a special characterization that is handier than the abstract definition given so far.

We already know by construction that $F_\Phi(\Omega) \subseteq \mathcal{N}_\Phi(\Omega)$ is dense in $\mathcal{N}_\Phi(\Omega)$ and that

$$\|f\|_\Phi = \|f\|_{\mathcal{N}_\Phi(\Omega)} \qquad \text{for all } f \in F_\Phi(\Omega).$$

These last two properties make the native space unique in the following sense.

Theorem 10.11 *Suppose that Φ is a symmetric positive definite kernel. Suppose further that $\mathcal{G}$ is a Hilbert space of functions $f : \Omega \to \mathbb{R}$ with reproducing kernel Φ. Then $\mathcal{G}$ is the native space $\mathcal{N}_\Phi(\Omega)$ and the inner products are the same.*

Proof From the second remark after the proof of Theorem 10.4 we know that $F_\Phi(\Omega) \subseteq \mathcal{G}$ and $\|f\|_{\mathcal{G}} = \|f\|_{\mathcal{N}_\Phi(\Omega)}$ for all $f \in F_\Phi(\Omega)$. Now take $f \in \mathcal{N}_\Phi(\Omega)$. Then there exists a

Cauchy sequence $\{f_n\} \subseteq F_\Phi(\Omega)$ converging to f in $\mathcal{N}_\Phi(\Omega)$ (remember that $R^{-1} f_n = f_n$). By Theorem 10.3 we know that f is given pointwise by $f(y) = \lim_{n\to\infty} f_n(y)$. However, f_n is also a Cauchy sequence in $\mathcal{G}$ that converges to a $g \in \mathcal{G}$. But now the reproducing-kernel property of $\mathcal{G}$ gives $g(x) = \lim_{n\to\infty} f_n(x)$. Thus $f = g \in \mathcal{G}$, which means that $\mathcal{N}_\Phi(\Omega) \subseteq \mathcal{G}$ with $\|f\|_{\mathcal{N}_\Phi(\Omega)} = \|f\|_{\mathcal{G}}$ for all $f \in \mathcal{N}_\Phi(\Omega)$. Next, suppose that $\mathcal{N}_\Phi(\Omega)$ does not equal $\mathcal{G}$. Since $\mathcal{N}_\Phi(\Omega)$ is closed we can then find an element $g \in \mathcal{G} \setminus \{0\}$ orthogonal to $\mathcal{N}_\Phi(\Omega)$. But this means that $g(y) = (g, \Phi(\cdot, y))_{\mathcal{G}} = 0$ for all $y \in \Omega$, which contradicts $g \neq 0$. Finally, because the norms are the same so are the inner products, by polarization. □

This uniqueness result allows us to give another characterization of the native space in the case where $\Omega = \mathbb{R}^d$ and Φ is translation invariant. This result uses Fourier transforms and shows that the native space actually consists of smooth functions.

Theorem 10.12 *Suppose that $\Phi \in C(\mathbb{R}^d) \cap L_1(\mathbb{R}^d)$ is a real-valued positive definite function. Define*

$$\mathcal{G} := \left\{ f \in L_2(\mathbb{R}^d) \cap C(\mathbb{R}^d) \; : \; \widehat{f} \Big/ \sqrt{\widehat{\Phi}} \in L_2(\mathbb{R}^d) \right\}$$

and equip this space with the bilinear form

$$(f, g)_{\mathcal{G}} := (2\pi)^{-d/2} \left(\widehat{f} / \sqrt{\widehat{\Phi}}, \widehat{g} / \sqrt{\widehat{\Phi}} \right)_{L_2(\mathbb{R}^d)} = (2\pi)^{-d/2} \int_{\mathbb{R}^d} \frac{\widehat{f}(\omega)\overline{\widehat{g}(\omega)}}{\widehat{\Phi}(\omega)} d\omega.$$

Then $\mathcal{G}$ is a real Hilbert space with inner product $(\cdot, \cdot)_{\mathcal{G}}$ and reproducing kernel $\Phi(\cdot - \cdot)$. Hence $\mathcal{G}$ is the native space of Φ on $\mathbb{R}^d$, i.e. $\mathcal{G} = \mathcal{N}_\Phi(\mathbb{R}^d)$, and both inner products coincide. In particular, every $f \in \mathcal{N}_\Phi(\mathbb{R}^d)$ can be recovered from its Fourier transform $\widehat{f} \in L_1(\mathbb{R}^d) \cap L_2(\mathbb{R}^d)$.

Note that we use the sloppy notation $\mathcal{N}_\Phi(\mathbb{R}^d)$ instead of the precise notation $\mathcal{N}_{\Phi(\cdot - \cdot)}(\mathbb{R}^d)$, for simplicity. This should cause no misunderstandings.

Proof From Corollary 6.12 we know that $\widehat{\Phi} \in L_1(\mathbb{R}^d)$. For $f \in \mathcal{G}$ this means in particular that $\widehat{f} \in L_1(\mathbb{R}^d)$ because

$$\int_{\mathbb{R}^d} |\widehat{f}(\omega)| d\omega \le \left(\int_{\mathbb{R}^d} \frac{|\widehat{f}(\omega)|^2}{\widehat{\Phi}(\omega)} d\omega \right)^{1/2} \left(\int_{\mathbb{R}^d} \widehat{\Phi}(\omega) d\omega \right)^{1/2}.$$

Hence Plancherel's theorem and the continuity of f and $(\widehat{f})^\vee$ allow us to recover f pointwise from its Fourier transform via

$$f(x) = (2\pi)^{-d/2} \int_{\mathbb{R}^d} \widehat{f}(\omega) e^{ix^T\omega} d\omega.$$

Since $(\cdot, \cdot)_{\mathcal{G}}$ is obviously $\mathbb{R}$-bilinear, it is an inner product if it is real-valued and positive definite. For a real-valued f the L_2-Fourier transform satisfies $\overline{\widehat{f}(x)} = \widehat{f}(-x)$ almost everywhere, so that

$$\begin{aligned}\int_{\mathbb{R}^d} \frac{\widehat{f}(\omega)\overline{\widehat{g}(\omega)}}{\widehat{\Phi}(\omega)} d\omega &= \int_{\omega_1>0} \left(\frac{\widehat{f}(\omega)\overline{\widehat{g}(\omega)}}{\widehat{\Phi}(\omega)} + \frac{\widehat{f}(-\omega)\overline{\widehat{g}(-\omega)}}{\widehat{\Phi}(-\omega)} \right) d\omega \\ &= \int_{\omega_1>0} \frac{\widehat{f}(\omega)\overline{\widehat{g}(\omega)} + \overline{\widehat{f}(\omega)}\widehat{g}(\omega)}{\widehat{\Phi}(\omega)} d\omega \\ &= 2\int_{\omega_1>0} \frac{\Re[\widehat{f}(\omega)\overline{\widehat{g}(\omega)}]}{\widehat{\Phi}(\omega)} d\omega \in \mathbb{R},\end{aligned}$$

proving that the bilinear form is indeed real-valued. As a weighted L_2-inner product, $(\cdot,\cdot)_{\mathcal{G}}$ is also positive definite.

Thus far we know that $\mathcal{G}$ is a pre-Hilbert space, and our next step is to prove that it is complete. For this reason suppose that $\{f_n\}$ is a Cauchy sequence in $\mathcal{G}$. This means that $\left\{\widehat{f_n}/\sqrt{\widehat{\Phi}}\right\}$ is a Cauchy sequence in $L_2(\mathbb{R}^d)$. Thus there exists a function $g \in L_2(\mathbb{R}^d)$ with $\widehat{f_n}/\sqrt{\widehat{\Phi}} \to g$ in $L_2(\mathbb{R}^d)$. From our initial assumptions we can conclude that $g\sqrt{\widehat{\Phi}} \in L_1(\mathbb{R}^d) \cap L_2(\mathbb{R}^d)$. Namely,

$$\int_{\mathbb{R}^d} \left| g(\omega)\sqrt{\widehat{\Phi}(\omega)} \right| d\omega \le \left(\int_{\mathbb{R}^d} |g(\omega)|^2 d\omega \right)^{1/2} \left(\int_{\mathbb{R}^d} \widehat{\Phi}(\omega) d\omega \right)^{1/2}$$

and

$$\int_{\mathbb{R}^d} \left| g(\omega)\sqrt{\widehat{\Phi}(\omega)} \right|^2 d\omega \le \|\widehat{\Phi}\|_{L_\infty(\mathbb{R}^d)} \|g\|^2_{L_2(\mathbb{R}^d)}.$$

Thus

$$f(x) := (2\pi)^{-d/2} \int_{\mathbb{R}^d} g(\omega)\sqrt{\widehat{\Phi}(\omega)} e^{ix^T\omega} d\omega, \qquad x \in \Omega,$$

is well defined, continuous, an element of $L_2(\mathbb{R}^d)$, and satisfies $\widehat{f}/\sqrt{\widehat{\Phi}} = g \in L_2(\mathbb{R}^d)$. Furthermore we have

$$\begin{aligned}|f(x) - f_n(x)| &\le (2\pi)^{-d/2} \int_{\mathbb{R}^d} \left| g(\omega)\sqrt{\widehat{\Phi}(\omega)} - \widehat{f_n}(\omega) \right| d\omega \\ &\le (2\pi)^{-d/2} \left\| g - \widehat{f_n}/\sqrt{\widehat{\Phi}} \right\|_{L_2(\mathbb{R}^d)} \|\widehat{\Phi}\|_{L_1(\mathbb{R}^d)},\end{aligned}$$

which means that f is also real-valued and hence that $f \in \mathcal{G}$. Finally,

$$\|f - f_n\|_{\mathcal{G}} = (2\pi)^{-d/4} \left\| \frac{\widehat{f}}{\sqrt{\widehat{\Phi}}} - \frac{\widehat{f_n}}{\sqrt{\widehat{\Phi}}} \right\|_{L_2(\mathbb{R}^d)} = (2\pi)^{-d/4} \left\| g - \frac{\widehat{f_n}}{\sqrt{\widehat{\Phi}}} \right\|_{L_2(\mathbb{R}^d)} \to 0$$

for $n \to \infty$. Thus $\mathcal{G}$ is complete. It remains to show that $\Phi(\cdot - \cdot)$ is the reproducing kernel of $\mathcal{G}$. First of all, Φ is bounded by $\Phi(0)$ and in $L_1(\mathbb{R}^d)$, so that it is also in $L_2(\mathbb{R}^d)$. Moreover,

$$\left\| \Phi(\cdot - y)^\wedge / \sqrt{\widehat{\Phi}} \right\|_{L_2(\mathbb{R}^d)} = \|\widehat{\Phi}\|_{L_1(\mathbb{R}^d)},$$

so that $\Phi(\cdot - y) \in \mathcal{G}$ for every $y \in \mathcal{G}$. The reproduction follows immediately from

$$\begin{aligned}(f, \Phi(\cdot - y))_{\mathcal{G}} &= (2\pi)^{-d/2} \int_{\mathbb{R}^d} \frac{\widehat{f}(\omega)\overline{\widehat{\Phi}(\omega)e^{-i\omega^T y}}}{\widehat{\Phi}(\omega)} d\omega \\ &= (2\pi)^{-d/2} \int_{\mathbb{R}^d} \widehat{f}(\omega)e^{i\omega^T y} d\omega \\ &= f(y).\end{aligned}$$

□

Because of the previous result, native spaces are in one way a generalization of Sobolev spaces. Let us clarify this. Remember that for $s > d/2$ the Sobolev space of order s is defined as

$$H^s(\mathbb{R}^d) = \left\{ f \in L_2(\mathbb{R}^d) \cap C(\mathbb{R}^d) \ : \ \widehat{f}(\cdot)(1 + \|\cdot\|_2^2)^{s/2} \in L_2(\mathbb{R}^d) \right\}.$$

Hence, if Φ has a Fourier transform that decays only algebraically then its native space is a Sobolev space.

Corollary 10.13 *Suppose that $\Phi \in L_1(\mathbb{R}^d) \cap C(\mathbb{R}^d)$ satisfies*

$$c_1(1 + \|\omega\|_2^2)^{-s} \le \widehat{\Phi}(\omega) \le c_2(1 + \|\omega\|_2^2)^{-s}, \qquad \omega \in \mathbb{R}^d$$

with $s > d/2$ and two positive constants $c_1 \le c_2$. Then the native space $\mathcal{N}_\Phi(\mathbb{R}^d)$ corresponding to Φ coincides with the Sobolev space $H^s(\mathbb{R}^d)$, and the native space norm and the Sobolev norm are equivalent.

10.3 Native spaces for conditionally positive definite kernels

As in the case of positive definite functions we generalize the notion of a conditionally positive definite function to a conditionally positive definite kernel.

Definition 10.14 *Suppose that $\mathcal{P}$ is a finite-dimensional subspace of $C(\Omega)$, $\Omega \subseteq \mathbb{R}^d$. A continuous symmetric kernel $\Phi : \Omega \times \Omega \to \mathbb{R}$ is said to be conditionally positive definite on Ω with respect to $\mathcal{P}$ if, for any N pairwise distinct centers $x_1, \ldots, x_N \in \Omega$ and all $\alpha \in \mathbb{R}^N \setminus \{0\}$ with*

$$\sum_{j=1}^{N} \alpha_j p(x_j) = 0 \tag{10.2}$$

for all $p \in \mathcal{P}$, the quadratic form

$$\sum_{j=1}^{N} \sum_{k=1}^{N} \alpha_j \alpha_k \Phi(x_j, x_k) \tag{10.3}$$

is positive.

The domain $\Omega \subseteq \mathbb{R}^d$ can still be quite arbitrary, except that now it should contain at least one $\mathcal{P}$-unisolvent subset. In contrast with the definition of a conditionally positive definite function we will allow a more general space $\mathcal{P}$. A conditionally positive definite function of order m is therefore a conditionally positive definite kernel with respect to $\mathcal{P} = \pi_{m-1}(\mathbb{R}^d)$. Again, we restrict ourselves here to real-valued kernels. Everything said about complex-valued kernels in the last section remains true.

We can proceed as in the case of a positive definite kernel to construct the native space of a conditionally positive kernel. Hence, we start with the linear space

$$F_\Phi(\Omega) := \left\{ \sum_{j=1}^N \alpha_j \Phi(\cdot, x_j) \; : \; N \in \mathbb{N},\, \alpha \in \mathbb{R}^N,\, x_1, \ldots, x_N \in \Omega, \right.$$
$$\left. \text{with } \sum_{j=1}^N \alpha_j p(x_j) = 0 \text{ for all } p \in \mathcal{P} \right\},$$

which becomes a pre-Hilbert space by introduction of the inner product

$$\left(\sum_{j=1}^N \alpha_j \Phi(\cdot, x_j), \sum_{k=1}^M \beta_k \Phi(\cdot, y_k) \right)_\Phi := \sum_{j=1}^N \sum_{k=1}^M \alpha_j \beta_k \Phi(x_j, y_k).$$

Note that the additional conditions in the definition of $F_\Phi(\Omega)$ ensure the definiteness of the inner product.

Continuing as in the case of positive definite kernels, we can form the Hilbert-space completion $\mathcal{F}_\Phi(\Omega)$ of $F_\Phi(\Omega)$ with respect to $(\cdot, \cdot)_\Phi$. But since $\Phi(\cdot, x)$ is in general not included in $F_\Phi(\Omega)$, we cannot use $f(x) = (f, \Phi(\cdot, x))_\Phi$ as in the case of positive definite kernels to derive that point evaluation functionals are continuous on $F_\Phi(\Omega)$ and thus on $\mathcal{F}_\Phi(\Omega)$. Hence we cannot interpret the abstract elements of $\mathcal{F}_\Phi(\Omega)$ as functions in this way. To arrive at an interpretation of the elements of $\mathcal{F}_\Phi(\Omega)$ as functions we have to make a detour.

To this end we choose a $\mathcal{P}$-unisolvent subset $\Xi = \{\xi_1, \ldots, \xi_Q\} \subseteq \Omega$ with $|\Xi| = \dim \mathcal{P} = Q$ elements and a Lagrange basis p_j, $1 \le j \le Q$, of $\mathcal{P}$ with respect to this set of centers. This choice will be fixed for the rest of this section. Next, we define a functional

$$\delta_{(x)} := \delta_x - \sum_{k=1}^Q p_k(x) \delta_{\xi_k}, \qquad x \in \Omega,$$

and a function

$$G(\cdot, x) := \delta_{(x)}^y \Phi(\cdot, y) = \Phi(\cdot, x) - \sum_{k=1}^Q p_k(x) \Phi(\cdot, \xi_k), \qquad x \in \Omega. \tag{10.4}$$

The functional obviously satisfies $\delta_{(x)}(p) = 0$ for all $p \in \mathcal{P}$; therefore it lies in $F_\Phi(\Omega)^*$ and so has a continuous extension to $\mathcal{F}_\Phi(\Omega)$. The function is obviously an element of $F_\Phi(\Omega)$. Furthermore, for $x \in \Omega$ we have the representation

$$\delta_{(x)}(f) = (f, G(\cdot, x))_\Phi, \qquad f \in F_\Phi(\Omega),$$

which extends to $\mathcal{F}_\Phi(\Omega)$ by continuity. Hence, if $\delta_{(x)}$ is now to play the role of δ_x we have to define the mapping

$$R : \mathcal{F}_\Phi(\Omega) \to C(\Omega), \qquad R(f)(x) := (f, G(\cdot, x))_\Phi.$$

The range of R is indeed a subset of $C(\Omega)$. By Cauchy–Schwarz we have again

$$|Rf(x) - Rf(y)| \le \|f\|_\Phi \|G(\cdot, x) - G(\cdot, y)\|_\Phi,$$

but this time with

$$\begin{aligned}\|G(\cdot, x) - G(\cdot, y)\|_\Phi^2 &= \Phi(x, x) + \Phi(y, y) - 2\Phi(x, y)\\ &\quad -2\sum_{j=1}^{Q}[p_j(x) - p_j(y)]\left[\Phi(x, \xi_j) - \Phi(y, \xi_j)\right]\\ &\quad +\sum_{j,k=1}^{Q}[p_k(x) - p_k(y)]\{p_j(x) - p_j(y)\}\Phi(\xi_j, \xi_k).\end{aligned}$$

Following the lines of the previous section, the next step is to show that R is injective.

Lemma 10.15 *The linear mapping $R : \mathcal{F}_\Phi(\Omega) \to C(\Omega)$ is injective.*

Proof $Rf = 0$ for $f \in \mathcal{F}_\Phi(\Omega)$ means that $(f, G(\cdot, x))_\Phi = 0$ for all $x \in \Omega$. Now, we choose an arbitrary $h = \sum_{j=1}^N \alpha_j \Phi(\cdot, x_j)$ from $F_\Phi(\Omega)$. For the coefficients of this element we have

$$\begin{aligned}\sum_{j=1}^{N}\alpha_j G(\cdot, x_j) &= \sum_{j=1}^{N}\alpha_j \Phi(\cdot, x_j) - \sum_{j=1}^{N}\alpha_j \sum_{k=1}^{Q}\Phi(\cdot, \xi_k)p_k(x_j)\\ &= h - \sum_{k=1}^{Q}\Phi(\cdot, \xi_k)\sum_{j=1}^{N}\alpha_j p_k(x_j)\\ &= h,\end{aligned}$$

showing that $(f, h)_\Phi = 0$ for all $h \in F_\Phi(\Omega)$. But because $\mathcal{F}_\Phi(\Omega)$ is the completion of $F_\Phi(\Omega)$ this means that $f = 0$. □

This allows us to interpret $\mathcal{F}_\Phi(\Omega)$ as a space of functions. But since $G(\cdot, x)$ and $\delta_{(x)}$ have the dispensable property $G(\cdot, \xi_k) = \delta_{(\xi_k)} \equiv 0$, $1 \le k \le Q$, we find that $Rf(\xi_k) = 0$, $1 \le k \le Q$, for all $f \in \mathcal{F}_\Phi(\Omega)$. However, $f(\xi_k)$ does not vanish for all $f \in F_\Phi(\Omega)$, so that it is not reasonable to take $R(\mathcal{F}_\Phi(\Omega))$ as the definition of the native space. The right choice comes from a closer look at how R acts on $F_\Phi(\Omega)$. Here, we find for $f = \sum \beta_j \Phi(\cdot, x_j)$ that

$$Rf(x) = (f, G(\cdot, x))_\Phi = f(x) - \sum_{k=1}^{Q} p_k(x) f(\xi_k) = f(x) - \Pi_{\mathcal{P}} f(x),$$

introducing the projection operator

$$\Pi_{\mathcal{P}} : C(\Omega) \to \mathcal{P}, \qquad \Pi_{\mathcal{P}}(f) = \sum_{k=1}^{Q} f(\xi_k) p_k. \tag{10.5}$$

Note that, with this operator, $G(\cdot, x)$ can also be written as $G(\cdot, x) = \Phi(\cdot, x) - \Pi_{\mathcal{P}}^x \Phi(\cdot, x)$.

Definition 10.16 *The native space corresponding to a symmetric kernel Φ that is conditionally positive definite on Ω with respect to $\mathcal{P}$ is defined by*

$$\mathcal{N}_\Phi(\Omega) := R(\mathcal{F}_\Phi(\Omega)) + \mathcal{P}.$$

The space is equipped with a semi-inner product via

$$(f, g)_{\mathcal{N}_\Phi(\Omega)} = (R^{-1}(f - \Pi_{\mathcal{P}} f),\ R^{-1}(g - \Pi_{\mathcal{P}} g))_\Phi.$$

Let us investigate this definition in more detail. First of all, the sum $R(\mathcal{F}_\Phi(\Omega)) + \mathcal{P}$ is direct because any element of $R(\mathcal{F}_\Phi(\Omega))$ vanishes on Ξ and this set is $\mathcal{P}$-unisolvent, which means that zero is the only element from $\mathcal{P}$ that vanishes there. From this fact we can also deduce that a general element $f = R(g) + p$ satisfies $f(\xi_k) = p(\xi_k)$, $1 \le k \le Q$, and hence can be written as $f = R(g) + \Pi_{\mathcal{P}}(f)$. This, again, means that $f - \Pi_{\mathcal{P}}(f)$ is an element of $R(\mathcal{F}_\Phi(\Omega))$ and that the semi-inner product is well defined. Obviously its null space is given by $\mathcal{P}$. Moreover, if $f \in F_\Phi(\Omega)$ then the definition ensures that $|f|_{\mathcal{N}_\Phi(\Omega)} = \|f\|_\Phi$.

Finally, we have the following Taylor expansion of $f \in \mathcal{N}_\Phi(\Omega)$, which can also be seen as the generalization of the reproducing-kernel property of positive definite kernels.

Theorem 10.17 *Suppose that $\Phi : \Omega \times \Omega \to \mathbb{R}$ is a symmetric kernel that is conditionally positive definite on Ω with respect to $\mathcal{P} \subseteq C(\Omega)$. Every $f \in \mathcal{N}_\Phi(\Omega)$ can be written as*

$$f(x) = \Pi_{\mathcal{P}} f(x) + (f, G(\cdot, x))_{\mathcal{N}_\Phi(\Omega)}$$

with the function G from (10.4) and the projector $\Pi_{\mathcal{P}}$ from (10.5).

Proof Since $G(\cdot, x) \in F_\Phi(\Omega)$ for every $x \in \Omega$ we have, as remarked earlier, $G(\cdot, x) = R^{-1}(G(\cdot, x) - \Pi_{\mathcal{P}} G(\cdot, x))$. From the definitions we can derive that

$$\begin{aligned} f(x) &= \Pi_{\mathcal{P}} f(x) + (R^{-1}(f - \Pi_{\mathcal{P}} f),\ G(\cdot, x))_\Phi \\ &= \Pi_{\mathcal{P}} f(x) + (R^{-1}(f - \Pi_{\mathcal{P}} f),\ R^{-1}(G(\cdot, x) - \Pi_{\mathcal{P}} G(\cdot, x)))_\Phi \\ &= \Pi_{\mathcal{P}} f(x) + (f, G(\cdot, x))_{\mathcal{N}_\Phi(\Omega)}. \end{aligned}$$

□

Note that in the case of a positive definite kernel both the definition of the native space in Definition 10.16 and the representation formula from Theorem 10.17 reduce to the definition and the reproducing-kernel formula given in Definition 10.9 and formula (10.1) since in this case $G(\cdot, x)$ reduces to Φ and $\mathcal{N}_\Phi(\Omega)$ to $R(\mathcal{F}_\Phi(\Omega))$.

The next theorem gives some more information on the connection with reproducing kernels. Even if $\mathcal{N}_\Phi(\Omega)$ does not have a reproducing kernel, the main part $R(\mathcal{F}_\Phi(\Omega))$ will do so.

Theorem 10.18 *The bilinear form* $(\cdot,\cdot)_{\mathcal{N}_\Phi(\Omega)}$ *is, on the space*

$$X_\Phi(\Omega) = \{f \in \mathcal{N}_\Phi(\Omega) : f(\xi_k) = 0, 1 \le k \le Q\} = R(\mathcal{F}_\Phi(\Omega)),$$

an inner product, which makes this space a Hilbert space. Moreover, this space has the reproducing kernel

$$\begin{aligned}\kappa(x,y) := \Phi(x,y) &- \sum_{k=1}^{Q} p_k(x)\Phi(\xi_k,y) - \sum_{\ell=1}^{Q} p_\ell(y)\Phi(x,\xi_\ell)\\ &+ \sum_{k=1}^{Q}\sum_{\ell=1}^{Q} p_k(x)p_\ell(y)\Phi(\xi_k,\xi_\ell),\end{aligned}$$

i.e. every $f \in X_\Phi(\Omega)$ *has the representation*

$$f(x) = (f, \kappa(\cdot,x))_{\mathcal{N}_\Phi(\Omega)}.$$

Proof We know that the linear mapping $R : \mathcal{F}_\Phi(\Omega) \to X_\Phi(\Omega)$ is isometric and bijective. Since $\mathcal{F}_\Phi(\Omega)$ is a Hilbert space, so is $X_\Phi(\Omega)$. In particular the bilinear form $(\cdot,\cdot)_{\mathcal{N}_\Phi(\Omega)}$ becomes an inner product on $X_\Phi(\Omega)$.

The kernel κ is a symmetric function and satisfies $\kappa(\cdot,y) = RG(\cdot,y)$ for all $y \in \Omega$. This shows that $\kappa(\cdot,y)$ is indeed an element of $X_\Phi(\Omega)$ for every $y \in \Omega$. Finally, for $f \in X_\Phi(\Omega)$ and $x \in \Omega$ we obtain

$$\begin{aligned}(f,\kappa(\cdot,x))_{\mathcal{N}_\Phi(\Omega)} &= (R^{-1}f, G(\cdot,x))_\Phi\\ &= \Pi_\mathcal{P} f(x) + (R^{-1}(f - \Pi_\mathcal{P} f), R^{-1}(G(\cdot,x) - \Pi_\mathcal{P} G(\cdot,x)))_\Phi\\ &= f(x)\end{aligned}$$

by Theorem 10.17. □

The previous theorem allows us to characterize native spaces for conditionally positive definite kernels in a way similar to the positive definite result in Theorem 10.11. As a consequence, we can derive a characterization based on (generalized) Fourier transforms comparable to the one in Theorem 10.12.

Proposition 10.19 *Suppose that* $\mathcal{G} \subseteq C(\Omega)$ *carries a semi-inner product* $(\cdot,\cdot)_\mathcal{G}$ *with null space* $\mathcal{P} \subseteq \mathcal{G}$ *such that* $\mathcal{G}_0 := \{g \in \mathcal{G} : g(\xi_k) = 0, 1 \le k \le Q\}$ *is a Hilbert space with reproducing kernel* κ. *Then* $\mathcal{G}$ *is the native space corresponding to* Φ *on* Ω.

Proof Since obviously $\mathcal{G} = \mathcal{G}_0 \oplus \mathcal{P}$ and $\mathcal{N}_\Phi(\Omega) = R(\mathcal{F}_\Phi(\Omega)) \oplus \mathcal{P}$ it suffices to show that $R(\mathcal{F}_\Phi(\Omega)) = \mathcal{G}_0$. This can be done in a way similar to that in the proof of Theorem 10.11.

First of all, for $f = \sum_{j=1}^N \alpha_j \Phi(\cdot, x_j) \in F_\Phi(\Omega)$ we have

$$Rf(x) = f(x) - \Pi_{\mathcal{P}} f(x) = \sum_{j=1}^N \alpha_j G(x, x_j) = \sum_{j=1}^N \alpha_j \kappa(x, x_j),$$

showing *that* $R(F_\Phi(\Omega)) \subseteq \mathcal{G}_0$. Moreover, for $f_1, f_2 \in R(F_\Phi(\Omega))$ it is true that $(f_1, f_2)_{\mathcal{N}_\Phi(\Omega)} = (f_1, f_2)_{\mathcal{G}}$.

For every $f \in \mathcal{F}_\Phi(\Omega)$ there exists a Cauchy sequence $(f_n) \subseteq R(F_\Phi(\Omega))$ satisfying in particular $f(x) = \lim_{n\to\infty} f_n(x)$ for all $x \in \Omega$. However, since f_n is also a Cauchy sequence in $\mathcal{G}_0$ there exists a limit $g \in \mathcal{G}_0$. From the continuity of the inner product we can derive again that $g(x) = \lim_{n\to\infty} f_n(x)$. Hence, $f \equiv g$. Finally, if $g \in \mathcal{G}_0$ is perpendicular to $R(\mathcal{F}_\Phi(\Omega))$ then all inner products $(g, \kappa(\cdot, x))_{\mathcal{G}}$, $x \in \Omega$, have to vanish. But this means that $g \equiv 0$. □

The native space carries only a semi-inner product. This semi-inner product has the null space $\mathcal{P}$. The following simple trick defines an inner product on $\mathcal{N}_\Phi(\Omega)$ and makes the native space a reproducing-kernel Hilbert space.

Theorem 10.20 *The native space $\mathcal{N}_\Phi(\Omega)$ corresponding to a conditionally positive definite kernel Φ carries the inner product*

$$(f, g) := (f, g)_{\mathcal{N}_\Phi(\Omega)} + \sum_{k=1}^Q f(\xi_k) g(\xi_k).$$

With this inner product $\mathcal{N}_\Phi(\Omega)$ becomes a reproducing-kernel Hilbert space with reproducing kernel

$$K(x, y) = \kappa(x, y) + \sum_{k=1}^Q p_k(x) p_k(y),$$

where κ is the kernel from Theorem 10.18.

Proof Obviously, the new inner product is symmetric, bilinear, and nonnegative. If

$$0 = (f, f) = (f, f)_{\mathcal{N}_\Phi(\Omega)} + \sum_{k=1}^Q |f(\xi_k)|^2$$

for an $f \in \mathcal{N}_\Phi(\Omega)$ then each summand has to be zero. But $(f, f)_{\mathcal{N}_\Phi(\Omega)} = 0$ means that $f \in \mathcal{P}$, and the additional information $f(\xi_k) = 0$ for $1 \le k \le Q$ coupled with the choice of the set Ξ leads to $f \equiv 0$. Hence, $(\cdot, \cdot)$ is positive definite. Let us come to the reproducing-kernel property. Since p_ℓ is a Lagrangian basis for Ξ and since $\kappa(\xi_k, \cdot) = 0$, we have

$$\begin{aligned}\sum_{k=1}^Q f(\xi_k) K(\xi_k, x) &= \sum_{k=1}^Q f(\xi_k) \kappa(\xi_k, x) + \sum_{k=1}^Q f(\xi_k) \sum_{\ell=1}^Q p_\ell(x) p_\ell(\xi_k) \\ &= \sum_{k=1}^Q f(\xi_k) p_k(x).\end{aligned}$$

Moreover, since $G(\cdot, x)$, $\kappa(\cdot, x)$, and $K(\cdot, x)$ differ only by a polynomial, Theorem 10.17 gives the representation

$$\begin{aligned} f(x) &= \Pi_{\mathcal{P}} f(x) + (f, \kappa(\cdot, x))_{\mathcal{N}_\Phi(\Omega)} \\ &= \sum_{k=1}^{Q} f(\xi_k) K(\xi_k, x) + (f, K(\cdot, x))_{\mathcal{N}_\Phi(\Omega)} \\ &= (f, K(\cdot, x)). \end{aligned}$$

□

In the case of positive definite functions, we have seen that the native space on $\mathbb{R}^d$ can be characterized by using Fourier transforms. Something similar is possible in the case of conditionally positive definite functions if we use generalized Fourier transforms instead.

Theorem 10.21 *Suppose that $\Phi \in C(\mathbb{R}^d)$ is an even conditionally positive definite function of order $m \in \mathbb{N}_0$. Suppose further that Φ has a generalized Fourier transform $\widehat{\Phi}$ of order m that is continuous on $\mathbb{R}^d \setminus \{0\}$. Let $\mathcal{G}$ be the real vector space consisting of all functions $f \in C(\mathbb{R}^d)$ that are slowly increasing and have a generalized Fourier transform $\widehat{f}$ of order $m/2$ that satisfies $\widehat{f}/\sqrt{\widehat{\Phi}} \in L_2(\mathbb{R}^d)$. Equip $\mathcal{G}$ with the symmetric bilinear form*

$$(f, g)_{\mathcal{G}} = (2\pi)^{-d/2} \int_{\mathbb{R}^d} \frac{\widehat{f}(\omega)\overline{\widehat{g}(\omega)}}{\widehat{\Phi}(\omega)} d\omega.$$

Then $\mathcal{G}$ is the native space corresponding to Φ, i.e. $\mathcal{G} = \mathcal{N}_\Phi(\mathbb{R}^d)$, and the semi-inner product $(\cdot, \cdot)_{\mathcal{N}_\Phi(\mathbb{R}^d)}$ coincides with the semi-inner product $(\cdot, \cdot)_{\mathcal{G}}$. Furthermore, every $f \in \mathcal{N}_\Phi(\Omega)$ has the representation

$$f(x) = \Pi_{\mathcal{P}} f(x) + (2\pi)^{-d/2} \int_{\mathbb{R}^d} \widehat{f}(\omega) \left(e^{ix^T\omega} - \sum_{k=1}^{Q} p_k(x) e^{i\xi_k^T\omega} \right) d\omega.$$

Proof Obviously $(\cdot, \cdot)_{\mathcal{G}}$ is $\mathbb{R}$-bilinear. It is also symmetric if it is real-valued. This can be shown in the same way as in the proof of Theorem 10.12 as long as we know that the generalized Fourier transform $\widehat{f}$ satisfies $\overline{\widehat{f}(\omega)} = \widehat{f}(-\omega)$ almost everywhere. Since this relation is satisfied for all real-valued test functions $\gamma \in \mathcal{S}_m$, the definition of a generalized Fourier transform and $\widehat{\gamma(-\cdot)}(\omega) = \widehat{\gamma}(-\omega)$ for such test functions ensures that

$$\begin{aligned} \int_{\mathbb{R}^d} \overline{\widehat{f}(\omega)} \gamma(\omega) d\omega &= \int_{\mathbb{R}^d} f(\omega) \overline{\widehat{\gamma}}(\omega) d\omega = \int_{\mathbb{R}^d} f(\omega) \widehat{\gamma}(-\omega) d\omega \\ &= \int_{\mathbb{R}^d} \widehat{f}(\omega) \gamma(-\omega) d\omega = \int_{\mathbb{R}^d} \widehat{f}(-\omega) \gamma(\omega) d\omega. \end{aligned}$$

The uniqueness of the generalized Fourier transform gives the stated relation.

Note that $\pi_{m-1}(\mathbb{R}^d) \subseteq \mathcal{G}$ is the null space of $(\cdot, \cdot)_{\mathcal{G}}$ by Proposition 8.10. With regard to Proposition 10.19 it remains to show that the space $\mathcal{G}_0 = \{g \in \mathcal{G} : g(\xi_k) = 0, 1 \le k \le Q\}$ is a Hilbert space of functions with reproducing kernel κ. This is done in three steps.

Step 1 $\kappa(\cdot, y)$ is an element of $\mathcal{G}_0$ for all $y \in \mathbb{R}^d$.
Obviously $\kappa(\xi_k, y) = 0$ for all $1 \le k \le Q$ and $y \in \mathbb{R}^d$. Next, $\kappa(\cdot, y)$ has the generalized Fourier transform

$$\kappa(\cdot, y)^\wedge(\omega) = \widehat{\Phi}(\omega)\left(e^{-iy^T\omega} - \sum_{k=1}^{Q} p_k(y)e^{-i\xi_k^T\omega}\right)$$

of order $m/2$. To see this we have to show that

$$\int_{\mathbb{R}^d} \kappa(\omega, y)\widehat{\gamma}(\omega)d\omega = \int_{\mathbb{R}^d} \kappa(\cdot, y)^\wedge(\omega)\gamma(\omega)d\omega$$

for every $\gamma \in \mathcal{S}_m$. By setting $\xi_0 = y$, $\alpha_0 = 1$, and $\alpha_k = -p_k(y)$, $1 \le k \le Q$, Lemma 8.11 yields

$$g(\omega) := \gamma(\omega)\left(e^{-iy^T\omega} - \sum_{k=1}^{Q} p_k(y)e^{-i\xi_k^T\omega}\right) = \gamma(\omega)\sum_{k=0}^{Q}\alpha_k e^{-i\xi_k^T\omega} \in \mathcal{S}_{2m}$$

for all $\gamma \in \mathcal{S}_m$. Moreover,

$$\widehat{g}(\omega) = \widehat{\gamma}(\omega + y) - \sum_{k=1}^{Q} p_k(y)\widehat{\gamma}(\omega + \xi_k).$$

Since $\kappa(\cdot, y) = G(\cdot, y) - \Pi_{\mathcal{P}} G(\cdot, y)$ this gives

$$\begin{aligned}
\int_{\mathbb{R}^d} \kappa(\omega, y)\widehat{\gamma}(\omega)d\omega &= \int_{\mathbb{R}^d}\left(\Phi(\omega - y) - \sum_{k=1}^{Q} p_k(y)\Phi(\omega - \xi_k)\right)\widehat{\gamma}(\omega)d\omega \\
&= \int_{\mathbb{R}^d} \Phi(\omega)\widehat{g}(\omega)d\omega = \int_{\mathbb{R}^d}\widehat{\Phi}(\omega)g(\omega)d\omega \\
&= \int_{\mathbb{R}^d}\widehat{\Phi}(\omega)\left(e^{-iy^T\omega} - \sum_{k=1}^{Q} p_k(y)e^{-i\xi_k^T\omega}\right)\gamma(\omega)d\omega.
\end{aligned}$$

Finally, $\kappa(\cdot, y)^\wedge/\sqrt{\widehat{\Phi}} \in L_2(\mathbb{R}^d)$ because

$$\begin{aligned}
\int_{\mathbb{R}^d}\frac{|\kappa(\cdot, y)^\wedge(\omega)|^2}{\widehat{\Phi}(\omega)}d\omega &= \int_{\mathbb{R}^d}\widehat{\Phi}(\omega)\left|e^{-iy^T\omega} - \sum_{k=1}^{Q} p_k(y)e^{-i\xi_k^T\omega}\right|^2 d\omega \\
&= \int_{\mathbb{R}^d}\widehat{\Phi}(\omega)\left|\sum_{k=0}^{Q}\alpha_k e^{-i\xi_k^T\omega}\right|^2 d\omega \\
&= (2\pi)^{d/2}\sum_{\ell,k=0}^{Q}\alpha_k\alpha_\ell\Phi(\xi_k - \xi_\ell) < \infty
\end{aligned}$$

by Corollary 8.13.

Step 2 κ is the reproducing kernel of $\mathcal{G}_0$.

To show this we use the test functions $g_\ell(x) = (\ell/\pi)^{d/2} e^{-\ell\|x\|_2^2}$ from Theorem 5.20 again. From our previous observations we know that the function

$$\gamma_\ell(\omega) := \left(e^{iy^T\omega} - \sum_{k=1}^{Q} p_k(\omega) e^{i\xi_k^T\omega} \right) \widehat{g}_\ell(\omega)$$

belongs to $\mathcal{S}_m$. Since $|\widehat{g}_\ell(\omega)| \le (2\pi)^{-d/2}$ for all $\omega \in \mathbb{R}^d$ and $\widehat{g}_\ell(\omega) \to (2\pi)^{-d/2}$ for $\ell \to \infty$ we can conclude that

$$\begin{aligned}
(f, \kappa(\cdot, y))_{\mathcal{G}} &= \lim_{\ell\to\infty} \int_{\mathbb{R}^d} \frac{\widehat{f}(\omega)\overline{\kappa(\cdot, y)^\wedge(\omega)}}{\widehat{\Phi}(\omega)} \widehat{g}_\ell(\omega) d\omega \\
&= \lim_{\ell\to\infty} \int_{\mathbb{R}^d} \widehat{f}(\omega) \left(e^{iy^T\omega} - \sum_{k=1}^{Q} p_k(y) e^{i\xi_k^T\omega} \right) \widehat{g}_\ell(\omega) d\omega \\
&= \lim_{\ell\to\infty} \int_{\mathbb{R}^d} \widehat{f}(\omega)\gamma_\ell(\omega) d\omega \\
&= \lim_{\ell\to\infty} \int_{\mathbb{R}^d} f(\omega)\widehat{\gamma}_\ell(\omega) d\omega \\
&= \lim_{\ell\to\infty} \int_{\mathbb{R}^d} f(\omega) \left(g_\ell(\omega - y) - \sum_{k=1}^{Q} p_k(y) g_\ell(\omega - \xi_k) \right) d\omega \\
&= f(y) - \sum_{k=1}^{Q} p_k(y) f(\xi_k) \\
&= f(y)
\end{aligned}$$

for all $f \in \mathcal{G}_0$.

Step 3 $\mathcal{G}_0$ is complete.

Suppose that $\{f_n\} \subseteq \mathcal{G}_0$ is a Cauchy sequence. Because

$$|f_n(x) - f_m(x)| \le \|f_n - f_m\|_{\mathcal{G}} \kappa(x, x)^{1/2}, \tag{10.6}$$

the sequence $\{f_n(x)\}$ is a Cauchy sequence in $\mathbb{R}$ and we can define a function f by $f(x) := \lim_{n\to\infty} f_n(x)$. On account of (10.6) the sequence f_n converges uniformly on compact subsets of $\mathbb{R}^d$ and hence the function f is continuous. Next, note that $\|f_n\|_{\mathcal{G}}$ is bounded since f_n is a Cauchy sequence in $\mathcal{G}_0$. From the fact that Φ is slowly increasing and from

$$\begin{aligned}
|f_n(x)|^2 &\le \|f_n\|_{\mathcal{G}}^2 \kappa(x, x) \\
&\le C \left(\Phi(0) - 2\sum_{k=1}^{Q} p_k(x)\Phi(x - \xi_k) + \sum_{j,k=1}^{Q} p_k(x) p_j(x) \Phi(\xi_j - \xi_k) \right),
\end{aligned}$$

we can deduce that f is also slowly increasing. Obviously f satisfies $f(\xi_k) = 0$ for $1 \le k \le Q$. Finally, since $\{f_n\}$ is a Cauchy sequence in $\mathcal{G}_0$, $\{\widehat{f}_n/\sqrt{\widehat{\Phi}}\}$ must be a Cauchy sequence in $L_2(\mathbb{R}^d)$ and must converge to a function $g \in L_2(\mathbb{R}^d)$. It remains to show that $g\sqrt{\widehat{\Phi}}$ is the generalized Fourier transform of f. To this end, choose $\gamma \in \mathcal{S}_m$. From

$$\begin{aligned}
&\int_{\mathbb{R}^d} \left| \left(\widehat{f}_n(\omega) - \sqrt{\widehat{\Phi}(\omega)} g(\omega) \right) \gamma(\omega) \right| d\omega \\
&= \int_{\mathbb{R}^d} \left| \frac{\widehat{f}_n(\omega)}{\sqrt{\widehat{\Phi}(\omega)}} - g(\omega) \right| \left| \sqrt{\widehat{\Phi}(\omega)} \gamma(\omega) \right| d\omega \\
&\le \left\| \frac{\widehat{f}_n}{\sqrt{\widehat{\Phi}}} - g \right\|_{L_2(\mathbb{R}^d)} \left(\int_{\mathbb{R}^d} \widehat{\Phi}(\omega) |\gamma(\omega)|^2 d\omega \right)^{1/2} \\
&\to 0 \qquad \text{for } n \to \infty,
\end{aligned}$$

we see that

$$\begin{aligned}
\int_{\mathbb{R}^d} f(\omega) \widehat{\gamma}(\omega) d\omega &= \lim_{n\to\infty} \int_{\mathbb{R}^d} f_n(\omega) \widehat{\gamma}(\omega) d\omega \\
&= \lim_{n\to\infty} \int_{\mathbb{R}^d} \widehat{f}_n(\omega) \gamma(\omega) d\omega \\
&= \int_{\mathbb{R}^d} \sqrt{\widehat{\Phi}(\omega)} g(\omega) \gamma(\omega) d\omega.
\end{aligned}$$

This, together with the obvious fact $\sqrt{\widehat{\Phi}}\, g \in L_2^{\mathrm{loc}}(\mathbb{R}^d \setminus \{0\})$ shows that the generalized Fourier transform of f is indeed $\sqrt{\widehat{\Phi}} g$. Obviously $\widehat{f}/\sqrt{\widehat{\Phi}} = g \in L_2(\mathbb{R}^d)$. Thus we have shown that $f \in \mathcal{G}_0$ and hence that $\mathcal{G}_0$ is complete. The given representation formula follows from the reproducing-kernel property and the Fourier transform of $\kappa(\cdot, y)$. □

10.4 Further characterizations of native spaces

With Theorems 10.12 and 10.21 we have already had two examples of equivalent representations of the native Hilbert space. In this section we want to give several other characterizations. We start with two characterizations for conditionally positive definite kernels, continuing to use the notation of the previous section. After that we will give a characterization valid only for positive definite kernels.

The first equivalent formulation we want to give is based on finitely supported linear functionals on $C(\Omega)$ that vanish on $\mathcal{P}$. To be more precise we form the set

$$L_{\mathcal{P}}(\Omega) := \left\{ \lambda_{N,\alpha,X} = \sum_{j=1}^{N} \alpha_j \delta_{x_j} \;:\; N \in \mathbb{N}, \alpha \in \mathbb{R}^N, x_1, \ldots, x_N \in \Omega, \right.$$
$$\left. \text{with } \lambda_{N,\alpha,X}(p) = 0 \text{ for all } p \in \mathcal{P} \right\}$$

and equip it with the inner product

$$(\lambda_{N,\alpha,X}, \lambda_{M,\beta,Y})_\Phi := \sum_{j=1}^{N} \sum_{k=1}^{M} \alpha_j \beta_k \Phi(x_j, y_k).$$

Obviously there is a one-to-one relation between $L_{\mathcal{P}}(\Omega)$ and $F_\Phi(\Omega)$, given simply by

$$L_{\mathcal{P}}(\Omega) \to F_\Phi(\Omega), \qquad \lambda \mapsto \lambda^x \Phi(\cdot, x),$$

where λ^x means action with respect to the variable x. This shows in particular that $\lambda(f) = (\lambda^x \Phi(\cdot, x), f)_\Phi$ for all $\lambda \in L_{\mathcal{P}}(\Omega)$ and all $f \in F_\Phi(\Omega)$, so that the norm defined on $L_{\mathcal{P}}$ gives the operator norm if $L_{\mathcal{P}}(\Omega)$ is interpreted as a subspace of the dual space of $F_\Phi(\Omega)$.

However, we do not want to exploit this relation explicitly any further; we shall use it whenever it is appropriate. Instead, we will look at the space of all functions on which these functionals are continuous.

Theorem 10.22 *Suppose that Φ is conditionally positive definite on Ω with respect to $\mathcal{P}$. Define*

$$\mathcal{G} = \{f \in C(\Omega) : |\lambda(f)| \le C_f \|\lambda\|_\Phi \text{ for all } \lambda \in L_{\mathcal{P}}(\Omega)\}.$$

This space carries the semi-norm

$$|f|_{\mathcal{G}} = \sup_{\substack{\lambda \in L_{\mathcal{P}}(\Omega) \\ \lambda \ne 0}} \frac{|\lambda(f)|}{\|\lambda\|_\Phi}.$$

Then we have $\mathcal{N}_\Phi(\Omega) = \mathcal{G}$ and both semi-norms are equal.

Proof Suppose that $f \in \mathcal{N}_\Phi(\Omega)$. To show that $f \in \mathcal{G}$ we first remark that for

$$G(\cdot, x) = \Phi(\cdot, x) - \sum_{k=1}^{Q} \Phi(\cdot, \xi_k) p_k(x)$$

and a general $\lambda \in L_{\mathcal{P}}(\Omega)$ we have

$$\lambda^x(G(\cdot, x)) = \lambda^x \Phi(\cdot, x) - \sum_{k=1}^{Q} \Phi(\cdot, \xi_k) \lambda(p_k) = \lambda^x \Phi(\cdot, x).$$

Using the reproduction formula $f(x) = \Pi_{\mathcal{P}} f(x) + (f, G(\cdot, x))_{\mathcal{N}_\Phi(\Omega)}$ given in Theorem 10.17 we find that

$$\begin{aligned} \lambda(f) &= \lambda(\Pi_{\mathcal{P}} f) + \lambda^x (f, G(\cdot, x))_{\mathcal{N}_\Phi(\Omega)} \\ &= (f, \lambda^x \Phi(\cdot, x))_{\mathcal{N}_\Phi(\Omega)} \\ &\le |f|_{\mathcal{N}_\Phi(\Omega)} \|\lambda\|_\Phi, \end{aligned}$$

and thus f is an element of $\mathcal{G}$. Furthermore we have established that $|f|_{\mathcal{G}} \le |f|_{\mathcal{N}_\Phi(\Omega)}$.

Now let us assume that f is an element of $\mathcal{G}$ and that we want to prove that f belongs to the native space. This f allows us to define a linear functional

$$F_f : F_\Phi(\Omega) \to \mathbb{R}, \qquad \lambda^x \Phi(\cdot, x) \mapsto \lambda(f),$$

which is continuous because of the definition of $\mathcal{G}$. Hence F_f has a continuous extension to $\mathcal{F}_\Phi(\Omega)$ and we can use Riesz' representation theorem to represent this extension by

$$F_f(g) = (g, Sf)_\Phi \qquad \text{for all } g \in \mathcal{F}_\Phi(\Omega).$$

Here Sf is the Riesz representer for this functional F_f. For proving $f \in \mathcal{N}_\Phi(\Omega)$ it suffices to show that f and $R(Sf)$ differ only by an element of $\mathcal{P}$. To see this, we use the definition $R(Sf)(x) = (Sf, G(\cdot, x))_\Phi$ and that for every $\mu \in L_\mathcal{P}(\Omega)$ we have $\mu^x G(\cdot, x) = \mu^x \Phi(\cdot, x)$. This gives

$$\begin{aligned}
\mu(f - RSf) &= \mu(f) - (Sf, \mu^x G(\cdot, x))_\Phi \\
&= \mu(f) - F_f(\mu^x G(\cdot, x)) \\
&= \mu(f) - F_f(\mu^x \Phi(\cdot, x)) \\
&= \mu(f) - \mu(f) \\
&= 0.
\end{aligned}$$

If we specify that $\mu = \delta_{(x)}$ then we end up with

$$f(x) = R(Sf)(x) + \sum_{k=1}^{Q} f(\xi_k) p_k(x).$$

Finally, since $Sf \in \mathcal{F}_\Phi(\Omega)$ we can choose a sequence $\lambda_j \in L_\mathcal{P}(\Omega)$ such that $\lambda_j^x \Phi(\cdot, x) \to Sf$ in $\mathcal{F}_\Phi(\Omega)$. Hence $\lambda_j(f) = (Sf, \lambda_j^x \Phi(\cdot, x))_\Phi \to \|Sf\|_\Phi^2$ and $\|\lambda_j\|_\Phi \to \|Sf\|_\Phi$ for $j \to \infty$. This allows us to make the bound

$$|f|_\mathcal{G} \geq \lim_{j\to\infty} \frac{|\lambda_j(f)|}{\|\lambda_j\|_\Phi} = \frac{\|Sf\|_\Phi^2}{\|Sf\|_\Phi} = |f|_{\mathcal{N}_\Phi(\Omega)}.$$

□

The previous result is not only interesting in its own right; it has also the following consequence.

Corollary 10.23 *The space $\mathcal{N}_\Phi(\Omega)$ is independent of the particular choice of $\Xi = \{\xi_1, \ldots, \xi_Q\}$. The semi-norms for any two different choices are equal.*

Our next characterization allows us to determine whether a function belongs to the native space, simply by looking at sequences of interpolants. Thus it is numerically applicable because it is based only on function values. For its proof we need a result that will also play an important role in a later chapter.

Lemma 10.24 *Suppose that $X = \{x_1, \ldots, x_N\} \subseteq \Omega$ is $\mathcal{P}$-unisolvent. Denote the unique interpolant, based on a conditionally positive definite kernel Φ and the set X, of a function $f \in \mathcal{N}_\Phi(\Omega)$ by $s_{f,X}$. Then we have*

$$(f - s_{f,X}, s)_{\mathcal{N}_\Phi(\Omega)} = 0$$

for every $s \in \text{span}\{\Phi(\cdot, x_j) : x_j \in X\} \cap F_\Phi(\Omega) + \mathcal{P}$. In particular, this gives

$$(f - s_{f,X}, s_{f,X})_{\mathcal{N}_\Phi(\Omega)} = 0.$$

Proof Any such function s can be written in the form $s = \lambda^x \Phi(\cdot, x) + \Pi_{\mathcal{P}}(s)$ with a certain linear functional $\lambda = \sum_{j=1}^N \alpha_j \delta_{x_j} \in L_{\mathcal{P}}(\Omega)$. The reproduction formula of Theorem 10.17 gives

$$(f - s_{f,X})(x) = \Pi_{\mathcal{P}}(f - s_{f,X})(x) + (f - s_{f,X},\, G(\cdot, x))_{\mathcal{N}_\Phi(\Omega)}, \qquad x \in \Omega.$$

If we use $\lambda^x G(\cdot, x) = \lambda^x \Phi(\cdot, x)$, the fact that λ vanishes on elements of $\mathcal{P}$, and the fact that $s_{f,X}$ interpolates f in x_j, yielding $\lambda(f - s_{f,X}) = 0$, we can conclude that

$$\begin{aligned} 0 &= \lambda(f - s_{f,X}) \\ &= \lambda(\Pi_{\mathcal{P}}(f - s_{f,X})) + (f - s_{f,X},\, \lambda^x \Phi(\cdot, x))_{\mathcal{N}_\Phi(\Omega)} \\ &= (f - s_{f,X},\, s)_{\mathcal{N}_\Phi(\Omega)}. \end{aligned}$$

Finally, $s_{f,X}$ satisfies the conditions imposed on s. □

For later reasons we state an immediate consequence.

Corollary 10.25 *In the situation of Lemma 10.24 we have the estimates* $|s_{f,X}|_{\mathcal{N}_\Phi(\Omega)} \le |f|_{\mathcal{N}_\Phi(\Omega)}$ *and* $|f - s_{f,X}|_{\mathcal{N}_\Phi(\Omega)} \le |f|_{\mathcal{N}_\Phi(\Omega)}$.

Proof According to Lemma 10.24 the functions $f - s_{f,X}$ and $s_{f,X}$ are mutually orthogonal. As a consequence we can deduce the Pythagorean law

$$|f - s_{f,X}|^2_{\mathcal{N}_\Phi(\Omega)} + |s_{f,X}|^2_{\mathcal{N}_\Phi(\Omega)} = |f|^2_{\mathcal{N}_\Phi(\Omega)},$$

which gives immediately both bounds. □

Theorem 10.26 *Let Φ be a conditionally positive definite kernel on Ω with respect to $\mathcal{P}$. Denote by $s_{f,X}$ the interpolant to a function $f \in C(\Omega)$ based on a $\mathcal{P}$-unisolvent X using Φ. Then f belongs to the native space $\mathcal{N}_\Phi(\Omega)$ if and only if there exists a constant c_f such that $|s_{f,X}|_{\mathcal{N}_\Phi(\Omega)} \le c_f$ for all $\mathcal{P}$-unisolvent $X \subseteq \Omega$. Moreover, in the case $f \in \mathcal{N}_\Phi(\Omega)$ the smallest possible constant c_f is given by $|f|_{\mathcal{N}_\Phi(\Omega)}$.*

Proof If $f \in \mathcal{N}_\Phi(\Omega)$, Corollary 10.25 shows that $|s_{f,X}|_{\mathcal{N}_\Phi(\Omega)} \le |f|_{\mathcal{N}_\Phi(\Omega)}$, which gives for such an f the upper bound $c_f \le |f|_{\mathcal{N}_\Phi(\Omega)}$ if c_f is the minimal choice. Next, let us assume $|s_{f,X}|_{\mathcal{N}_\Phi(\Omega)} \le c_f$ for all $\mathcal{P}$-unisolvent $X \subseteq \Omega$. For an arbitrary

$$\lambda_{N,\alpha,X} = \sum_{j=1}^N \alpha_j \delta_{x_j} \in L_{\mathcal{P}}(\Omega),$$

we choose a $\mathcal{P}$-unisolvent set $Y \supseteq X$ and let $s_{f,Y}$ be the interpolant on this set Y to f. Then $s_{f,Y}$ belongs to $\mathcal{N}_\Phi(\Omega)$ and we have $\lambda_{N,\alpha,X}(f - s_{f,Y}) = 0$. Thus we can make the estimate

$$\begin{aligned} |\lambda_{N,\alpha,X}(f)| &\le |\lambda_{N,\alpha,X}(f - s_{f,Y})| + |\lambda_{N,\alpha,X}(s_{f,Y})| \\ &\le \|\lambda_{N,\alpha,X}\|_\Phi \, |s_{f,Y}|_{\mathcal{N}_\Phi(\Omega)} \\ &\le c_f \|\lambda_{N,\alpha,X}\|_\Phi. \end{aligned}$$

As this holds for all $\lambda_{N,\alpha,X}$ we have $f \in \mathcal{N}_\Phi(\Omega)$ and $|f|_{\mathcal{N}_\Phi(\Omega)} \le c_f$ by Theorem 10.22. □

Our next characterization is only for positive definite and not for conditionally positive definite kernels. Moreover, we assume the set $\Omega \subseteq \mathbb{R}^d$ to be compact. We need some preparatory results on embeddings.

Lemma 10.27 *Suppose that $\Omega \subseteq \mathbb{R}^d$ is compact and Φ is a symmetric positive definite kernel on Ω. Then the native space $\mathcal{N}_\Phi(\Omega)$ has a continuous linear embedding into $L_2(\Omega)$.*

Proof Since Φ is the reproducing kernel of its native space we have

$$|f(x)|^2 = |(f, \Phi(\cdot, x))|^2_{\mathcal{N}_\Phi(\Omega)} \le \|f\|^2_{\mathcal{N}_\Phi(\Omega)} \|\Phi(\cdot, x)\|^2_{\mathcal{N}_\Phi(\Omega)} = \|f\|^2_{\mathcal{N}_\Phi(\Omega)} \Phi(x, x)$$

for every $f \in \mathcal{N}_\Phi(\Omega)$ and $x \in \Omega$. This implies that $\|f\|_{L_2(\Omega)} \le C\|f\|_{\mathcal{N}_\Phi(\Omega)}$ with $C^2 = \int_\Omega \Phi(x, x)dx$. The latter integral is finite because Φ is continuous and Ω is compact. □

Now we introduce the integral operator $T : L_2(\Omega) \to L_2(\Omega)$, defined by

$$Tv(x) := \int_\Omega \Phi(x, y)v(y)dy, \qquad v \in L_2(\Omega), \quad x \in \Omega. \tag{10.7}$$

Obviously Tv is continuous. But it is also an element of the native space.

Proposition 10.28 *Suppose that Φ is a symmetric positive definite kernel of the compact set $\Omega \subseteq \mathbb{R}^d$. Then the integral operator T maps $L_2(\Omega)$ continuously into the native space $\mathcal{N}_\Phi(\Omega)$. It is the adjoint of the embedding operator of the native space $\mathcal{N}_\Phi(\Omega)$ into $L_2(\Omega)$, i.e. it satisfies*

$$(f, v)_{L_2(\Omega)} = (f, Tv)_{\mathcal{N}_\Phi(\Omega)}, \qquad f \in \mathcal{N}_\Phi(\Omega), \quad v \in L_2(\Omega). \tag{10.8}$$

The range of T is dense in $\mathcal{N}_\Phi(\Omega)$.

Proof We use the characterization of Theorem 10.22 to show that $Tv \in \mathcal{N}_\Phi(\Omega)$. Hence we will pick an arbitrary $\lambda \in L(\Omega)$ (recall that we do not have any side conditions), and we see that

$$|\lambda(Tv)| \le \|v\|_{L_2(\Omega)} \|\lambda^x \Phi(\cdot, x)\|_{L_2(\Omega)}, \le C\|v\|_{L_2(\Omega)} \|\lambda\|_\Phi,$$

by Lemma 10.27. This gives $\|Tv\|_{\mathcal{N}_\Phi(\Omega)} \le C\|v\|_{L_2(\Omega)}$. To prove (10.8) we start with an $f \in F_\Phi(\Omega)$ and find, by the reproducing-kernel property, that

$$\begin{aligned}(f, v)_{L_2(\Omega)} &= \sum_{j=1}^N \alpha_j \int_\Omega \Phi(x, x_j)v(x)dx = \sum_{j=1}^N \alpha_j Tv(x_j)\\ &= \sum_{j=1}^N \alpha_j (Tv, \Phi(\cdot, x_j))_{\mathcal{N}_\Phi(\Omega)} = (f, Tv)_{\mathcal{N}_\Phi(\Omega)}\end{aligned}$$

for all $v \in L_2(\Omega)$. Since $F_\Phi(\Omega)$ is dense in $\mathcal{N}_\Phi(\Omega)$ and $\mathcal{N}_\Phi(\Omega)$ is continuously embedded into $L_2(\Omega)$ the general case follows by continuous extension. The final statement is a consequence of the general properties of adjoint mappings. The closure of the range of the operator T is the orthogonal complement of the kernel of its adjoint operator. But this is the whole space. □

It is well known that in our situation the operator $T : L_2(\Omega) \to L_2(\Omega)$ is a compact operator. Moreover, it satisfies

$$(Tv, v)_{L_2(\Omega)} = (Tv, Tv)_{\mathcal{N}_\Phi(\Omega)} \geq 0$$

for all $v \in L_2(\Omega)$. For such an operator, Mercer's theorem (see Pogorzelski [154] for example) guarantees the existence of a countable set of positive eigenvalues $\rho_1 \geq \rho_2 \geq \cdots > 0$ and continuous eigenfunctions $\{\varphi_n\}_{n\in\mathbb{N}} \subseteq L_2(\Omega)$ such that $T\varphi_n = \rho_n \varphi_n$. Furthermore, $\{\phi_n\}$ is an orthonormal basis for $L_2(\Omega)$ and the kernel Φ possesses the absolutely and uniformly convergent representation

$$\Phi(x, y) = \sum_{n=1}^{\infty} \rho_n \varphi_n(x) \varphi_n(y).$$

This allows us to derive our final characterization for native spaces.

Theorem 10.29 *Suppose Φ is a symmetric positive definite kernel on a compact set $\Omega \subseteq \mathbb{R}^d$. Then its native space is given by*

$$\mathcal{N}_\Phi(\Omega) = \left\{ f \in L_2(\Omega) : \sum_{n=1}^{\infty} \frac{1}{\rho_n} |(f, \varphi_n)_{L_2(\Omega)}|^2 < \infty \right\} \tag{10.9}$$

and the inner product has the representation

$$(f, g)_{\mathcal{N}_\Phi} = \sum_{n=1}^{\infty} \frac{1}{\rho_n} (f, \varphi_n)_{L_2(\Omega)} (g, \varphi_n)_{L_2(\Omega)}, \qquad f, g \in \mathcal{N}_\Phi(\Omega). \tag{10.10}$$

Proof Denote the set on the right-hand side of (10.9) by $\mathcal{G}$ and the inner product on the right-hand side of (10.10) by $(\cdot, \cdot)_\mathcal{G}$. We start by showing that $\mathcal{G} \subseteq \mathcal{N}_\Phi(\Omega)$ and $\|f\|_{\mathcal{N}_\Phi(\Omega)} \leq \|f\|_\mathcal{G}$ for all $f \in \mathcal{G}$. If $f \in \mathcal{G}$ is given, f is continuous because

$$\begin{aligned}
\sum_{n=1}^{\infty} |(f, \varphi_n)_{L_2(\Omega)} \varphi_n(x)| &\leq \left(\sum_{n=1}^{\infty} \frac{|(f, \varphi_n)_{L_2(\Omega)}|^2}{\rho_n} \right)^{1/2} \left(\sum_{n=1}^{\infty} \rho_n |\varphi_n(x)|^2 \right)^{1/2} \\
&= \|f\|_\mathcal{G} \sqrt{\Phi(x, x)}.
\end{aligned}$$

Moreover, for $\lambda = \sum_{j=1}^{N} \alpha_j \delta_{x_j} \in L(\Omega)$ we find that

$$\begin{aligned}
|\lambda(f)| &\leq \left(\sum_{n=1}^{\infty} \frac{1}{\rho_n} |(f, \varphi_n)_{L_2(\Omega)}|^2 \right)^{1/2} \left(\sum_{n=1}^{\infty} \rho_n |\lambda(\varphi_n)|^2 \right)^{1/2} \\
&= \|f\|_\mathcal{G} \left(\sum_{j,k=1}^{N} \alpha_j \alpha_k \sum_{n=1}^{\infty} \rho_n \varphi_n(x_j) \varphi_n(x_k) \right)^{1/2} \\
&= \|f\|_\mathcal{G} \left(\sum_{j,k=1}^{N} \alpha_j \alpha_k \Phi(x_j, x_k) \right)^{1/2} \\
&= \|f\|_G \|\lambda\|_\Phi,
\end{aligned}$$

which leads to the desired result by Theorem 10.22.

It remains to show that $\mathcal{N}_\Phi(\Omega) \subseteq \mathcal{G}$ and that $\|f\|_{\mathcal{G}} \le \|f\|_{\mathcal{N}_\Phi(\Omega)}$ for all $f \in \mathcal{N}_\Phi(\Omega)$. To achieve this, we start by looking at the dense subset $T(L_2(\Omega)) \subseteq \mathcal{N}_\Phi(\Omega)$. For an element $f = Tv$, $v \in L_2(\Omega)$, we can conclude from the L_2 expansion of v that $f = \sum_{n=1}^\infty \rho_n(v, \varphi_n)_{L_2(\Omega)}\varphi_n$, so that $(f, \varphi_n)_{L_2(\Omega)} = \rho_n(v, \varphi_n)_{L_2(\Omega)}$. This allows us to calculate its native space norm:

$$\begin{aligned}\|f\|^2_{\mathcal{N}_\Phi(\Omega)} = (v, Tv)_{L_2(\Omega)} &= \sum_{n=1}^{\infty}(v, \varphi_n)_{L_2(\Omega)}(f, \varphi_n)_{L_2(\Omega)}\\ &= \sum_{n=1}^{\infty} \frac{1}{\rho_n}|(f, \varphi_n)_{L_2(\Omega)}|^2,\end{aligned}$$

so that $\|\cdot\|_{\mathcal{N}_\Phi(\Omega)}$ and $\|\cdot\|_{\mathcal{G}}$ are the same on $T(L_2(\Omega))$. For an arbitrary $f \in \mathcal{N}_\Phi(\Omega)$ we choose a sequence $\{f_j\} \subseteq T(L_2(\Omega))$ with $\|f - f_j\|_{\mathcal{N}_\Phi(\Omega)} \to 0$ for $j \to \infty$. For $N, j \in \mathbb{N}$ we have the bound

$$\sum_{n=1}^{N} \frac{1}{\rho_n}|(f_j, \varphi_n)_{L_2(\Omega)}|^2 \le \|f_j\|_{\mathcal{N}_\Phi(\Omega)}.$$

Since f_j converges to f the sum is uniformly bounded in j and N. Using the fact that the native space is continuously embedded in $L_2(\Omega)$ and letting j tend to infinity gives therefore

$$\sum_{n=1}^{N} \frac{1}{\rho_n}|(f, \varphi_n)_{L_2(\Omega)}|^2 \le \|f\|_{\mathcal{N}_\Phi(\Omega)}.$$

Hence, we can let N tend to infinity also, which shows that $f \in \mathcal{G}$ and $\|f\|_{\mathcal{G}} \le \|f\|_{\mathcal{N}_\Phi(\Omega)}$; this actually establishes norm equality. □

Picard's theorem on the range of a compact integral operator gives also

Corollary 10.30 *Suppose that Φ is a symmetric positive definite kernel on a compact set $\Omega \subseteq \mathbb{R}^d$. Then the range of the integral operator (10.7) is given by*

$$T(L_2(\Omega)) = \left\{ f \in L_2(\Omega) : \sum_{n=1}^{\infty} \frac{|(f, \varphi_n)|^2_{L_2(\Omega)}}{\rho_n^2} < \infty \right\}.$$

This space will play a particular role in the context of improved error estimates for radial basis function interpolants in Section 11.5.

10.5 Special cases of native spaces

In this section we want to take a closer look at the native spaces of two instances of basis functions. In the first instance we investigate the compactly supported functions of Section 9.4. The other class of functions is provided by certain thin-plate splines and powers. It will turn out that the native spaces are Sobolev and Beppo Levi spaces, respectively.

Let us start with the compactly supported functions $\Phi_{d,k} = \phi_{d,k}(\|\cdot\|_2)$ from Chapter 9. We know that such a function has a classical radial Fourier transform $\widehat{\Phi}_{d,k} = \mathcal{F}_d\phi_{d,k}(\|\cdot\|_2)$.

It is our goal to show that this Fourier transform decays as $(1 + \|\cdot\|_2)^{-d-2k-1}$, which means by Corollary 10.13 that the associated native space is the Sobolev space $H^s(\mathbb{R}^d)$ with $s = k + d/2 + 1/2$.

We start our investigation by restricting ourselves to the odd-dimensional case. To this end let us set $d = 2n + 1$ and $m = n + k$. With this we derive

$$\begin{aligned}\mathcal{F}_d\phi_{d,k}(r) &= \mathcal{F}_{d+2k}\phi_{\lfloor d/2\rfloor+k+1}(r) = \mathcal{F}_{2m+1}\phi_{m+1}(r)\\ &= r^{-3m-2}\int_0^r (r-t)^{m+1}t^{m+1/2}J_{m-1/2}(t)dt.\end{aligned}$$

Thus we can use the representation (6.9) for bounding the Fourier transform.

Lemma 10.31 *For every $m \in \mathbb{N}_0$ there exists a constant C_m such that*

$$\mathcal{F}_{2m+1}\phi_{m+1}(r) \le C_m r^{-2m-2}$$

for all $r > 0$.

Proof By Lemma 6.19 we know that $\mathcal{F}_{2m+1}\phi_{m+1}(r) = B_m f_m(r) r^{-3m-2}$ with a certain constant B_m and a nonnegative function f_m defined in that lemma. Thus, it suffices to show that $f_m(r) \le Cr^m$. We now show by induction that, more precisely, $f_m(r) \le 2^{m+1}r^m/m!$. In the case $m = 0$ we have $f_0(r) = 1 - \cos r$, which obviously satisfies $f_0(r) \le 2$. Now suppose that everything is settled for $m \ge 0$. Then

$$f_{m+1}(r) = \int_0^r f_m(t)f_0(r-t)dt \le \int_0^r \frac{2^{m+1}}{m!}t^m 2dt = \frac{2^{m+2}}{(m+1)!}r^{m+1},$$

which completes our proof. □

We want to point out that the constant C_m of the last lemma is given explicitly. Moreover, even if this bound is valid for $r > 0$ it is only of interest for large r. For r close to zero we know that $\mathcal{F}_{2m+1}\phi_{m+1}$ is bounded by a constant.

Next we turn to the lower bounds on $\mathcal{F}_{2m+1}\phi_{m+1}$. Since $\mathcal{F}_1\phi_1$ coincides, up to a constant, with $(1 - \cos r)/r^2$ there is no chance of getting a lower bound for this function. But in all other cases, i.e. $m \ge 1$, it is possible.

Lemma 10.32 *For every $m \in \mathbb{N}$ there exist constants $r_m, c_m > 0$ such that*

$$\mathcal{F}_{2m+1}\phi_{m+1}(r) \ge c_m r^{-2m-2}$$

for all $r \ge r_m$. Moreover, $\mathcal{F}_{2m+1}\phi_{m+1}$ is strictly positive on $[0, \infty)$.

Proof The proof is by induction on m. Again we use the representation $\mathcal{F}_{2m+1}\phi_{m+1}(r) = B_m f_m(r) r^{-3m-2}$ of Lemma 6.19 and concentrate on showing that $f_m(r) \ge c_m r^m$ for $r \ge r_m$. If $m = 1$ it is easy to compute that

$$f_1(r) = \int_0^r f_0(t)f_0(r-t)dt = r + \tfrac{1}{2}r\cos r - \tfrac{3}{2}\sin r \ge \tfrac{1}{2}r - \tfrac{3}{2} \ge \tfrac{1}{4}r$$

if $r \geq 6$. Hence we have found r_1 and c_1. Now suppose that our statement is true for $m \geq 1$. Then for $m+1$ we have

$$\begin{aligned} f_{m+1}(r) &= \int_0^r f_m(t)[1-\cos(r-t)]\,dt \\ &\geq \int_{r_m}^r c_m t^m [1-\cos(r-t)]\,dt \\ &= \frac{c_m}{m+1} r^{m+1} - \frac{c_m}{m+1} r_m^{m+1} - c_m \int_{r_m}^r t^m \cos(r-t)\,dt \end{aligned}$$

for $r \geq r_m$. We use integration by parts to bound the last integral via

$$\begin{aligned} \left| \int_{r_m}^r t^m \cos(r-t)\,dt \right| &\leq |r_m^m \sin(r-r_m)| + m \left| \int_{r_m}^r t^{m-1} \sin(r-t)\,dt \right| \\ &\leq r_m^m + m \int_{r_m}^r t^{m-1} dt \\ &= r_m^m + r^m - r_m^m = r^m. \end{aligned}$$

Hence

$$f_{m+1}(r) \geq \frac{c_m}{m+1} r^{m+1} - c_m r^m - \frac{c_m}{m+1} r_m^{m+1} \geq \frac{c_m}{2(m+1)} r^{m+1}$$

for sufficiently large $r \geq r_{m+1}$.

Since ϕ_{m+1} is nonnegative and nonvanishing, $\mathcal{F}_{2m+1}\phi_{m+1}$ is positive definite on $\mathbb{R}^{2m+1}$ according to Corollary 6.9. This means in particular that $\mathcal{F}_{2m+1}\phi_{m+1}(0) > 0$ by Theorem 6.2. Furthermore, the function f_1 is the (Laplace-)convolution of two nonnegative functions with isolated zeros. Thus it has to be positive on $(0, \infty)$. Finally f_m, $m \geq 2$, is the (Laplace-)convolution of a positive function with $1 - \cos r$ and, therefore, also positive on $(0, \infty)$. □

This finishes our investigations in the odd-dimensional case. Next we turn to even space dimensions. To this end we set $d = 2n$, $m = n + k$, to get

$$\mathcal{F}_d \phi_{d,k}(r) = \mathcal{F}_{2m} \phi_{m+1}(r) = r^{-3m-1} \int_0^r (r-t)^{m+1} t^m J_{m-1}(t)\,dt.$$

In the odd-dimensional case the functions f_n defined by Laplace-transform conforming convolution were an appropriate tool for determining the Fourier transform of ϕ_{m+1}. Something similar is true in the even-dimensional case. Let us define

$$\begin{aligned} g_0(r) &= \int_0^r J_0(t)dt, \\ g_m(r) &= \int_0^r f_{m-1}(r-t) g_0(t)dt, \qquad m \geq 1, \end{aligned}$$

with f_m from Lemma 6.19. Then the result analogous to Lemma 6.19 becomes

Lemma 10.33 *Let $A_m = 2^m(m+1)!\,\Gamma(m+1/2)/\sqrt{\pi}$ for $m \in \mathbb{N}$. Then*

$$\mathcal{F}_{2m}\phi_{m+1}(r) = A_m r^{-3m-1} g_m(r).$$

Proof The proof is similar to the proof of Lemma 6.19. On the one hand, the Fourier transform can be written as $\mathcal{F}_{2m}\phi_{m+1}(r) = r^{-3m-1}h(r)$ with $h(r) = \int_0^r h_1(r-t)h_2(t)\,dt$, $h_1(t) = t^{m+1}$ and $h_2(t) = t^m J_{m-1}(t)$. The Laplace transform of h_1 is given by $\mathcal{L}h_1(r) = (m+1)!\,r^{-m-2}$. Setting $\nu = m-1 \ge 0$ in Lemma 5.7 yields

$$\mathcal{L}h_2(r) = \frac{2^m\Gamma(m+1/2)r}{\sqrt{\pi}(1+r^2)^{m+1/2}},$$

so that the Laplace transform of h becomes

$$\mathcal{L}h(r) = \frac{A_m}{r^{m+1}(1+r^2)^{m+1/2}}.$$

On the other hand, we know by Lemma 5.8 that

$$\int_0^\infty J_0(t)e^{-rt}dt = \frac{1}{(1+r^2)^{1/2}}$$

so that

$$\mathcal{L}g_0(r) = \int_0^\infty \int_0^t J_0(s)ds\, e^{-rt}dt = \frac{1}{r}\int_0^\infty J_0(s)e^{-rs}ds = \frac{1}{r(1+r^2)^{1/2}}.$$

Moreover, f_{m-1} has Laplace transform $\mathcal{L}f_{m-1}(r) = r^{-m}(1+r^2)^{-m}$, as shown in the proof of Lemma 6.19. This gives

$$\mathcal{L}g_m(r) = \frac{1}{r^{m+1}(1+r^2)^{m+1/2}}$$

and $h = A_m g_m$ by uniqueness, which completes the proof. □

This iterative representation of the Fourier transform $\mathcal{F}_{2m}\phi_{m+1}$ together with the bounds on f_m allow us to find upper and lower bounds for $\mathcal{F}_{2m}\phi_{m+1}$.

Lemma 10.34 *For every $m \in \mathbb{N}$ there exists a constant $C_m > 0$ such that*

$$\mathcal{F}_{2m}\phi_{m+1}(r) \le C_m r^{-2m-1}$$

for all $r > 0$. Moreover, for $m \ge 2$ there exist an $r_m > 0$ and a $c_m > 0$ such that

$$\mathcal{F}_{2m}\phi_{m+1}(r) \ge c_m r^{-2m-1}$$

for all $r \ge r_m$. Finally, $\mathcal{F}_{2m}\phi_{m+1}$ is strictly positive on $[0,\infty)$.

Proof By Lemma 5.9 we know that $\lim_{t\to\infty} g_0(t) = 1$. Thus there exists a $t_0 > 0$ such that $1/2 \le g_0(t) \le 3/2$ for all $t \ge t_0$. Since g_0 is also continuous this means in particular that g_0 is bounded. Moreover, from the proof of Lemma 10.31 we know that $f_{m-1}(t) \le Ct^{m-1}$, which gives

$$|g_m(r)| \le \int_0^r |f_{m-1}(t)||g_0(r-t)|dt \le C \int_0^r t^{m-1}dt = Cr^m.$$

Since $\mathcal{F}_{2m}\phi_{m+1}(r) = A_m r^{-3m-1} g_m(r)$ this proves the upper bound.

For the lower bound we first choose an $r_0 > 0$ such that $g_0(r) \ge 1/2$ and $f_{m-1}(r) \ge Cr^{m-1}$ for all $r \ge r_0/2$. The latter was done in the proof of Lemma 10.32. Since both g_0 and f_{m-1} are nonnegative we can obtain the estimate

$$g_m(r) \ge \int_{r_0/2}^{r/2} f_{m-1}(t) g_0(r-t)dt \ge C \int_{r_0/2}^{r/2} t^{m-1}dt \ge Cr^m$$

for sufficiently large $r \ge r_0$. The use of $\mathcal{F}_{2m}\phi_{m+1}(r) = A_m r^{-3m-1} g_m(r)$ finishes the proof for this part.

Finally, since $g_0(r) > 0$ for all $r > 0$ and $f_{m-1}(r) > 0$ or $f_{m-1}(r) = 1 - \cos r$ if $m \ge 2$ or $m = 1$ respectively, g_m has to be positive on $(0, \infty)$. Since ϕ_{m+1} is nonnegative, $\mathcal{F}_{2m}\phi_{m+1}$ is positive definite on $\mathbb{R}^{2m}$, showing that $\mathcal{F}_{2m}\phi_{m+1}$ has to be positive at zero and hence everywhere. □

Now that we have complete control over the Fourier transform of $\phi_{d,k}$ we can state and proof our main result.

Theorem 10.35 *Let $\Phi_{d,k} = \phi_{d,k}(\|\cdot\|_2)$ denote the compactly supported radial basis function of minimal degree that is positive definite on $\mathbb{R}^d$ and in C^{2k}. Let $d \ge 3$ if $k = 0$. Then there exist constants $c_1, c_2 > 0$ depending only on d and k such that*

$$c_1(1 + \|\omega\|_2)^{-d-2k-1} \le \widehat{\Phi}_{d,k}(\omega) \le c_2(1 + \|\omega\|_2)^{-d-2k-1}$$

for all $\omega \in \mathbb{R}^d$. This means in particular that

$$\mathcal{N}_{\Phi_{d,k}}(\mathbb{R}^d) = H^{d/2+k+1/2}(\mathbb{R}^d),$$

i.e. the native space for these basis functions is a classical Sobolev space.

Proof The preceding results show that both upper and lower bounds are valid for sufficiently large arguments $r = \|\omega\|_2$, say $\|\omega\|_2 \ge r_0$. But as $\widehat{\Phi}_{d,k}$ is a continuous and positive function on $\mathbb{R}^d$ the bounds have to hold with possibly worse constants on the whole of $\mathbb{R}^d$. Finally, the native space is the Sobolev space by Corollary 10.13. □

After investigating the native spaces for the compactly supported functions of minimal degree we now turn to another famous class of radial basis functions, the thin-plate splines. To be more precise we want to characterize the native spaces of the functions $\Phi_{d,\ell} := \phi_{d,\ell}(\|\cdot\|_2)$ with $\ell > d/2$ and

$$\phi_{d,\ell}(r) := \begin{cases} \dfrac{\Gamma(d/2-\ell)}{2^{2\ell}\pi^{d/2}(\ell-1)!} r^{2\ell-d}, & \text{for } d \text{ odd}, \\[2ex] \dfrac{(-1)^{\ell+(d-2)/2}}{2^{2\ell-1}\pi^{d/2}(\ell-1)!(\ell-d/2)!} r^{2\ell-d}\log r & \text{for } d \text{ even}. \end{cases} \tag{10.11}$$

From Theorems 8.16 and 8.17 we know that $\Phi_{d,\ell}$ has a generalized Fourier transform

$$\widehat{\Phi}_{d,\ell}(\omega) = (2\pi)^{-d/2}\|\omega\|_2^{-2\ell}$$

of order $m = \ell - \lceil d/2 \rceil + 1$, so that $\Phi_{d,\ell}$ is conditionally positive definite of order m. In contrast with our earlier convention but in accordance with Proposition 8.2 we will consider $\Phi_{d,\ell}$ as a conditionally positive definite function of order ℓ and its generalized Fourier transform as of order ℓ for the rest of this section.

The reason for choosing the constant factor in this way is given by the simple structure of the Fourier transform, which leads to the fact that $\Phi_{d,\ell}$ is a fundamental solution of the iterated Laplacian. Remember that the Laplacian operator is defined to be $\Delta := \sum_{j=1}^d \partial^2/\partial x_j^2$ and the iterated Laplacian to be $\Delta^\ell := \Delta\Delta^{\ell-1}$.

Theorem 10.36 *Let $d, \ell \in \mathbb{N}$ with $\ell > d/2$. If $\Phi_{d,\ell} := \phi_{d,\ell}(\|\cdot\|_2)$ with the univariate function $\phi_{d,\ell}$ from (10.11) then*

$$(-1)^\ell \int_{\mathbb{R}^d} \Phi_{d,\ell}(\omega)\Delta^\ell g(x-\omega)d\omega = g(x)$$

for all $g \in \mathcal{S}$ and $x \in \mathbb{R}^d$.

Proof Define γ by its Fourier transform $\widehat{\gamma}(\omega) := \Delta^\ell g(x-\omega)$. Then γ is given by

$$\begin{aligned}\gamma(\omega) &= (2\pi)^{-d/2} \int_{\mathbb{R}^d} \Delta^\ell g(x-\eta)e^{i\eta^T\omega}d\eta \\ &= e^{ix^T\omega}(\Delta^\ell g)^\wedge(\omega) \\ &= (-1)^\ell e^{ix^T\omega}\|\omega\|_2^{2\ell}\,\widehat{g}(\omega),\end{aligned}$$

showing that $\gamma \in \mathcal{S}_{2\ell}$ for every $g \in \mathcal{S}$ and $x \in \mathbb{R}^d$. Hence, we can invoke the theory on generalized Fourier transforms to derive

$$(-1)^\ell \int_{\mathbb{R}^d} \Phi_{d,\ell}(\omega)\Delta^\ell g(x-\omega)d\omega = (2\pi)^{-d/2}\int_{\mathbb{R}^d} e^{ix^T\omega}\widehat{g}(\omega)d\omega = g(x),$$

using the special form of the Fourier transform of $\Phi_{d,\ell}$ mentioned earlier. □

Next we introduce Beppo Levi spaces. To this end we have to define the generalized derivative of a continuous function.

Definition 10.37 *Let $f \in L_1^{\mathrm{loc}}(\mathbb{R}^d)$ and $\alpha \in \mathbb{N}^d$ be given. A function $f_\alpha \in L_1^{\mathrm{loc}}(\mathbb{R}^d)$ is the generalized (or weak) derivative of f of order α if*

$$\int_{\mathbb{R}^d} f(x)D^\alpha\gamma(x)dx = (-1)^{|\alpha|}\int_{\mathbb{R}^d} f_\alpha(x)\gamma(x)dx \qquad (10.12)$$

is satisfied for all $\gamma \in C_0^\infty(\mathbb{R}^d)$. We will use the notation $D^\alpha f := f_\alpha$ again.

For $\ell > d/2$, the linear space

$$\mathrm{BL}_\ell(\mathbb{R}^d) := \{f \in C(\mathbb{R}^d) : D^\alpha f \in L_2(\mathbb{R}^d) \text{ for all } |\alpha| = \ell\}$$

equipped with the inner product

$$(f, g)_{\mathrm{BL}_\ell(\mathbb{R}^d)} := \sum_{|\alpha|=\ell} \frac{\ell!}{\alpha!}(D^\alpha f, D^\alpha g)_{L_2(\mathbb{R}^d)}$$

is called the Beppo Levi space on $\mathbb{R}^d$ of order ℓ.

Beppo Levi spaces can be introduced in a much more general way. The most general version starts with $\mathcal{D}_\Omega = C_0^\infty(\Omega)$, $\Omega \subseteq \mathbb{R}^d$, and its dual $\mathcal{D}'_\Omega$, the set of distributions. The advanced reader will know what type of continuity is meant in the definition of the dual space. Next, one chooses a separable complete space E of functions defined on Ω and defines the Beppo Levi space to be

$$\mathrm{BL}_\ell(E) := \{f \in \mathcal{D}'_\Omega : D^\alpha f \in E \text{ for all } |\alpha| = \ell\}.$$

In our specific situation it is possible to show that the two definitions coincide, i.e. $\mathrm{BL}_\ell(\mathbb{R}^d) = \mathrm{BL}_\ell(L_2(\mathbb{R}^d))$. Details may be found in the papers by Deny and Lions [45], Duchon [47], and Light and Wayne [108] and the other sources on Beppo Levi spaces cited in the references.

The choice of weights $\ell!/\alpha!$ in the definition is motivated by expressing $\|x\|_2^{2\ell}$ as $\|x\|_2^{2\ell} = \sum_{|\alpha|=\ell} \ell! x^{2\alpha}/\alpha!$. This also means that we can express the iterated Laplacian by $\Delta^\ell = \sum_{|\alpha|=\ell} \ell! D^{2\alpha}/\alpha!$. Both will be important later on.

The rest of this section is devoted to showing that the native space of $\Phi_{d,\ell}$ is the Beppo Levi space $\mathrm{BL}_\ell(\mathbb{R}^d)$. We start by showing that the null space of the semi-inner product is the space of polynomials of degree less than ℓ. Clearly, $\pi_{\ell-1}(\mathbb{R}^d)$ is in the null space of $(\cdot, \cdot)_{\mathrm{BL}_\ell(\mathbb{R}^d)}$. It remains to prove that they are actually the same.

Lemma 10.38 *Suppose that $f \in \mathrm{BL}_\ell(\mathbb{R}^d)$, $\ell > d/2$, satisfies $D^\alpha f = 0$ for all $|\alpha| = \ell$. Then f is a polynomial of degree less than ℓ.*

Proof We use approximation by convolution. Let $g \in C_0^\infty(\mathbb{R}^d)$ be nonnegative and even, having integral one. Set $g_n := n^d g(n\,\cdot)$ as usual. Then we know from Theorem 5.22 that $f * g_n \in C^\infty(\mathbb{R}^d)$ and $D^\alpha(f * g_n) = f * (D^\alpha g_n)$. An application of the definition of the generalized derivative gives immediately $D^\alpha(f * g_n) = (D^\alpha f) * g_n = 0$ for $|\alpha| = \ell$. Hence for all $n \in \mathbb{N}$ and all $|\alpha| = \ell$ the C^∞-functions $D^\alpha(f * g_n)$ are zero, implying that $f * g_n \in \pi_{\ell-1}(\mathbb{R}^d)$ for all n. Moreover, $f * g_n(x)$ tends to $f(x)$ as $n \to \infty$ for all $x \in \mathbb{R}^d$. But if we fix $x \in \mathbb{R}^d$ the latter convergence means the convergence of the coefficients of the polynomials $f * g_n \in \pi_{\ell-1}(\mathbb{R}^d)$. Thus f is also a polynomial of degree less than ℓ. □

The next step is to show that the native space of the thin-plate splines $\Phi_{d,\ell}$ is contained in the Beppo Levi space $BL_\ell(\mathbb{R}^d)$, and that on $\mathcal{N}_{\Phi_{d,\ell}}(\mathbb{R}^d)$ both semi-inner products are equal. To this end we will use the Fourier transform representation of $\mathcal{N}_{\Phi_{d,\ell}}(\mathbb{R}^d)$ given in Theorem 10.21.

Proposition 10.39 *For $\ell > d/2$ let $\Phi_{d,\ell} = \phi_{d,\ell}(\|\cdot\|_2)$ be the thin-plate spline defined in (10.11). If $\Phi_{d,\ell}$ is considered to be a conditionally positive definite function of order ℓ*

then the associated native space is contained in the Beppo Levi space of order ℓ, i.e. $\mathcal{N}_{\Phi_{d,\ell}}(\mathbb{R}^d) \equiv \mathcal{N}_{\Phi_{d,\ell},\pi_{\ell-1}(\mathbb{R}^d)}(\mathbb{R}^d) \subseteq \mathrm{BL}_\ell(\mathbb{R}^d)$, and the semi-inner products are the same on this subspace.

Proof Let q_α be the monomial $q_\alpha(x) = x^\alpha$, $\alpha \in \mathbb{N}_0^d$. Suppose that $f \in \mathcal{N}_{\Phi_{d,\ell}}(\mathbb{R}^d)$ is given. Then f possesses a generalized Fourier transform $\widehat{f}$ of order $\ell/2$ with $\omega \mapsto \widehat{f}(\omega)\|\omega\|_2^{\ell} \in L_2(\mathbb{R}^d)$. This means that the function $\widehat{f}q_\alpha$ is in $L_2(\mathbb{R}^d)$ for all $|\alpha| = \ell$. Hence we can define

$$f_\alpha(\omega) := (\widehat{f}q_\alpha(i\cdot))^\vee(\omega) \in L_2(\mathbb{R}^d)$$

for $|\alpha| = \ell$, using the inverse L_2-Fourier transform. Since f is real-valued, so is f_α. Since $q_\alpha\gamma^\vee \in \mathcal{S}_\ell$ for $\gamma \in C_0^\infty(\mathbb{R}^d)$ we find that

$$\begin{aligned}
\int_{\mathbb{R}^d} f_\alpha(\omega)\gamma(\omega)d\omega &= \int_{\mathbb{R}^d} (\widehat{f}q_\alpha(i\cdot))^\vee(\omega)\gamma(\omega)d\omega \\
&= \int_{\mathbb{R}^d} \widehat{f}(\omega)(i\omega)^\alpha\gamma^\vee(\omega)d\omega \\
&= (-1)^{|\alpha|}\int_{\mathbb{R}^d} \widehat{f}(\omega)(D^\alpha\gamma)^\vee(\omega)d\omega \\
&= (-1)^{|\alpha|}\int_{\mathbb{R}^d} f(x)D^\alpha\gamma(x)dx,
\end{aligned}$$

showing that f_α is the generalized derivative of f. Hence, the native space is contained in the Beppo Levi space.

Finally, for $f, g \in \mathcal{N}_{\Phi_{d,\ell}}(\mathbb{R}^d)$ we have

$$\begin{aligned}
(f, g)_{\mathrm{BL}_\ell(\mathbb{R}^d)} &= \sum_{|\alpha|=\ell} \frac{\ell!}{\alpha!}\int_{\mathbb{R}^d} f_\alpha(x)g_\alpha(x)dx \\
&= \sum_{|\alpha|=\ell} \frac{\ell!}{\alpha!}\int_{\mathbb{R}^d} \widehat{f_\alpha}(\omega)\overline{\widehat{g_\alpha}(\omega)}d\omega \\
&= \sum_{|\alpha|=\ell} \frac{\ell!}{\alpha!}\int_{\mathbb{R}^d} \widehat{f}(\omega)\overline{\widehat{g}(\omega)}(i\omega)^\alpha(-i\omega)^\alpha d\omega \\
&= \int_{\mathbb{R}^d} \widehat{f}(\omega)\overline{\widehat{g}(\omega)}\|\omega\|_2^{2\ell}d\omega \\
&= (2\pi)^{-d/2}\int_{\mathbb{R}^d} \frac{\widehat{f}(\omega)\overline{\widehat{g}(\omega)}}{\widehat{\Phi}_{d,\ell}(\omega)}d\omega \\
&= (f, g)_{\mathcal{N}_{\Phi_{d,\ell}}(\mathbb{R}^d)}.
\end{aligned}$$

□

It remains to show that the inclusion is actually an identity. One might be tempted to define a generalized Fourier transform $\widehat{f}$ for a function f from the Beppo Levi space by

$$\widehat{f}(\omega) := \frac{\widehat{f_\alpha}(\omega)}{(i\omega)^\alpha}.$$

Unfortunately, it is not at all simple to prove that this is indeed the generalized Fourier transform in our sense. The reason for this is that $\gamma(\omega)/(i\omega)^\alpha$ is not even continuous at zero for $\gamma \in \mathcal{S}_\ell$.

Hence, instead of proving that the Beppo Levi space is a subspace of the native space we will show that every function from the Beppo Levi space that is orthogonal to all functions from the native space with respect to the semi-inner product of the Beppo Levi space is actually a polynomial of degree less than ℓ.

The first step in this direction is to reformulate Theorem 10.36 in a way appropriate for Beppo Levi functions. To this end we want to employ density results. But since the Beppo Levi space is only equipped with a semi-inner product, we have to be more precise about what we understand by density.

Theorem 10.40 *Let $\ell > d/2$. Then the set $C_0^\infty(\mathbb{R}^d)$ is dense in $\mathrm{BL}_\ell(\mathbb{R}^d)$. To be more precise, for every $f \in \mathrm{BL}_\ell(\mathbb{R}^d)$, every compact subset K of $\mathbb{R}^d$, and every $\epsilon > 0$ there exists a function $g \in C_0^\infty(\mathbb{R}^d)$ such that*

(1) $\|f - g\|_{L_\infty(K)} < \epsilon$,
(2) $\|D^\alpha f - D^\alpha g\|_{L_2(\mathbb{R}^d)} < \epsilon$ for all $|\alpha| = \ell$.

Proof In the first step we will show that the set $C^\infty(\mathbb{R}^d) \cap \mathrm{BL}_\ell(\mathbb{R}^d)$ is dense in $\mathrm{BL}_\ell(\mathbb{R}^d)$ in the sense specified in the theorem. This follows immediately from approximation by convolution. If we use $f * g_n$, where $\{g_n\}$ is a sequence from Theorem 5.22 then we know from this theorem that $f * g_n \in C^\infty(\mathbb{R}^d)$ and that property (1) is satisfied for sufficiently large n. Moreover, since $D^\alpha(f * g_n) = (D^\alpha f) * g_n$ and $D^\alpha f \in L_2(\mathbb{R}^d)$ for $|\alpha| = \ell$, the same theorem tells us that $f * g_n \in \mathrm{BL}_\ell(\mathbb{R}^d)$ and that the second property is also satisfied for sufficiently large n.

Hence, it remains to show that the functions of $C^\infty(\mathbb{R}^d) \cap \mathrm{BL}_\ell(\mathbb{R}^d)$ can be approximated by $C_0^\infty(\mathbb{R}^d)$ functions in the stated way. Let us assume that $f \in C^\infty(\mathbb{R}^d) \cap \mathrm{BL}_\ell(\mathbb{R}^d)$ is given. We choose a function $\psi \in C_0^\infty(\mathbb{R}^d)$, which is identically one on $\|x\|_2 \le 1$, identically zero on $\|x\|_2 \ge 2$, and has maximum absolute value one, and set $f_k := \psi(\cdot/k) f \in C_0^\infty(\mathbb{R}^d)$. Then Leibniz' rule gives

$$D^\alpha f_k(x) = \sum_{0 \ne \beta \le \alpha} \binom{\alpha}{\beta} \frac{1}{k^{|\beta|}} D^\beta \psi(x/k) D^{\alpha-\beta} f(x) + \psi(x/k) D^\alpha f(x),$$

so that

$$\begin{aligned}\|D^\alpha f - D^\alpha f_k\|_{L_2(\mathbb{R}^d)} &\le \frac{1}{k} \sum_{0 \ne \beta \le \alpha} \binom{\alpha}{\beta} \|D^\beta \psi\|_{L_\infty(\mathbb{R}^d)} \|D^{\alpha-\beta} f\|_{L_2(\mathbb{R}^d)} \\ &\quad + \left(\int_{\|x\|_2 > k} |D^\alpha f(x)|^2 dx \right)^{1/2}.\end{aligned}$$

The last expression clearly tends to zero as $k \to \infty$, which settles the second property, while the first one is obvious. □

This density result allows us to draw some very important conclusions.

Theorem 10.41 *Suppose that* $\lambda = \sum_{j=1}^{N} \lambda_j \delta_{x_j}$ *is an element of* $L_{\pi_{\ell-1}(\mathbb{R}^d)}$, *i.e.* $\lambda(p) = 0$ *for all* $p \in \pi_{\ell-1}(\mathbb{R}^d)$. *Set* $f_\lambda = \lambda^y \Phi_{d,\ell}(\cdot - y) = \sum \lambda_j \Phi_{d,\ell}(\cdot - x_j)$. *Then for every* $f \in \mathrm{BL}_\ell(\mathbb{R}^d)$, $\ell > d/2$, *and every* $x \in \mathbb{R}^d$ *we have the representation*

$$\sum_{j=1}^{N} \lambda_j f(x - x_j) = (f, f_\lambda(x - \cdot))_{\mathrm{BL}_\ell(\mathbb{R}^d)} \tag{10.13}$$

$$= \sum_{|\alpha|=\ell} \frac{\ell!}{\alpha!} \int_{\mathbb{R}^d} D^\alpha f(y) D^\alpha f_\lambda(x - y) dy.$$

Proof One consequence of Proposition 10.39 is that f_λ, which is an element of the native space, is contained in the Beppo Levi space. Hence $D^\alpha f_\lambda$ is an element of $L_2(\mathbb{R}^d)$. Since also $D^\alpha f \in L_2(\mathbb{R}^d)$ by definition, it follows easily from Lemma 5.21 that the right-hand side of (10.13) is a continuous function.

Next, let us first assume that $f \in C_0^\infty(\mathbb{R}^d)$. Then the definition of generalized derivatives, the choice of coefficients in the semi-inner product, and Theorem 10.36 give

$$\begin{aligned}
(-1)^\ell (f, f_\lambda(x - \cdot))_{\mathrm{BL}_\ell(\mathbb{R}^d)} &= \sum_{|\alpha|=\ell} \frac{\ell!}{\alpha!} \int_{\mathbb{R}^d} D^\alpha f(x - y) D^\alpha f_\lambda(y) dy \\
&= \sum_{|\alpha|=\ell} \frac{\ell!}{\alpha!} \int_{\mathbb{R}^d} D^{2\alpha} f(x - y) f_\lambda(y) dy \\
&= \int_{\mathbb{R}^d} \Delta^\ell f(x - y) f_\lambda(y) dy \\
&= \sum_{j=1}^{N} \lambda_j \int_{\mathbb{R}^d} \Phi_{d,\ell}(y - x_j) \Delta^\ell f(x - y) dy \\
&= (-1)^\ell \sum_{j=1}^{N} \lambda_j f(x - x_j),
\end{aligned}$$

proving the result in this case. For a general $f \in \mathrm{BL}_\ell(\mathbb{R}^d)$ we fix $x \in \mathbb{R}^d$ and choose a compact set $K \subseteq \mathbb{R}^d$ such that $x - x_j \in K$ for $1 \le j \le N$. For an arbitrary $\epsilon > 0$ we choose a $g \in C_0^\infty(\mathbb{R}^d)$ according to Theorem 10.40. Then two applications of the triangle inequality show that the absolute value of the difference in the two sides of (10.13) can be bounded by $\epsilon \left(\sum_{j=1}^{N} |\lambda_j| + |f_\lambda|_{\mathrm{BL}_\ell(\mathbb{R}^d)} \right)$, which tends to zero with $\epsilon \to 0$. □

This is the major step in proving several things. For example we can now readily conclude that every function f from the Beppo Levi space is slowly increasing; we will derive a representation formula for f and finally we will show that such a function has to be in the native space of the thin-plate splines.

To achieve all these goals we use a now familiar concept. Suppose that $\Xi = \{\xi_1, \ldots, \xi_Q\} \subseteq \mathbb{R}^d$ is $\pi_{\ell-1}(\mathbb{R}^d)$-unisolvent and $p_1, \ldots, p_Q$ is a Lagrange basis of $\pi_{\ell-1}(\mathbb{R}^d)$

with respect to Ξ. Next we define for a fixed $x \in \mathbb{R}^d$ the functional

$$\lambda := \delta_{-x} - \sum_{k=1}^{Q} p_k(x)\delta_{-\xi_k},$$

which is very similar to the functional $\delta_{(x)}$ employed earlier. To see that λ annihilates polynomials we can simply apply it to the basis $q_k := p_k(-\cdot)$. Denote by Π_ℓ the projection $\Pi_\ell = \Pi_{\pi_{\ell-1}(\mathbb{R}^d)}$.

Theorem 10.42 *Let $\ell > d/2$. Every function $f \in \mathrm{BL}_\ell(\mathbb{R}^d)$ has the representation*

$$f(x) = \Pi_\ell f(x) + (f, G(\cdot, x))_{\mathrm{BL}_\ell(\mathbb{R}^d)}, \qquad x \in \mathbb{R}^d.$$

Here G is the function (10.4) for the basis functions specified in (10.11). This representation means in particular that f is slowly increasing.

Proof The special choice of λ made in the paragraph before this theorem together with Theorem 10.41 shows that, for $\omega \in \mathbb{R}^d$,

$$f(\omega + x) = \sum_{j=1}^{Q} p_j(x) f(\omega + \xi_j) + (f, f_\lambda(\omega - \cdot))_{\mathrm{BL}_\ell(\mathbb{R}^d)} \tag{10.14}$$

with

$$f_\lambda(\omega) = \Phi_{d,\ell}(\omega + x) - \sum_{j=1}^{Q} p_j(x)\Phi_{d,\ell}(\omega + \xi_j).$$

But the definition of G, the fact that $\Phi_{d,\ell}$ is even, and the fact that the generalized derivative of $\Phi_{d,\ell}$ coincides with the usual one outside zero allow us to conclude that $D_1^\alpha G(\omega, x) = (-1)^{|\alpha|} D^\alpha f_\lambda(-\omega)$. Hence setting $\omega = 0$ in (10.14) gives the stated representation.

Finally, since the Beppo Levi semi-norm and the native space semi-norm coincide on the native space we see that the Beppo Levi semi-norm of $G(\cdot, x)$ grows at most as a polynomial in x and so does f. □

Theorem 10.43 *For $\ell > d/2$ let $\Phi_{d,\ell} = \phi_{d,\ell}(\|\cdot\|_2)$ be the thin-plate spline defined in (10.11). If $\Phi_{d,\ell}$ is considered to be a conditionally positive definite function of order ℓ then the associated native space is the Beppo Levi space of order ℓ, i.e. $\mathcal{N}_{\Phi_{d,\ell}}(\mathbb{R}^d) \equiv \mathcal{N}_{\Phi_{d,\ell},\pi_{\ell-1}(\mathbb{R}^d)}(\mathbb{R}^d) = \mathrm{BL}_\ell(\mathbb{R}^d)$ and the semi-inner products are the same.*

Proof We know already that the native space is contained in the Beppo Levi space. Moreover, since the semi-norms coincide on this subspace the native space is a complete subspace, meaning that every Cauchy sequence has a (not necessarily unique) limit. This in turn means that if the native space is not the whole Beppo Levi space then there must be an element $f \in \mathrm{BL}_\ell(\mathbb{R}^d)$ that is orthogonal to the native space. The representation formula stated in Theorem 10.42 now gives $f \in \pi_{\ell-1}(\mathbb{R}^d)$. □

10.6 An embedding theorem

Since in several cases the native space of a conditionally positive definite kernel does not coincide with a classical function space, it is important to know properties such as the smoothness of the functions in the native space in advance, given only information about the kernel itself. By construction we know already that $\mathcal{N}_\Phi(\Omega) \subseteq C(\Omega)$. Now we want to see how the smoothness of the kernel is inherited by the native space.

To this end, we use the forward differences already introduced in (7.4), with a slightly different notation,

$$\Delta_{k,h} f(r) := \sum_{j=0}^{k} (-1)^{k-j} \binom{k}{j} f(r + jh),$$

and their multivariate versions

$$\Delta_{\alpha,h} f(x) := \Delta_{\alpha_1,h}^{\mathrm{x}_1} \cdots \Delta_{\alpha_d,h}^{\mathrm{x}_d} f(x)$$

for $\alpha = (\alpha_1, \ldots, \alpha_d)^T \in \mathbb{N}_0^d$ and $x = (\mathrm{x}_1, \ldots, \mathrm{x}_d)^T \in \mathbb{R}^d$. Here Δ^{x_j} means that Δ acts with respect to the x_j-variable. From the property of the univariate forward difference, it obviously follows that

$$\lim_{h \to 0} h^{-|\alpha|} \Delta_{\alpha,h} f(x) = D^\alpha(x)$$

if f is $|\alpha|$-times continuously differentiable around x.

Lemma 10.44 *Suppose that $\Omega \subseteq \mathbb{R}^d$ is open and that $\Phi \in C^{2k}(\Omega \times \Omega)$ is a conditionally positive definite symmetric kernel with respect to $\mathcal{P} \subseteq C^k(\Omega)$. Then the function $G(\cdot, \cdot)$ from (10.4) is k-times continuously differentiable with respect to the second argument, and for every $x \in \Omega$ and every $\alpha \in \mathbb{N}_0^d$ with $|\alpha| \le k$ the function $D_2^\alpha G(\cdot, x)$ is in $\mathcal{N}_\Phi(\Omega)$. Here D_2^α means that we differentiate with respect to the second argument.*

Proof Obviously G possesses k continuous derivatives with respect to its second argument. Moreover, $D_2^\alpha G(\cdot, x)$ is in $C^k(\Omega)$ as a function of the first argument for all $|\alpha| \le k$ and all $x \in \Omega$.

Fix $\alpha \in \mathbb{N}_0^d$ with $|\alpha| \le k$ and $x \in \Omega$. Define the function $\varphi_n := \Delta_{\alpha,1/n,2} G(\cdot, x)$. Here, the additional 2 in the subscript means that Δ acts with respect to the second argument of G. Since $G(\cdot, x) \in F_\Phi(\Omega)$ for every $x \in \Omega$, we also have $\varphi_n \in F_\Phi(\Omega)$. Hence we have a representation of the form $\varphi_n = \lambda_n^y \Phi(\cdot, y)$ with $\lambda_n \in L_\mathcal{P}(\Omega)$. The definition also ensures that $\varphi_n(y) = \lambda_n^u \Phi(y, u) \to D_2^\alpha G(y, x)$. Now, φ_n is a Cauchy sequence in $F_\Phi(\Omega)$. Because

$$(\varphi_n, \varphi_m)_\Phi = \lambda_n^u \lambda_m^v \Phi(u, v) \to D_1^\alpha D_2^\alpha G(x, x) =: c$$

for $m, n \to \infty$, we have

$$\|\varphi_n - \varphi_m\|_\Phi^2 = \|\varphi_n\|_\Phi^2 + \|\varphi_m\|_\Phi^2 - 2(\varphi_n, \varphi_m)_\Phi \to c + c - 2c = 0$$

for $m, n \to \infty$. Thus there exists a $\varphi \in \mathcal{F}_\Phi(\Omega)$ with $\|\varphi - \varphi_n\|_\Phi \to 0$ for $n \to \infty$. For this element we make the computation

$$\begin{aligned} R\varphi(y) &= (\varphi, G(\cdot, y))_\Phi \\ &= \lim_{n\to\infty} (\varphi_n, G(\cdot, y))_\Phi \\ &= \lim_{n\to\infty} (\varphi_n(y) - \Pi_\mathcal{P}\varphi_n(y)) \\ &= D_2^\alpha G(y, x) - \Pi_\mathcal{P}^u \left(D_2^\alpha G(u, x)\right)|_{u=x}, \end{aligned}$$

showing that $D_2^\alpha G(\cdot, x)$ indeed belongs to $\mathcal{N}_\Phi(\Omega)$. □

With this lemma at hand it is easy to prove the smoothness of the functions belonging to the native space of a smooth basis function.

Theorem 10.45 *Suppose that $\Omega \subseteq \mathbb{R}^d$ is open and that $\Phi \in C^{2k}(\Omega \times \Omega)$ is a conditionally positive definite kernel with respect to $\mathcal{P} \subseteq C^k(\Omega)$; then $\mathcal{N}_\Phi(\Omega) \subseteq C^k(\Omega)$ and, for every $f \in \mathcal{N}_\Phi(\Omega)$, every $\alpha \in \mathbb{N}_0^d$ with $|\alpha| \le k$, and every $x \in \Omega$, we have the representation*

$$D^\alpha f(x) = D^\alpha \Pi_\mathcal{P} f(x) + \left(f, D_2^\alpha G(\cdot, x)\right)_{\mathcal{N}_\Phi(\Omega)}. \tag{10.15}$$

Proof We will show (10.15) by induction on $|\alpha|$. This will obviously prove the existence and continuity of the derivatives. For $|\alpha| = 0$, formula (10.15) obviously coincides with the representation in Theorem 10.17. For $|\alpha| > 0$ we can assume that $\alpha_1 > 0$. Hence, with $\beta = (\alpha_1 - 1, \alpha_2, \dots, \alpha_d)^T$ we find that

$$\begin{aligned} D^\alpha f(x) &= \lim_{h\to 0} \frac{1}{h}\left[D^\beta f(x + he_1) - D^\beta f(x)\right] \\ &= \lim_{h\to 0} \frac{1}{h}\left[D^\beta(\Pi_\mathcal{P} f)(x + he_1) - D^\beta(\Pi_\mathcal{P} f)(x)\right] \\ &\quad + \lim_{h\to 0}\left(f, \frac{1}{h}\left[D_2^\beta G(\cdot, x + he_1) - D_2^\beta G(\cdot, x)\right]\right)_{\mathcal{N}_\Phi(\Omega)} \\ &= D^\alpha(\Pi_\mathcal{P} f)(x) + \left(f, D_2^\alpha G(\cdot, x)\right)_{\mathcal{N}_\Phi(\Omega)}, \end{aligned}$$

using the fact that the derivatives of $G(\cdot, x)$ exist by Lemma 10.44. As usual e_1 denotes the first unit vector in $\mathbb{R}^d$. □

In the situation where Φ is a positive definite function (i.e. a translation-invariant kernel) which is in $L_1(\mathbb{R}^d)$ and which has a Fourier transform that decays like $(1 + \|\cdot\|_2^2)^{-s}$ we know by Corollary 10.13 that the native space is actually the Sobolev space $H^s(\mathbb{R}^d)$. If $s > k + d/2$ then the Fourier inversion formula guarantees that $\Phi \in C^{2k}(\mathbb{R}^d)$ and Theorem 10.45 shows that $H^s(\mathbb{R}^d) \subseteq C^k(\mathbb{R}^d)$, which is Sobolev's embedding theorem.

10.7 Restriction and extension

In this section we want to investigate how the native space $\mathcal{N}_\Phi(\Omega)$ depends on the region Ω. To do this we have to be more careful about the notation, even if sometimes it seems excessive.

Let us assume that we are dealing with two regions that satisfy $\Omega_1 \subseteq \Omega_2 \subseteq \mathbb{R}^d$. We are now interested in the questions whether the functions from $\mathcal{N}_\Phi(\Omega_1)$ have an extension to Ω_2 and whether the restrictions of the functions from $\mathcal{N}_\Phi(\Omega_2)$ to Ω_1 lie in $\mathcal{N}_\Phi(\Omega_1)$. Of course, both should be true and we shall prove the results in this section.

The crucial point in everything we do here is that we assume that the set Ξ is already contained in Ω_1, that $\mathcal{P} \subseteq C(\Omega_2)$, and that $\Phi \in C(\Omega_2 \times \Omega_2)$ is a conditionally positive definite kernel with respect to $\mathcal{P}$ on Ω_2. In the case of a positive definite kernel we need only the last assumption, that Φ is positive definite on the larger set.

Theorem 10.46 *Each function $f \in \mathcal{N}_\Phi(\Omega_1)$ has a natural extension to a function $Ef \in \mathcal{N}_\Phi(\Omega_2)$. Furthermore, $|Ef|_{\mathcal{N}_\Phi(\Omega_2)} = |f|_{\mathcal{N}_\Phi(\Omega_1)}$.*

Proof Since $\Omega_1 \subseteq \Omega_2$ we have a natural extension $\epsilon : F_\Phi(\Omega_1) \to F_\Phi(\Omega_2)$, simply by evaluation of a function $f \in F_\Phi(\Omega_1)$ at points from Ω_2 also. Since the norm $\|f\|_{\Phi,\Omega_1}$ depends only on the centers and coefficients of f, we have obviously $\|\epsilon f\|_{\Phi,\Omega_2} = \|f\|_{\Phi,\Omega_1}$. Hence ϵ is an isometric embedding that has an continuous extension $\epsilon : \mathcal{F}_\Phi(\Omega_1) \to \mathcal{F}_\Phi(\Omega_2)$. This allows us to construct the extension operator $E : \mathcal{N}_\Phi(\Omega_1) \to \mathcal{N}_\Phi(\Omega_2)$ in the following way. Every $f \in \mathcal{N}_\Phi(\Omega_1)$ has the representation $f(x) = \Pi_\mathcal{P} f(x) + R_{\Omega_1}(\tilde{f})(x)$, with $\tilde{f} \in \mathcal{F}_\Phi(\Omega_1)$. For this f and $x \in \Omega_2$ we define

$$Ef(x) = \Pi_\mathcal{P} f(x) + R_{\Omega_2}(\epsilon \tilde{f})(x).$$

The function $\Pi_\mathcal{P} f$ has an obvious extension to Ω_2. This is why we did not use different notation. Moreover, for $x \in \Omega_1$ we have

$$\begin{aligned} R_{\Omega_2}(\epsilon \tilde{f})(x) &= (\epsilon \tilde{f}, G_{\Omega_2}(\cdot, x))_{\Phi,\Omega_2} = (\epsilon \tilde{f}, \epsilon G_{\Omega_1}(\cdot, x))_{\Phi,\Omega_2} \\ &= (\tilde{f}, G_{\Omega_1}(\cdot, x))_{\Phi,\Omega_1}, \end{aligned}$$

showing that $Ef(x) = f(x)$ for $f \in \mathcal{N}_\Phi(\Omega_1)$ and $x \in \Omega_1$. Finally, for two functions $f, g \in \mathcal{N}_\Phi(\Omega_1)$ the identities

$$(Ef, Eg)_{\mathcal{N}_\Phi(\Omega_2)} = (\epsilon \tilde{f}, \epsilon \tilde{g})_{\Phi,\Omega_2} = (\tilde{f}, \tilde{g})_{\Phi,\Omega_1} = (f, g)_{\mathcal{N}_\Phi(\Omega_1)}$$

show that E is isometric. □

Now let us turn to the restriction of a function $f \in \mathcal{N}_\Phi(\Omega_2)$ to Ω_1. By Theorem 10.26, $f|\Omega_1 \in \mathcal{N}_\Phi(\Omega_1)$ if there exists a constant c_f such that $|\lambda(f|\Omega_1)| \le c_f \|\lambda\|_{\Phi,\Omega_1}$ for all $\lambda \in L_\mathcal{P}(\Omega_1)$. Since obviously $L_\mathcal{P}(\Omega_1) \subseteq L_\mathcal{P}(\Omega_2)$ it is true that there exists a constant c_f with

$$|\lambda(f|\Omega_1)| \le c_f \|\lambda\|_{\Phi,\Omega_2} = c_f \|\lambda\|_{\Phi,\Omega_1},$$

giving $f|\Omega_1 \in \mathcal{N}_{\Phi(\Omega_1)}$. Finally,

$$\|f|\Omega_1\|_{\mathcal{N}_\Phi(\Omega_1)} = \sup_{\substack{\lambda \in L_\mathcal{P}(\Omega_1) \\ \lambda \ne 0}} \frac{|\lambda(f)|}{\|\lambda\|_{\Phi,\Omega_2}} \le \sup_{\substack{\lambda \in L_\mathcal{P}(\Omega_2) \\ \lambda \ne 0}} \frac{|\lambda(f)|}{\|\lambda\|_{\Phi,\Omega_2}} = \|f\|_{\mathcal{N}_\Phi(\Omega_2)}$$

finishes the proof of our next theorem.

Theorem 10.47 *The restriction $f|\Omega_1$ of any function $f \in \mathcal{N}_\Phi(\Omega_2)$ is contained in $\mathcal{N}_\Phi(\Omega_1)$ with a semi-norm that is less than or equal to the semi-norm of f.*

The concept of restriction and extension has an interesting implication in the case of Sobolev spaces. We already know that the native space over the entire $\mathbb{R}^d$ of a basis function with algebraically decaying Fourier transform is a Sobolev space. Now we are able to show that this is also true for "nice" regions $\Omega \subseteq \mathbb{R}^d$. To this end, let us recall that the Sobolev space $H^k(\Omega)$, $k \in \mathbb{N}$, for a measurable set Ω can be introduced using weak derivatives. It consists of all functions $f \in L_2(\Omega)$ having weak derivatives in $L_2(\Omega)$ of order $|\alpha| \le k$. The norm on $H^k(\Omega)$ is then given by $\|u\|^2_{H^k(\Omega)} = \sum_{|\alpha|\le k} \|D^\alpha u\|^2_{L_2(\Omega)}$. In the case $\Omega = \mathbb{R}^d$ it is known that this norm is equivalent to the norm previously defined by Fourier transformation.

Corollary 10.48 *Suppose that $\Phi \in L_1(\mathbb{R}^d)$ has a Fourier transform that decays as $(1 + \|\cdot\|_2^2)^{-k}$, $k \in \mathbb{N}$, $k > d/2$. Suppose that $\Omega \subseteq \mathbb{R}^d$ has a Lipschitz boundary. Then $\mathcal{N}_\Phi(\Omega) = H^k(\Omega)$ with equivalent norms.*

Proof Every $f \in \mathcal{N}_\Phi(\Omega)$ has an extension $Ef \in \mathcal{N}_\Phi(\mathbb{R}^d) = H^k(\mathbb{R}^d)$. Thus we have $f = Ef|\Omega \in H^k(\Omega)$ and $\|f\|_{H^k(\Omega)} \le \|Ef\|_{H^k(\mathbb{R}^d)} \le c\|Ef\|_{\mathcal{N}_\Phi(\mathbb{R}^d)} = c\|f\|_{\mathcal{N}_\Phi(\Omega)}$. However, for a Lipschitz-bounded region it is well known (see Brenner and Scott [31]) that every function $f \in H^k(\Omega)$ has an extension $\tilde{E}f \in H^k(\mathbb{R}^d) = \mathcal{N}_\Phi(\mathbb{R}^d)$ satisfying $\|\tilde{E}f\|_{H^k(\mathbb{R}^d)} \le C\|f\|_{H^k(\Omega)}$. Thus $f = \tilde{E}f|\Omega \in \mathcal{N}_\Phi(\Omega)$ and

$$\|f\|_{\mathcal{N}_\Phi(\Omega)} \le \|\tilde{E}f\|_{\mathcal{N}_\Phi(\mathbb{R}^d)} \le c\|\tilde{E}f\|_{H^k(\mathbb{R}^d)} \le c\|f\|_{H^k(\Omega)}.$$

□

Note that we also have extensions for functions from $\mathcal{N}_\Phi(\Omega)$ for more general regions Ω, including regions with corners and even finite regions. This stands in sharp contrast with the Sobolev case, where regions exists that do not allow an extension to $\mathbb{R}^d$. In this sense we have found another reason why native spaces are generalizations of classical Sobolev spaces.

10.8 Notes and comments

The concept of reproducing-kernel Hilbert spaces is well established in numerical analysis. Apparently the first deep investigation goes back to Aronszajn [2] in 1950. Another good source is the book [129] of Meschkowski from 1962. More recently, in particular in the context of radial basis functions, the overview articles [167, 168] from Schaback and the inventive work [112, 113] by Madych and Nelson have given much help in clarifying the theory. Another valuable resource on native spaces is the diploma thesis [98] by Klein.

The results on the native spaces of compactly supported functions were initially given by the present author [191, 192].

There is a huge number of publications on Beppo Levi spaces, which turn out to be the native spaces for thin-plate splines. The interested reader might have a look at the work of

Duchon [47–49], of Meinguet [122–126], of Deny and Lions [45], and of Mizuta [135, 136]. Despite these numerous publications, the approach given in the present text seems to be new.

The fact that the smoothness of a given kernel is inherited by its native space, as pointed out in the Section 10.6, will be of some importance later on, when we try to solve partial differential equations using radial basis functions. We will also see that the radial basis function interpolant approximates not only the function but also its derivatives.

The extension theorems presented here are rather simple but sufficient in many situations. But when the best approximation order has to be found, it seems that deeper results are necessary. First steps in this direction can be found in Light and Vail [107].

11
Error estimates for radial basis function interpolation

The goal of this chapter is to derive error estimates for the interpolation process based on (conditionally) positive definite kernels. As in the case of classical univariate spline interpolation, it is possible to show that convergence takes place not only for the function itself but also for its derivatives. The error estimates are again expressed in terms of the fill distance

$$h_{X,\Omega} = \sup_{x\in\Omega} \min_{x_j \in X} \|x - x_j\|_2,$$

so that convergence is studied for $h_{X,\Omega} \to 0$. We will concentrate on error estimates for functions coming from the associated native space of the basis function of interest.

11.1 Power function and first estimates

In this section we will be concerned with estimating the difference or *error* between an (unknown) function f coming from the native Hilbert space $\mathcal{N}_\Phi(\Omega)$ of a (conditionally) positive definite kernel Φ and its interpolant $s_{f,X}$. Once again, we will assume the kernel to be real-valued and symmetric throughout the entire chapter. The starting point for error estimates is to rewrite the interpolant in its Lagrangian form. To this end we use the following notation. Let $A = (\Phi(x_i, x_j)) \in \mathbb{R}^{N\times N}$ and $P = (p_j(x_i)) \in \mathbb{R}^{N\times Q}$ where $p_1, \dots, p_Q$ form a basis of $\mathcal{P}$. Furthermore, let $R(x) = (\Phi(x, x_1), \dots, \Phi(x, x_N))^T \in \mathbb{R}^N$ and $S(x) = (p_1(x), \dots, p_Q(x))^T \in \mathbb{R}^Q$. Finally, let $e^{(j)} \in \mathbb{R}^N$ denote the jth unit vector.

If $X = \{x_1, \dots, x_N\}$ is $\mathcal{P}$-unisolvent then the linear system

$$\begin{pmatrix} A & P \\ P^T & 0 \end{pmatrix} \begin{pmatrix} \alpha^{(j)} \\ \beta^{(j)} \end{pmatrix} = \begin{pmatrix} e^{(j)} \\ 0 \end{pmatrix}$$

is uniquely solvable. The associated functions

$$u_j^* = \sum_{i=1}^N \alpha_i^{(j)} \Phi(\cdot, x_i) + \sum_{k=1}^Q \beta_k^{(j)} p_k$$

obviously satisfy $u_j^*(x_i) = \delta_{ij}$ and belong to the space

$$V_X := \mathcal{P} + \left\{ \sum_{j=1}^N \alpha_j \Phi(\cdot, x_j) : \sum_{j=1}^N \alpha_j p(x_j) = 0,\ p \in \mathcal{P} \right\}.$$

Since every function f from V_X is uniquely determined by $f|X$, we must have $f = \sum f(x_j)u_j^*$ for such a function. This gives the first part of the following theorem.

Theorem 11.1 *Suppose that Φ is a conditionally positive definite kernel with respect to $\mathcal{P}$ on $\Omega \subseteq \mathbb{R}^d$. Suppose that $X = \{x_1, \ldots, x_N\} \subseteq \Omega$ is $\mathcal{P}$-unisolvent. Then there exist functions $u_j^* \in V_X$ such that $u_j^*(x_k) = \delta_{jk}$. Moreover, there exist functions v_j^*, $1 \le j \le Q$, such that*

$$\begin{pmatrix} A & P \\ P^T & 0 \end{pmatrix} \begin{pmatrix} u^*(x) \\ v^*(x) \end{pmatrix} = \begin{pmatrix} R(x) \\ S(x) \end{pmatrix}. \tag{11.1}$$

Proof It remains to prove the existence of $v^*(x)$, so that $u^*(x)$ and $v^*(x)$ together satisfy (11.1). Since $\mathcal{P} \subseteq V_X$, we must have $p = \sum p(x_j)u_j^*$ for all $p \in \mathcal{P}$, or equivalently $P^T u^*(x) = S(x)$. Hence we are left with showing that $Au^*(x) - R(x) \in P(\mathbb{R}^Q)$, because this guarantees the existence of $v^*(x)$. As (11.1) has a unique solution, this finishes the proof. Since the orthogonal complement of $P(\mathbb{R}^Q)$ is given by the null space of P^T, it suffices to show that $\gamma^T(Au^*(x) - R(x)) = 0$ for all $\gamma \in \mathbb{R}^N$ with $P^T\gamma = 0$. But $P^T\gamma = 0$ means that γ is admissible, i.e. $\gamma^T R(x) \in V_X$. This means in particular that

$$\gamma^T R(x) = \sum_{j=1}^N u_j^*(x)\gamma^T R(x_j) = \sum_{j=1}^N u_j^*(x) \sum_{i=1}^N \gamma_j \Phi(x_i, x_j) = \gamma^T A u^*(x)$$

or, equivalently, that $\gamma^T(Au^*(x) - R(x)) = 0$. □

Note that the functions $v_j^*(x)$ have the remarkable property $v_k^*(x_\ell) = 0$. As a consequence of Theorem 11.1, we are now able to rewrite an interpolant as

$$s_{f,X}(x) = \sum_{j=1}^N f(x_j)u_j^*(x), \tag{11.2}$$

which will be very useful later on. Furthermore, we see that the function $s_{f,X}$ is as smooth as the functions u_j^* and these functions inherit by (11.1) the smoothness of Φ with respect to the first argument and that of $\mathcal{P}$. Thus if Φ is in C^k with respect to the first argument and $\mathcal{P} \subseteq C^k(\Omega)$ then so is $s_{f,X}$. Of course, this also follows immediately from the standard representation of $s_{f,X}$.

In what follows, we will write $D^\alpha R(x)$ for $(D_1^\alpha \Phi(x, x_1), \ldots, D_1^\alpha \Phi(x, x_N))^T$, where D_1^α again denotes the derivative with respect to the first argument. We use $D^\alpha S(x)$ in the same way, component-wise.

A formal differentiation of (11.1) then gives

$$\begin{pmatrix} A & P \\ P^T & 0 \end{pmatrix} \begin{pmatrix} D^\alpha u^*(x) \\ D^\alpha v^*(x) \end{pmatrix} = \begin{pmatrix} D^\alpha R(x) \\ D^\alpha S(x) \end{pmatrix}. \tag{11.3}$$

Under the assumptions that $\Phi \in C^{2k}(\Omega \times \Omega)$, that $\mathcal{P} \subseteq C^k(\Omega)$, and that $\Omega \subseteq \mathbb{R}^d$ is open, we know that both $f \in \mathcal{N}_\Phi(\Omega)$ and $s_{f,X}$ are in $C^k(\Omega)$. Thus it seems to be natural to ask for error bounds not only on $f - s_{f,X}$ but also on the derivatives $D^\alpha(f - s_{f,X})$ for $|\alpha| \le k$.

Definition 11.2 *Suppose that $\Omega \subseteq \mathbb{R}^d$ is open and that $\Phi \in C^{2k}(\Omega \times \Omega)$ is a conditionally positive definite kernel on Ω with respect to $\mathcal{P} \subseteq C^k(\Omega)$. If $X = \{x_1, \dots, x_N\} \subseteq \Omega$ is $\mathcal{P}$-unisolvent then for every $x \in \Omega$ and $\alpha \in \mathbb{N}_0^d$ with $|\alpha| \le k$ the power function is defined by*

$$\left[P_{\Phi,X}^{(\alpha)}(x)\right]^2 := D_1^\alpha D_2^\alpha \Phi(x,x) - 2\sum_{j=1}^N D^\alpha u_j^*(x) D_1^\alpha \Phi(x,x_j)$$
$$+ \sum_{i,j=1}^N D^\alpha u_i^*(x) D^\alpha u_j^*(x) \Phi(x_i,x_j).$$

This function plays an important role in our estimates, as we shall see very soon, but first we will have another look at the power function. If then we keep x, X, Φ, and α fixed then we can replace the constant vector $D^\alpha u^*(x) \in \mathbb{R}^N$ by an arbitrary vector $u \in \mathbb{R}^N$. Thus let us define the quadratic form $\mathcal{Q} : \mathbb{R}^N \to \mathbb{R}$ by

$$\mathcal{Q}(u) = D_1^\alpha D_2^\alpha \Phi(x,x) - 2\sum_{j=1}^N u_j D_1^\alpha \Phi(x,x_j)$$
$$+ \sum_{i,j=1}^N u_i u_j \Phi(x_i,x_j), \qquad u \in \mathbb{R}^N.$$

If necessary, we will also write $\mathcal{Q}_\Phi(u) = \mathcal{Q}(u)$. With this definition the power function becomes

$$\left[P_{\Phi,X}^{(\alpha)}(x)\right]^2 = \mathcal{Q}(D^\alpha u^*(x)),$$

and we will exploit this fact later on. But to do this we need a different representation of the quadratic form $\mathcal{Q}$.

Lemma 11.3 *Suppose that $\Phi \in C^{2k}(\Omega \times \Omega)$ is a conditionally positive definite kernel with respect to $\mathcal{P} \subseteq C^k(\Omega)$. Fix $x \in \Omega$. Now suppose that $u^{(\alpha)} \in \mathbb{R}^N$ is a vector that satisfies $\sum_j u_j^{(\alpha)} p(x_j) = D^\alpha p(x)$ for all $p \in \mathcal{P}$. Then the quadratic form $\mathcal{Q}$ has the representation*

$$\mathcal{Q}(u^{(\alpha)}) = \left| D_2^\alpha G(\cdot,x) - \sum_{j=1}^N u_j^{(\alpha)} G(\cdot,x_j) \right|^2_{\mathcal{N}_\Phi(\Omega)}, \tag{11.4}$$

where G is the modified kernel from (10.4).

Proof The proof involves some simple, straightforward, but unfortunately also lengthy and tedious computations. The right-hand side of (11.4) can be expressed as

$$\left|D_2^\alpha G(\cdot,x)\right|^2_{\mathcal{N}_\Phi(\Omega)} - 2\sum_{j=1}^N u_j^{(\alpha)} \left(D_2^\alpha G(\cdot,x), G(\cdot,x_j)\right)_{\mathcal{N}_\Phi(\Omega)}$$
$$+ \sum_{i,j=1}^N u_i^{(\alpha)} u_j^{(\alpha)} \left(G(\cdot,x_i), G(\cdot,x_j)\right)_{\mathcal{N}_\Phi(\Omega)}.$$

Thus we have to compute these three types of inner products. Since $G(\cdot,x) =$

$\Phi(\cdot, x) - \sum_{n=1}^{Q} p_n(x)\Phi(\cdot, \xi_n)$ we have immediately

$$\left(G(\cdot, x_i), G(\cdot, x_j)\right)_{\mathcal{N}_\Phi(\Omega)} = \Phi(x_i, x_j) + \sum_{n,\ell=1}^{Q} p_n(x_i)p_\ell(x_j)\Phi(\xi_\ell, \xi_n)$$
$$- \sum_{n=1}^{Q} p_n(x_i)\Phi(x_j, \xi_n) - \sum_{\ell=1}^{Q} p_\ell(x_j)\Phi(\xi_\ell, x_i).$$

Moreover, using $\sum u_j^{(\alpha)} p(x_j) = D^\alpha p(x)$ gives

$$\sum_{i,j=1}^{N} u_i^{(\alpha)} u_j^{(\alpha)} \left(G(\cdot, x_i), G(\cdot, x_j)\right)_{\mathcal{N}_\Phi(\Omega)}$$
$$= \sum_{i,j=1}^{N} u_i^{(\alpha)} u_j^{(\alpha)} \Phi(x_i, x_j) - 2\sum_{j=1}^{N}\sum_{n=1}^{Q} D^\alpha p_n(x) u_j^{(\alpha)} \Phi(x_j, \xi_n)$$
$$+ \sum_{n,\ell=1}^{Q} D^\alpha p_n(x) D^\alpha p_\ell(x) \Phi(\xi_\ell, \xi_n).$$

Next, from Lemma 10.44 we know that $D_2^\alpha G(\cdot, x)$ is in the native space and has therefore a representation

$$D_2^\alpha G(y, x) = \sum_{\ell=1}^{Q} D_2^\alpha G(\xi_\ell, x) p_\ell(y) + (D_2^\alpha G(\cdot, x), G(\cdot, y))_{\mathcal{N}_\Phi(\Omega)}$$

by Theorem 10.17. This allows us to compute the second term in our initial sum. The definition of $G(\cdot, \cdot)$ and the reproduction property of the coefficients $u_j^{(\alpha)}$ yield

$$\sum_{j=1}^{N} u_j^{(\alpha)} \left(D_2^\alpha G(\cdot, x), G(\cdot, x_j)\right)_{\mathcal{N}_\Phi(\Omega)}$$
$$= \sum_{j=1}^{N} u_j^{(\alpha)} D_2^\alpha \Phi(x_j, x) - \sum_{j=1}^{N}\sum_{n=1}^{Q} u_j^{(\alpha)} D^\alpha p_n(x) \Phi(x_j, \xi_n)$$
$$- \sum_{\ell=1}^{Q} D^\alpha p_\ell(x) D_2^\alpha \Phi(\xi_\ell, x) + \sum_{n,\ell=1}^{Q} D^\alpha p_\ell(x) D^\alpha p_n(x) \Phi(\xi_\ell, \xi_n).$$

Finally, since $D_2^\alpha G(\cdot, x) \in \mathcal{N}_\Phi(\Omega) \subseteq C^k(\Omega)$, Theorem 10.45 allows the representation

$$D_1^\alpha D_2^\alpha G(y, x) = D^\alpha(\Pi_{\mathcal{P}} D_2^\alpha G(\cdot, x))(y) + \left(D_2^\alpha G(\cdot, x), D_2^\alpha G(\cdot, y)\right)_{\mathcal{N}_\Phi(\Omega)}.$$

Hence, after some manipulation we get

$$\left(D_2^\alpha G(\cdot, x), D_2^\alpha G(\cdot, x)\right)_{\mathcal{N}_\Phi(\Omega)}$$
$$= D_1^\alpha D_2^\alpha \Phi(x, x) - \sum_{n=1}^{Q} D^\alpha p_n(x) D_1^\alpha \Phi(x, \xi_n) - \sum_{\ell=1}^{Q} D^\alpha p_\ell(x) D_2^\alpha \Phi(\xi_\ell, x)$$
$$+ \sum_{n,\ell=1}^{Q} D^\alpha p_\ell(x) D^\alpha p_n(x) \Phi(\xi_\ell, \xi_n).$$

Summing up all the terms and using $D_1^\alpha \Phi(x, y) = D_2^\alpha \Phi(y, x)$, we derive the stated result. □

The reader should have noticed that the technical intrincacy in the proof of the preceding lemma was caused by two things. If, however, one uses a positive definite kernel and is only interested in the error of the pure function and not its derivatives, one immediately has the representation

$$\begin{aligned}\mathcal{Q}(u) &= \left\| \Phi(\cdot, x) - \sum_{j=1}^N u_j \Phi(\cdot, x_j) \right\|_{\mathcal{N}_\Phi(\Omega)}^2 \\ &= \Phi(x, x) - 2 \sum_{j=1}^N u_j \Phi(x, x_j) + \sum_{i,j=1}^N u_i u_j \Phi(x_i, x_j).\end{aligned}$$

After this preparatory step it is easy to show how the power function is involved in finding error estimates for our approximation scheme.

Theorem 11.4 *Let $\Omega \subseteq \mathbb{R}^d$ be open. Suppose that $\Phi \subseteq C^{2k}(\Omega \times \Omega)$ is a conditionally positive definite kernel on Ω with respect to $\mathcal{P} \subseteq C^k(\Omega)$. Suppose further that $X = \{x_1, \dots, x_N\} \subseteq \Omega$ is $\mathcal{P}$-unisolvent. Denote the interpolant of $f \in \mathcal{N}_\Phi(\Omega)$ by $s_{f,X}$. Then for every $x \in \Omega$ and every $\alpha \in \mathbb{N}_0^d$ with $|\alpha| \le k$ the error between f and its interpolant can be bounded by*

$$|D^\alpha f(x) - D^\alpha s_{f,X}(x)| \le P_{\Phi,X}^{(\alpha)}(x) |f|_{\mathcal{N}_\Phi(\Omega)}.$$

Proof Using representation (11.2), the Taylor formula from Theorem 10.17, and the reproduction property of the coefficients, which follows from (11.3), we see that

$$\begin{aligned}D^\alpha s_{f,X}(x) &= \sum_{j=1}^N f(x_j) D^\alpha u_j^*(x) \\ &= \sum_{j=1}^N D^\alpha u_j^*(x) \left[\Pi_{\mathcal{P}} f(x_j) + (f, G(\cdot, x_j))_{\mathcal{N}_\Phi(\Omega)} \right] \\ &= D^\alpha(\Pi_{\mathcal{P}} f)(x) + \left(f, \sum_{j=1}^N D^\alpha u_j^*(x) G(\cdot, x_j) \right)_{\mathcal{N}_\Phi(\Omega)}.\end{aligned}$$

However, Theorem 10.45 allows us to write

$$D^\alpha f(x) = D^\alpha(\Pi_{\mathcal{P}} f)(x) + (f, D_2^\alpha G(\cdot, x))_{\mathcal{N}_\Phi(\Omega)},$$

which, together with the previous equation, yields

$$\begin{aligned}|D^\alpha(f - s_{f,X})(x)| &= \left| \left(f, D_2^\alpha G(\cdot, x) - \sum_{j=1}^N D^\alpha u_j^*(x) G(\cdot, x_j) \right)_{\mathcal{N}_\Phi(\Omega)} \right| \\ &\le |f|_{\mathcal{N}_\Phi(\Omega)} P_{\Phi,X}^{(\alpha)}(x)\end{aligned}$$

by Lemma 11.3. □

This theorem allows us to split the error between the unknown function f from the native space and its interpolant into two terms, one term (the power function) being independent of f and the other independent of the centers X. Our further investigation of the error will be done by bounding the power function in an appropriate way. Therefore, we regard the power function again as a function of the coefficients $u_j^*(x)$.

Theorem 11.5 *Let $\Omega \subseteq \mathbb{R}^d$ be open. Suppose that $\Phi \in C^{2k}(\Omega \times \Omega)$ is a conditionally positive definite kernel on Ω with respect to $\mathcal{P} \subseteq C^k(\Omega)$. Suppose further that $X = \{x_1, \dots, x_N\} \subseteq \Omega$ is $\mathcal{P}$-unisolvent. Define for $x \in \Omega$ and $\alpha \in \mathbb{N}_0^d$ with $|\alpha| \le k$ the function $\mathcal{Q} : \mathbb{R}^N \to \mathbb{R}$:*

$$\mathcal{Q}(u) := D_1^\alpha D_2^\alpha \Phi(x, x) - 2 \sum_{j=1}^N u_j D_1^\alpha \Phi(x, x_j) + \sum_{i=1}^N \sum_{j=1}^N u_i u_j \Phi(x_i, x_j).$$

The minimum of this function on the set

$$M = \left\{ u \in \mathbb{R}^N : \sum_{j=1}^N u_j p(x_j) = D^\alpha p(x) \, \text{for all } p \in \mathcal{P} \right\}$$

is given by the vector $D^\alpha u^*(x)$, where $u^*(x)$ is found in Theorem 11.1:

$$\mathcal{Q}(D^\alpha u^*(x)) \le \mathcal{Q}(u) \qquad \text{for all } u \in M.$$

Proof If we adopt the notation of the paragraph before Theorem 11.1 we see that the function $\mathcal{Q}$ takes the form $\mathcal{Q}(u) = D_1^\alpha D_2^\alpha \Phi(x, x) - 2u^T D^\alpha R(x) + u^T A u$ and has to be minimized over $M = \{u : P^T u = D^\alpha S(x)\}$. Since M is nonempty and A is positive definite on $M_0 := \{u : P^T u = 0\}$, Lemma 4.2 yields that this quadratic minimization problem has a unique solution that can be computed using Lagrange multipliers. Doing so, we derive the equations

$$\begin{aligned} Au + Pv &= D^\alpha R(x), \\ P^T u &= D^\alpha S(x), \end{aligned}$$

where v denotes the Lagrange multiplier. This system is uniquely solved by the functions $u = D^\alpha u^*(x)$ and $v = D^\alpha v^*(x)$. □

Having this minimal property in mind, the idea is to bounding the power function by plugging into $\mathcal{Q}$ an appropriate vector $\widetilde{u}^{(\alpha)}(x)$ instead of the optimal vector $D^\alpha u^*(x)$.

11.2 Error estimates in terms of the fill distance

As indicated in the last section we now want to bound the power function by replacing the optimal vector $D^\alpha u^*(x)$ appropriately.

We will do this first for conditionally positive definite functions and then for arbitrary symmetric conditionally positive definite kernels $\Phi \in C^{2k}(\Omega \times \Omega)$. In both situations we assume the general finite-dimensional subspace $\mathcal{P}$ to be $\pi_{m-1}(\mathbb{R}^d)$.

In the following, the region $\Omega \subseteq \mathbb{R}^d$ is always assumed to be open. But this is only necessary for estimates on the derivatives. In the non-derivative case Ω has only to satisfy an interior cone condition. Moreover, if Ω is not open, the estimates on the derivatives hold in every interior point.

For our estimates we will employ local polynomial reproductions as we have studied them in Chapter 4, in particular in the form of Theorem 3.14. But here we need a more general version covering also derivatives. Again, norming sets are the key ingredient. To use them, we first have to derive a Bernstein inequality for multivariate polynomials.

Proposition 11.6 *Suppose that $\Omega \subseteq \mathbb{R}^d$ is bounded and satisfies an interior cone condition with radius $r > 0$ and angle θ. If $p \in \pi_\ell(\mathbb{R}^d)$ and $\alpha \in \mathbb{N}_0$ is a multi-index for which $|\alpha| \le \ell$ then*

$$\|D^\alpha p\|_{L_\infty(\Omega)} \le \left(\frac{2\ell^2}{r\sin\theta}\right)^{|\alpha|} \|p\|_{L_\infty(\Omega)}.$$

Proof Obviously the result is true if $\alpha = 0$ or $D^\alpha p = 0$. Hence let us assume that ∇p is not identically zero. The maximum of $\|\nabla p(x)\|_2$ over $\overline{\Omega}$ occurs at some point $x_M \in \overline{\Omega}$. Obviously, the maximum is positive. Let $\eta = \nabla p(x_M)/\|\nabla p(x_M)\|_2$. Because $x_M \in \overline{\Omega}$, the cone condition, which holds for $\overline{\Omega}$ as well as Ω, implies that x_M is the vertex of a cone $C \subseteq \overline{\Omega}$ having radius r, axis along a direction ξ, and angle θ. We may adjust the sign of p so that $\eta^T\xi \ge 0$. By looking at the intersection of the cone C with a plane containing ξ and η, we see that there is a unit vector ζ pointing into the cone and satisfying $\eta^T\zeta \ge \cos(\pi/2 - \theta) = \sin\theta$. It follows that

$$\|\nabla p(x_M)\|_2 = \frac{\partial p}{\partial \eta}(x_M) \le \csc\theta \frac{\partial p}{\partial \zeta}(x_M).$$

However, for $t \in \mathbb{R}$, $\tilde{p}(t) := p(x_M + t\zeta)$ is in $\pi_\ell(\mathbb{R})$. In particular, it obeys the usual Bernstein inequality on $0 \le t \le r$:

$$|\tilde{p}'(t)| \le \frac{2\ell^2}{r} \max_{t\in[0,r]} |\tilde{p}(t)| \le \frac{2\ell^2}{r}\|p\|_{L_\infty(\Omega)}.$$

Since $\tilde{p}'(0) = (\partial p/\partial\zeta)(x_M)$, we have for all $x \in \overline{\Omega}$

$$\|\nabla p(x)\|_2 \le \|\nabla p(x_M)\|_2 \le \csc\theta \frac{\partial p}{\partial \zeta}(x_M) \le \frac{2\ell^2}{r\sin\theta}\|p\|_{L_\infty(\Omega)}.$$

Noting that $|(\partial p/\partial x_j)(x)| \le \|\nabla p(x)\|_2$ and, keeping track of polynomial degrees as we differentiate, we obtain the stated result. □

With this result, Theorem 3.4 together with Theorem 3.8 immediately yields the global version of a polynomial reproduction.

Proposition 11.7 *Let $p \in \pi_\ell(\mathbb{R}^d)$ and let Ω be bounded, satisfying an interior cone condition with radius $r > 0$ and angle θ. Suppose that $h > 0$ and the set $X = \{x_1, \dots, x_N\} \subseteq \Omega$ satisfy*

(1) $h \le \dfrac{r \sin\theta}{4(1+\sin\theta)\ell^2}$,
(2) for every $B(x,h) \subseteq \Omega$ *there is a center* $x_j \in X \cap B(x,h)$;

then for any multi-index α *with* $|\alpha| \le \ell$ *there exist real numbers* $a_j^\alpha(x)$ *such that*

$$D^\alpha p(x) = \sum_{j=1}^N a_j^\alpha(x) p(x_j)$$

for all $p \in \pi_\ell(\mathbb{R}^d)$. *Moreover,*

$$\sum_{j=1}^N |a_j^\alpha(x)| \le 2\left(\frac{2\ell^2}{r\sin\theta}\right)^{|\alpha|}.$$

As pointed out after the proof of Theorem 3.8, the second condition is automatically satisfied if h is the fill distance $h_{X,\Omega}$. Using again the fact that a cone satisfies a cone condition itself, we can proceed as in Section 3.3 to derive the following local version.

Theorem 11.8 *Suppose that* $\Omega \subseteq \mathbb{R}^d$ *is bounded and satisfies an interior cone condition. Let* $\ell \in \mathbb{N}_0$ *and* $\alpha \in \mathbb{N}_0^d$ *with* $|\alpha| \le \ell$. *Then there exist constants* $h_0, c_1^{(\alpha)}, c_2^{(\alpha)} > 0$ *such that for all* $X = \{x_1, \ldots, x_N\} \subseteq \Omega$ *with* $h_{X,\Omega} \le h_0$ *and every* $x \in \Omega$ *there exist numbers* $\widetilde{u}_1^{(\alpha)}(x), \ldots, \widetilde{u}_N^{(\alpha)}(x)$ *with*

(1) $\sum_{j=1}^N \widetilde{u}_j^{(\alpha)}(x) p(x_j) = D^\alpha p(x)$ *for all* $p \in \pi_\ell(\mathbb{R}^d)$,
(2) $\sum_{j=1}^N |\widetilde{u}_j^{(\alpha)}(x)| \le c_1^{(\alpha)} h_{X,\Omega}^{-|\alpha|}$,
(3) $\widetilde{u}_j^{(\alpha)}(x) = 0$, *if* $\|x - x_j\|_2 > c_2^{(\alpha)} h_{X,\Omega}$.

Note that the construction ensures also the following important property. Given $x \in \Omega$, at most those $\widetilde{u}_j^{(\alpha)}(x)$ that belong to a center x_j in the cone associated with x are nonzero. Hence all line segments that connect one of these x_i with either x or another x_j are also contained in Ω. This allows us to apply Taylor's formula later on.

Our first main result deals with (conditionally) positive definite functions. After it, we will also deal with (conditionally) positive definite kernels. To treat the case of a (conditionally) positive definite function we have to remark that an even function $\Phi : \mathbb{R}^d \to \mathbb{R}$ that is in $C^k(\mathbb{R}^d)$ gives rise to a symmetric kernel $\Phi(\cdot - \cdot)$ that is in $C^{2\lfloor k/2 \rfloor}(\mathbb{R}^d \times \mathbb{R}^d)$. Moreover, we do not have to restrict ourselves to open regions Ω, even in the case of derivatives, because any function $f \in \mathcal{N}_\Phi(\Omega)$ is the restriction of a function from $\mathcal{N}_\Phi(\mathbb{R}^d) \subseteq C^{\lfloor k/2 \rfloor}(\mathbb{R}^d)$.

Theorem 11.9 *Suppose that* $\Phi \in C^k(\mathbb{R}^d)$ *is conditionally positive definite of order* m. *Suppose further that* $\Omega \subseteq \mathbb{R}^d$ *is bounded and satisfies an interior cone condition. Fix* $\ell \ge m-1$. *For* $\alpha \in \mathbb{N}_0^d$ *with* $|\alpha| \le k/2$ *and* $X = \{x_1, \ldots, x_N\} \subseteq \Omega$ *satisfying* $h_{X,\Omega} \le h_0$ *the power function can be bounded:*

$$\begin{aligned}\left[P_{\Phi,X}^{(\alpha)}(x)\right]^2 &\le |D^{2\alpha}\Phi(0) - D^{2\alpha} p(0)| \\ &\quad + 2c_1^{(\alpha)} h_{X,\Omega}^{-|\alpha|} \|D^\alpha \Phi - D^\alpha p\|_{L_\infty(B(0,\, c_2^{(\alpha)} h_{X,\Omega}))} \\ &\quad + [c_1^{(\alpha)}]^2 h_{X,\Omega}^{-2|\alpha|} \|\Phi - p\|_{L_\infty(B(0,\, 2c_2^{(\alpha)} h_{X,\Omega}))},\end{aligned} \tag{11.5}$$

where p is an arbitrary polynomial from $\pi_\ell(\mathbb{R}^d)$ and the constants h_0, $c_1^{(\alpha)}$, $c_2^{(\alpha)}$ come from Theorem 11.8.

Proof Let us introduce the notation

$$\Delta_0 := |D^{2\alpha}\Phi(0) - D^{2\alpha}p(0)|,$$
$$\Delta_1 := \|D^\alpha\Phi - D^\alpha p\|_{L_\infty(B(0,\, c_2^{(\alpha)}h_{X,\Omega}))}$$
$$\Delta_2 := \|\Phi - p\|_{L_\infty(B(0,\, 2c_2^{(\alpha)}h_{X,\Omega}))}.$$

The polynomial reproduction property of the functions $\widetilde{u}^{(\alpha)}$ from Theorem 11.8 gives $\sum_{j=1}^N \widetilde{u}_j^{(\alpha)} D^\alpha p(x - x_j) = (-1)^{|\alpha|} D^\alpha p(0)$ and, when applied twice,

$$\sum_{i,j,=1}^N \widetilde{u}_i^{(\alpha)}(x)\widetilde{u}_j^{(\alpha)}(x)p(x_i - x_j) = (-1)^{|\alpha|}\sum_{i=1}^N \widetilde{u}_i^{(\alpha)}(x)D^\alpha p(x_i - x)$$
$$= (-1)^{|\alpha|}D^{2\alpha}p(0).$$

Hence, if we rewrite the quadratic form $\mathcal{Q}$ with Φ replaced by p, we find that $\mathcal{Q}_p(\widetilde{u}^{(\alpha)}) = 0$. Thus we can bound the power function:

$$\left[P_{\Phi,X}^{(\alpha)}(x)\right]^2 \le \mathcal{Q}_\Phi(\widetilde{u}^{(\alpha)}) - \mathcal{Q}_p(\widetilde{u}^{(\alpha)}) = \mathcal{Q}_{\Phi-p}(\widetilde{u}^{(\alpha)})$$
$$\le \Delta_0 + \Delta_1\sum_{j=1}^N \left|\widetilde{u}_j^{(\alpha)}(x)\right| + \Delta_2\sum_{i,j=1}^N \left|\widetilde{u}_i^{(\alpha)}(x)\right|\left|\widetilde{u}_j^{(\alpha)}(x)\right|$$
$$\le \Delta_0 + 2c_1^{(\alpha)}h_{X,\Omega}^{-|\alpha|}\Delta_1 + \left[c_1^{(\alpha)}\right]^2 h_{X,\Omega}^{-2|\alpha|}\Delta_2.$$

This holds for any $p \in \pi_\ell(\mathbb{R}^d)$. In the preceding estimates we have in particular employed the other two features of the coefficients. □

Note that if one is interested only in function values, the error estimate of the last theorem becomes

$$P_{\Phi,X}^2(x) \le \left(1 + c_1^{(0)}\right)^2 \|\Phi - p\|_{L_\infty(B(0,\, 2c_2^{(0)}h_{X,\Omega}))}.$$

Furthermore, in the case of a radial function $\Phi = \phi(\|\cdot\|_2)$ one can use univariate polynomials $p \in \pi_n(\mathbb{R})$ as multivariate polynomials via $p(\|\cdot\|_2^2) \in \pi_{2n}(\mathbb{R}^d)$, which leads to the estimates

$$P_{\Phi,X}^2(x) \le \left(1 + c_1^{(0)}\right)^2 \max_{0\le r\le 2c_2^{(0)}h_{X,\Omega}} |\phi(r) - p(r^2)|$$
$$= \left(1 + c_1^{(0)}\right)^2 \max_{0\le s\le [2c_2^{(0)}]^2 h_{X,\Omega}^2} |\phi(\sqrt{s}) - p(s)|. \tag{11.6}$$

Theorem 11.9 allows us to state our first generic error estimate.

Definition 11.10 *The space $C_\nu^k(\mathbb{R}^d)$ is defined to consist of all functions $f \in C^k(\mathbb{R}^d)$ whose derivatives of order k satisfy $D^\alpha f(x) = \mathcal{O}(\|x\|_2^\nu)$ for $\|x\|_2 \to 0$.*

Theorem 11.11 *Suppose that $\Phi \in C_\nu^k(\mathbb{R}^d)$ is conditionally positive definite of order m. Suppose further that $\Omega \subseteq \mathbb{R}^d$ is bounded and satisfies an interior cone condition. For $\alpha \in \mathbb{N}_0^d$ with $|\alpha| \le k/2$ and $X = \{x_1, \dots, x_N\} \subseteq \Omega$ satisfying $h_{X,\Omega} \le h_0$ we have the error bound*

$$\|D^\alpha f - D^\alpha s_{f,X}\|_{L_\infty(\Omega)} \le C h_{X,\Omega}^{(k+\nu)/2-|\alpha|} |f|_{\mathcal{N}_\Phi(\Omega)}.$$

Proof We fix $\ell \ge \max\{m-1, k-1\}$ and take p as the Taylor polynomial of Φ of degree $k-1$, i.e. $p(x) = \sum_{|\beta|<k} D^\beta \Phi(0) x^\beta / \beta!$. Since $\Phi \in C_\nu^k(\mathbb{R}^d)$, we immediately get for any $|\gamma| \le k$

$$|D^\gamma \Phi(x) - D^\gamma p(x)| \le \sum_{|\beta|=k-|\gamma|} \frac{|D^{\beta+\gamma}\Phi(\xi)|}{\beta!} |x^\beta| \le C h_{X,\Omega}^{k-|\gamma|+\nu},$$

provided that $\|x\|_2 \le c h_{X,\Omega}$. Inserting the corresponding results for $\gamma = 0, \alpha, 2\alpha$ into (11.5) gives the desired result. □

Remark 11.12 *It is worthwhile to note that the error estimate of the preceeding theorem holds for every $h \le h_0$ which satisfies the condition that every ball $B(x, h) \subseteq \Omega$ contains at least one point from X. Moreover, $h_0 = r/C_2$ with C_2 from Theorem 3.14. Finally, the constant C depends on Ω only via its cone condition angle θ.*

We end this section by discussing the kernel case.

Theorem 11.13 *Let $\Omega \subseteq \mathbb{R}^d$ be open and bounded, satisfying an interior cone condition. Suppose that $\Phi \in C^{2k}(\Omega \times \Omega)$ is conditionally positive definite with respect to $\pi_{m-1}(\mathbb{R}^d)$. Denote the interpolant to $f \in \mathcal{N}_\Phi(\Omega)$ that is based on the $\pi_{m-1}(\mathbb{R}^d)$-unisolvent set $X = \{x_1, \dots, x_N\}$ by $s_{f,X}$. Fix $\alpha \in \mathbb{N}_0^d$ with $|\alpha| \le k$. Then there exist constants $h_0, C > 0$ such that*

$$|D^\alpha f(x) - D^\alpha s_{f,X}(x)| \le C C_\Phi(x)^{1/2} h_{X,\Omega}^{k-|\alpha|} |f|_{\mathcal{N}_\Phi(\Omega)}, \qquad x \in \Omega,$$

if $h_{X,\Omega} \le h_0$. The number $C_\Phi(x)$ is defined by

$$C_\Phi(x) := \max_{\substack{\beta,\nu \in \mathbb{N}_0^d \\ |\beta|+|\nu|=2k}} \max_{z,w \in \Omega \cap B(x,\, c_2^{(\alpha)} h_{X,\Omega})} \left| D_1^\beta D_2^\nu \Phi(z, w) \right|.$$

and the constant C is independent of x, f, and Φ.

Proof We will make use in the following of two Taylor expansions. In both cases we keep the first argument $w \in \Omega$ fixed and expand the function with respect to its second argument around w. Moreover, we have to ensure that the line segment between $w \in \Omega$ and $z \in \Omega$ is also contained in Ω. Fix $\nu \in \mathbb{N}_0^d$ with $|\nu| \le k$. The first expansion is

$$D_1^\nu \Phi(w, z) = \sum_{|\beta|<2k-|\nu|} \frac{D_2^\beta D_1^\nu \Phi(w, w)}{\beta!} (z - w)^\beta + R(w, z, \nu),$$

with remainder

$$R(w, z, \nu) := \sum_{|\beta|=2k-|\nu|} \frac{D_2^\beta D_1^\nu \Phi(w, \xi_{w,z}^\nu)}{\beta!}(z-w)^\beta.$$

Here, $\xi_{w,z}^\nu$ is a point on the line segment between w and z. Similarly, we have the second expansion,

$$D_2^\nu \Phi(w, z) = \sum_{|\beta|<2k-|\nu|} \frac{D_2^{\beta+\nu}\Phi(w, w)}{\beta!}(z-w)^\beta + S(w, z, \nu).$$

This time the remainder takes the form

$$S(w, z, \nu) := \sum_{|\beta|=2k-|\nu|} \frac{D_2^{\beta+\nu}\Phi(w, \eta_{w,z}^\nu)}{\beta!}(z-w)^\beta$$

with $\eta_{w,z}^\nu$ on the line segment between w and z.

We have to bound the power function to achieve the desired result. To this end we use the vector $u := \widetilde{u}^{(\alpha)}(x)$ from Theorem 11.8 with an $\ell \ge \max\{2k-1, m-1\}$. Moreover, by the remarks made after Theorem 11.8, we know that for a fixed $x \in \Omega$ all x_j relevant to the construction are contained in the cone associated with x. Hence all line segments between x_i and x or x_i and another x_j for these x_j are contained in Ω. The power function can be bounded by $[P_{\Phi,X}^{(\alpha)}(x)]^2 \le \mathcal{Q}(u)$, and the latter is given by

$$\mathcal{Q}(u) = D_1^\alpha D_2^\alpha \Phi(x, x) - 2\sum_j u_j D_1^\alpha \Phi(x, x_j) + \sum_{i,j} u_i u_j \Phi(x_i, x_j).$$

The summation is always over only those indices j with $u_j \ne 0$. The first Taylor expansion used twice gives

$$\begin{aligned}\mathcal{Q}(u) = {} & D_1^\alpha D_2^\alpha \Phi(x, x) \\ & - 2\sum_j u_j \left(\sum_{|\beta|<2k-|\alpha|} \frac{D_2^\beta D_1^\alpha \Phi(x, x)}{\beta!}(x_j - x)^\beta + R(x, x_j, \alpha) \right) \\ & + \sum_{i,j} u_i u_j \left(\sum_{|\beta|<2k} \frac{D_2^\beta \Phi(x_i, x_i)}{\beta!}(x_j - x_i)^\beta + R(x_i, x_j, 0) \right).\end{aligned}$$

An application of the reproduction property of the coefficient vector u together with an application of the second Taylor expansion yields

$$\begin{aligned}\mathcal{Q}(u) = {} & D_1^\alpha D_2^\alpha \Phi(x, x) - 2D_2^\alpha D_1^\alpha \Phi(x, x) - 2\sum_j u_j R(x, x_j, \alpha) \\ & + \sum_i u_i \sum_{|\beta|<2k-|\alpha|} \frac{D_2^{\alpha+\beta}\Phi(x_i, x_i)}{\beta!}(x - x_i)^\beta + \sum_{i,j} u_i u_j R(x_i, x_j, 0) \\ = {} & -D_1^\alpha D_2^\alpha \Phi(x, x) - \sum_j u_j \left(2R(x, x_j, \alpha) - \sum_i u_i R(x_i, x_j, 0) \right) \\ & + \sum_i u_i \left[D_2^\alpha \Phi(x_i, x) - S(x_i, x, \alpha) \right].\end{aligned}$$

Since $D_2^\alpha \Phi(x_i, x) = D_1^\alpha \Phi(x, x_i)$, a final application of the first Taylor formula and another application of the reproduction property of the coefficient vector leads finally to

$$Q(u) = -\sum_j u_j \left(R(x, x_j, \alpha) + S(x_j, x, \alpha) - \sum_i u_i R(x_i, x_j, 0) \right).$$

However, we know that $\sum_j |u_j| \le c_1^{(\alpha)} h_{X,\Omega}^{-|\alpha|}$. Moreover, because $\|x - x_j\|_2 \le c_2^{(\alpha)} h_{X,\Omega}$ and $\|x_i - x_j\|_2 \le 2c_2^{(\alpha)} h_{X,\Omega}$ we see that the first two remainder terms can be bounded by $CC_\Phi(x) h_{X,\Omega}^{2k-|\alpha|}$ and the last term also by $CC_\Phi(x) h_{X,\Omega}^{2k-|\alpha|}$, using the bound on the ℓ_1-norm of the coefficients again. This gives for the power function $P_{\Phi,X}^{(\alpha)}(x) \le CC_\Phi(x)^{1/2} h_{X,\Omega}^{k-|\alpha|}$ and hence the desired result. □

The reason for the special treatment of the number C_Φ is that in many cases it allows an improvement over the $\mathcal{O}(h^{k-|\alpha|})$ order, as we will see very soon. Moreover, if all derivatives of Φ of order $2k$ are continuous on $\overline{\Omega} \times \overline{\Omega}$ then $C_\Phi(x)$ is uniformly bounded on Ω.

Moreover, the assumption that X is $\pi_{m-1}(\mathbb{R}^d)$-unisolvent is automatically satisfied if $h_{X,\Omega} \le h_0$. The latter condition was mainly made to allow unique polynomial interpolation in a subset of X for polynomials of degree at most $\ell \ge m - 1$.

Finally, in the case of a function Φ, the number $C_\Phi(x)$ has the form

$$C_\Phi(x) = \max_{|\beta|=2k} \|D^\beta \Phi\|_{L_\infty(B(0,\, 2c_2^{(\alpha)} h_{X,\Omega}))} \tag{11.7}$$

and is obviously independent of x.

11.3 Estimates for popular basis functions

It is time to apply the general result of Theorem 11.11 to those basis functions that have accompanied us so far. Our first application deals with basis functions of infinite smoothness. Without a closer look at the involved constants we immediately get an arbitrary convergence order.

Theorem 11.14 *Let Φ be one of the Gaussians or the (inverse) multiquadrics. Suppose that Φ is conditionally positive definite of order m. Suppose further that $\Omega \subseteq \mathbb{R}^d$ is bounded and satisfies an interior cone condition. Denote the radial basis function interpolant to $f \in \mathcal{N}_\Phi(\Omega)$ based on Φ and $X = \{x_1, \dots, x_N\}$ by $s_{f,X}$. Fix $\alpha \in \mathbb{N}_0^d$. For every $\ell \in \mathbb{N}$ with $\ell \ge \max\{|\alpha|, m-1\}$ there exist constants $h_0(\ell), C_\ell > 0$ such that*

$$|D^\alpha f(x) - D^\alpha s_{f,X}(x)| \le C_\ell h_{X,\Omega}^{\ell-|\alpha|} |f|_{\mathcal{N}_\Phi(\Omega)}$$

for all $x \in \Omega$, provided that $h_{X,\Omega} \le h_0(\ell)$.

To derive spectral convergence orders it is crucial to study how the constants $h_0(\ell)$ and C_ℓ depend on ℓ. We will discuss this in more detail in Section 11.4.

In the case of basis functions with a finite number of continuous derivatives it is important to know the exact Hölder class $C_\nu^k(\mathbb{R}^d)$ to which they belong.

Lemma 11.15 *Let $\Phi(x) = \|x\|_2^\beta$ with $\beta > 0$, $\beta \notin 2\mathbb{N}$. For every $\alpha \in \mathbb{N}_0^d$ there exists a homogenous polynomial $p_{\alpha,\beta} \in \pi_{|\alpha|}(\mathbb{R}^d)$ such that*

$$D^\alpha \Phi(x) = p_{\alpha,\beta}(x)\|x\|_2^{\beta-2|\alpha|}$$

for every $x \neq 0$. In particular, there exists a constant c_α such that $|D^\alpha \Phi(x)| \le c_\alpha \|x\|_2^{\beta-|\alpha|}$ for every $x \neq 0$, showing that $\Phi \in C^{\lceil\beta\rceil-1}_{\beta-\lceil\beta\rceil-1}(\mathbb{R}^d)$.

Proof The proof is by induction on the length of α. For $|\alpha| = 0$ there is nothing to show. Now assume that $|\alpha| \ge 1$. Without restriction we can assume $\alpha_1 \ge 1$. Define $\gamma = (\alpha_1 - 1, \alpha_2, \dots, \alpha_d)^T$. Then there exists a homogeneous polynomial $p_{\gamma,\beta}$ of degree $|\gamma|$ such that

$$\begin{aligned} D^\alpha \Phi(x) &= \frac{\partial}{\partial x_1} D^\gamma \Phi(x) \\ &= \frac{\partial}{\partial x_1}\left[p_{\gamma,\beta}(x)\|x\|_2^{\beta-2|\gamma|}\right] \\ &= \left(\frac{\partial p_{\gamma,\beta}}{\partial x_1}(x)\|x\|_2^2 + (\beta - 2|\gamma|)x_1 p_{\gamma,\beta}(x)\right)\|x\|_2^{\beta-2|\gamma|-2} \\ &=: p_{\alpha,\beta}(x)\|x\|_2^{\beta-2|\alpha|}. \end{aligned}$$

The polynomial $p_{\alpha,\beta}$ is indeed a homogenous polynomial of degree $|\alpha|$, because the derivative of a homogenous polynomial of degree ℓ is a homogenous polynomial of degree $\ell - 1$, and the product of two homogenous polynomials of degree ℓ and k is a homogenous polynomial of degree $\ell + k$. □

The following theorem is an immediate consequence.

Theorem 11.16 *Suppose that $\Omega \subseteq \mathbb{R}^d$ is bounded and satisfies an interior cone condition. Let $\Phi(x) = (-1)^{\lceil\beta/2\rceil}\|x\|_2^\beta$, $\beta > 0$, $\beta \notin 2\mathbb{N}$. Denote the interpolant of a function $f \in \mathcal{N}_\Phi(\Omega)$ based on this basis function and the set of centers $X = \{x_1, \dots, x_N\} \subseteq \Omega$ by $s_{f,X}$. Then there exist constants $h_0, C > 0$ such that*

$$|D^\alpha f(x) - D^\alpha s_{f,X}(x)| \le C h_{X,\Omega}^{\beta/2-|\alpha|}|f|_{\mathcal{N}_\Phi(\Omega)}, \qquad x \in \Omega,$$

for all α with $|\alpha| \le (\lceil\beta\rceil - 1)/2$ provided that $h_{X,\Omega} \le h_0$.

The next family of functions at which we want to look are the compactly supported functions $\Phi_{d,k} = \phi_{d,k}(\|\cdot\|_2)$ constructed in Chapter 9.

Theorem 11.17 *Let $\Phi_{d,k} = \phi_{d,k}(\|\cdot\|_2)$ be the functions from Theorem 9.13. Suppose that $\Omega \subseteq \mathbb{R}^d$ is bounded and satisfies an interior cone condition. Denote the radial basis function interpolant of $f \in \mathcal{N}_{\Phi_{d,k}}(\Omega)$ based on $\Phi_{d,k}$ and $X = \{x_1, \dots, x_N\} \subseteq \Omega$ by $s_{f,X}$. Then there exist constants $C, h_0 > 0$ such that*

$$|D^\alpha f(x) - D^\alpha s_{f,X}(x)| \le C h_{X,\Omega}^{k+1/2-|\alpha|}\|f\|_{\mathcal{N}_\Phi(\Omega)}$$

for every $\alpha \in \mathbb{N}_0^d$ with $|\alpha| \le k$ and every $x \in \Omega$, provided that $h_{X,\Omega} \le h_0$.

Proof From the estimates on the Fourier transform of $\Phi_{d,k}$ we already know that $\Phi_{d,k} \in C^{2k}(\mathbb{R}^d)$. Moreover, we know that $\phi_{d,k} \in C^{2k}(\mathbb{R})$ by construction. Hence, since $\phi_{d,k}$ is also a polynomial on $0 \le r \le 1$ of degree $\ell := \lfloor d/2 \rfloor + 3k + 1$, this means that exactly the first k odd coefficients of this polynomial must vanish. Thus we have

$$\phi_{d,k}(r) = a_0 + a_2 r^2 + \cdots + a_{2k} r^{2k} + a_{2k+1} r^{2k+1} + \cdots + a_\ell r^\ell$$

for $0 \le r \le 1$ with certain coefficients a_j. Taking $p(k) = a_0 + a_2\|x\|_2^2 + \cdots + a_{2k}\|x\|_2^{2k}$ shows $|D^\beta(\Phi_{d,k} - p)(x)| \le C_\beta \|x\|_2^{2k+1-|\beta|}$ for $|\beta| \le 2k$ by Lemma 11.15. Thus Theorem 11.9 gives the stated result. □

Once again, we point out that convergence for the compactly supported basis functions is gained by keeping the support radius fixed while the fill distance tends to zero. This means in particular that the advantage of the compact support gets more and more lost. In the end, the compactly supported basis functions act like globally supported ones. In contrast with this *nonstationary* setting, one could be tempted to use a *stationary* approach, i.e. to choose the support radius proportional to the fill distance. In this setting the bandwidth of the interpolation matrices is approximately constant, at least for quasi-uniform data sets. The price for this stationary setting is that we cannot conclude convergence from estimates on the power function. To be more precise, suppose we scale a basis function Φ with support in the unit ball $B(0, 1)$ by $\Phi_\delta = \Phi(\cdot/\delta)$. Then it is easy to see that the power function scales as $P_{\Phi_\delta,X}(x) = P_{\Phi,X/\delta}(x/\delta)$ with $X/\delta = \{x_1/\delta, \dots, x_N/\delta\}$. Since obviously we also have $h_{X/\delta,\Omega} = h_{X,\delta\Omega}/\delta$ we can see that

$$P_{\Phi_\delta,X}(x) \le C \left(\frac{h_{X,\delta\Omega}}{\delta}\right)^{k+1/2},$$

which will not tend to zero if δ is chosen proportional to h. Of course, we have to take into account that the native space norm to Φ_δ varies also with δ. But numerical examples show that the interpolation error does not tend to zero for $h \to 0$. Nonetheless, the error goes down to a certain threshold and remains constant afterwards. This effect is sometimes called *approximate approximation* (see for example Maz'ya and Schmidt [119]) and needs further investigation in this context.

The distinction between stationary and nonstationary settings plays a particularly important role for interpolation on a grid. For example, in [155] Powell takes the stationary point of view and shows that the "Gaussian fares badly", since Gaussian interpolation does not even provide uniform convergence. However, in [156] Ron makes a more general investigation, which covers the nonstationary setting also, and he concludes that spectral convergence holds for Gaussian interpolation. Hence when reading articles on approximation on a grid one should always keep this distinction in mind.

If only one interpolant has to be computed, which in general will be the case in applications (in contrast with numerical testing), the choice of the right support radius of a compactly supported basis function is thus a delicate question. In a later chapter we will introduce numerical methods that try to take advantage of the compact support and yield convergence nonetheless.

Our final example deals with the approximation power of thin-plate splines.

Lemma 11.18 *Let $\Phi(x) = \|x\|_2^{2k} \log \|x\|_2$ with $k \in \mathbb{N}$. For every $\alpha \in \mathbb{N}_0^d$ there exist homogenous polynomials $p_{\alpha,k}, q_{\alpha,k} \in \pi_{|\alpha|}(\mathbb{R}^d)$ such that*

$$D^\alpha \Phi(x) = (p_{\alpha,k}(x) + q_{\alpha,k}(x) \log \|x\|_2)\|x\|_2^{2k-2|\alpha|}$$

for every $x \neq 0$. In particular, there exists a constant c_α such that $|D^\alpha \Phi(x)| \leq c_\alpha(1 + \log \|x\|_2)\|x\|_2^{2k-|\alpha|}$ for every $x \neq 0$, showing that $\Phi \in C^{2k-1}(\mathbb{R}^d)$.

Proof The proof is again by induction on the length of α. For $|\alpha| = 0$ there is nothing to show. Now assume $|\alpha| \geq 1$. Without restriction we assume again $\alpha_1 \geq 1$. Define $\gamma = (\alpha_1 - 1, \alpha_2, \dots, \alpha_d)^T$. Then there exist homogeneous polynomials $p_{\gamma,k}, q_{\gamma,k}$ of degree $|\gamma|$ such that

$$\begin{aligned}
D^\alpha \Phi(x) &= \frac{\partial}{\partial x_1} D^\gamma \Phi(x) \\
&= \frac{\partial}{\partial x_1} \left[\left(p_{\gamma,k}(x) + q_{\gamma,k}(x) \log \|x\|_2 \right) \|x\|_2^{2k-2|\gamma|} \right] \\
&= \left[\left(\frac{\partial p_{\gamma,k}}{\partial x_1}(x)\|x\|_2^2 + q_{\gamma,k}(x)x_1 + 2(k - |\gamma|)p_{\gamma,k}(x)x_1 \right) \right. \\
&\quad \left. + \left(\frac{\partial q_{\gamma,k}}{\partial x_1}(x)\|x\|_2^2 + 2(k - |\gamma|)x_1 q_{\gamma,k}(x) \right) \log \|x\|_2 \right] \|x\|_2^{2(k-|\alpha|)} \\
&=: \left(p_{\alpha,k}(x) + q_{\alpha,k}(x) \log \|x\|_2 \right) \|x\|_2^{2k-2|\alpha|}.
\end{aligned}$$

The polynomials $p_{\alpha,k}, q_{\alpha,k}$ are homogenous by the same arguments as those given in the proof of Lemma 11.15. □

If we proceed as in the case $\phi(r) = r^\beta$ to get estimates for thin-plate splines, we could bound the derivatives of order $2k - 1$ by

$$|D^\alpha \Phi(x)| \leq C(1 + \log \|x\|_2)\|x\|_2 \leq \frac{C}{\epsilon}\|x\|_2^{1-\epsilon}$$

for every $\epsilon > 0$. This gives the bound

$$\left[P_{\Phi,X}^{(\alpha)}(x) \right]^2 \leq \frac{C_\alpha}{\epsilon} h^{2k-\epsilon-2|\alpha|},$$

which is not the full order we have in mind. Hence the obvious application of Theorem 11.11 is not appropriate in this case. The reason for this is mainly the use of the Taylor polynomials. Instead, we have to use Theorem 11.9 directly.

Theorem 11.19 *Suppose that $\Omega \subseteq \mathbb{R}^d$ is bounded and satisfies an interior cone condition. Let $\Phi(x) = (-1)^{k+1}\|x\|_2^{2k} \log \|x\|_2$. Denote the interpolant of a function $f \in \mathcal{N}_\Phi(\Omega)$ based on this basis function and the set of centers $X = \{x_1, \dots, x_N\} \subseteq \Omega$ by $s_{f,X}$. Then there exist constants $h_0, C > 0$ such that*

$$|D^\alpha f(x) - D^\alpha s_{f,X}(x)| \leq C h_{X,\Omega}^{k-|\alpha|} |f|_{\mathcal{N}_\Phi(\Omega)}$$

for all $x \in \Omega$ and all α with $|\alpha| \leq k - 1$, provided that $h_{X,\Omega} \leq h_0$.

Proof Set $h = h_{X,\Omega}$. Let us denote the right-hand side of (11.5) by $F(p, h)$. Next we define the bijective map $T : \pi_{2k}(\mathbb{R}^d) \to \pi_{2k}(\mathbb{R}^d)$ by

$$Tp(x) = h^{-2k} p(hx) - \|x\|_2^{2k} \log h.$$

Since $\Phi(x/h) = h^{2k}\Phi(x) - h^{-2k}\|x\|_2^{2k} \log h$ we have

$$\Phi(x) - p(x) = h^{2k} \left(\Phi(x/h) - Tp(x/h)\right)$$

for every $p \in \pi_{2k}(\mathbb{R}^d)$, giving

$$D^\beta \Phi(x) - D^\beta p(x) = h^{2k-|\beta|} \left(D^\beta \Phi(x/h) - D^\beta T p(x/h)\right)$$

for every $\beta \in \mathbb{N}_0^d$ with $|\beta| \le 2k - 1$. This means in particular that

$$D^{2\alpha}\Phi(0) - D^{2\alpha} p(0) = h^{2k-2|\alpha|}(D^{2\alpha}\Phi(0) - D^{2\alpha} T p(0))$$

and, moreover,

$$\begin{aligned}
&\|D^\alpha \Phi - D^\alpha p\|_{L_\infty(B(0,c_2^{(\alpha)}h))} \\
&\quad = \sup_{\|x\|_2 \le c_2^{(\alpha)} h} |D^\alpha \Phi(x) - D^\alpha p(x)| \\
&\quad = h^{2k-|\alpha|} \sup_{\|x\|_2 \le c_2^{(\alpha)} h} |D^\alpha \Phi(x/h) - D^\alpha T p(x/h)| \\
&\quad = h^{2k-|\alpha|} \sup_{\|x\|_2 \le c_2^{(\alpha)}} |D^\alpha \Phi(x) - D^\alpha T p(x)| \\
&\quad = h^{2k-|\alpha|} \|D^\alpha \Phi - D^\alpha T p\|_{L_\infty(B(0,c_2^{(\alpha)}))}.
\end{aligned}$$

Finally, by setting $\ell = \max\{m - 1, 2k\}$ in Theorem 11.9 we obtain

$$\begin{aligned}
\left[P_{\Phi,X}^{(\alpha)}(x)\right]^2 &\le \inf_{p \in \pi_{2k}(\mathbb{R}^d)} F(p, h) \\
&= h^{2k-2|\alpha|} \inf_{p \in \pi_{2k}(\mathbb{R}^d)} F(Tp, 1) \\
&= h^{2k-2|\alpha|} \inf_{p \in \pi_{2k}(\mathbb{R}^d)} F(p, 1).
\end{aligned}$$

The last infimum is a constant, which proves the stated bound. □

Let us summarize our results on bounding the power function in the following way. For every basis function we have found a function F such that

$$P_{\Phi,X}^2(x) \le F(h_{X,\Omega}), \qquad x \in \Omega.$$

The basis functions together with their functions F are collected in Table 11.1. The function F is given only up to a constant that may depend on Ω, d, and Φ but not on X. The spectral convergence results for Gaussians and (inverse) multiquadrics will be obtained in Section 11.4.

Table 11.1 *Upper bounds on $P^2_{\Phi,X}$ in terms of h*

	$\Phi(x) = \phi(r), r = \|x\|_2$	$F(h)$
Gaussians	$e^{-\alpha r^2}, \quad \alpha > 0$	$e^{-c\|\log h\|/h}$
multiquadrics (MQs)	$(-1)^{\lceil\beta\rceil}(c^2+r^2)^\beta, \quad \beta > 0, \beta \notin \mathbb{N}$	$e^{-\tilde{c}/h}$
inverse MQs	$(c^2+r^2)^\beta, \quad \beta < 0$	$e^{-\tilde{c}/h}$
powers	$(-1)^{\lceil\beta/2\rceil}r^\beta, \quad \beta > 0, \beta \notin 2\mathbb{N}$	h^β
thin-plate splines	$(-1)^{k+1}r^{2k}\log r, \quad k \in \mathbb{N}$	h^{2k}
compactly supported functons	$\phi_{d,k}(r)$	h^{2k+1}

11.4 Spectral convergence for Gaussians and (inverse) multiquadrics

In Theorem 11.14 we saw that interpolation by Gaussians and (inverse) multiquadrics leads to arbitrarily high algebraic approximation orders for functions from the associated native space, i.e.

$$\|f - s_{f,X}\|_{L_\infty(\Omega)} \le C_\ell h^\ell_{X,\Omega}|f|_{\mathcal{N}_\Phi(\Omega)}$$

holds for every $\ell \in \mathbb{N}$. It is now our goal to conclude also even spectral convergence orders, i.e.

$$\|f - s_{f,X}\|_{L_\infty(\Omega)} \le e^{-c/h_{X,\Omega}}|f|_{\mathcal{N}_\Phi(\Omega)} \tag{11.8}$$

with a certain constant $c > 0$. To this end we have to study how the constants h_0, c_1, c_2 depend on the polynomial degree ℓ. For example, if we use the result on local polynomial reproduction given in Theorem 3.14, which was based on norming sets, we know that a possible choice is $h_0 = c_0/\ell^2$, $c_1 = 2$, $c_2 = \tilde{c}_2\ell^2$, where $c_0, \tilde{c}_2$ are constants independent of ℓ. The analysis to come will show that this allows error estimates of the form

$$\|f - s_{f,X}\|_{L_\infty(\Omega)} \le Ce^{-c/\sqrt{h_{X,\Omega}}}|f|_{\mathcal{N}_\Phi(\Omega)}. \tag{11.9}$$

To gain the full spectral order given in (11.8), however, we need a local polynomial reproduction that allows h_0 and c_2 to depend linearly on $1/\ell$ and ℓ, respectively. To achieve this we have to sacrifice something, namely the uniform bound on c_1. We will see that it does not really matter if this constant grows even exponentially in ℓ.

The following result specifies what we have just stated. Unfortunately, it does not seem to have such an elegant proof as the other results on local polynomial reproduction that we have encountered so far. The only known proof is based on some deeper results from algebraic geometry and the theory of nondegenerate points. To give the proof here would exceed the scope of this text. The interested reader should have a look at Madych and Nelson [114] for details. But let us point out again that the somewhat weaker estimates given in (11.9) follow easily from Theorem 3.14 and the proof of Theorem 11.22 below.

Proposition 11.20 *Define $\gamma_1 = 2$ and $\gamma_d = 2d(1 + \gamma_{d-1})$ for $d = 2, 3, \ldots$. Let ℓ and q be positive integers with $q \ge \gamma_d(\ell + 1)$. Let Ω be a cube in $\mathbb{R}^d$. Subdivide Ω into q^d identical subcubes. If $X \subseteq \Omega$ is a set of $N \ge q^d$ points such that each subcube contains at least one of these points then, for all $p \in \pi_\ell(\mathbb{R}^d)$,*

$$\|p\|_{L_\infty(\Omega)} \le e^{2d\gamma_d(\ell+1)} \|p\|_{L_\infty(X)}.$$

This result allows us to state a new version of a local polynomial reproducing process. For simplicity we restrict ourselves to cubes $W(x_0, R) = \{x \in \mathbb{R}^d : \|x - x_0\|_\infty \le R\}$.

Theorem 11.21 *Let $\Omega = W(x_0, R)$ be a cube in $\mathbb{R}^d$. There exist constants $c_0, c_2 > 0$ depending only on Ω such that for every $\ell \in \mathbb{N}$ and every $X = \{x_1, \ldots, x_N\} \subseteq \Omega$ with $h_{X,\Omega} \le c_0/\ell$ we can find functions $u_j : \Omega \to \mathbb{R}$ satisfying*

(1) $\sum_{j=1}^N u_j(x)p(x_j) = p(x)$ for all $x \in \Omega$ and all $p \in \pi_\ell(\mathbb{R}^d)$,
(2) $\sum_{j=1}^N |u_j(x)| \le e^{2d\gamma_d(\ell+1)}$ for all $x \in \Omega$,
(3) $u_j(x) = 0$ if $\|x - x_j\|_2 > c_2 \ell h_{X,\Omega}$.

The constant γ_d is defined in Proposition 11.20.

Proof Let $q := \gamma_d(\ell + 1)$ and $h := h_{X,\Omega}$. Define

$$h_0 := \frac{2R}{3q} = \frac{2R}{3\gamma_d(\ell + 1)} =: \frac{c_0}{\ell}.$$

If $h \le h_0$ then we can find for every $x \in \Omega$ a cube W_x of side length $3hq$ that is completely contained in Ω and has x as one of its corners. This cube can be divided into q^d subcubes of side length $3h$. Hence, the interior of each of these subcubes contains a ball with radius h, namely the ball centered at the center of the subcube. But this means that each of the subcubes contains at least one point from X. Proposition 11.20 gives, with $Y = X \cap W_x = \{y_1, \ldots, y_M\}$,

$$\|p\|_{L_\infty(W_x)} \le e^{2d\gamma_d(\ell+1)} \|p\|_{L_\infty(Y)}, \qquad p \in \pi_\ell(\mathbb{R}^d).$$

This allows us to invoke Theorem 3.4 to find functions $u_j^x : W_x \to \mathbb{R}$, $1 \le j \le M$, such that

$$\sum_{j=1}^M u_j^x(y) p(y_j) = p(y), \qquad y \in W_x, \quad p \in \pi_\ell(\mathbb{R}^d),$$

and

$$\sum_{j=1}^M |u_j^x(y)| \le e^{2dq}, \qquad y \in W_x.$$

Hence, if we define the functions $u_k : \Omega \to \mathbb{R}$, $1 \le k \le N$, by

$$u_k(x) := \begin{cases} u_j^x(x), & \text{if } x_k = y_j, \\ 0 & \text{otherwise,} \end{cases}$$

we see that the first two stated properties are satisfied. For the third property we only have to remark that $u_k(x) \neq 0$ means that $x_k = y_j$ for a certain index j. But since x and y_j lie both in the cube W_x they have a separation at most $\sqrt{d}q3h =: c_2\ell h$. □

With this result at hand it is easy to establish spectral convergence for Gaussians and multiquadric-like functions.

Theorem 11.22 *Let Ω be a cube in $\mathbb{R}^d$. Suppose that $\Phi = \phi(\|\cdot\|_2)$ is a conditionally positive definite function such that $f := \phi(\sqrt{\cdot})$ satisfies $|f^{(\ell)}(r)| \leq \ell! M^\ell$ for all integers $\ell \geq \ell_0$ and all $r \in [0, \infty)$, where $M > 0$ is a fixed constant. Then there exists a constant $c > 0$ such that the error between a function $f \in \mathcal{N}_\Phi(\Omega)$ and its interpolant $s_{f,X}$ can be bounded by*

$$\|f - s_{f,X}\|_{L_\infty(\Omega)} \leq e^{-c/h_{X,\Omega}} |f|_{\mathcal{N}_\Phi(\Omega)} \tag{11.10}$$

for all data sites X with sufficiently small $h_{X,\Omega}$.

Moreover, if f satisfies even $|f^{(\ell)}(r)| \leq M^\ell$, the error bound can be improved to

$$\|f - s_{f,X}\|_{L_\infty(\Omega)} \leq e^{c \log(h_{X,\Omega})/h_{X,\Omega}} |f|_{\mathcal{N}_\Phi(\Omega)}, \tag{11.11}$$

whenever $h_{X,\Omega}$ is sufficiently small.

Proof From the estimate (11.6) we have the bound

$$P^2_{\Phi,X}(x) \leq [1 + c_1(2n)]^2 \|f - p\|_{L_\infty[0,\, 4[c_2(2n)]^2 h^2]}$$

for $x \in \Omega$ if $h = h_{X,\Omega}$ and $p \in \pi_n(\mathbb{R})$. We have already indicated that the constants c_1, c_2 from the local polynomial reproduction depend on $\ell = 2n$. If $h \leq c_0/(2n)$ then we can use the constants $c_1(2n) = e^{2d\gamma_d(2n+1)}$, and $c_2(2n) = 2c_2 n$ from Theorem 11.21. Moreover, if we take p to be the Taylor polynomial of f around zero of degree n then the first assumption made on f gives

$$|f(t) - p(t)| \leq \frac{|f^{(n+1)}(\xi)|}{(n+1)!} t^{n+1} \leq (Mt)^{n+1}.$$

Hence, we can further estimate that

$$\begin{aligned}
P^2_{\Phi,X}(x) &\leq \left(1 + e^{2d\gamma_d(2n+1)}\right)^2 \|f - p\|_{L_\infty[0, 16c_2^2 n^2 h^2]} \\
&\leq 4e^{4d\gamma_d(2n+1)} (C_1 n^2 h^2)^{n+1} \\
&\leq e^{\log 4 + 4d\gamma_d(2n+1)} (C_1 n^2 h^2)^{n+1} \\
&\leq \left(e^{C_2}\right)^{n+1} (C_1 n^2 h^2)^{n+1} \\
&= (C_3 n^2 h^2)^{n+1},
\end{aligned}$$

where the C_i denote the appropriate constants. The latter estimate holds uniformly of course, for $x \in \Omega$. Next, define $c_4 := \min\{c_0/2, 1/\sqrt{eC_3}\}$. Then we choose n so that $c_4/(n+1) \leq h \leq c_4/n$, which gives

$$P^2_{\Phi,X}(x) \leq e^{-(n+1)} \leq e^{-c_4/h}.$$

This establishes (11.10) with $c = c_4/2$. To see (11.11) we just have to remark that the second assumption on f now leads to

$$P^2_{\Phi,X}(x) \leq \frac{(C_3 n^2 h^2)^{n+1}}{(n+1)!}.$$

Stirling's formula from Proposition 5.2 yields

$$\frac{1}{(n+1)!} \leq \left(\frac{e}{n+1}\right)^{n+1},$$

so that

$$P^2_{\Phi,X}(x) \leq (eC_3 n h^2)^{n+1}.$$

Hence, if we now define $c_4 := \min\{c_0/2, 1/(eC_3)\}$ and choose n in the same manner as before we see that

$$P^2_{\Phi,X}(x) \leq h^{n+1} \leq h^{c_4/h},$$

which finishes the proof in this case. □

It is easy to apply this theorem to Gaussians and multiquadrics. For example, if $\phi(r) = e^{-\alpha r^2}$, $\alpha > 0$, we have $f(r) = e^{-\alpha r}$ and $f^{(\ell)}(r) = (-1)^\ell \alpha^\ell e^{-\alpha r}$, $\ell \in \mathbb{N}_0$, so that the second assumption on f holds with $M = \alpha$ in this case. Thus, for Gaussians we have error bounds of the form (11.11).

In the case $\phi(r) = (1 + r^2)^\beta$ with $\beta < 0$ or $\beta > 0$, $\beta \notin \mathbb{N}$, we have $f(r) = (1 + r)^\beta$, so that

$$f^{(\ell)}(r) = \beta(\beta - 1)\cdots(\beta - \ell + 1)(1 + r)^{\beta-\ell}.$$

From

$$\left|\frac{\beta - j}{j+1}\right| = \left|\frac{j + 1 - (\beta + 1)}{j+1}\right| \leq 1 + \frac{|\beta + 1|}{j+1} \leq 1 + |\beta + 1| =: M$$

we can conclude that

$$|f^{(\ell)}(r)| = \ell! \frac{|\beta|}{1} \frac{|\beta - 1|}{2} \cdots \frac{|\beta - \ell + 1|}{\ell} \leq \ell! M^\ell$$

provided $\ell \geq \lceil \beta \rceil$. Hence, (inverse) multiquadrics satisfy the first assumption on f, which leads to error bounds of the form (11.10).

11.5 Improved error estimates

After establishing the basic theory on error estimates for interpolation by radial basis functions we now turn to the question how the previously derived approximation orders can be improved. There are at least two ways. The first way restricts the space of functions to be interpolated by assuming more smoothness than the native space provides. This is rather natural and the reader might think of the theory on univariate splines. The second way weakens the norm in which the error is estimated. Replacing the L_∞-norm by a weaker L_p-norm should result in a better approximation order. The methods we have in mind lead to an algebraic improvement of this order; hence they are only interesting in the case of basis kernels that have finite smoothness. For basis functions such as multiquadrics and Gaussians, where the convergence order is spectral, they are almost pointless. The latter improvement type will be discussed in the next section.

The basic idea of the first improvement technique is the following. If $f \in \mathcal{N}_\Phi(\Omega)$ is given and $s_{f,X}$ denotes its interpolant then the function $g := f - s_{f,X}$ is also a member of the native space. Moreover, since it vanishes on X it has the zero function as its unique interpolant. This means that the standard error estimate from Theorem 11.4 becomes

$$|f(x) - s_{f,X}(x)| \le P_{\Phi,X}(x)|f - s_{f,X}|_{\mathcal{N}_\Phi(\Omega)} \tag{11.12}$$

and we will squeeze out the additional $h_{X,\Omega}$ terms of $|f - s_{f,X}|_{\mathcal{N}_\Phi(\Omega)}$ by means of certain assumptions on f.

The easiest way to achieve this is for positive definite kernels $\Phi : \Omega \times \Omega \to \mathbb{R}$ defined on a compact set $\Omega \subseteq \mathbb{R}^d$. In (10.7) we introduced an integral operator

$$Tv(x) = \int_\Omega \Phi(x, y)v(y)dy$$

that maps $L_2(\Omega)$-functions to functions from the native space. In Corollary 10.30 we described its range in terms of its eigenfunctions, when considered as an operator from $L_2(\Omega)$ to $L_2(\Omega)$.

Theorem 11.23 *Suppose that Φ is a symmetric positive definite kernel on a compact set $\Omega \subseteq \mathbb{R}^d$. Then for every $f \in T(L_2(\Omega))$ we have*

$$|f(x) - s_{f,X}(x)| \le P_{\Phi,X}(x)\|P_{\Phi,X}\|_{L_2(\Omega)}\|T^{-1}f\|_{L_2(\Omega)}, \qquad x \in \Omega.$$

Proof Let $f = Tv$, $v \in L_2(\Omega)$. Taking the $L_2(\Omega)$-norm of (11.12) yields

$$\|f - s_{f,X}\|_{L_2(\Omega)} \le \|P_{\Phi,X}\|_{L_2(\Omega)}\|f - s_{f,X}\|_{\mathcal{N}_\Phi(\Omega)}.$$

Using the orthogonality relation from Lemma 10.24 together with (10.8) leads to

$$\begin{aligned}
\|f - s_{f,X}\|^2_{\mathcal{N}_\Phi(\Omega)} &= (f - s_{f,X}, f)_{\mathcal{N}_\Phi(\Omega)} \\
&= (f - s_{f,X}, Tv)_{\mathcal{N}_\Phi(\Omega)} \\
&= (f - s_{f,X}, v)_{L_2(\Omega)} \\
&\le \|f - s_{f,X}\|_{L_2(\Omega)}\|v\|_{L_2(\Omega)} \\
&\le \|P_{\Phi,X}\|_{L_2(\Omega)}\|f - s_{f,X}\|_{\mathcal{N}_\Phi(\Omega)}\|v\|_{L_2(\Omega)}.
\end{aligned}$$

Canceling one $\|f - s_{f,X}\|_{\mathcal{N}_\Phi(\Omega)}$ factor and inserting the result back into (11.12) proves the result. □

This result means for example in the case of the compactly supported functions $\phi_{d,k}$ that the L_∞-order can be improved to $2k+1$ from $k+1/2$ provided that the functions come from the restricted space. But techniques that we that will learn soon allow an additional improvement to $2k+1+d/2$.

Instead of treating the general case of conditionally positive definite kernels we will discuss only the most important example, namely the thin-plate splines $\phi_{d,\ell}$ defined in (10.11). This time, the approximated functions come from the Sobolev space $H^{2\ell}(\mathbb{R}^d)$, which is the intersection of all Beppo Levi spaces $\mathrm{BL}_k(\mathbb{R}^d)$ with $k \le 2\ell$.

Theorem 11.24 *Suppose that $\Phi = \Phi_{d,\ell}$ denotes the thin-plate spline with $\ell > d/2$, considered as being conditionally positive definite of order ℓ. Let $\Omega \subseteq \mathbb{R}^d$ be bounded and satisfy an interior cone condition. Then for every $f \in H^{2\ell}(\mathbb{R}^d)$ with support in Ω we have*

$$\begin{aligned} |f(x) - s_{f,X}(x)| &\le P_{\Phi,X}(x)\|P_{\Phi,X}\|_{L_2(\Omega)}\|\Delta^\ell f\|_{L_2(\Omega)}, \qquad x \in \Omega, \\ &\le C h_{X,\Omega}^{2\ell-d}\|\Delta^\ell f\|_{L_2(\Omega)}, \end{aligned}$$

where the last inequality holds for all sufficiently dense sets X.

Proof The proof is based on the same ideas as the previous one. We only have to replace the estimate on the native space norm. Remember that if we use the thin-plate splines as conditionally positive definite functions of order ℓ, their native space is the Beppo Levi space $\mathrm{BL}_\ell(\mathbb{R}^d)$. This time we use

$$\begin{aligned} |f - s_{f,X}|^2_{\mathrm{BL}_\ell(\mathbb{R}^d)} &= (f - s_{f,X}, f)_{\mathrm{BL}(\mathbb{R}^d)} \\ &= \sum_{|\alpha|=\ell} \frac{\ell!}{\alpha!} \int_{\mathbb{R}^d} D^\alpha(f - s_{f,X})(x) D^\alpha f(x)dx \\ &= \sum_{|\alpha|=\ell} \frac{\ell!}{\alpha!}(-1)^\ell \int_{\mathbb{R}^d} (f - s_{f,X})(x) D^{2\alpha} f(x)dx \\ &= (-1)^\ell \int_{\mathbb{R}^d} (f - s_{f,X})(x)\Delta^\ell f(x)dx \\ &= (-1)^\ell \int_\Omega (f - s_{f,X})(x)\Delta^\ell f(x)dx, \end{aligned}$$

so that

$$|f - s_{f,X}|^2_{\mathrm{BL}_\ell(\mathbb{R}^d)} \le \|f - s_{f,X}\|_{L_2(\Omega)}\|\Delta^\ell f\|_{L_2(\Omega)}.$$

The partial integration we have just carried out can be justified using density arguments similar to those employed in the proof of Theorem 10.40. Remember that f is compactly supported with support in Ω. □

This result can further be improved if the $L_2(\Omega)$-error is estimated more dexterously. To this end a localization is necessary, which we will describe in the next section in a more general setting.

11.6 Sobolev bounds for functions with scattered zeros

Suppose that the native space $\mathcal{N}_\Phi(\Omega)$ of a radial basis function Φ is a Sobolev space $H^k(\Omega)$ or that it is continuously embedded into such a Sobolev space. In this situation, we can derive more accurate error estimates than before. These estimates use the whole range of L_p-norms and do not need the power function approach at all. They are based only upon the fact that the error $u := f - s_{f,X}$ is a function from the Sobolev space which has zeros at X and which is, by Corollary 10.25, bounded in the native space norm by the native space norm of f.

Since these results hold in a more general setting, we first introduce for $1 \le p \le \infty$ the Sobolev space $W_p^k(\Omega)$ as the set of all functions $f \in L_p(\Omega)$ having weak derivatives $D^\alpha f$ also in $L_p(\Omega)$ for all $|\alpha| \le k$. We will then use the Sobolev semi-norms (which we have also called Beppo Levi semi-norms)

$$|u|_{W_p^k(\Omega)}^p := \sum_{|\alpha|=k} \|D^\alpha u\|_{L_p(\Omega)}^p, \qquad k \in \mathbb{N}_0, \quad 1 \le p < \infty.$$

The case $p = \infty$ is defined in an obvious manner, replacing the sum by the maximum. The full norm on $W_p^k(\Omega)$ is then given by summing up all semi-norms, i.e. $\|u\|_{W_p^k(\Omega)}^p = \sum_{j=0}^k |u|_{W_p^j(\Omega)}^p$. For an application to the compactly supported basis functions from Chapter 9 we also have to deal with fractional Sobolev spaces. One way to introduce them is by interpolation theory. Here, we use the direct approach and define for $1 \le p < \infty$, $k \in \mathbb{N}_0$, and $0 < s < 1$,

$$|u|_{W_p^{k+s}(\Omega)} := \left(\sum_{|\alpha|=k} \int_\Omega \int_\Omega \frac{|D^\alpha u(x) - D^\alpha u(y)|^p}{\|x-y\|_2^{d+ps}} dx dy \right)^{1/p}$$

$$\|u\|_{W_p^{k+s}(\Omega)} := \left(\|u\|_{W_p^k(\Omega)}^p + |u|_{W_p^{k+s}(\Omega)}^p \right)^{1/p}.$$

Now suppose that $u \in W_p^{k+s}(\Omega)$ vanishes on $X \subseteq \Omega$ and that $k > d/p - m$, so that $W_p^k(\Omega) \subseteq C^m(\Omega)$ by Sobolev's embedding theorem. We wish to establish the following result:

$$|u|_{W_q^m(\Omega)} \le c h_{X,\Omega}^{k+s-|\alpha|-d(1/p-1/q)_+} |u|_{W_p^{k+s}(\Omega)}, \tag{11.13}$$

which can be used immediately for our radial basis function interpolation process. Since the proof of (11.13) involves a certain technicality, we will now outline its ideas for the special case $p = q = 2$. The first step is to cover Ω by "nice" local patches $\mathcal{D}$ of diameter $\mathcal{O}(h_{X,\Omega})$. The term "nice" will be explained very soon. Then on each patch $\mathcal{D}$ we approximate u by a polynomial $p \in \pi_k(\mathbb{R}^d)$ that is an averaged Taylor polynomial:

$$|u|_{W_2^m(\mathcal{D})} \le |u - p|_{W_2^m(\mathcal{D})} + |p|_{W_2^m(\mathcal{D})}. \tag{11.14}$$

The derivatives $D^\alpha p$, $|\alpha| = m$, of p can be expressed using Proposition 11.7 as

$$D^\alpha p(x) = \sum_{j=1}^{N} a_j^{(\alpha)}(x) p(x_j) = \sum_{j=1}^{N} a_j^{(\alpha)}(x)[p(x_j) - u(x_j)],$$

since $u|X = 0$. Moreover, we know that the coefficient vector in this representation can be bounded by $Ch_{X,\Omega}^{-|\alpha|}$ with a constant that depends only on the region geometry and the polynomial degree (and the constant factor between the local cone radius and $h_{X,\Omega}$). Inserting this into (11.14) yields

$$|u|_{W_2^m(\mathcal{D})} \le |u - p|_{W_2^m(\mathcal{D})} + Ch_{X,\Omega}^{-m+d/2} |u - p|_{L_\infty(\mathcal{D})},$$

where the additional $h_{X,\Omega}^{d/2}$ comes from the volume of $\mathcal{D}$. Hence we have reduced everything to a local polynomial approximation problem. When this is solved, the local estimates are put together to form a global one. If the patches do not overlap too much, the sum of the local Sobolev norms is equivalent to the Sobolev norm on Ω. We now start the detailed discussion.

Definition 11.25 *A domain $\mathcal{D}$ is said to be star-shaped with respect to a ball $B(x_c, \rho) := \{x \in \mathbb{R}^d : \|x - x_c\|_2 \le \rho\}$ if, for every $x \in \mathcal{D}$, the closed convex hull of $\{x\} \cup B$ is contained in $\mathcal{D}$. If $\mathcal{D}$ is bounded then the chunkiness parameter γ is defined to be the ratio of the diameter $d_{\mathcal{D}}$ of $\mathcal{D}$ to the radius $\rho_{\max}$ of the largest ball relative to which $\mathcal{D}$ is star-shaped.*

A bounded domain $\mathcal{D}$ is contained in a ball $B(x_c, R)$. Throughout the rest of this section we want to assume that $\rho_{\max}/2 \le \rho \le \rho_{\max}$, so that we have the obvious chain of inequalities $\rho_{\max}/2 \le \rho \le \rho_{\max} \le d_D \le 2R$. Hence, the chunkiness parameter satisfies

$$\frac{1}{2} \le \frac{d_{\mathcal{D}}}{2\rho} \le \gamma = \frac{d_{\mathcal{D}}}{\rho_{\max}} \le \frac{2R}{\rho}. \tag{11.15}$$

Such domains satisfy a simple interior cone condition.

Proposition 11.26 *If $\mathcal{D}$ is bounded, star-shaped with respect to $B(x_c, \rho)$, and contained in $B(x_c, R)$ then $\mathcal{D}$ satisfies an interior cone condition with radius ρ and angle $\vartheta = 2 \arcsin[\rho/(2R)]$.*

Proof It is easy to check that when $x \in B(x_c, \rho)$ the cone condition is satisfied if the central axis of the cone is directed along a diameter of the ball $B(x_c, \rho)$. If x is outside that ball then we consider the convex hull of x and the intersection of the sphere $S(x, \|x - x_c\|_2) = \{y \in \mathbb{R}^d : \|y - x\|_2 = \|x_c - x\|_2\}$ with $B(x_c, \rho)$. This is a cone and, because $\mathcal{D}$ is star-shaped with respect to $B(x_c, \rho)$, it is contained in $\mathcal{D}$. Its radius is the distance from x to x_c. To find its angle ϑ, we consider a triangle formed by x, x_c, and any point on y in the intersection of $S(x, \|x - x_c\|_2)$ and the sphere $S(x_c, \rho)$. This is an isosceles triangle, since $\|x_c - x\|_2 = \|y - x\|_2$. The angle $\angle x_c x y = \vartheta$; the side opposite this angle has length ρ. A little trigonometry then gives us $\|x_c - x\|_2 \sin(\vartheta/2) = \rho/2$. Consequently, we have $\vartheta = 2 \arcsin[\rho/(2\|x_c - x\|_2)]$. Moreover, since $\mathcal{D} \subseteq B(x_c, R)$, we also have $\|x_c - x\|_2 \le R$.

Thus $\vartheta \geq 2\arcsin[\rho/(2R)]$. Finally, $\rho \leq \|x - x_c\|_2$ implies that the cone with vertex x, axis along $x_c - x$, and angle $\vartheta = 2\arcsin[\rho/(2R)]$ is contained in $\mathcal{D}$. □

After setting up the geometry of our local patches we introduce the approximating polynomial. As stated before, this is an averaged Taylor polynomial:

$$Q_k u(x) := \sum_{|\alpha|<k} \frac{1}{\alpha!} \int_{B_\rho} D^\alpha u(y)(x-y)^\alpha \phi_\rho(y)dy,$$

where $\phi \in C_0^\infty(\mathbb{R}^d)$ is chosen such that $\phi_\rho = \rho^{-d}\phi(\cdot/\rho)$ is supported in $B_\rho = B(0,\rho)$ and forms an approximation of the identity.

Our first result deals with the case of integer-order Sobolev spaces. We will omit the proof here; it can be found in the book [31] of Brenner and Scott.

Proposition 11.27 *Let $1 < p < \infty$ and $k > d/p$ or $p = 1$ and $k \geq d/p$. Then there exists a constant $C > 0$ depending only on k, d, p such that, for every $u \in W_p^k(\mathcal{D})$,*

$$\|u - Q_k u\|_{L_\infty(\mathcal{D})} \leq C(1+\gamma)^d d_{\mathcal{D}}^{k-d/p} |u|_{W_p^k(\mathcal{D})}.$$

Here γ denotes the chunkiness parameter for $\mathcal{D}$.

We are now going to extend this result to fractional-order Sobolev spaces $W_p^{k+s}(\mathcal{D})$. To this end, we first investigate the action of Q_{k+1} on a function from $W_p^k(\mathcal{D})$, which is at least well defined.

Lemma 11.28 *For $1 < p < \infty$ and $k > d/p$ or $p = 1$ and $k \geq d$, if $u \in W_p^k(\mathcal{D})$ and $P \in \pi_k(\mathbb{R}^d)$ then*

$$\|u - Q_{k+1}u\|_{L_\infty(\mathcal{D})} \leq C(1+\gamma)^d d_{\mathcal{D}}^{k-d/p} |u - P|_{W_p^k(\mathcal{D})}, \tag{11.16}$$

where C depends only on k, d, p.

Proof Since $P \in \pi_k(\mathbb{R})$ we have $Q_{k+1}P = P$. Hence, if we set $v := u - P$ then we get

$$\begin{aligned} u - Q_{k+1}u &= v - Q_{k+1}v \\ &= v - Q_k v - \sum_{|\alpha|=k} \frac{1}{\alpha!} \int_{B_\rho} D^\alpha v(y)(\cdot - y)^\alpha \phi_\rho(y)dy. \end{aligned}$$

The term $v - Q_k v$ can be bounded using Proposition 11.27, while the rest enjoys the upper bound

$$\begin{aligned} C d_{\mathcal{D}}^k \rho^{-d} \max_{|\alpha|=k} \int_{B_\rho} |D^\alpha v(y)| dy &\leq C d_{\mathcal{D}}^k \rho^{-d} \operatorname{vol}(B_\rho)^{1-1/p} |v|_{W_p^k(\mathcal{D})} \\ &\leq C(d_{\mathcal{D}}/\rho)^{d/p} d_{\mathcal{D}}^{k-d/p} |v|_{W_p^k(\mathcal{D})}. \end{aligned}$$

Finally, $d_{\mathcal{D}}/\rho \leq 2\gamma$ and $\gamma^{d/p} \leq (1+\gamma)^d$ give the desired result. □

Now we are able to derive our result for fractional-order Sobolev spaces. Such a result could also be proved by using interpolation theory on the operator Q_{k+1}, but here we will prove it directly.

Proposition 11.29 *Let $0 < s \le 1$ and $m \in \mathbb{N}$. Let $1 < p < \infty$ and $k > m + d/p$, or $p = 1$ and $k \ge m + d$. For $u \in W_p^{k+s}(\mathcal{D})$ we have*

$$\|u - Q_{k+1}u\|_{W_\infty^m(\mathcal{D})} \le C(1+\gamma)^{d(1+1/p)} d_{\mathcal{D}}^{k+s-m-d/p} |u|_{W_p^{k+s}(\mathcal{D})}$$

with a constant $C > 0$ depending only on k, d, p.

Proof We start with the situation $m = 0$. The case $s = 1$ follows immediately from Proposition 11.27. Hence, we might assume that $0 < s < 1$. In Lemma 11.28 let $P = Q_{k+1}u \in \pi_k(\mathbb{R}^d)$. The identity

$$D^\alpha Q_{k+1}u = Q_{k+1-|\alpha|} D^\alpha u, \tag{11.17}$$

which is easily established, holds for $|\alpha| \le k$. In particular, if we take $|\alpha| = k$ then we have

$$D^\alpha Q_{k+1}u = Q_1 D^\alpha u = \int_{B_\rho} \phi_\rho(y) D^\alpha u(y) dy,$$

which is of course a constant. Since $\int_{B_\rho} \phi_\rho(y)dy = 1$, we can use standard arguments, $\|x - y\|_2 \le d_{\mathcal{D}}$, bounds on ϕ_ρ, and Hölder's inequality, to obtain the estimate

$$\begin{aligned}
|D^\alpha(u - Q_{k+1}u)(x)| &\le \int_{B_\rho} \phi_\rho(y)\, |D^\alpha u(x) - D^\alpha u(y)|\, dy \\
&\le \int_{B_\rho} \phi_\rho(y) \|x-y\|_2^{s+d/p} \frac{|D^\alpha u(x) - D^\alpha u(y)|}{\|x-y\|_2^{s+d/p}} dy \\
&\le C\rho^{-d} d_{\mathcal{D}}^{s+d/p} \int_{B_\rho} \frac{|D^\alpha u(x) - D^\alpha u(y)|}{\|x-y\|_2^{s+d/p}} dy \\
&\le C d_{\mathcal{D}}^{s+d/p} \rho^{-d/p} \left\| \frac{D^\alpha u(x) - D^\alpha u}{\|x - \cdot\|_2^{s+d/p}} \right\|_{L^p(\mathcal{D})}.
\end{aligned}$$

Raising both sides to the power p, integrating over $\mathcal{D}$, and summing over all $|\alpha| = k$ gives

$$|u - P|_{W_p^k(\mathcal{D})}^p \le C^p d_{\mathcal{D}}^{sp+d} \rho^{-d} |u|_{W_p^{k+s}(\mathcal{D})}^p.$$

A final application of $d_{\mathcal{D}}/\rho \le 2\gamma$ and taking the pth root of both sides gives

$$|u - P|_{W_p^k(\mathcal{D})} \le C d_{\mathcal{D}}^s \gamma^{d/p} |u|_{W_p^{k+s}(\mathcal{D})}.$$

Inserting this into the bound of Lemma 11.28 yields the result for $m = 0$. The general case $m > 0$ follows from this, from (11.17), and from the relation $|D^\alpha u|_{W_p^{k-|\alpha|}(\mathcal{D})} \le |u|_{W_p^k(\mathcal{D})}$. □

So far we have been concerned with polynomial approximation in Sobolev spaces over small regions. We have not used the fact that our functions vanish on a discrete subset X at all. We will employ this now in the way that we have already pointed out in the introductory part of this section. Hence, our first estimate is

$$|u|_{W^m_\infty(\mathcal{D})} \le |u - Q_{k+1}u|_{W^m_\infty(\mathcal{D})} + |Q_{k+1}u|_{W^m_\infty(\mathcal{D})}. \tag{11.18}$$

Our star-shaped domain $\mathcal{D}$ satisfies a cone condition with radius $\rho > 0$ and angle $\vartheta = 2\arcsin[\rho/(2R)]$. Thus, by Proposition 11.29 and the fact that the chunkiness parameter satisfies $\gamma \le d_{\mathcal{D}}/r \le \csc(\vartheta/2)$, the first term can be bounded by

$$|u - Q_{k+1}u|_{W^m_\infty(\mathcal{D})} \le C d_{\mathcal{D}}^{k+s-m-d/p}|u|_{W^{k+s}_p(\mathcal{D})}$$

with a constant C depending only on k, d, p, and ϑ. To bound the second term in (11.18) we assume that the conditions in Proposition 11.7 are satisfied, so that by that proposition we have the representation

$$D^\alpha p(x) = \sum_{j=1}^N a_j^\alpha(x) p(x_j), \qquad x \in \mathcal{D},$$

for all $p \in \pi_k(\mathbb{R}^d)$, with certain coefficients satisfying

$$\sum_{j=1}^N |a_j^\alpha(x)| \le 2\left(\frac{2k^2}{\rho\sin\vartheta}\right)^{|\alpha|}, \qquad x \in \mathcal{D}.$$

This together with $u|X = 0$ allows us to estimate for $|\alpha| = m$

$$\begin{aligned}|D^\alpha(Q_{k+1}u)(x)| &\le \sum_{j=1}^N |a_j^\alpha(x)||u(x_j) - Q_{k+1}u(x_j)|\\ &\le C\rho^{-m} d_{\mathcal{D}}^{k+s-d/p}|u|_{W^{k+s}_p(\mathcal{D})}\\ &\le C d_{\mathcal{D}}^{k+s-m-d/p}|u|_{W^{k+s}_p(\mathcal{D})},\end{aligned}$$

with a generic constant $C > 0$ that depends only on d, p, k, m, and ϑ. To derive the last estimate we have once again used the upper bound $d_{\mathcal{D}}/\rho \le \csc(\vartheta/2)$. Putting these two bounds together gives the first part of the following proposition.

Proposition 11.30 *Let k be a positive integer, $1 \le p < \infty$, $0 < s \le 1$, $1 \le q \le \infty$, and let $m \in \mathbb{N}_0$ satisfy $k > m + d/p$, or, for $p = 1$, $k \ge m + d$. Also, let $X \subseteq \mathcal{D}$ be a discrete set that satisfies the two conditions in Proposition 11.7 with an $h > 0$. If $u \in W_p^{k+s}(\mathcal{D})$ satisfies $u|_X = 0$ then*

$$|u|_{W^m_q(\mathcal{D})} \le C d_{\mathcal{D}}^{k+s-m+d(1/q-1/p)}|u|_{W^{k+s}_p(\mathcal{D})}.$$

Here, the constant C depends only on k, d, p, m, and the angle ϑ corresponding to the cone condition that $\mathcal{D}$ satisfies.

Proof The remaining case $q < \infty$ follows from

$$|u|_{W^m_q(\mathcal{D})} \le C\,\mathrm{vol}(D)^{1/q}|u|_{W^m_\infty(\mathcal{D})} = C d_{\mathcal{D}}^{d/q}|u|_{W^m_\infty(\mathcal{D})},$$

where the first C is $\#\{\alpha \in \mathbb{N}_0^d : |\alpha| = m\} = \mathcal{O}(m^{d-1})$ and the second involves also the volume of the unit ball in $\mathbb{R}^d$. □

This concludes our local estimates. It is important to notice that the constants depend on the local domain $\mathcal{D}$ only via the angle ϑ and the radius ρ of the cone condition.

The next step is to cover our global region Ω, which is supposed to be bounded and to satisfy a cone condition with radius r and angle θ, using domains that are star-shaped. To this end, we introduce the following quantities. Let $h = h_{X,\Omega}$ be the fill distance and

$$\begin{aligned}
\vartheta &:= 2 \arcsin\left(\frac{\sin\theta}{4(1+\sin\theta)}\right),\\
Q(k,\theta) &:= \frac{\sin\theta \sin\vartheta}{8k^2(1+\sin\theta)(1+\sin\vartheta)},\\
R &:= Q(k,\theta)^{-1}h,\\
\rho &:= \frac{\sin\theta}{2(1+\sin\theta)}R.
\end{aligned}$$

With these settings we define the sets

$$T_\rho := \left\{t \in (2\rho/\sqrt{d})\mathbb{Z}^d : B(t,\rho) \subseteq \Omega\right\}$$

and

$$\mathcal{D}_t = \{x \in \Omega : \operatorname{co}(\{x\} \cup B(t,\rho)) \subseteq \Omega \cap B(t,R)\}, \qquad t \in T_\rho,$$

where $\operatorname{co}(A)$ denotes the closed convex hull of the set A.

Lemma 11.31 *With the quantities just introduced, suppose that the fill distance $h = h_{X,\Omega}$ satisfies $h \le Q(k,\theta)r$. Then the following hold true:*

(1) each $\mathcal{D}_t$ is star-shaped with respect to the ball $B(t,\rho)$ and satisfies $B(t,\rho) \subseteq \mathcal{D}_t \subseteq \Omega \cap B(t,R)$;
(2) each $\mathcal{D}_t$ satisfies a cone condition with angle ϑ and radius ρ;
(3) $\Omega = \bigcup_{t \in T_\rho} \mathcal{D}_t$ and $d_{\mathcal{D}_t} \le 2R = 2h/Q(k,\theta)$;
(4) $\sum_{t \in T_\rho} \chi_{\mathcal{D}_t} \le M_1$;
(5) $\#T_\rho \le M_2 \rho^{-d}$.

Here χ_B denotes the characteristic function of the set B and M_1, M_2 are constants depending only on k, θ, d.

Proof Obviously the first property is automatically satisfied for all $\rho > 0$. Hence, by Proposition 11.26, $\mathcal{D}_t$ satisfies a cone condition with radius ρ and angle

$$2\arcsin\left(\frac{\rho}{2R}\right) = 2\arcsin\left(\frac{R\sin\theta}{2(1+\sin\theta)}\frac{1}{2R}\right) = \vartheta.$$

Moreover, its diameter is bounded by $d_{\mathcal{D}_t} \le 2R = 2h/Q(k,\theta)$. Next note that our assumption on h gives $R \le r$, so that Ω also satisfies a cone condition with radius R and angle θ. Hence, if an arbitrary $x \in \Omega$ is given then we can find a cone $C(x) = C(x,\xi,\theta,R) \subseteq \Omega$. The definition of ρ and Lemma 3.7 ensure that the ball $B(y, 2\rho)$ centered at $y = x + (2\rho/\sin\theta)\xi$ is contained in Ω. For y we can choose a point $t \in (2\rho/\sqrt{d})\mathbb{Z}^d$ with $\|y - t\|_2 \le \rho$ which gives that the ball $B(t, \rho) \subseteq B(y, 2\rho)$ is also contained in Ω. Hence $t \in T_\rho$ and, since $C(x)$ is convex and since $\|x - t\|_2 \le R$, we additionally have $\mathrm{co}(\{x\} \cup B(t, \rho)) \subseteq \Omega \cap B(t, R)$, so that $x \in \mathcal{D}_t$. This shows the third property.

For the fourth property note that $\mathcal{D}_t \subseteq B(t, R)$ is contained in the cube $W(t, R)$ and that this cube contains at most

$$M_1 := \left(\frac{2\sqrt{d}(1 + \sin\theta)}{\sin\theta} + 1 \right)^d$$

points from $(2\rho/\sqrt{d})\mathbb{Z}^d$. The last property is justified in the same way, since Ω is bounded. □

Now that we have the local sets we can formulate and proof our main result of this section.

Theorem 11.32 *Suppose Ω is bounded and satisfies an interior cone condition. Let k be a positive integer, $0 < s \le 1$, $1 \le p < \infty$, $1 \le q \le \infty$, and let $m \in \mathbb{N}_0$ satisfy $k > m + d/p$, for $p > 1$ or, for $p = 1$, $k \ge m + d$. Also, let $X \subseteq \Omega$ be a discrete set with mesh norm h satisfying $h \le Q(k, \theta)r$. If $u \in W_p^{k+s}(\Omega)$ satisfies $u|_X = 0$ then*

$$|u|_{W_q^m(\Omega)} \le C h^{k+s-m-d(1/p-1/q)_+} |u|_{W_p^{k+s}(\Omega)}, \tag{11.19}$$

where $(x)_+ = x$ if $x \ge 0$ and is 0 otherwise.

Proof We will use the notation introduced in the paragraph before Lemma 11.31. First of all note that, since $h \le Q(k, \theta)r$, Lemma 11.31 is applicable. Furthermore, our definition of ρ, R, and $Q(k, \theta)$ establish

$$h = \frac{\rho \sin\vartheta}{4k^2(1 + \sin\vartheta)},$$

which allows us to employ Proposition 11.30. The lemma and proposition just mentioned immediately establish the result in the case $q = \infty$. For $1 \le q < \infty$, however, the decomposition of Ω implies that we have

$$\begin{aligned}
|u|^q_{W_q^m(\Omega)} &= \sum_{|\alpha|=m} \int_\Omega |D^\alpha u(x)|^q dx \\
&\le \sum_{t \in T_\rho} \sum_{|\alpha|=m} \int_{\mathcal{D}_t} |D^\alpha u(x)|^q dx = \sum_{t \in T_\rho} |u|^q_{W_q^m(\mathcal{D}_t)} \\
&\le (\#T_r)^{q(1/q-1/p)_+} \left(\sum_{t \in T_\rho} |u|^p_{W_q^m(\mathcal{D}_t)} \right)^{q/p},
\end{aligned}$$

where the last bound follows from standard inequalities relating the p-norm and the q-norm on finite-dimensional spaces. Proposition 11.30 together with $d_{\mathcal{D}_t} \le 2R = 2Q(k, \theta)^{-1}h$

gives the bound

$$\left(\sum_{t\in T_\rho} |u|^p_{W^m_q(\mathcal{D}_t)}\right)^{1/p} \le C h^{k+s-m+d(1/q-1/p)} \left(\sum_{t\in T_r} |u|^p_{W^{k+s}_p(\mathcal{D}_t)}\right)^{1/p}.$$

Here, it has been essential that all involved constants depend only on the cone condition, which is the same for all $\mathcal{D}_t$. Now using Lemma 11.31 again yields

$$\begin{aligned}\sum_{t\in T_\rho} |u|^p_{W^{k+s}_p(\mathcal{D}_t)} &\le \sum_{|\alpha|=k} \int_\Omega \sum_{t\in T_\rho} \chi_{\mathcal{D}_t}(x) \int_{\mathcal{D}_t} \frac{|D^\alpha u(x) - D^\alpha u(y)|^p}{\|x-y\|_2^{d+sp}} dy dx \\ &\le M_1 \sum_{|\alpha|=k} \int_\Omega \int_\Omega \frac{|D^\alpha u(x) - D^\alpha u(y)|^p}{\|x-y\|_2^{d+sp}} dy dx \\ &\le M_1 |u|^p_{W^{k+s}_p(\Omega)}.\end{aligned}$$

A final application of Lemma 11.31 together with

$$\rho = \frac{\sin\theta}{2Q(k,\theta)(1+\sin\theta)} h$$

shows that $\#T_\rho \le C h^{-d}$. Putting all these things together and taking

$$d\left(\frac{1}{q} - \frac{1}{p}\right) - d\left(\frac{1}{q} - \frac{1}{p}\right)_+ = -d\left(\frac{1}{p} - \frac{1}{q}\right)_+$$

into account establishes the desired result. □

We end this section by applying the last theorem to radial basis functions of compact support and to thin-plate splines.

Let $\Phi_{d,k} = \phi_{d,k}(\|\cdot\|_2)$ be the compactly supported basis functions from Definition 9.11. If $\Omega \subseteq \mathbb{R}^d$ has a Lipschitz boundary, we know by Theorem 10.35 and Corollary 10.48, which can be extended to the case of fractional Sobolev spaces (see the discussion in [31]), that the associated native space is norm-equivalent to the Sobolev space $H^{k+(d+1)/2}(\Omega) = W_2^{k+(d+1)/2}(\Omega)$. We treat this case in the more general situation where $\Phi \in L_1(\mathbb{R}^d)$ has a Fourier transform that satisfies

$$c_1(1+\|\omega\|_2^2)^{-\tau} \le \widehat{\Phi}(\omega) \le c_2(1+\|\omega\|_2^2)^{-\tau}, \qquad \omega \in \mathbb{R}^d, \tag{11.20}$$

with $\tau > d/2$.

Corollary 11.33 *Suppose that $\Omega \subseteq \mathbb{R}^d$ is bounded, has a Lipschitz boundary, and satisfies an interior cone condition with radius r and angle θ. Let $X \subseteq \Omega$ be a given discrete set of centers and $s_{f,X}$ be the interpolant. Suppose that Φ satisfies (11.20) with $\tau = k+s$, where k is a positive integer and $0 < s \le 1$. If $m \in \mathbb{N}_0$ satisfies $k > m + d/2$ then the error between $f \in W_2^\tau(\Omega)$ and its interpolant $s_{f,X}$ can be bounded by*

$$|f - s_{f,X}|_{W^m_q(\Omega)} \le C h_{X,\Omega}^{\tau-m-d(1/2-1/q)_+} \|f\|_{W_2^\tau(\Omega)}$$

for all sufficiently dense sets X.

Next, we want to apply these results to the thin-plate splines $\Phi_{d,\ell} = \phi_{d,\ell}(\|\cdot\|_2)$ from (10.11). We know by Theorem 10.43 that the global native space $\mathcal{N}_{\Phi_{d,\ell}}(\mathbb{R}^d)$ is the Beppo Levi space $\mathrm{BL}_\ell(\mathbb{R}^d)$. To apply Theorem 11.32, we need an extension operator from $\mathrm{BL}_\ell(\Omega)$ to $\mathrm{BL}_\ell(\mathbb{R}^d)$.

Lemma 11.34 *Suppose that $\Omega \subseteq \mathbb{R}^d$ is open and bounded and satisfies an interior cone condition. For every $f \in \mathrm{BL}_\ell(\mathbb{R}^d)$, $\ell > d/2$, there exists a unique function $f^\Omega \in \mathrm{BL}_\ell(\mathbb{R}^d)$ with $f^\Omega|\Omega = f$ and*

$$|f^\Omega|_{\mathrm{BL}_\ell(\mathbb{R}^d)} = \min\{|g|_{\mathrm{BL}_\ell(\mathbb{R}^d)} : g \in \mathrm{BL}_\ell(\mathbb{R}^d) \quad \text{and} \quad g|\Omega = f|\Omega\}.$$

Proof Fix a $\pi_\ell(\mathbb{R}^d)$-unisolvent set $\Xi = \{\xi_1, \ldots, \xi_Q\} \subseteq \Omega$ and introduce the inner product

$$(f, g)_{\mathbb{R}^d} := (f, g)_{\mathrm{BL}_\ell(\mathbb{R}^d)} + \sum_{j=1}^{Q} f(\xi_j)g(\xi_j).$$

With this inner product, $\mathrm{BL}_\ell(\mathbb{R}^d)$ becomes a reproducing-kernel Hilbert space (see Theorem 10.20). Moreover, since all relevant functions coincide with f when restricted to Ξ, the minimization problem is equivalent to minimizing the norm $\|\cdot\|_{\mathbb{R}^d}$ on

$$V_f = \{g \in \mathrm{BL}_\ell(\mathbb{R}^d) : g|\Omega = f|\Omega\}.$$

But this set is obviously nonempty since it contains f, it is convex, and it is closed. The last follows from the reproducing-kernel property. If $\{g_n\} \subseteq V_f$ converges to $g \in \mathrm{BL}_\ell(\mathbb{R}^d)$ then the reproducing kernel gives also pointwise convergence, i.e. $g_n(x) \to g(x)$, $x \in \mathbb{R}^d$. This means $g|\Omega = f$.

Moreover, the minimization problem amounts to nothing other than finding the best approximation from V_f to 0. Because of the properties of V_f just stated, this is uniquely solvable. □

Lemma 11.35 *Let $\ell > d/2$ and $\Omega \subseteq \mathbb{R}^d$ be open and bounded, satisfying an interior cone condition. For every $f \in H^\ell(\Omega)$ there exists a unique $|\cdot|_{\mathrm{BL}_\ell(\mathbb{R}^d)}$ minimal extension $f^\Omega \in \mathrm{BL}_\ell(\mathbb{R}^d)$. Moreover, this extension is continuous, i.e. there exists a constant $K > 0$ such that*

$$|f^\Omega|_{\mathrm{BL}_\ell(\mathbb{R}^d)} \le K|f|_{\mathrm{BL}_\ell(\Omega)}.$$

Proof Since $\Omega \subseteq \mathbb{R}^d$ satisfies the cone condition there is a continuous extension operator from $H^\ell(\Omega)$ to $H^\ell(\mathbb{R}^d)$, meaning that there exists a constant $C > 0$ such that we can find for every $f \in H^\ell(\Omega)$ a function $\widetilde{f} \in H^\ell(\mathbb{R}^d)$ with $\widetilde{f}|\Omega = f$ and $\|\widetilde{f}\|_{H^\ell(\mathbb{R}^d)} \le C\|f\|_{H^\ell(\Omega)}$. Since obviously $H^\ell(\mathbb{R}^d) \subseteq \mathrm{BL}_\ell(\mathbb{R}^d)$, Lemma 11.34 gives us a function $f^\Omega \in \mathrm{BL}_\ell(\mathbb{R}^d)$ which coincides on Ω with f and which has a minimal Beppo Levi semi-norm amongst all such functions. The uniqueness follows from Lemma 11.34 and the fact that all possible extensions $\widetilde{f}$ of f coincide with f on Ω.

By the proof of that lemma we even know that f^Ω has minimal $\|\cdot\|_{\mathbb{R}^d}$-norm with $\|f\|_{\mathbb{R}^d}^2 := |f|^2_{\mathrm{BL}_\ell(\mathbb{R}^d)} + \sum |f(\xi_j)|^2$. Hence, using the Sobolev embedding theorem, we have

$$\|f^\Omega\|_{\mathbb{R}^d} \le \|\widetilde{f}\|_{\mathbb{R}^d} \le C\|\widetilde{f}\|_{H^\ell(\mathbb{R}^d)} \le C\|f\|_{H^\ell(\Omega)}$$

with some generic constant $C > 0$. If we can show that $\|\cdot\|_\Omega$ defined by $\|f\|_\Omega^2 = |f|_{\mathrm{BL}_\ell(\Omega)} + \sum |f(\xi_j)|^2$ is equivalent to $\|\cdot\|_{H^\ell(\Omega)}$ on $H^\ell(\Omega)$ then we can immediately derive $\|f^\Omega\|_{\mathbb{R}^d} \le C\|f\|_\Omega$ and hence, since $\Xi \subseteq \Omega$, also $|f^\Omega|_{\mathrm{BL}_\ell(\mathbb{R}^d)} \le C|f|_{\mathrm{BL}_\ell(\Omega)}$. To show norm equivalence we use standard arguments from the theory of Sobolev spaces. First of all, $\|\cdot\|_\Omega$ can obviously be bounded by a constant times $\|\cdot\|_{H^\ell(\Omega)}$, again by the Sobolev embedding theorem. Unfortunately this is not the inequality we need. To prove the other inequality let us assume that it is wrong. Then we can find a sequence $\{\varphi_n\}_{n\in\mathbb{N}}$ with $\|\varphi_n\|_{H^\ell(\Omega)} = 1$ and

$$1 = \|\varphi_n\|^2_{H^\ell(\Omega)} > n\left(\sum_{|\alpha|=\ell} \frac{\ell!}{\alpha!}\|D^\alpha\varphi_n\|^2_{L_2(\Omega)} + \sum_{j=1}^{Q} |\varphi_n(\xi_j)|^2\right). \tag{11.21}$$

From this we can deduce that $\|D^\alpha\varphi_n\|_{L_2(\Omega)} \to 0$ for $n \to \infty$ provided that $|\alpha| = \ell$.

Since Ω satisfies an interior cone condition, $H^\ell(\Omega)$ must be compactly embedded in $H^{\ell-1}(\Omega)$. This means that $\{\varphi_n\}$ is relatively compact in $H^{\ell-1}(\Omega)$. Hence there exists a convergent subsequence. For simplicity we call this subsequence $\{\varphi_n\}$ again, i.e. $\varphi_n \to \varphi$ in $H^{\ell-1}(\Omega)$ with $\varphi \in H^{\ell-1}(\Omega)$. But since $\|D^\alpha\varphi_n\|_{L_2(\Omega)} \to 0$ if $|\alpha| = \ell$, $\{\varphi_n\}$ is a Cauchy sequence, even in $H^\ell(\Omega)$, that converges to an element, say $\widetilde{\varphi} \in H^\ell(\Omega)$. Since it also converges in $H^{\ell-1}(\Omega)$ to φ we must have $\varphi = \widetilde{\varphi} \in H^\ell(\Omega)$ and $\varphi_n \to \varphi$ in $H^\ell(\Omega)$. Moreover, we can conclude that $D^\alpha\varphi = 0$ for all $|\alpha| = \ell$. As in the proof of Lemma 10.38 we see that φ coincides on Ω with a polynomial from $\pi_{\ell-1}(\mathbb{R}^d)$. By Sobolev's embedding theorem again, we find that $\varphi_n(x) \to \varphi(x)$, $x \in \Omega$. This means in conjunction with (11.21) that $\sum |\varphi(\xi_j)|^2 = 0$ and thus $\varphi(\xi_j) = 0$ for all $1 \le j \le Q$. Since Ξ is $\pi_{\ell-1}(\mathbb{R}^d)$-unisolvent, we can conclude that $\varphi = 0$, which contradicts $\|\varphi\|_{H^\ell(\Omega)} = \lim_{n\to\infty} \|\varphi_n\|_{H^\ell(\Omega)} = 1$. □

Theorem 11.36 *Let $\ell > m + d/2$. Suppose that $\Omega \subseteq \mathbb{R}^d$ is open and bounded and satisfies an interior cone condition. Consider the thin-plate splines $\Phi_{d,\ell}$ as conditionally positive definite of order ℓ. Then the error between $f \in H^\ell(\Omega)$ and its interpolant $s_{f,X}$ can be bounded by*

$$|f - s_{f,X}|_{W_p^m(\Omega)} \le C h_{X,\Omega}^{\ell-m-d(1/2-1/p)_+} |f|_{\mathrm{BL}_\ell(\Omega)}$$

for $1 \le p \le \infty$. Finally, if $f \in H^{2\ell}(\Omega)$ has compact support in Ω, then we have the improved bound

$$|f(x) - s_{f,X}(x)| \le C h_{X,\Omega}^{2\ell-d/2} \|\Delta^\ell f\|_{L_2(\Omega)}, \qquad x \in \Omega.$$

Proof The first estimate is obviously true in the case $p = \infty$. Hence we can assume that $1 \le p < \infty$ for the first two estimates.

According to Lemma 11.35 we can extend $f \in H^\ell(\Omega)$ to $f^\Omega \in \mathrm{BL}_\ell(\mathbb{R}^d)$. Then the interpolant to f based on $X \subseteq \Omega$ coincides with the interpolant to f^Ω. The density result in Theorem 10.40 allows us to apply Theorem 11.32 in this situation also, yielding

$$|f - s_{f,X}|_{W_p^m(\Omega)} \le C h_{X,\Omega}^{\ell-m-d(1/2-1/p)_+} |f^\Omega - s_{f^\Omega,X}|_{\mathrm{BL}_\ell(\Omega)}.$$

Moreover, by Corollary 10.25 we have $|f^\Omega - s_{f^\Omega,X}|_{\mathrm{BL}_\ell(\Omega)} \le |f^\Omega|_{\mathrm{BL}_\ell(\Omega)}$ and, following Lemma 11.35, the latter may be bounded by $K|f|_{\mathrm{BL}_\ell(\Omega)}$.

Finally, if $f \in H^{2\ell}(\Omega)$ has compact support in Ω then we obviously have $f^\Omega = f$. Hence, the error estimate just established gives for $p = 2$

$$|f - s_{f,X}|_{L_2(\mathbb{R}^d)} \le C h_{X,\Omega}^{\ell} |f - s_{f,X}|_{\mathrm{BL}_\ell(\mathbb{R}^d)}.$$

Using this in the proof of Theorem 11.24 yields the final error bound. □

Note that the condition for $f \in H^{2\ell}(\Omega)$ to have compact support in Ω can be weakened by assuming that certain normal derivatives vanish.

Thin-plate splines are probably the most examined and best understood basis functions. Nonetheless, there are still some important open problems. More details about this are given, among other things, in the next section.

11.7 Notes and comments

Nowadays there is common agreement upon the fact that error estimates of the interpolation process using radial basis functions should first of all take place in the native space. The material presented in the first three sections of this chapter is based upon the pioneering work of Duchon [49], Madych and Nelson [111–113], and Wu and Schaback [204]. But the number of publications in this particular field is steadily increasing and the interested reader should have a look at the bibliography. Theorem 11.13 borrows ideas from Levesley and Ragozin [103].

Let us point out that the seminal paper [114] by Madych and Nelson is so far the only one (except the somewhat weaker version [195] by the present author, which is based on the same ideas) that establishes spectral convergence orders for Gaussians and (inverse) multiquadrics. But we also want to emphasize that this paper does not prove a spectral order of the form e^{-c/h^2} for the Gaussians, as is sometimes suggested in other publications.

The experienced reader has probably noticed that the improved estimates derived in Section 11.5 borrow ideas from classical spline theory. As in that case, rather simple Hilbert space arguments are used in Theorem 11.23 to double the approximation order; see also Schaback [166, 168]. The estimates on the L_p-norm use a trick that has become known as *Duchon's localization trick* [49], where they appear in the context of thin-plate spline approximation for the first time. The general version given here comes from Narcowich *et al.* [149]. However, in the case of spline approximation it is well known that the optimal order cannot be achieved using only Hilbert space arguments; this should also be true in the case of radial basis function approximation. Moreover, there is quite a gap in the approximation orders that can be realized if the data sites form a regular grid or if they are truly scattered and finitely many. Let us discuss this in more detail in the case of thin-plate splines. Let $\gamma_p = \min\{\ell, \ell - d/2 + d/p\}$ be the L_p-approximation order as derived in the first case mentioned in Theorem 11.36. It is known (see Buhmann [33]) that for sufficiently smooth functions the L_∞-order is 2ℓ, if the data sites form a regular infinite grid of grid size h. The same is true if the data sites are a finite grid on $[0, 1]^d$ and the error is measured in any compact subset of $(0, 1)^d$ (see Bejancu [22]). However, it is also known that if the L_∞-order is larger than 2ℓ then the approximated function f becomes "trivial" in the sense that it has to satisfy of $\Delta^\ell f = 0$ on Ω, so that the saturation order for these functions is 2ℓ (see Schaback and Wendland [171]).

In the papers [90–94] Johnson showed that the L_p-order for smooth functions does not exceed $\ell + 1/p$ for $1 \le p \le \infty$ and he improved it to $\gamma_p + 1/p$ for $1 \le p \le 2$, so that most is known about this case. He also showed that the L_∞-order is 2ℓ except for a boundary layer of size $\mathcal{O}(h|\log h|)$.

Some people still argue that, on the one hand, the native space, particularly for Gaussians and (inverse) multiqadrics, is rather small since the Fourier transform of a function from one of these spaces has to decay exponentially fast. This is beyond doubt true but, on the other hand, Shannon's famous sampling theorem holds in its original form for an even smaller class of functions, namely only band-limited ones, and nobody would argue the importance of this theorem. Nonetheless, there is some progress in escaping the native space and extending the error estimates to larger function spaces. The first result in this direction came from Schaback [165], where interpolation was replaced by approximation. Yoon [207, 208] approached interpolation using Schaback's ideas but had to work with scaled functions. The most recent results are those of Brownlee and Light [32] for thin-plate splines and Narcowich and Ward [144] for more general basis functions.

12
Stability

In this chapter we will be concerned with the stability of the radial basis function interpolation process. Let us introduce the subject with an example.

We consider the following one-dimensional interpolation problem. The data sites are given by $x_j = j/n \in [0,1]$, $0 \le j \le n$, and the basis function is the inverse multiquadric $\phi(r) = 1/\sqrt{1+r^2}$. From the previous chapter we know that the sequence of interpolants s_n to a function from the native space of ϕ converges to f as e^{-cn} and this convergence comes from estimates on the power function.

To compute the interpolant, however, we have to invert the interpolation matrices $A_n = (\phi((i-j)/n))$. Unfortunately, the smallest eigenvalue $\lambda_{\min}(n)$ tends to zero as fast as $P^2_{\Phi,X}$. To illustrate this behavior we have plotted both $-\log \lambda_{\min}(n)$, the negative logarithm of the smallest eigenvalue of A_n, and $-\log \|P^2_{\Phi,X}\|_{L_\infty[0,1]}$ in Figure 12.1. For a more mathematical investigation let us make the following definition. Assume that $\Phi : \Omega \times \Omega \to \mathbb{R}$ is a conditionally positive definite kernel with respect to $\mathcal{P}$. For $X = \{x_1, \dots, x_N\} \subseteq \Omega$ and a basis $p_1, \dots, p_Q$ of $\mathcal{P}$ we set $P = P_X = (p_j(x_i)) \in \mathbb{R}^{N \times Q}$. This allows us to express the side conditions (10.2) from Definition 10.14 by $P_X^T \alpha = 0$. With this notation we define

$$\lambda_{\min}(A_{\Phi,X}) = \inf_{\alpha \in \mathbb{R}^N \setminus \{0\}, P_X^T \alpha = 0} \frac{\alpha^T A_{\Phi,X} \alpha}{\alpha^T \alpha}, \tag{12.1}$$

where $A_{\Phi,X}$ denotes the usual interpolation matrix. Since the quadratic form (10.3) is positive on the set of all $\alpha \in \mathbb{R}^N$ with $P^T\alpha = 0$ we necessarily have $\lambda_{\min}(A_{\Phi,X}) > 0$.

Now why is $\lambda_{\min}(A_{\Phi,X})$ important for the stability of the interpolation process? Obviously, if Φ is an unconditionally positive definite kernel then $A_{\Phi,X}$ is a positive definite matrix and $\lambda_{\min}(A_{\Phi,X})$ is its smallest eigenvalue. But even in the case of a conditionally positive definite kernel this number is crucial. For example, for the interpolation of data $f|X$ we have to solve a system like (8.10). This shows that $\alpha^T A_{\Phi,X}\alpha = \alpha^T(f|X)$ and hence

$$\|\alpha\|_2 \le \frac{1}{\lambda_{\min}(A_{\Phi,X})} \|f|X\|_2.$$

So, we know more about the accuracy of our solution vector α if we know more about lower bounds for $\lambda_{\min}(A_{\Phi,X})$.

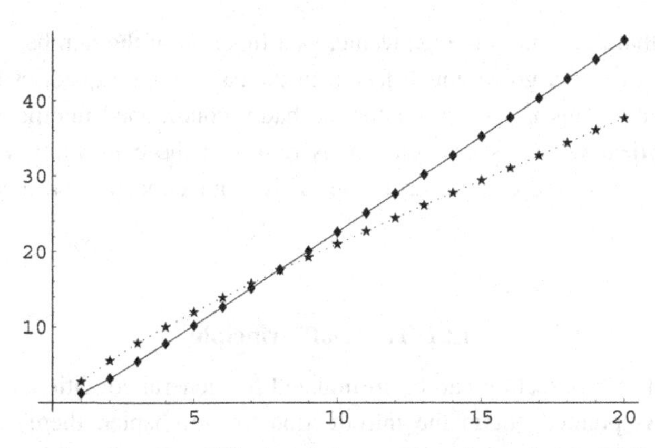

Fig. 12.1 The power function (dotted line) and the smallest eigenvalue (solid line) for integer n, $1 \le n \le 20$.

Moreover let us briefly discuss the condition number as a whole. If Φ is a symmetric positive definite kernel, $\lambda_{\min}(A_{\Phi,X})$ coincides with the norm $\|A_{\Phi,X}^{-1}\|_2$ and the condition number of $A_{\Phi,X}$ is given by

$$\operatorname{cond}(A_{\Phi,X}) = \|A_{\Phi,X}\|_2 \|A_{\Phi,X}^{-1}\|_2 = \frac{\lambda_{\max}(A_{\Phi,X})}{\lambda_{\min}(A_{\Phi,X})},$$

where $\lambda_{\max}$ denotes the maximum eigenvalue. The condition number of an interpolation matrix gives information on the numerical stability of the interpolation process. Hence we actually have to investigate both the maximum and the minimum eigenvalue. Fortunately $\lambda_{\max}$ behaves nicely compared to the smallest eigenvalue; to see this, we invoke Gershgorin's theorem, which gives for our matrix an index $j \in \{1, \dots, N\}$ with

$$|\lambda_{\max}(A_{\Phi,X}) - \Phi(x_j, x_j)| \le \sum_{\substack{k=1 \\ k \ne j}}^{N} |\Phi(x_j, x_k)|,$$

so that

$$\lambda_{\max}(A_{\Phi,X}) \le N \|\Phi(\cdot, \cdot)\|_{L_\infty(X \times X)},$$

which becomes, in the case of a positive definite function,

$$\lambda_{\max}(A_{\Phi,X}) \le N\Phi(0).$$

Hence if X is quasi-uniformly distributed then $\lambda_{\max}(A_{\Phi,X})$ grows at most like $h_{X,\Omega}^{-d}$, and this upper bound can also be established, in the case of the Gaussians and the compactly supported radial functions, for non-quasi-uniform data sets. Even if this upper bound appears already to be worse than expected, numerical tests show that the maximum eigenvalue indeed causes no problems. The same is true in the case of a conditionally positive definite function.

We will see that the minimum eigenvalue, as a function of the number of data sites or their separation distance, grows much faster, in the case of multiquadrics and Gaussians even exponentially. This is the reason for the badly conditioned interpolation matrices. But before we deal with this problem let us return to the connection with the power function mentioned in the opening example. This connection will be made in the next section.

12.1 Trade-off principle

Our main result in this section can be formulated for general conditionally positive definite kernels. As pointed out in the introduction to the chapter, there is a connection between the smallest eigenvalue of the main part of the interpolation matrix given by $A_{\Phi,X} = (\Phi(x_i, x_j))_{1\le i,j\le N}$ and the squared power function $P^2_{\Phi,X}(x)$. But since the power function depends on the point $x \in \Omega$ at which it is evaluated, and the smallest eigenvalue of $A_{\Phi,X}$ does not, the setting has to be slightly modified. This is done by adding the point x to the set of centers X. Let us define $x_0 = x$.

Theorem 12.1 *If $u_j^*(x)$, $1 \le j \le N$, denotes the cardinal functions from Theorem 11.1 then we have for all $x \notin X$*

$$[\lambda_{\min}(A_{\Phi,X\cup\{x\}})]^{-1} P^2_{\Phi,X}(x) \ge 1 + \sum_{j=1}^{N} [u_j^*(x)]^2.$$

Proof Define $u_0^*(x) = -1$ and $x_0 = x$. Then, the definition of the power function immediately gives

$$P^2_{\Phi,X}(x) = \sum_{j,k=0}^{N} u_j^*(x)u_k^*(x)\Phi(x_j, x_k) \ge \lambda_{\min}(A_{\Phi,X\cup\{x\}}) \sum_{j=0}^{N} [u_j^*(x)]^2.$$

□

This theorem can be interpreted in at least two different ways. On the one hand we have

$$\lambda_{\min}(A_{\Phi,X}) \le \min_{1\le k\le N} P^2_{\Phi,X\setminus\{x_k\}}(x_k),$$

giving both lower bounds for the power function and upper bounds for the eigenvalue. On the other hand we have

$$1 + \sum_{j=1}^{N} |u_j^*(x)|^2 \le \frac{P^2_{\Phi,X}(x)}{\lambda_{\min}(A_{\Phi,X\cup\{x\}})}, \qquad x \notin X,$$

which is an upper bound for the Lebesgue functions.

12.2 Lower bounds for $\lambda_{\min}$

For $\lambda_{\min}$ we use the following idea. Suppose that Φ is the conditionally positive definite kernel of interest. Suppose further that there exists a positive definite kernel Ψ such that

$$\sum_{j,k=1}^{N} \alpha_j \alpha_k \Phi(x_j, x_k) \geq \sum_{j,k=1}^{N} \alpha_j \alpha_k \Psi(x_j, x_k) \geq \lambda \|\alpha\|_2.$$

Then λ is obviously a lower bound for $\lambda_{\min}$. But what do we gain from this? Now we have to do the estimates for Ψ instead of Φ and of course Ψ has to depend on Φ. So how does Ψ depend on Φ? Before we answer these questions, we want to discuss the terms in which the lower bounds have to be expressed.

The approximation error between function and interpolant was expressed in terms of the fill distance $h_{X,\Omega}$, because the fill distance is a measure of how well the centers X cover the region Ω. But it is not a good measure of the stability. A point set X might have quite a big fill distance and the interpolation process is nonetheless badly conditioned. The reason for this is simply that only two points from X have to be very close. Thus it would seem to be more natural to express lower bounds on $\lambda_{\min}$ in terms of the separation distance, which has already been defined to be

$$q_X := \tfrac{1}{2} \min_{i \neq j} \|x_i - x_j\|_2.$$

The separation distance gives the maximum radius $r > 0$ such that all balls $\{x \in \mathbb{R}^d : \|x - x_j\|_2 < r\}$ are disjoint.

From now on our analysis will concentrate on real-valued conditionally positive definite functions that possess a positive generalized Fourier transform. Then by Corollary 8.13 we can express a typical quadratic form via

$$\sum_{j,k=1}^{N} \alpha_j \alpha_k \Phi(x_j - x_k) = (2\pi)^{-d/2} \int_{\mathbb{R}^d} \left| \sum_{j=1}^{N} \alpha_j e^{i\omega^T x_j} \right|^2 \widehat{\Phi}(\omega) d\omega.$$

Thus an appropriate Ψ is given if its Fourier transform $\widehat{\Psi}$ satisfies $\widehat{\Phi}(\omega) \geq \widehat{\Psi}(\omega)$ and is of order less than or equal to the order of $\widehat{\Phi}$.

Lemma 12.2 *Let χ_M be the characteristic function of $B(0, M)$, $M > 0$, i.e. $\chi_M(x) = 1$ if $\|x\|_2 \leq M$ and $\chi_M(x) = 0$ otherwise. Then*

$$\widehat{\chi_M}(x) = (\chi_M)^\vee(x) = M^{d/2} \|x\|_2^{-d/2} J_{d/2}(M \|x\|_2),$$

where J_ν is a Bessel function of the first kind.

Proof As χ_M is a radial function, its Fourier transform and inverse Fourier transform coincide and can be computed via Theorem 5.26 as

$$\widehat{\chi}_M(x) = \|x\|_2^{-(d-2)/2} \int_0^M t^{d/2} J_{(d-2)/2}(\|x\|_2 t) dt.$$

Using the definition of J_ν, the dominated convergence theorem and the multiplication property of the Γ-function give

$$\begin{aligned}\widehat{\chi}_M(x) &= \|x\|_2^{-(d-2)/2} \int_0^M \sum_{m=0}^{\infty} \frac{(-1)^m \, (\|x\|_2 t/2)^{2m+(d-2)/2}}{m!\,\Gamma(m+d/2)} t^{d/2} dt \\ &= \|x\|_2^{-d/2+1} \sum_{m=0}^{\infty} \frac{(-1)^m \, (\|x\|_2/2)^{2m+(d-2)/2}}{m!\,\Gamma(m+d/2)} \int_0^M t^{2m+d-1} dt \\ &= \|x\|_2^{-d/2+1} \sum_{m=0}^{\infty} \frac{(-1)^m \, (\|x\|_2/2)^{2m+(d-2)/2}}{m!\,\Gamma(m+d/2)} \frac{1}{d+2m} M^{2m+d} \\ &= M^{d/2} \|x\|_2^{-d/2} J_{d/2}(M\|x\|_2).\end{aligned}$$

□

This function is the key ingredient in finding the function Ψ.

Theorem 12.3 *Let Φ be an even conditionally positive definite function that possesses a positive Fourier transform $\widehat{\Phi} \in C(\mathbb{R}^d \setminus \{0\})$. With the function*

$$\varphi_0(M) := \inf_{\|\omega\|_2 \le 2M} \widehat{\Phi}(\omega)$$

a lower bound on $\lambda_{\min}$ is given by

$$\lambda_{\min}(A_{\Phi,X}) \ge \frac{\varphi_0(M)}{2\Gamma(d/2+1)} \left(\frac{M}{2^{3/2}}\right)^d$$

for any $M > 0$ satisfying

$$M \ge \frac{12}{q_X} \left(\frac{\pi \Gamma^2(d/2+1)}{9}\right)^{1/(d+1)} \tag{12.2}$$

or, a fortiori,

$$M \ge \frac{6.38d}{q_X}. \tag{12.3}$$

Proof Let us define Ψ by its Fourier transform as

$$\widehat{\Psi}(\omega) \equiv \widehat{\Psi}_M(\omega) := \frac{\varphi_0(M)\Gamma(d/2+1)}{2^d M^d \pi^{d/2}} (\chi_M * \chi_M)(\omega),$$

where $f * g$ denotes the convolution from Theorem 5.16. Then $\widehat{\Psi} \ge 0$ has support in $B(0, 2M)$, which shows that $\widehat{\Phi}(\omega) \ge \widehat{\Psi}(\omega)$ for $\|\omega\|_2 > 2M$. For $\|\omega\|_2 \le 2M$ note that

$$\widehat{\Psi}(\omega) \le \frac{\varphi_0(M)\Gamma(d/2+1)}{2^d M^d \pi^{d/2}} \operatorname{vol}(B(0, 2M)) \le \varphi_0(M) \le \widehat{\Phi}(\omega).$$

This shows that Ψ is a good candidate and that we have to bound the quadratic form for Ψ. This is done directly. First note that

$$\begin{aligned}\Psi_M(x) &= \frac{\varphi_0(M)\Gamma(d/2+1)}{2^d M^d \pi^{d/2}}(\chi_M * \chi_M)^\vee(x)\\ &= \frac{\varphi_0(M)\Gamma(d/2+1)}{2^d M^d \pi^{d/2}}(2\pi)^{d/2}|\widehat{\chi}_M(x)|^2\\ &= \frac{\varphi_0(M)\Gamma(d/2+1)}{2^{d/2}}\|x\|_2^{-d} J_{d/2}^2(M\|x\|_2).\end{aligned}$$

Next we use

$$\begin{aligned}\sum_{j,k=1}^N \alpha_j\alpha_k\Psi_M(x_j-x_k) &\geq \|\alpha\|_2^2\Psi_M(0) - \sum_{j\neq k}|\alpha_j||\alpha_k||\Psi_M(x_j-x_k)|\\ &\geq \|\alpha\|_2^2\Psi_M(0) - \frac{1}{2}\sum_{j\neq k}(|\alpha_j|^2+|\alpha_k|^2)|\Psi_M(x_j-x_k)|\\ &\geq \|\alpha\|_2^2\left(\Psi_M(0) - \max_{1\leq j\leq N}\sum_{\substack{k=1\\k\neq j}}^N|\Psi_M(x_j-x_k)|\right).\end{aligned}$$

By Proposition 5.6 we know that

$$\Psi_M(0) = \frac{\varphi_0(M)}{\Gamma(d/2+1)}\left(\frac{M}{2^{3/2}}\right)^d.$$

Hence, the stated bound on $\lambda_{\min}$ is $\Psi_M(0)/2$ and it remains to show that

$$\max_{1\leq j\leq N}\sum_{\substack{k=1\\k\neq j}}^N|\Psi_M(x_j-x_k)| \leq \frac{1}{2}\Psi_M(0)$$

for the chosen M. To this end we can assume that the maximum is taken for $x_1=0$, i.e. that

$$\max_{1\leq j\leq N}\sum_{\substack{k=1\\k\neq j}}^N|\Psi_M(x_j-x_k)| = \sum_{k=2}^N|\Psi_M(x_k)|.$$

Now the trick is to count the points in a different way. If we define

$$E_n = \{x\in\mathbb{R}^d : nq_X \leq \|x\|_2 < (n+1)q_X\}$$

then we see that every $x_j, 2\leq j\leq N$, is contained in exactly one of the $E_n, n\geq 1$. Moreover, since every ball $B(x_j, q_X)$ around x_j with radius q_X is essentially disjoint from a ball around $x_k \neq x_j$ with the same radius and since all these balls with center in E_n are contained in

$$\{x\in\mathbb{R}^d : (n-1)q_X \leq \|x\|_2 \leq (n+2)q_X\},$$

we can estimate the number of centers in E_n by comparing the volumes to get

$$\#\{x_j\in E_n\} \leq (n+2)^d - (n-1)^d \leq 3^d n^{d-1},$$

where the last inequality is easily shown by induction on d. From Proposition 5.6 we find

$$\begin{aligned}|\Psi_M(x)| &\leq \frac{\varphi_0(M)\Gamma(d/2+1)2^{d/2+2}}{M\pi}\|x\|_2^{-(d+1)}\\ &= \Psi_M(0)\frac{\Gamma^2(d/2+1)}{\pi}\left(\frac{4}{M\|x\|_2}\right)^{d+1},\end{aligned}$$

which allows us to bound $\Psi_M(x)$ on E_n:

$$|\psi_M(x)| \leq \Psi_M(0)\frac{\Gamma^2(d/2+1)}{\pi}\left(\frac{4}{Mnq_X}\right)^{d+1}, \qquad x \in E_n.$$

Thus if we use $\sum_{n=1}^{\infty} n^{-2} = \pi^2/6$ we get the bound

$$\begin{aligned}\sum_{k=2}^{N}|\Psi_M(x_k)| &\leq \sum_{n=1}^{\infty}\#\{x_j \in E_n\}\sup_{x\in E_n}|\Psi_M(x)|\\ &\leq \Psi_M(0)\frac{\Gamma^2(d/2+1)}{\pi}\left(\frac{4}{Mq_X}\right)^{d+1}3^d\sum_{n=1}^{\infty}n^{-2}\\ &= \Psi_M(0)\frac{\Gamma^2(d/2+1)\pi}{18}\left(\frac{12}{Mq_X}\right)^{d+1}\\ &\leq \frac{1}{2}\Psi_M(0),\end{aligned}$$

where the last inequality obviously holds for all M satisfying (12.2). To see that (12.3) implies (12.2), remember Stirling's formula from Proposition 5.2, which gives us here

$$\frac{\pi}{9}\Gamma^2(d/2+1) \leq \frac{\pi^2}{9}d^{d+1}(2e)^{-d}e^{1/(3d)}$$

and

$$\begin{aligned}\left(\frac{\pi}{9}\Gamma^2(d/2+1)\right)^{1/(d+1)} &\leq d\left(\frac{\pi^2}{9}\right)^{1/(d+1)}(2e)^{-d/(d+1)}e^{1/[3d(d+1)]}\\ &\leq d\frac{\pi}{3\sqrt{2e}}e^{1/6} \leq 0.531d,\end{aligned}$$

so that (12.3) is indeed sufficient. □

Our next step is to apply this result to our collection of different basis functions. To this end let us introduce the constants

$$M_d = 12\left(\frac{\pi\Gamma^2(d/2+1)}{9}\right)^{1/(d+1)} \quad \text{and} \quad C_d = \frac{1}{2\Gamma(d/2+1)}\left(\frac{M_d}{2^{3/2}}\right)^d$$

so that the bound becomes

$$\lambda_{\min}(A_{\Phi,X}) \geq C_d\varphi_0(M_d/q_X)q_X^{-d}.$$

Of course, instead of using the exact bound for M in the definition above we could also define $M_d = 6.38d$.

Our first example will be the Gaussian $\Phi(x) = e^{-\alpha\|x\|_2^2}$, $\alpha > 0$. From Theorems 5.18 and 5.16 we know that the Fourier transform of the Gaussian is given by

$$\widehat{\Phi}(\omega) = (2\alpha)^{-d/2} e^{-\|x\|_2^2/(4\alpha)},$$

which is clearly decreasing. Thus the infimum takes the value

$$\varphi_0(M) = (2\alpha)^{-d/2} e^{-M^2/\alpha},$$

giving

Corollary 12.4 *For interpolation with $\Phi(x) = e^{-\alpha\|x\|_2^2}$, the minimal eigenvalue of the interpolation matrix can be bounded by*

$$\begin{aligned}\lambda_{\min}(A_{\Phi,X}) &\geq C_d (2\alpha)^{-d/2} e^{-M_d^2/(q_X^2\alpha)} q_X^{-d} \\ &\geq C_d (2\alpha)^{-d/2} e^{-40.71d^2/(q_X^2\alpha)} q_X^{-d}.\end{aligned}$$

Next let us have a look at (inverse) multiquadrics. We know from Theorem 8.15 that $\Phi(x) = (c^2 + \|x\|_2^2)^\beta$, $\beta \in \mathbb{R}\backslash\mathbb{N}_0$, has up to a sign the generalized Fourier transform

$$\widehat{\Phi}(\omega) = \frac{2^{1+\beta}}{\Gamma(-\beta)} \left(\frac{\|\omega\|_2}{c}\right)^{-\beta-d/2} K_{d/2+\beta}(c\|\omega\|_2), \qquad \omega \neq 0.$$

Moreover, we know from Corollary 5.12 that $r \mapsto r^{-\beta-d/2} K_{d/2+\beta}(r)$ is nonincreasing and that

$$K_{d/2+\beta}(r) \geq C(d,\beta)\frac{e^{-r}}{\sqrt{r}}$$

for $r \geq 1$ with an explicitly known constant $C(d,\beta)$. The restriction $r \geq 1$ is only necessary if $|d + 2\beta| < 1$. In any case we have

$$\varphi_0(M) \geq \widetilde{C}(d,c,\beta)\frac{e^{-2cM}}{M^{\beta+(d+1)/2}}.$$

Corollary 12.5 *Using $\Phi(x) = (c^2 + \|x\|_2^2)^\beta$, $\beta \in \mathbb{R}\backslash\mathbb{N}_0$, as the basis function results in*

$$\lambda_{\min}(A_{\Phi,X}) \geq C(d,\beta,c) q_X^{\beta-d/2+1/2} e^{-2cM_d/q_X}$$

with an explicitly known constant $C(d,\beta,c)$.

After this more complicated example we turn to functions of finite smoothness. Thin-plate splines, powers, and compactly supported functions of minimal degree can all be treated in the same way. Let us start with thin-plate splines $\Phi(x) = (-1)^{k+1}\|x\|_2^{2k}\log\|x\|_2$. According to Theorem 8.17 they have the generalized Fourier transform

$$\widehat{\Phi}(\omega) = c_k \|\omega\|_2^{-d-2k}$$

with a constant c_k specified in that theorem. Hence our theory yields for thin-plate splines

Corollary 12.6 *In the case where $\Phi(x) = (-1)^{k+1}\|x\|_2^{2k}\log\|x\|_2$, we can bound the minimum eigenvalue $\lambda_{\min}(A_{\Phi,X})$ as follows:*

$$\lambda_{\min}(A_{\Phi,X}) \geq C_d c_k (2M_d)^{-d-2k} q_X^{2k}.$$

Next, the powers $\Phi(x) = (-1)^{\lceil \beta/2 \rceil}\|x\|_2^{\beta}$, $x \in \mathbb{R}^d$, with $\beta > 0$, $\beta \notin 2\mathbb{N}$, have the generalized Fourier transform $\widehat{\Phi}(\omega) = c_\beta \|\omega\|_2^{-d-\beta}$ by Theorem 8.16. Again, the constant c_β is specified in the theorem. This Fourier transform leads to

Corollary 12.7 *In the case where $\Phi(x) = (-1)^{\lceil \beta/2 \rceil}\|x\|_2^{\beta}$, $\beta > 0$, $\beta \notin 2\mathbb{N}$, we have*

$$\lambda_{\min}(A_{\Phi,X}) \geq C_d c_\beta (2M_d)^{-d-\beta} q_X^{\beta}.$$

Finally, let us have a look at the compactly supported radial basis functions $\Phi_{d,k} = \phi_{d,k}(\|\cdot\|_2)$ of minimal degree defined in Definition 9.11. Even if we do not know the Fourier transform of these functions explicitly, we know from Theorem 10.35 that

$$\widehat{\Phi}_{d,k}(\omega) \geq C\|\omega\|_2^{-d-2k-1}$$

for sufficiently large $\|\omega\|_2$, with the possible exception of $d = 1, 2$ in the case $k = 0$. The constant C depends only on d and k. Since $\widehat{\Phi}_{d,k}$ is continuous and positive on $\mathbb{R}^d$ we obtain

Corollary 12.8 *In the case of the compactly supported radial basis functions of minimal degree of Section 9.4, the smallest eigenvalue of the interpolation matrix can be bounded as follows:*

$$\lambda_{\min}(A_{\Phi,X}) \geq C q_X^{2k+1}.$$

As in the case of error estimates we summarize our results in the following form. For every basis function we have found a function G such that

$$\lambda_{\min}(A_{\Phi,X}) \geq G(q_X).$$

Table 12.1 contains the functions G for the various basis functions Φ up to a constant factor that depends only on Φ and d but not on X.

Let us come back to the trade-off principle. If we use the function G that we have just introduced and the function F from Table 11.1, which gave a bound on the squared power function, we see that

$$G(q_X) \leq \lambda_{\min}(A_{X,\Phi}) \leq P^2_{\Phi,X\setminus\{x_j\}}(x_j) \leq F(h_{X\setminus\{x_j\},\Omega})$$

for every $x_j \in X$. Hence, if X and $X \setminus \{x_j\}$ are quasi-uniform then the separation distance q_X and the fill distance $h_{X\setminus\{x_j\},\Omega}$ are the same size. Since in the case of basis functions of finite smoothness the functions F and G differ only by a constant factor and have the same exponent, this means in particular that the estimates of both the upper bounds for the power function and the lower bounds for the smallest eigenvalue are sharp concerning the order. We see also that in the case of Gaussians there is a substantial gap between G and F, while

Table 12.1 *Lower bounds on $\lambda_{\min}$ in terms of q*

	$\Phi(x) = \phi(r)$, $r = \|x\|_2$	$G(q)$
Gaussians	$e^{-\alpha r^2}$, $\alpha > 0$	$q^{-d} e^{-M_d^2/(\alpha q^2)}$
		$q^{-d} e^{-40.71 d^2/(\alpha q^2)}$
multiquadrics	$(-1)^{\lceil \beta/2 \rceil}(c^2 + r^2)^\beta$,	$q^{\beta-(d-1)/2} e^{-2M_d c/q}$
inverse MQ	$\beta \in \mathbb{R} \setminus \mathbb{N}_0$	$q^{\beta-(d-1)/2} e^{-12.76 dc/q}$
powers	$(-1)^{\lceil \beta/2 \rceil} r^\beta$,	q^β
	$\beta > 0$, $\beta \notin 2\mathbb{N}$	
thin-plate splines	$(-1)^{k+1} r^{2k} \log r$,	q^{2k}
	$k \in \mathbb{N}$	
compactly supported functions	$\phi_{d,k}(r)$	q^{2k+1}

in the case of the (inverse) multiquadrics, better results concerning the involved constants are all that is necessary.

12.3 Change of basis

We have seen that expressing the radial basis function interpolant $s_{f,X}$ of a function f in the standard basis can lead to badly conditioned interpolation matrices. The condition number depends more on the separation distance than on the number N of centers.

Of course, if in particular the basis function is positive definite, leading to a positive definite interpolation matrix, we have all the preconditioning methods known from classical linear algebra to hand. The most promising methods seem to be the preconditioned conjugate gradient method and the incomplete Cholesky factorization. But since these methods are described in good books on numerical linear algebra we will skip the details here. There is a lack of preconditioning methods specially tailored to the radial basis function situation. The few existing methods seem to be inferior even to the classical preconditioner just mentioned.

In the first place, a bad condition number is a result of the naturally chosen basis, namely $\Phi(\cdot, x_1), \ldots, \Phi(\cdot, x_N)$ (plus a basis for $\mathcal{P}$), and we might be interested in finding a better basis for the subspace

$$V_X := \operatorname{span}\{\Phi(\cdot, x_j) : x_j \in X\} + \mathcal{P}.$$

This is a well-known idea in approximation theory. For example, the success of using splines in the univariate setting goes hand in hand with finding the B-spline basis. Obviously, if we choose the cardinal basis $\{u_j^*\}$ as a basis for V_X then the interpolation matrix becomes the identity matrix. Unfortunately, finding the cardinal basis is at least as difficult as solving the linear problem itself.

In this section we want to discuss other bases for this finite-dimensional space. Particularly in the case of thin-plate splines we will see that it is possible to find a basis which leads to an interpolation matrix that is independent of the separation distance q_X but dependent on the number of centers.

We will restrict ourselves here to the case of conditionally positive definite kernels. But the reader should keep in mind that a positive definite kernel can also act as a conditionally positive definite kernel.

In Section 10.3 we introduced two further kernels associated with the initial kernel Φ. Remember that we had previously chosen a set $\Xi = \{\xi_1, \dots, \xi_Q\} \subseteq \Omega$ that is $\mathcal{P}$-unisolvent and a cardinal basis $p_1, \dots, p_Q$ for $\mathcal{P}$ satisfying $p_\ell(\xi_k) = \delta_{\ell,k}$. Then we defined the kernels

$$\kappa(x, y) := \Phi(x, y) - \sum_{k=1}^{Q} p_k(x)\Phi(\xi_k, y) - \sum_{\ell=1}^{Q} p_\ell(y)\Phi(x, \xi_\ell)$$
$$+ \sum_{\ell=1}^{Q}\sum_{k=1}^{Q} p_k(x)p_\ell(y)\Phi(\xi_k, \xi_\ell) \tag{12.4}$$

and

$$K(x, y) = \kappa(x, y) + \sum_{\ell=1}^{Q} p_\ell(x)p_\ell(y).$$

Another way of describing the relation between K and κ is by the projection operator $\Pi_{\mathcal{P}} f = \sum_{\ell=1}^{Q} f(\xi_\ell)p_\ell$:

$$\Pi_{\mathcal{P}} K(\cdot, y) = \sum_{\ell=1}^{Q} \kappa(\xi_\ell, y)p_\ell + \sum_{\ell=1}^{Q}\sum_{k=1}^{Q} p_k(\xi_\ell)p_k(y)p_\ell = \sum_{\ell=1}^{Q} p_\ell(y)p_\ell,$$

which implies that

$$\kappa(\cdot, y) = K(\cdot, y) - \Pi_{\mathcal{P}} K(\cdot, y). \tag{12.5}$$

Note that if $\Xi \subseteq X$ then it is true that $K(\cdot, x_j)$ and $\kappa(\cdot, x_j)$ both lie in V_X. Thus the question to be answered here is whether we can use $\{K(\cdot, x_1), \dots, K(\cdot, x_N)\}$ or $\{\kappa(\cdot, x_1), \dots, \kappa(\cdot, x_N)\}$ as a basis for V_X. Obviously, the second family is doomed to fail since $\kappa(\cdot, \xi_k) = 0$ for $1 \le k \le Q$. Thus in this situation we have at least to add $\mathcal{P}$ again.

Theorem 12.9 *The kernel $K : \Omega \times \Omega \to \mathbb{R}$ is positive definite on Ω. Moreover, if $\widetilde{\Omega} = \Omega \backslash \Xi$ then $\kappa : \widetilde{\Omega} \times \widetilde{\Omega} \to \mathbb{R}$ is positive definite on $\widetilde{\Omega}$. Both kernels are conditionally positive definite with respect to $\mathcal{P}$ on Ω.*

Proof First of all, if a set of distinct points $X = \{x_1, \dots, x_N\} \subseteq \Omega$ is given and if we have an $\alpha \in \mathbb{R}^N$ that satisfies

$$\sum_{j=1}^{N} \alpha_j p(x_j) = 0 \qquad \text{for } p \in \mathcal{P} \tag{12.6}$$

then obviously

$$\sum_{i,j=1}^{N} \alpha_i \alpha_j K(x_i, x_j) = \sum_{i,j=1}^{N} \alpha_i \alpha_j \kappa(x_i, x_j) = \sum_{i,j=1}^{N} \alpha_i \alpha_j \Phi(x_i, x_j),$$

showing that both K and κ are conditionally positive definite on Ω with respect to $\mathcal{P}$. Let us have a closer look at K for arbitrary $\alpha \in \mathbb{R}^N$. From Theorem 10.20 we know that K is the reproducing kernel for the native space $\mathcal{N}_\Phi(\Omega)$ with respect to the inner product

$$(f, g) = (f, g)_{\mathcal{N}_\Phi(\Omega)} + \sum_{\ell=1}^{Q} f(\xi_\ell) g(\xi_\ell).$$

Together with Theorem 10.3 this means that

$$\sum_{i,j=1}^{N} \alpha_i \alpha_j K(x_i, x_j) = \left\| \sum_{j=1}^{N} \alpha_j K(\cdot, x_j) \right\|^2 \geq 0,$$

showing K to be at least positive semi-definite. But since we have a norm now the quadratic form is zero if and only if $\sum_{j=1}^{N} \alpha_j K(x, x_j) = 0$ for all $x \in \Omega$. Setting $x = \xi_\ell$ and using $\kappa(\xi_\ell, \cdot) = 0$ shows that actually α satisfies (12.6). Thus the first part of our proof gives $\alpha = 0$.

Now let us look at κ. We start with centers $X = \{x_1, \ldots, x_N\} \subseteq \widetilde{\Omega}$, which means that $X \cap \Xi = \emptyset$. Thus $Y = X \cup \Xi$ consists of $N + Q$ distinct points. Let $y_j = x_j$ for $1 \leq j \leq N$ and $y_{N+j} = \xi_j$ for $1 \leq j \leq Q$. Next suppose that $\alpha \in \mathbb{R}^N \setminus \{0\}$ is given. If we define $\beta \in \mathbb{R}^{N+Q}$ by $\beta_j = \alpha_j$ for $1 \leq j \leq N$ and $\beta_{N+j} = -\sum_{i=1}^{N} \alpha_i p_j(x_i)$ for $1 \leq j \leq Q$ then we have

$$\sum_{j=1}^{N+Q} \beta_j p_k(y_j) = \sum_{j=1}^{N} \alpha_j p_k(x_j) - \sum_{\ell=1}^{Q} \sum_{i=1}^{N} \alpha_i p_\ell(x_i) p_k(\xi_\ell) = 0$$

for $1 \leq k \leq Q$. Thus β satisfies (12.6) for Y instead of X. Moreover, it is now easy to see that

$$\sum_{i,j=1}^{N} \alpha_i \alpha_j \kappa(x_i, x_j) = \sum_{i,j=1}^{N+Q} \beta_i \beta_j \Phi(y_i, y_j) > 0,$$

which proves the result for κ. □

Thus we can restate our initial interpolation problem in two new ways. Let us start with the simpler one.

Corollary 12.10 *If $\Xi \subseteq X$ then the interpolant $s_{f,X}$ can be written as*

$$s_{f,X} = \sum_{j=1}^{N} \alpha_j K(\cdot, x_j),$$

where the coefficients are determined by $s_{f,X}(x_j) = f_j$, $1 \leq j \leq N$.

When using κ we have to be more careful, since $\Xi \subseteq X$ does not lead to linearly independent functions $\kappa(\cdot, x_j)$. But we need $\Xi \subseteq X$ to ensure that we get the same interpolant. So assume that $x_j = \xi_j$ for $1 \le j \le Q$. Then we know at least that the matrix

$$C = (\kappa(x_i, x_j))_{Q+1 \le i, j \le N}$$

is positive definite. Or, in other words, the family $\{\kappa(\cdot, x_j) : Q+1 \le j \le N\}$ is linearly independent. Since $\kappa(x_j, \cdot) = 0$ for $1 \le j \le Q$, we immediately have that $\{\kappa(\cdot, x_j) : Q+1 \le j \le N\} \cup \{p_k : 1 \le k \le Q\}$ is a basis for V_X.

Thus we can restate the interpolation problem using this basis.

Corollary 12.11 *If $\Xi \subseteq X$ satisfies $x_j = \xi_j$ for $1 \le j \le Q$ then the interpolant can be written as*

$$s_{f,X}(x) = \sum_{j=1}^{Q} \beta_k p_k(x) + \sum_{j=Q+1}^{N} \alpha_j \kappa(x, x_j)$$

and the coefficients are again determined by $s_{f,X}(x_j) = f(x_j)$, $1 \le j \le N$.

Since the $\{p_\ell\}$ form a Lagrangian basis for Ξ and since κ vanishes if one of its arguments is an element from Ξ, the interpolation conditions lead to the matrix equation

$$\begin{pmatrix} I & O \\ \widetilde{P} & C \end{pmatrix} \begin{pmatrix} \beta \\ \alpha \end{pmatrix} = f|X$$

with the $\mathbb{R}^{Q \times Q}$ identity matrix I and the $\mathbb{R}^{(N-Q) \times Q}$ matrix $\widetilde{P} = (p_j(x_i))$, where i runs over the indices starting with $i = Q+1$.

Note that the first row means that $\beta = (f(x_1), \ldots, f(x_Q))^T$ and so solving the system reduces to solving

$$C\alpha = \widetilde{f} - \widetilde{P}\beta$$

with $\widetilde{f} = (f(x_{Q+1}), \ldots, f(x_N))^T$, which can be solved using a solver for positive definite matrices.

Let us do an example to see how the three different approaches work. To this end we choose the basis function to be $\Phi(x, y) = \|x - y\|_2^2 \log \|x - y\|_2$, which is conditionally positive definite on $\mathbb{R}^d \times \mathbb{R}^d$ of order $m = 2$. Thus if we treat the two-dimensional case, $\mathcal{P}$ is the space of linear polynomials in $\mathbb{R}^2$ having dimension $Q = 3$.

Our simple example uses lattice points on $[0, 1]^2$, and we will choose $\Xi = \{(0, 0)^T, (1, 0)^T, (0, 1)^T\}$ with the associated Lagrange basis $p_1(x) = 1 - x_1 - x_2$, $p_2(x) = x_1$, and $p_3(x) = x_2$. Table 12.2 contains the results for different spacings. The matrix $\widetilde{A}$ is the standard matrix using the standard basis for V_X as it appears for example in (8.10).

It can be seen that the condition numbers are roughly of the same size, even slightly worse, in the case of the new kernels. But in this example we changed the separation distance and the number of points. If we are only interested in the effect of the separation distance we have to keep the number of points fixed. This is of interest in itself, for reasons that will

Table 12.2 *Condition numbers in the ℓ_2-norm*

Spacing h	Conventional matrix $\widetilde{A}$	Reproducing-kernel matrix K	Homogeneous matrix C
1/8	3.5158×10^3	1.8390×10^4	7.5838×10^3
1/16	3.8938×10^4	2.6514×10^5	1.1086×10^5
1/32	5.1363×10^5	4.0007×10^6	1.6864×10^6
1/64	7.6183×10^6	6.2029×10^7	2.6264×10^7

Table 12.3 *Condition numbers for a fixed number of centers*

Scale parameter α	Conventional matrix $\widetilde{A}$	Reproducing-kernel matrix K	Homogeneous matrix C
0.001	2.4349×10^8	8.4635×10^8	5.4938×10^3
0.01	2.4364×10^6	8.4640×10^6	5.4938×10^3
0.1	2.5179×10^4	8.5134×10^4	5.4938×10^3
1.0	3.6458×10^2	1.3660×10^3	5.4938×10^3
10	1.8742×10^6	1.2609×10^3	5.4938×10^3
100	1.1520×10^{11}	1.1396×10^5	5.4938×10^3
1000	3.5478×10^{15}	1.1386×10^7	5.4938×10^3

become clear later, when we want to solve a large RBF system by splitting it up into a couple of smaller systems.

Thus our next model problem is again thin-plate spline interpolation, but this time on a uniform 5×5 grid in $[0, \alpha]^2$ with spacing $q_X = \alpha/4$. Table 12.3 shows the results.

Now the situation has completely changed. The condition number of the matrix C is independent of the scaling α. We will spend the rest of this section explaining this behavior.

We start with a lemma that is an improved version of a technique we have already used, in the proof of Theorem 11.9.

Lemma 12.12 *Let $\Xi = \{\xi_1, \ldots, \xi_Q\}$ be a $\pi_{m-1}(\mathbb{R}^d)$-unisolvent set and let $p_1, \ldots, p_Q \in \pi_{m-1}(\mathbb{R}^d)$ be such that $p_k(\xi_\ell) = \delta_{k,\ell}$. Then for any $p \in \pi_{2m-1}(\mathbb{R}^d)$ it is true that*

$$0 = p(x-y) - \sum_{k=1}^{Q} p_k(y)p(x-\xi_k) - \sum_{\ell=1}^{Q} p_\ell(x)p(\xi_\ell - y)$$

$$+ \sum_{k,\ell=1}^{Q} p_\ell(x)p_k(y)p(\xi_\ell - \xi_k).$$

Proof Obviously it suffices to prove the identity for the monomials $q_\alpha(x) = x^\alpha$, $\alpha \in \mathbb{N}_0^d$, $|\alpha| \le 2m - 1$. The binomial theorem allows us to write

$$q_\alpha(x-y) = \sum_{0 \le \beta \le \alpha} a_\beta q_\beta(y) q_{\alpha-\beta}(x).$$

Thus to investigate the expression on the right-hand side of the statement in the lemma for $p = q_\alpha$, we have to investigate

$$q_{\alpha-\beta}(x)q_\beta(y) - \sum_{k=1}^{Q} p_k(y)q_\beta(\xi_k)q_{\alpha-\beta}(x) - \sum_{\ell=1}^{Q} p_\ell(x)q_{\alpha-\beta}(\xi_\ell)q_\beta(y)$$

$$+ \sum_{\ell,k=1}^{Q} p_\ell(x)p_k(y)q_{\alpha-\beta}(\xi_\ell)q_\beta(\xi_k)$$

$$= q_{\alpha-\beta}(x)q_\beta(y) - \sum_{k=1}^{Q} p_k(y)q_\beta(\xi_k)q_{\alpha-\beta}(x)$$

$$- \sum_{\ell=1}^{Q} p_\ell(x)q_{\alpha-\beta}(\xi_\ell)\left(q_\beta(y) - \sum_{k=1}^{Q} p_k(y)q_\beta(\xi_k)\right)$$

$$= q_{\alpha-\beta}(x)\left[q_\beta(y) - \Pi q_\beta(y)\right] - \sum_{k=1}^{Q} p_k(x)\left[q_\beta(y) - \Pi q_\beta(y)\right] q_{\alpha-\beta}(\xi_k)$$

$$= \left[q_{\alpha-\beta}(x) - \Pi q_{\alpha-\beta}(x)\right]\left[q_\beta(y) - \Pi q_\beta(y)\right],$$

where we have utilized the projection operator $\Pi f = \sum_{k=1}^{Q} f(\xi_k)p_k$ again. Now, since $|\alpha| \le 2m-1$, either $|\alpha - \beta| \le m-1$ or $|\beta| \le m-1$ for every β. Hence, the last expression is zero for all $\beta \in \mathbb{N}_0^d$ with $\beta \le \alpha$. But this means that the expression given in the lemma becomes

$$\sum_{\beta \le \alpha} a_\beta \left(q_{\alpha-\beta}(x) - \Pi q_{\alpha-\beta}(x)\right)\left(q_\beta(y) - \Pi q_\beta(y)\right) = 0$$

for $p = q_\alpha$, thus proving the lemma. □

This lemma was needed to show that under certain circumstances the kernel κ is in a certain way homogeneous.

Theorem 12.13 *Suppose that the symmetric function $\Phi \in C(\mathbb{R}^d \times \mathbb{R}^d)$ satisfies $\Phi(hx, hy) = h^\lambda \Phi(x, y) + q_h(x-y)$ for all $h > 0$ and $x, y \in \mathbb{R}^d$, where $\lambda \in \mathbb{R}$ and $q_h \in \pi_{2m-1}(\mathbb{R}^d)$. Let $\Xi = \{\xi_1, \dots, \xi_Q\}$ be unisolvent for $\pi_{m-1}(\mathbb{R}^d)$ with associated Lagrange basis $p_1, \dots, p_Q$. Let κ be the kernel (12.4) and κ^h for $h > 0$ be the kernel κ for the set $h\Xi = \{h\xi_1, \dots, h\xi_Q\}$ and the Lagrange functions $p_1^h, \dots, p_Q^h$ associated with this set. Then $\kappa^h(hx, hy) = h^\lambda \kappa(x, y)$ for all $x, y \in \mathbb{R}$.*

Proof Since obviously $p_k^h(x) = p_k(x/h)$, we have

$$\kappa^h(hx, hy) = \Phi(hx, hy) - \sum_{k=1}^{Q} p_k^h(hx)\Phi(h\xi_k, hy) - \sum_{\ell=1}^{Q} p_\ell^h(hy)\Phi(hx, h\xi_\ell)$$

$$+ \sum_{\ell,k=1}^{Q} p_\ell^h(hy)p_k^h(hx)\Phi(h\xi_k, h\xi_\ell)$$

$$
\begin{aligned}
&= h^\lambda \left(\Phi(x,y) - \sum_{k=1}^{Q} p_k(x)\Phi(\xi_k, y) - \sum_{\ell=1}^{Q} p_\ell(y)\Phi(x,\xi_\ell) \right. \\
&\qquad \left. + \sum_{k,\ell=1}^{Q} p_\ell(y)p_k(x)\Phi(\xi_k,\xi_\ell) \right) \\
&\quad + q_h(x-y) - \sum_{k=1}^{Q} p_k(x)q_h(\xi_k - y) - \sum_{\ell=1}^{Q} p_\ell(y)q_h(x-\xi_\ell) \\
&\quad + \sum_{k,\ell=1}^{Q} p_k(x)p_\ell(y)q_h(\xi_k - \xi_\ell) \\
&= h^\lambda \kappa(x,y).
\end{aligned}
$$

□

This theorem gives the reason why the condition number of the matrix C is independent of the scaling parameter. To see this we only have to show that the conditions of the theorem are met by thin-plate splines. But before that let us record

Corollary 12.14 *Suppose that $\Phi \in C(\mathbb{R}^d \times \mathbb{R}^d)$ is a conditionally positive definite kernel with respect to $\pi_{m-1}(\mathbb{R}^d)$ and satisfies the assumptions of the last theorem. Suppose further that $X = \{x_1, \ldots, x_N\}$ is disjoint with Ξ. Let $C = (\kappa(x_j, x_k))$ and $C^h = (\kappa^h(hx_j, hx_k))$; then both matrices have the same condition number with respect to the ℓ_2-norm.*

Proof Since both C and C^h are positive definite matrices, we can compute their minimal and maximal eigenvalues:

$$
\begin{aligned}
\lambda_{\min}(C^h) &= \min_{\|\alpha\|_2=1} \sum_{j,k=1}^{N} \alpha_j \alpha_k \kappa^h(hx_j, hx_k) \\
&= h^\lambda \min_{\|\alpha\|_2=1} \sum_{j,k=1}^{N} \alpha_j \alpha_k \kappa(x_j, x_k) \\
&= h^\lambda \lambda_{\min}(C).
\end{aligned}
$$

Replacing "min" by "max" shows that $\lambda_{\max}(C^h) = h^\lambda \lambda_{\max}(C)$. Hence both matrices have the same condition number. □

As mentioned before we will conclude this section by giving two examples.

Proposition 12.15 *The kernel $\Phi(x,y) = (-1)^{\lceil \beta/2 \rceil} \|x-y\|_2^\beta$, $\beta > 0$, $\beta \notin 2\mathbb{N}$, is a function that is conditionally positive definite of order $m = \lceil \beta/2 \rceil$ and satisfies the condition of Theorem 12.13 with $\lambda = \beta$ and $q_h \equiv 0$. The kernel $\Phi(x,y) = (-1)^{k+1}\|x-y\|_2^{2k} \log \|x-y\|_2$ is conditionally positive definite of order $m = k+1$ and satisfies the condition of Theorem 12.13 with $\lambda = 2k$ and a polynomial $q_h \in \pi_{2k}(\mathbb{R}^d) \subseteq \pi_{2m-1}(\mathbb{R}^d)$.*

Proof Everything said about the first kernel is obviously true, so let us turn to the second kernel, the thin-plate splines. There we have

$$\begin{aligned}\Phi(hx, hy) &= (-1)^{k+1} h^{2k} \|x - y\|_2^{2k} (\log h + \log \|x - y\|) \\ &= h^{2k} \Phi(x, y) + (-1)^{k+1} h^{2k} \log h \|x - y\|_2^{2k}.\end{aligned}$$

□

12.4 Notes and comments

The idea of expressing the condition number in terms of the so-called separation distance was initiated by Ball [8] and extended by Narcowich and Ward in [143–145, 147]. Their approach is based on the Schoenberg and Bernstein–Hausdorff–Widder theory on the one hand and on Micchelli's theorem on the other, so that it is restricted to radial functions that are (conditionally) positive definite on every $\mathbb{R}^d$. But their idea was so powerful that Schaback [163] could easily extend it to the general case of conditionally positive definite functions by using representations of Bochner's type.

The trade-off principle, also sometimes called the uncertainty relation, was first discovered in this field by Schaback [163]. The third section of the chapter was based on ideas of Beatson *et al.* [15].

13
Optimal recovery

So far, we have dealt with the following simple interpolation or approximation problem. An in general unknown function f is specified only at certain points $X = \{x_1, \ldots, x_N\}$, and we are interested in recovering the function f on a region Ω that is well covered by the centers X. In a later chapter we will concentrate on more general problems. But let us stick to this particular one a little longer. Why should we use (conditionally) positive definite kernels for recovering f?

We have already learnt that recovering f is a difficult task and that radial basis functions are a powerful tool for doing this. In particular, they can be used (at least theoretically – we come back to the numerical treatment in a later chapter) with truly scattered data and in every dimension. Moreover, positive definite functions appeared quite naturally in the context of reproducing-kernel Hilbert spaces.

But this is not the end of the story. Interpolants based on (conditionally) positive definite kernels are optimal in several other ways and the present chapter is devoted to this subject.

13.1 Minimal properties of radial basis functions

Let us start with best approximation. We have seen that the native space $\mathcal{N}_\Phi(\Omega)$ corresponding to a (conditionally) positive definite kernel Φ is an adequate function space. The interpolant $s_{f,X}$ is one candidate that uses the given information about f on X, but of course not the only one. More precisely, any function s from the space

$$V_X = \left\{ s = \sum_{j=1}^{N} \alpha_j \Phi(\cdot, x_j) + p : p \in \mathcal{P} \text{ and } \sum_{j=1}^{N} \alpha_j q(x_j) = 0 \text{ for all } q \in \mathcal{P} \right\} \quad (13.1)$$

can be considered.

Theorem 13.1 *Suppose that $\Phi \in C(\Omega \times \Omega)$ is a conditionally positive definite kernel with respect to the finite-dimensional space $\mathcal{P} \subseteq C(\Omega)$. Suppose further that X is $\mathcal{P}$-unisolvent and that $f \in \mathcal{N}_\Phi(\Omega)$ is known only at $X = \{x_1, \ldots, x_N\} \subseteq \Omega$. Then the interpolant $s_{f,X}$ is the best approximation to f from (13.1) with respect to the native space* *(semi-)norm, i.e.*

$$|f - s_{f,X}|_{\mathcal{N}_\Phi(\Omega)} \le |f - s|_{\mathcal{N}_\Phi(\Omega)}$$

for all $s \in V_X$. Hence, $s_{f,X}$ is the orthogonal projection of f onto V_X.

Proof By Lemma 10.24 we know that $(f - s_{f,X}, s)_{\mathcal{N}_\Phi(\Omega)} = 0$ for all $s \in V_X$. But this is the characterization of the best approximation in Hilbert spaces. The argument also holds in the case of a symmetric bilinear form with nonzero kernel. □

Note that in the case of a conditionally positive definite function the best approximation is uniquely determined only up to an element from $\mathcal{P}$. We can avoid this nonuniqueness by going over to the inner product

$$(f, g) = \sum_{k=1}^{Q} f(\xi_k)g(\xi_k) + (f, g)_{\mathcal{N}_\Phi(\Omega)}$$

with a $\mathcal{P}$-unisolvent subset $\Xi = \{\xi_1, \dots, \xi_Q\} \subseteq \Omega$; see Theorem 10.20.

But this is not yet the whole story. Recalling the minimal properties of splines, we know that they minimize an energy functional under all interpolatory functions from a Sobolev space. The same is true here.

Theorem 13.2 *Suppose that $\Phi \in C(\Omega \times \Omega)$ is a conditionally positive definite kernel with respect to the finite-dimensional space $\mathcal{P} \subseteq C(\Omega)$. Suppose further that X is $\mathcal{P}$-unisolvent and that values $f_1, \dots, f_N$ are given. Then the interpolant $s_{f,X}$ has minimal (semi-)norm $|\cdot|_{\mathcal{N}_\Phi(\Omega)}$ under all functions $s \in \mathcal{N}_\Phi(\Omega)$ that interpolate the data $\{f_j\}$ at the centers X, i.e.*

$$|s_{f,X}|_{\mathcal{N}_\Phi(\Omega)} = \min\{|s|_{\mathcal{N}_\Phi(\Omega)} : s \in \mathcal{N}_\Phi(\Omega) \text{ with } s(x_j) = f_j, 1 \le j \le N\}.$$

Proof The interpolant has a representation $s_{f,X} = \lambda^x \Phi(\cdot, x) + q$, where $\lambda = \sum_{j=1}^N \alpha_j \delta_{x_j}$ satisfies $\lambda(p) = 0$ for all $p \in \mathcal{P}$ and where $q \in \mathcal{P}$. Moreover, any $s \in \mathcal{N}_\Phi(\Omega)$ can be expressed by

$$s(x) = \Pi_{\mathcal{P}} s(x) + (s, G(\cdot, x))_{\mathcal{N}_\Phi(\Omega)},$$

where G is the function from (10.4). Since $\lambda^x \Phi(\cdot, x) = \lambda^x G(\cdot, x)$ we have

$$\begin{aligned}
(v, s_{f,X})_{\mathcal{N}_\Phi(\Omega)} &= \left(v, \sum_{j=1}^N \alpha_j \Phi(\cdot, x)\right)_{\mathcal{N}_\Phi(\Omega)} \\
&= (v, \lambda^x \Phi(\cdot, x))_{\mathcal{N}_\Phi(\Omega)} \\
&= \lambda(\Pi_{\mathcal{P}} v) + (v, \lambda^x G(\cdot, x))_{\mathcal{N}_\Phi(\Omega)} \\
&= \lambda(v) = \sum_{j=1}^N \alpha_j v(x_j) = 0
\end{aligned}$$

for every $v \in \mathcal{N}_\Phi(\Omega)$ with $v(x_j) = 0$ for $1 \le j \le N$. But this means that

$$|s_{f,X}|^2_{\mathcal{N}_\Phi(\Omega)} = (s_{f,X}, s_{f,X} - s + s)_{\mathcal{N}_\Phi(\Omega)} = (s_{f,X}, s)_{\mathcal{N}_\Phi(\Omega)}$$
$$\le |s_{f,X}|_{\mathcal{N}_\Phi(\Omega)} |s|_{\mathcal{N}_\Phi(\Omega)}$$

for all $s \in \mathcal{N}_\Phi(\Omega)$ with $s(x_j) = f_j$ for $1 \le j \le N$. □

It is interesting to see what this implies for the thin-plate splines, in particular for those specified in (10.11). For example, in the bivariate case we could consider the function $\phi(r) = r^2 \log r$, which is conditionally positive definite of order 2. Hence, the radial basis function interpolant

$$s_{f,X}(x) = \sum_{j=1}^{N} c_j \phi(\|x - x_j\|_2) + p(x),$$

where p is a bivariate linear polynomial, minimizes the semi-norm

$$|f|_{\mathrm{BL}_2(\mathbb{R}^2)} = \left(\int_{\mathbb{R}^d} \left| \frac{\partial^2 f}{\partial x_1^2}(x) \right|^2 + 2 \left| \frac{\partial^2 f}{\partial x_1 \partial x_2}(x) \right|^2 + \left| \frac{\partial^2 f}{\partial x_2^2}(x) \right|^2 dx \right)^{1/2}$$

under all interpolants from the Beppo Levi space $\mathrm{BL}_2(\mathbb{R}^2)$. This was the initial starting point for investigating thin-plate splines. Note that for space dimensions $d \ge 3$ one actually has to consider the thin-plate splines (10.11) as functions of order ℓ instead of order $m = \ell - \lceil d/2 \rceil + 1$, in order to state the minimization problem in the Beppo Levi space $BL_\ell(\mathbb{R}^d)$. For example, the function $\phi(r) = r$ is conditionally positive definite of order $m = 1$. Hence, only a constant has to be added to the radial sum to guarantee a unique interpolant. But if the interpolant should minimize the Beppo Levi semi-norm $|\cdot|_{BL_2(\mathbb{R}^3)}$ under all Beppo Levi functions it has to be formed with an additional linear polynomial.

Finally, let us come to the third minimal property of the radial basis function interpolant. This one is connected with the power function.

Remember that we can rewrite the interpolant $s_{f,X}$ using the cardinal functions $\{u_j^*\}$ as

$$s_{f,X}(x) = \sum_{j=1}^{N} u_j^*(x) f(x_j).$$

Thus we could ask the question whether these coefficients $\{u_j^*(x)\}$ are the best possible, or, equivalently, what is the solution of

$$\inf_{u \in \mathbb{R}^N : \sum u_j p(x_j) = p(x), p \in \mathcal{P}} \quad \sup_{f \in \mathcal{N}_{\Phi(\Omega)}, |f|_{\mathcal{N}_\Phi(\Omega)} = 1} \left| f(x) - \sum_{j=1}^{N} u_j f(x_j) \right|. \tag{13.2}$$

Theorem 13.3 *Suppose that $\Phi \in C(\Omega \times \Omega)$ is a conditionally positive definite kernel with respect to the finite-dimensional space $\mathcal{P} \subseteq C(\Omega)$. Suppose further that X is $\mathcal{P}$-unisolvent and $x \in \Omega$ is fixed. Then the solution vector to (13.2) is given by the cardinal functions*

$u_j^*(x)$, $1 \le j \le N$, *i.e. for* $f \in \mathcal{N}_\Phi(\Omega)$ *we have*

$$|f(x) - s_{f,X}(x)| \le \left| f(x) - \sum_{j=1}^N u_j f(x_j) \right|$$

for all choices $u_1, \ldots, u_N \in \mathbb{R}$ *with* $\sum u_j p(x_j) = p(x)$, $p \in \mathcal{P}$.

Proof Using again the representation for $f \in \mathcal{N}_\Phi(\Omega)$ from Theorem 10.17 leads us to

$$\begin{aligned} f(x) - \sum_{j=1}^N u_j f(x_j) &= \Pi_{\mathcal{P}} f(x) + (f, G(\cdot, x))_{\mathcal{N}_\Phi(\Omega)} \\ &\quad - \sum_{j=1}^N u_j \Pi_{\mathcal{P}} f(x_j) - \sum_{j=1}^N u_j (f, G(\cdot, x_j))_{\mathcal{N}_\Phi(\Omega)} \\ &= \left(f, G(\cdot, x) - \sum_{j=1}^N u_j G(\cdot, x_j) \right)_{\mathcal{N}_\Phi(\Omega)}. \end{aligned}$$

This shows that the norm of the pointwise error functional is given by

$$\sup_{f \in \mathcal{N}_\Phi(\Omega), |f|_{\mathcal{N}_\Phi(\Omega)}=1} \left| f(x) - \sum_{j=1}^N u_j f(x_j) \right| = \left| G(\cdot, x) - \sum_{j=1}^N u_j G(\cdot, x_j) \right|_{\mathcal{N}_\Phi(\Omega)} = \mathcal{Q}^{1/2}(u),$$

where $\mathcal{Q}$ is the quadratic form (here for $\alpha = 0$) from Lemma 11.3 and Theorem 11.5. But Theorem 11.5 simply states that $\mathcal{Q}(u^*(x)) \le \mathcal{Q}(u)$ for all admissible $u \in \mathbb{R}^N$. □

Obviously, a stronger result holds as well when derivatives are included. We leave the details, which are simple, to the reader.

Later on, the results of this section will be generalized to a setting where functionals other than point evaluations are involved. The more general setting can be described as follows. Suppose that we know of an unknown function $f \in \mathcal{N}_\Phi(\Omega)$ only the values $\lambda_j(f)$, $1 \le j \le N$, where λ_j is a continuous linear functional on the native space. Suppose further that we have another functional λ and that we are interested in finding an unknown value $\lambda(f)$ using only the values $\lambda_j(f)$. We will call such a problem a generalized interpolation problem. It will be discussed extensively in Chapter 16.

13.2 Abstract optimal recovery

The idea of optimally reconstructing functions can be put in a much more general framework, which we will now describe. Moreover, we will show how this general framework applies to radial basis function interpolation.

Let $\mathbb{U}$, $\mathbb{V}$, and $\mathbb{W}$ be three normed linear spaces. Let K be a subset of $\mathbb{U}$. We assume that we have some information on the elements of K, given by the linear mapping $\mathcal{J} : \mathbb{U} \to \mathbb{V}$. The mapping $\mathcal{J}$ is called the *information operator*. Moreover, we have another linear operator $T : \mathbb{U} \to \mathbb{W}$, which we want to call the *target operator*.

Our task is to reconstruct for every $x \in K$ the target $Tx \in \mathbb{W}$ from the information $\mathcal{J}x \in \mathbb{V}$. This can be done by a mapping $A : \mathcal{J}(K) \subseteq \mathbb{V} \to \mathbb{W}$, which cannot be linear. Such a mapping A is in this context called an *algorithm*. The whole situation can be visualized in the following diagram.

$$\begin{array}{ccc} K \subseteq \mathbb{U} & \xrightarrow{\mathcal{J}} & \mathcal{J}(K) \subseteq \mathbb{V} \\ \Big\downarrow T & \swarrow A & \\ \mathbb{W} & & \end{array}$$

To determine the usefulness of an algorithm A we measure its error by

$$E(A) := \sup_{x \in K} \|A(\mathcal{J}x) - Tx\|_{\mathbb{W}},$$

and the entire problem has an intrinsic error defined by

$$E^* = \inf_A E(A).$$

Now it should be clear what an optimal algorithm has to do.

Definition 13.4 *A mapping $A : \mathcal{J}(K) \to \mathbb{W}$ is called an optimal algorithm for the problem just described, if $E(A) = E^*$.*

Next, we derive a sufficient condition for an optimal algorithm. To this end we assume that K is symmetric, i.e. $x \in K$ implies $-x \in K$.

Theorem 13.5 *Suppose that K is symmetric. If there exists a mapping $F : \mathcal{J}(K) \to \mathbb{U}$ such that for $x \in K$*

(1) $x - F\mathcal{J}x \in K$,
(2) $\mathcal{J}(x - F\mathcal{J}x) = 0$

then $TF : \mathcal{J}(K) \to \mathbb{W}$ is optimal.

Proof First of all, the symmetry of K and the linearity of $\mathcal{J}$ show that we have for $x \in K$ with $\mathcal{J}x = 0$ also $\mathcal{J}(-x) = -\mathcal{J}(x) = 0$. This gives, for an arbitrary algorithm A and such an x,

$$\begin{aligned} \|Tx\| &= \tfrac{1}{2}\|Tx - A(0) + Tx + A(0)\| \\ &\le \tfrac{1}{2}\big[\|Tx - A(0)\| + \|Tx + A(0)\|\big] \\ &\le \max\{\|A(0) - Tx\|, \|A(0) + Tx\|\} \\ &= \max\{\|A(\mathcal{J}x) - Tx\|, \|A(\mathcal{J}(-x)) - T(-x)\|\} \\ &\le E(A). \end{aligned}$$

Hence we have established on the one hand the inequality

$$\sup\{\|Tx\| : x \in K, \mathcal{J}x = 0\} \le E^*.$$

On the other hand, the existence of an F with the stated properties now shows that

$$E(TF) = \sup_{x \in K} \|T(F\mathcal{J}x - x)\| \le \sup\{\|Tz\| : z \in K, \mathcal{J}z = 0\} \le E^*,$$

which means that TF is indeed an optimal algorithm. □

At first sight the previous result seems to be of only limited use, since the optimal algorithm TF includes the target operator. The following two examples show that this is in fact not the case.

In the first example, we choose $\mathbb{U} = \mathbb{W} = \mathcal{N}_\Phi(\Omega)$ as the native space of a (conditionally) positive definite kernel Φ and $T : \mathcal{N}_\Phi(\Omega) \to \mathcal{N}_\Phi(\Omega)$ as the identity mapping $T = id$. Moreover, we set $\mathbb{V} = \mathbb{R}^N$ and define $\mathcal{J} : \mathcal{N}_\Phi(\Omega) \to \mathbb{R}^N$ by $\mathcal{J}f = (f(x_1), \dots, f(x_N))^T$ for a given fixed set $\{x_1, \dots, x_N\} \subseteq \Omega$. Finally, K is chosen to be the unit ball in $\mathcal{N}_\Phi(\Omega)$, i.e. $K = \{f \in \mathcal{N}_\Phi(\Omega) : |f|_{\mathcal{N}_\Phi(\Omega)} = 1\}$. In this setting an algorithm $A : \mathcal{J}(K) \to \mathcal{N}_\Phi(\Omega)$ is optimal if it minimizes

$$E(A) = \sup_{|f|_{\mathcal{N}_\Phi(\Omega)} \le 1} |A(f|X) - f|_{\mathcal{N}_\Phi(\Omega)}. \tag{13.3}$$

To apply Theorem 13.5, we define $F : \mathbb{R}^N \to \mathcal{N}_\Phi(\Omega)$ to be the interpolation mapping, i.e. $F(f_1, \dots f_N) = s_{\{f_j\},X}$. Then the interpolation condition is equivalent to $\mathcal{J}F\mathcal{J}f = \mathcal{J}f$ for all $f \in \mathcal{N}_\Phi(\Omega)$, which is the second condition in Theorem 13.5. The first condition can be concluded from Theorem 13.1. Since $s_{f,X}$ is the best approximation to f, we in particularly have

$$|f - F\mathcal{J}f|_{\mathcal{N}_\Phi(\Omega)} \le |f|_{\mathcal{N}_\Phi(\Omega)} \le 1.$$

Hence by Theorem 13.5 interpolation is in this sense optimal.

Corollary 13.6 *Among all mappings $A : \mathcal{N}_\Phi(\Omega)|X \to \mathcal{N}_\Phi(\Omega)$, interpolation in X is optimal in the sense that it minimizes (13.3).*

The result obviously remains true in the following more general setting. Whenever an interpolatory operator provides a best-approximation property, it gives an optimal algorithm to the associated problem.

Our second example goes in the same direction. This time we choose $\mathbb{U}$, $K \subseteq \mathbb{U}$, $\mathbb{V}$, and $\mathcal{J}$ as in the last example, but we let $\mathbb{W}$ be $\mathbb{R}$ and $T : \mathcal{N}_\Phi(\Omega) \to \mathbb{R}$ be defined by $Tf = f(x)$ for a fixed $x \in \Omega$. This means that we now want to minimize

$$E(A) = \sup_{|f|_{\mathcal{N}_\Phi(\Omega)} \le 1} |A(f|X) - f(x)|. \tag{13.4}$$

Since the setting is almost the same as in the first example, we have in particular again that with $F(f|X) = s_{f,X}$ the conditions of Theorem 13.5 are satisfied. Hence, interpolation is again optimal.

Corollary 13.7 *Among all mappings $A : \mathcal{N}_\Phi(\Omega)|X \to \mathbb{R}$ the one that evaluates the interpolant $s_{f,X}$ at x is optimal in the sense that it minimizes (13.4).*

Obviously, one can easily think of several other examples and we encourage the reader to do so.

For simplicity, we have restricted ourselves here to a situation of exact information. It is possible to include the fact that $\mathcal{J}x$ is known only up to an error of magnitude $\epsilon > 0$ by redefining the error of an algorithm as

$$E(A, \epsilon) := \sup_{x \in K} \sup_{y \in \mathbb{V}: \|\mathcal{J}x - y\| \le \epsilon} \|Ay - Tx\|,$$

but this would lead us beyond the scope of this book.

13.3 Notes and comments

In this chapter we have recovered many of the striking properties of univariate splines in the context of multivariate approximation by (conditionally) positive definite kernels.

Deeper insights into the theory of optimal recovery can be gained from the review articles [73] by Golomb and Weinberger and [134] by Micchelli and Rivlin and the literature therein.

14
Data structures

So far we have dealt mainly with the theoretical background of scattered data approximation and interpolation. Except for the moving least squares approximation we have not discussed an efficient implementation of our theory. But from the investigation of the condition number of straightforward interpolation matrices we know that special techniques have to be developed to get an efficient but still accurate approximation method. Several approaches will be discussed in the next chapter.

Crucial to all these methods is the choice of the underlying data structure. It is worth reflecting for a while on how the centers can be stored in a computer most successfully. Thus, in this chapter we will discuss different ways of representing the centers $X = \{x_1, \dots, x_N\} \subseteq \Omega \subseteq \mathbb{R}^d$.

Let us start by collecting possible requests about the set of points X that the data structure should be able to answer efficiently.

The first question every user has to answer is whether all points should be kept in the main memory of the computer or whether the number of points is so large that it has to be stored on a hard disk or other external device. The difference is that in the latter case a reasonable ordering of the data points reduces the number of disk accesses, resulting in a dramatic reduction in run time. There is also an improvement in the first situation by a reasonable ordering, because of the cache of the computer, but it is not dramatic. We will concentrate on the situation where all points can be kept in the main memory of the computer.

The second question the user has to answer is whether all points are given in advance, which would allow us to build the data structure using the knowledge of all the points, or whether the points are given one by one, which would force us to build the data structure point by point. We will concentrate here on the former case. Nevertheless, each of our data structures allows us to add points after the initial build-up. But adding too many points in this way might lead to a degenerate data structure that in turn results in bad performance. In this situation it might be better to rebuild the whole data structure if more than a certain number of points has been inserted afterwards. We want to point out that there are certain applications for which a point-by-point insertion is necessary, for example, if greedy methods for solving partial differential equations numerically are employed. But, as said before, we will concentrate in the following on the most important case, where all points are given in advance.

Here, then, are the queries that the data structure should at least be able to answer efficiently.

- **Nearest neighbor search** Given a point $x \in \Omega$, what is the nearest neighbor of x from X? The nearest neighbor is defined to be the point from X that has minimal distance to x among all points from X. Of course, the nearest neighbor does not need to be unique.
- **ℓ nearest neighbors search** More generally, given a point $x \in \Omega$ and a number $\ell \in \mathbb{N}$, what are the ℓ nearest neighbors of x from X?
- **Range search** Given a region R, what are the points $X \cap R$? We have to answer this question in particular in the situation where R is a rectangle or a ball. In this situation we can hope for better results than in general.

While a range search has already appeared in the context of the moving least squares approximation, the nearest-neighbor problem comes into play naturally if we want to compute the separation distance q_X for a given data set. If it is possible to compute the nearest neighbor of a point x in constant time then we can compute q_X in $\mathcal{O}(N)$, which is favorably cheap when compared with the naive approach of computing all distances $\|x_j - x_k\|_2$ having $\mathcal{O}(N^2)$ complexity.

Some algorithms introduced in the next chapter cover the region Ω by a finite number of overlapping regions Ω_j, i.e. $\Omega = \cup_{j=1}^{M} \Omega_j$ such that every point $x \in \Omega$ is contained in at most $K > 0$ regions Ω_j. Here, $K > 0$ is an integer independent of x but depending on Ω and on the covering $\{\Omega_j\}$. This is a kind of a dual problem to the problems we have encountered so far. Thus we need a data structure that is able to solve the problem, as follows.

- **Containment query** Given $x \in \Omega$ report all regions Ω_j such that $x \in \Omega_j$.

Before we start looking at specific data structures let us discuss the *brute force method*. The naive approach would simply store the points X in an N-dimensional array of d-dimensional vectors and try to manage with that. Obviously, this would need $\mathcal{O}(dN)$ space and $\mathcal{O}(dN)$ time to built the "data structure"; these are the minimum necessary. Unfortunately, it is also obvious that a nearest-neighbor query can only be answered by testing all points in X, which costs $\mathcal{O}(N)$ time. The same is true for range search and containment query; both have an $\mathcal{O}(N)$ complexity in time. We have already seen in the context of moving least squares that this is unacceptable.

14.1 The fixed-grid method

Our first data structure is based on the simple idea of covering the region Ω by disjoint cubes of equal side length. For every cube we list the indices of the centers contained in it. It turns out that this method is very efficient for quasi-uniform data sets. Before we go more into details let us review some important properties of quasi-uniform data sets.

Let $\Omega \subseteq \mathbb{R}^d$ be bounded. Fix a quasi-uniformity constant $c_{\mathrm{qu}} > 0$. Then we say that a set of pairwise distinct centers $X = \{x_1, \ldots, x_N\} \subseteq \Omega$ is quasi-uniform with respect to c_{qu} if

$$q_X \leq h_{X,\Omega} \leq c_{\mathrm{qu}} q_X,$$

where q_X and $h_{X,\Omega}$ are the usual separation and fill distance, respectively. To require $q_X \le h_{X,\Omega}$ instead of $c_1 q_X \le h_{X,\Omega}$ is not a restriction if Ω satisfies an interior cone condition with radius $r > 0$. If $q_X \le r$ we can find for an $x_1 \in X$ a point $y \in \Omega$ with $\|y - x_1\|_2 = q_X$. Hence

$$\|y - x_j\|_2 \ge \|x_j - x_1\|_2 - \|x_1 - y\|_2 \ge 2q_X - q_X = q_X$$

for any other $x_j \in X$. But this means that $h_{X,\Omega} \ge q_X$.

The nice thing about quasi-uniform sets is that not only are q_X and $h_{X,\Omega}$ equivalent in the sense that their fractions can be bounded from above and below by constants but also both are equivalent to $N^{-1/d}$.

Proposition 14.1 *Let $\Omega \subseteq \mathbb{R}^d$ be bounded and measurable. Suppose that $X = \{x_1, \dots, x_N\} \subseteq \Omega$ is quasi-uniform with respect to $c_{\mathrm{qu}} > 0$. Then there exist constants $c_1, c_2 > 0$ depending only on the space dimension d, on Ω, and on c_{qu} such that*

$$c_1 N^{-1/d} \le h_{X,\Omega} \le c_2 N^{-1/d}.$$

Proof By definition of $h = h_{X,\Omega}$ we have

$$\Omega \subseteq \bigcup_{j=1}^{N} B(x_j, h),$$

hence a comparison of the volumes gives

$$\mathrm{vol}(\Omega) \le N h^d \frac{\pi^{d/2}}{\Gamma(d/2+1)}$$

or $h \ge c_1 N^{-1/d}$. For the second inequality note that since Ω is bounded there exists a ball $B(x_0, R)$ with $\Omega \subseteq B(x_0, R)$. Thus we have

$$\bigcup_{j=1}^{N} B(x_j, q_X) \subseteq B(x_0, R + q_X).$$

The balls $B(x_j, q_X)$ are essentially disjoint by definition of q_X. Hence $N q_X^d \le (R + q_X)^d$ or $N \le (1 + R/q_X)^d \le (2R)^d q_X^{-d}$, if $q_X \le R$. Note that $q_X > R$ does not make sense because in that situation X consists only of one point. The quasi-uniformity of X now leads to $N \le (2Rc_{\mathrm{qu}})^d h^{-d}$, showing also that $h \le c_2 N^{-1/d}$. □

Obviously, the equivalence of q_X and $h_{X,\Omega}$ and the equivalence of $h_{X,\Omega}$ and $N^{-1/d}$ lead to the equivalence of q_X and $N^{-1/d}$. The next result is similar to a result that we achieved for moving least squares.

Corollary 14.2 *Suppose that $X = \{x_1, \dots, x_N\} \subseteq \Omega$ is quasi-uniform. For any cube $W = W(y, c_s N^{-1/d})$ of side length $2c_s N^{-1/d}$ the number of centers in $X \cap W$ is bounded by a constant that is independent of N.*

Proof We argue as in the second part of the proof of Proposition 14.1. Let $Y = \{y_1, \ldots y_M\} = X \cap W$. Then the inclusion

$$\bigcup_{j=1}^{M} B(y_j, q_X) \subseteq B(y, \sqrt{d}c_s N^{-1/d} + q_X) \subseteq B(y, cq_X)$$

shows that $M \le c^d$, where c depends only on c_{qu}, Ω, d. □

Let us come back to the initial problem of defining a good data structure. The first step is to find the bounding box for Ω. In many applications Ω is not known in advance. Moreover, building the data structure is a problem relating just to the point set X and not to Ω. Hence, we will look for the bounding box BB for X instead Ω. Let us denote the coordinates of the points by $x_j = (\mathrm{x}_{j,1}, \ldots, \mathrm{x}_{j,d})^T$.

Algorithm 1 Building the grid structure

Input: Data sites $x_1, \ldots, x_N \in \mathbb{R}^d$.
Output: For each grid cell a list of those points contained in it.

Find the bounding box BB for the data points.
Define a grid consisting of $\lfloor N^{1/d} \rfloor^d$ equally sized cells on Q.
for *every grid cell* **do**
  Initialize a list of indices for the points contained in that cell.

for $1 \le j \le N$ **do**
  Determine the cell C with $x_j \in C$.
  Store j in the list for cell C.

Algorithm 1 describes how the grid structure can be built. The most difficult part is to organize the indices for the grid. This can be done in the following way. Choose a small $\epsilon > 0$ and let

$$\alpha_k := \min_{1 \le j \le N} \mathrm{x}_{j,k} - \epsilon \qquad \text{and} \qquad \beta_k := \max_{1 \le j \le N} \mathrm{x}_{j,k}$$

for every $1 \le k \le d$. These numbers essentially describe the bounding box BB for the data set X. The lower bound is slightly smaller than expected, to ensure that in what follows all indices are between 1 and $\lfloor N^{1/d} \rfloor$.

To find out in which box a point $x = (\mathrm{x}_1, \ldots, \mathrm{x}_d)^T \in \mathbb{R}^d$ is contained, we compute the index vector $(I(\mathrm{x}_1), \ldots, I(\mathrm{x}_d))^T$ with

$$I(\mathrm{x}_k) := \left\lceil \frac{\mathrm{x}_k - \alpha_k}{\beta_k - \alpha_k} \lfloor N^{1/d} \rfloor \right\rceil.$$

Then this index vector has to be mapped to a unique index, which can be done by defining

$$I(x) := \sum_{k=1}^{d} [I(x_k) - 1] \lfloor N^{1/d} \rfloor^{k-1}.$$

Let us analyze the resources needed by this algorithm. We need $\mathcal{O}(dN)$ space to store N points from $\mathbb{R}^d$. The list of cells needs another $\mathcal{O}(N)$ space for the pointers to the lists of indices. Finally, since every point is reported only once in the lists of indices, we need $\mathcal{O}(N)$ space for storing all indices. Thus, summing up, we need $\mathcal{O}(dN)$ space to build the entire data structure.

Next, let us turn to the computational complexity. In the first step of the algorithm we have to compute the maximum and minimum of each coordinate. This can be done in $\mathcal{O}(dN)$ time. In the second step we have to build the data structure for the cells. This means we have to define an array of size $\lfloor N^{1/d} \rfloor^d$ whose entries are lists of indices. Hence this can be done in $\mathcal{O}(N)$ time. Finally, in the third step we have to find for each center x_j its box. The computation of its index costs $\mathcal{O}(d)$ time, so that we need $\mathcal{O}(dN)$ time for the third step.

Note that we have not made use of the fact that X is quasi-uniform at all. We have derived the following result.

Theorem 14.3 *Algorithm 1 needs $\mathcal{O}(dN)$ time and $\mathcal{O}(dN)$ space to build a grid data structure for N points in $\mathbb{R}^d$.*

The left-hand part of Figure 14.1 shows a typical example of a gridded set of centers.

With this data structure given, it is now time to see how it helps to answer our requests. Let us start with the range query. In Figure 14.1 the algorithm for an ellipsoidal region is demonstrated. The precise formulation of the procedure is given in Algorithm 2.

Any analysis of the computational complexity has to consider the size of the query region. In general, bounds on the necessary time are given in terms of N and of the number of points in the query region. Here, we want to take a different point of view. We are mainly interested in query regions R that have size $\mathcal{O}(N^{-1/d})$. If this is the case then there

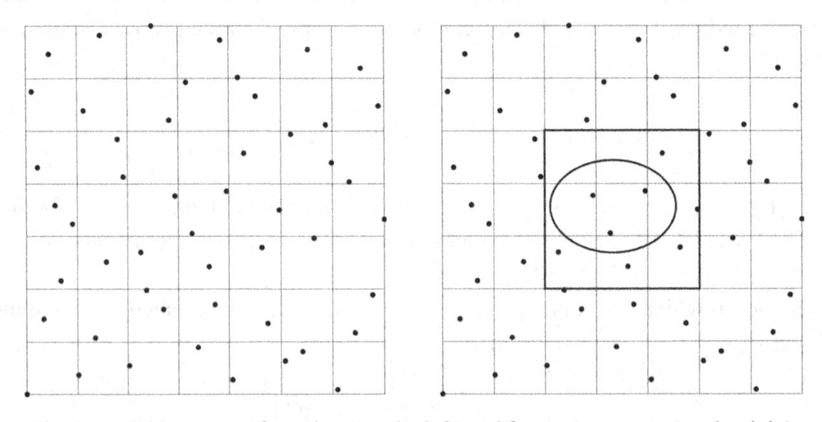

Fig. 14.1 Grid structure for points (on the left) and for a range query (on the right).

Algorithm 2 Range query

Input: Query region $R \subseteq \mathbb{R}^d$.
Output: All points that lie in R.

Compute the indices of all cells W with $R \cap W \neq \emptyset$.
If this is too expensive then compute first the bounding box B of R and then all indices of cells W with $B \cap W \neq \emptyset$.
Traverse all cells found and report those points contained in R.

exists a cube $W = W(y, c_1 N^{-1/d})$ containing R. Since our cells are cubes with side length $\mathcal{O}(N^{-1/d})$, the number of cells that have common points with W and hence with R is bounded by $\mathcal{O}(c^d)$, i.e. it is constant in terms of the number N of points but exponential in terms of space dimension. Moreover, if X is quasi-uniform then the number of points in each cell is also bounded by $\mathcal{O}(c^d)$ by Corollary 14.2. Hence we have

Proposition 14.4 *Let $X = \{x_1, \dots, x_N\} \subseteq \Omega$ be quasi-uniform. Suppose that the query region R is of size $\mathcal{O}(N^{-1/d})$. If its bounding box can be computed in constant time and if it is also possible to decide in constant time whether a point belongs to R, then Algorithm 2 needs constant time in terms of N and exponential time in terms of d to report all points contained in R.*

We finish this section by investigating the nearest neighbor problem. To look for the nearest neighbor of x, we simply locate the box that contains x and compute the distance of all points x_j in that box. Moreover, we have to check the points in the neighboring boxes. To make this procedure more precise, we will introduce the concept of surrounding boxes. For a given box

$$B = [\gamma_1, \delta_1] \times \cdots \times [\gamma_d, \delta_d]$$

we define the kth surrounding box to be

$$S_k(B) = \{[\gamma_1 + h_1\alpha_1, \delta_1 + h_1\alpha_1] \times \cdots \times [\gamma_d + h_d\alpha_d, \delta_d + h_d\alpha_d] : \|\alpha\|_\infty = k\},$$

where $h_j = \delta_j - \gamma_j$ and α is in $\mathbb{Z}^d$. In particular, $S_0(B)$ contains only B itself. See Figure 14.2 for an illustration.

With this definition we can formulate the algorithm for the nearest neighbor search as in Algorithm 3. Remember, that we already have the bounding box BB of X.

If X is quasi-uniform then we know that each box of the grid contains only a constant number of centers. Moreover, the size of the grid boxes is proportional to $h_{X,\Omega}$. Hence the algorithm terminates after a constant number of steps.

Proposition 14.5 *Let $X = \{x_1, \dots, x_N\} \subseteq \Omega$ be quasi-uniform. For every $x \in \mathbb{R}^d$, Algorithm 3 reports the nearest neighbor from X to x in constant time in terms of N.*

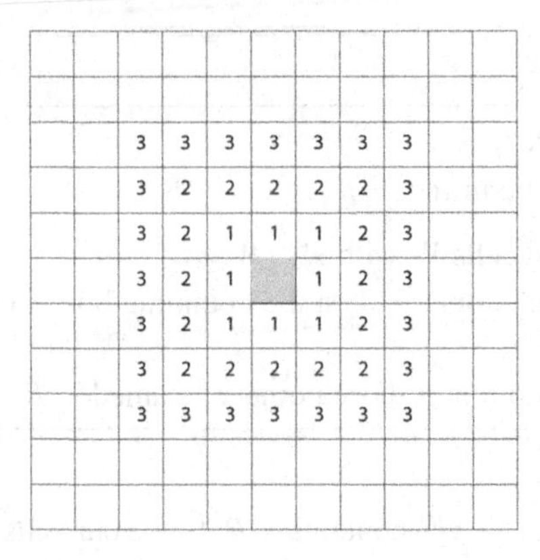

Fig. 14.2 Typical surroundings of a two-dimensional box.

Algorithm 3 Nearest neighbor search

Input: $x \in \mathbb{R}^d$ query point.
Output: A point $x_j \in X$ closest to x.

Find the box B with $x \in B$.
Initialize $dist$ by infinity.
for $k = 0, 1, 2, \ldots$ **do**
  for *every* $W \in S_k(B)$ **do**
    Compute the distance of each $x_j \in W \cap BB$ to x, store the closest so far, and update $dist$.
  Stop if $dist < k \min_{1 \le j \le d} h_j$.

A generalization to the ℓ nearest neighbors problem is obvious. The time taken to report those neighbors depends also on the number ℓ.

The fixed-grid method can also be seen in a different light. It turns out to be an efficient data structure for the containment query. Let us describe this in more detail. Suppose that Ω is an axis-parallel cube. Then we take the covering $\{\Omega_j\}$ to be cubes centered at a grid $h\mathbb{Z}^d$ of side length $c_s h$ that have common points with Ω. On the boundary of Ω we might not need all the small cubes. Hence, an offset $1 \le k \le c_s$ determines how much our covering overlaps Ω. Figure 14.3 demonstrates the idea. It shows the fine $h\mathbb{Z}^d$ grid on Ω and some of the macro-boxes Ω_j.

Obviously, for every $x \in \Omega$ those boxes Ω_j that contain x can be determined in constant time. Moreover, only a constant number of these boxes overlap, i.e. each x is contained only in a constant number of boxes and the constant is independent of x. Finally, if X is a quasi-uniform data set with fill distance h then the number of points from X in one of the

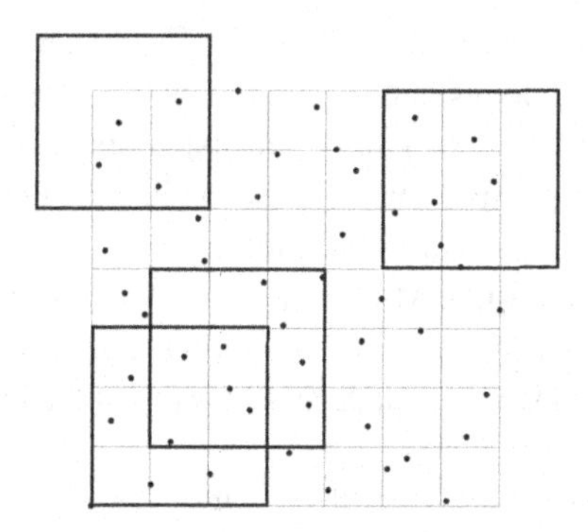

Fig. 14.3 Covering a region with boxes.

boxes Ω_j is bounded by a uniform constant. If Ω is not a box, then the general idea still holds. Only at the boundary do the boxes have to be adjusted.

14.2 *kd*-Trees

In the last section we divided the bounding box of the centers X into equally sized subboxes and collected all points for each of these subboxes. If the centers fail to be quasi-uniform (which means numerically that the quasi-uniformity constant c_{qu} is huge), we have a large number of empty boxes, or boxes that contain a large number of points, or both. In this situation it seems to be favorable to divide the bounding box in a different way. One possibility is to use kd-trees. The name initiated from work in a k-dimensional space, hence, in our setting these trees should be called dd-trees. But we will keep the commonly used name.

The kd-tree data structure is based on a recursive subdivision of space into disjoint rectangular regions called boxes. Each node of the tree is associated with such a box B and with a set of data points that are contained in this box. The root node of the tree is associated with the bounding box that contains all the data points. Consider an arbitrary node in the tree. As long as the number of data points associated with this node is greater than a small quantity, called the *bucket size*, the box is split into two boxes by an axis-orthogonal hyperplane that intersects this box. There are a number of different splitting rules, which determine how this hyperplane might be selected. We will discuss them in detail later on. The two resulting boxes are the cells associated with the two children of this node. The data points lying in the original box are split between these two children, depending on which side of the splitting hyperplane they are. Points lying on the hyperplane itself may be associated with either child (according to the dictates of the splitting rule). When the number of points associated with the current box falls below the bucket size then the resulting node is declared a leaf, and these points are stored with the node.

Thus, in addition to the data points themselves, a kd-tree is specified by two additional parameters, the bucket size and a splitting rule. The tree itself is a binary tree with two different types of nodes, splitting nodes and leaves. The root node, is in principle a splitting node (unless the number of points is less than the bucket size), which contains additional information such as the number of points, the space dimension, and the bucket size. A splitting node stores the integer splitting dimension indicating the coordinate axis orthogonal to the cutting hyperplane. It also stores the splitting value where the hyperplane intersects the axis of the splitting dimension. Moreover, it contains two pointers, one for each child (corresponding to the low and high sides of the cutting plane). A leaf node stores the number of points that are associated with this node and an array of the indices of the points lying in the associated box.

Next, let us discuss possible splitting rules. To this end we denote by Y the current subset of data points in the current box B. Let $B(Y) \subseteq B$ be the bounding box of Y. For the root node, Y is equal to X, the set of all points, and $B = B(X)$ is the bounding box of the centers. The aspect ratio of a box is the ratio of its longest and shortest side lengths. Given a dimension, the point spread of Y along this dimension is the difference between the largest and the smallest coordinate of all points in Y for this dimension. Finally, for a set A of n numbers, the median is the number m that partitions A into two subsets, one with $\lfloor n/2 \rfloor$ elements that are not greater than m and the other with $\lceil n/2 \rceil$ elements that are not less than m. Typical splitting rules are as follows.

- **Standard *kd*-tree splitting rule** The splitting dimension is the dimension of the maximum spread of Y. The splitting value is the median of the coordinates of Y along this dimension.
- **Cyclic splitting rule** This splitting rule works in the same way as the standard rule, but the splitting dimension is not chosen by the maximum spread. Instead all coordinates are chosen one after the other in a cyclic way.
- **Midpoint splitting rule** The splitting dimension is the dimension of the longest side of B. The splitting value is the midpoint of this side of the box. If there is more than one dimension with the longest side choose the dimension with the widest point spread.
- **Sliding midpoint rule** The splitting dimension and splitting value are chosen as in the case of the midpoint splitting rule, provided that points from Y lie on both sides of the cutting plane. If this is not the case then the cutting plane is moved from the empty side towards the first point of Y. The coordinate along the splitting dimension of this point now determines the splitting value. The point itself is put into the former empty (child-)box while all other points from Y remain in the other box.

Obviously, several other splitting rules are possible, but those mentioned above are commonly used. Before we analyze the complexity of building a kd-tree using the standard splitting rule, we want to point out that on the one hand this rule may lead to boxes with arbitrarily high aspect ratios. On the other hand, the midpoint splitting rule produces boxes with bounded aspect ratios but it may lead to trivial splits. These are splits where one of the child boxes does not contain a point at all. Hence the depth and size of a tree can be arbitrarily large, even exceeding $\mathcal{O}(N)$. A possible way out is the sliding midpoint rule. It cannot produce trivial splits and therefore both depth and size are bounded by $\mathcal{O}(N)$. It may also produce boxes of high aspect ratio but this seems to be less likely. A more thorough analysis of the complexity will follow. Figure 14.4 shows a decomposition of space using

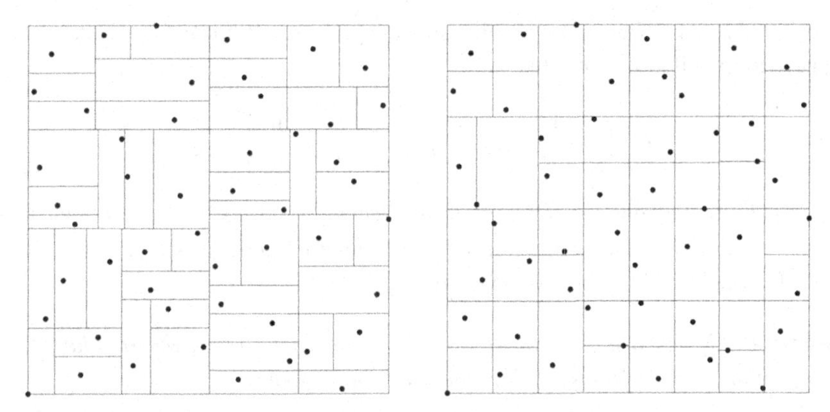

Fig. 14.4 A *kd*-tree decomposition with standard splitting rule (on the left) and sliding midpoint splitting rule (on the right).

the standard splitting rule (on the left) and the sliding midpoint rule (on the right). In both cases the bucket size has been chosen as one.

Assuming that the splitting rule is chosen, that the bounding box BB of X is at hand, and that b denotes the bucket size, Algorithm 4 describes the procedure of building a *kd*-tree.

Algorithm 4 Build *kd*-tree

Input: Data points $X = \{x_1, \ldots, x_N\} \subseteq \mathbb{R}^d$, box B.
Output: Root of a *kd*-tree.

if *X contains at most b points* **then**
- Return a leaf storing the indices of these points.

else
- Determine splitting dimension and value and split the current bounding box B and its points into two subboxes B_1, B_2 and two subsets of points Y_1, Y_2, respectively.
- Apply this algorithm to Y_1, B_1, resulting in a pointer ν_1.
- Apply this algorithm to Y_2, B_2, resulting in a pointer ν_2.
- Create a splitting node ν storing the splitting dimension, the splitting value, the left child ν_1, and the right child ν_2.
- Return ν.

Let us analyze this algorithm when the standard splitting rule is employed. The most expensive step in each recursive call is to find the splitting value. Median finding can be done in linear time but the necessary algorithms are rather complicated. It appears to be easier to presort the set of points on each coordinate in a preprocessing step, which can be done in $\mathcal{O}(dN \log N)$ time and needs additional $\mathcal{O}(dN)$ space at least temporarily to keep the sorted lists. Then it is easy to find the median in linear time. It is also easy to construct the two sorted lists for the recursive calls from the given list in linear time. Hence, the time $T(N)$ needed to build the tree satisfies

$$T(N) = \begin{cases} \mathcal{O}(1) & \text{if } N \le b, \\ \mathcal{O}(N) + 2T(\lceil N/2 \rceil) & \text{if } N > b \end{cases}$$

which on solving the equation turns out to be $\mathcal{O}(N \log N)$. Finally, because a kd-tree is a binary tree with $\mathcal{O}(N)$ leaves and because every internal node uses $\mathcal{O}(1)$ space the total amount of space is $\mathcal{O}(N)$. Thus we have proven

Theorem 14.6 *To construct the kd-tree for N points in $\mathbb{R}^d$ using the standard splitting rule, Algorithm 4 needs $\mathcal{O}(dN \log N)$ time and $\mathcal{O}(dN)$ space.*

A consequence of this result is that a kd-tree built with the standard splitting rule has $\mathcal{O}(\log N)$ depth. As pointed out earlier, the other splitting rules do not guarantee this feature. Nonetheless, if the points are quasi-uniformly distributed then all splitting rules have roughly the same complexity and need the same space.

Algorithm 5 Range query

Input: Query region $R \subseteq \mathbb{R}^d$ and root ν of kd-tree.
Output: All points contained in R.

```
if ν is a leaf then
    Report all points stored at ν that are in R.
else
    Let ν1 and ν2 denote the left and right child of ν, respectively and let B1 and B2 denote
        their associated boxes.
    for i = 1, 2 do
        if Bi is fully contained in R then
            Report all points in the tree rooted at νi.
        else
            if R intersects Bi then
                Apply this algorithm to R and νi.
```

Let us now turn to the range query problem. A possible solution is described in Algorithm 5. The main test in this algorithm is whether R intersects the box $B(\nu)$ associated with a node ν. If this is rather complicated then it is better to replace R by its bounding box, to apply Algorithm 5 to the bounding box, and to test the reported points again in a final step.

Note that it is not necessary to compute the associated cell $B(\nu)$ each time. The current cell is maintained through the recursive calls using only the information on the splitting value $s(\nu)$ and the splitting dimension $d(\nu)$. For example, the cell for the left child ν_1 of a node ν is

$$B(\nu_1) = B(\nu) \cap \{x \in \mathbb{R}^d : x_{d(\nu)} \le s(\nu)\}.$$

The complexity analysis of the range query shows that Algorithm 5 needs slightly more time to answer a range query problem than Algorithm 2, if the data points are quasi-uniform and if the range has size $\mathcal{O}(h)$. This is a natural consequence of the tree structure.

Proposition 14.7 *Let $X = \{x_1, \ldots, x_N\} \subseteq \mathbb{R}^d$ be quasi-uniform in its bounding box BB. Suppose that the query region R is of size $\mathcal{O}(N^{-1/d})$. If its bounding box can be computed in constant time and if it is also possible to decide in constant time whether a point belongs to R, then Algorithm 5 needs $\mathcal{O}(\log N)$ time to report all points contained in R.*

Proof If X is quasi-uniform in its bounding box then each splitting rule divides the current box into two boxes that have a side length in the direction of the spitting dimension that is roughly half the side length of the original box. Hence no coordinate is preferred by the splitting rule and the boxes associated with the leaves have side lengths $\mathcal{O}(h_{X,BB})$ in each coordinate. Since the query region has also size $\mathcal{O}(h_{X,BB})$, only a constant number of leaf boxes has to be investigated. Finally, since the depth of the tree is $\mathcal{O}(\log N)$, the complexity for reporting all points contained in R is $\mathcal{O}(\log N)$. □

Next, let us have a look at the nearest-neighbor problem. There are different ways of implementing a strategy for this problem using a kd-data structure. The standard procedure is given in Algorithm 6.

Algorithm 6 Nearest neighbor search

Input: Query point x and root ν of kd-tree.
Output: A point $x_j \in X$ that is closest to x.

Initialize the distance *dist* as infinity and reserve space to store the nearest neighbor *nn*.
if *ν is a leaf* **then**
  Compute the distance of each point stored in ν to x.
  if *the current distance is less than dist* **then**
    Update *dist* and *nn*.
if *ν is a splitting node* **then**
  Visit that child of ν whose associated box contains x.
  The other child has only to be visited if its associated box has a distance to x that is less than the current distance *dist*.
Return *nn*.

The generalization to the ℓ nearest neighbors problem is obvious again. The same is true for the analysis in the case of quasi-uniform points. In view of the facts that each box associated with a leaf has size $\mathcal{O}(h)$ and that two parallel splitting hyperplanes have separation distance at least $\mathcal{O}(h)$, only a constant number of leaves will be visited. Since the tree has depth $\mathcal{O}(\log N)$, we can make the following proposition:

Proposition 14.8 *Let $X = \{x_1, \ldots, x_N\} \subseteq \mathbb{R}^d$ be quasi-uniform in its bounding box BB. For every $x \in \mathbb{R}^d$, Algorithm 6 needs $\mathcal{O}(\log N)$ time to find the nearest neighbor of x.*

Since our result on building the *kd*-tree does not depend on quasi-uniformity it is interesting to see how our algorithms behave if X is not quasi-uniform. There is a very large number of results addressing this problem. Here, we want to give only one of them. To this end we assume that the *kd*-tree has been built using the cyclic splitting rule.

Proposition 14.9 *A range query with an axis-parallel rectangle R can be performed in $\mathcal{O}(N^{1-1/d} + k)$ time, where N is the number of points in the kd-tree and k is the number of reported points.*

Proof Since each point contained in the query region is dealt with exactly once in a "report all points" call of Algorithm 5, the time necessary to report all relevant points is linear in the number k of reported points.

Hence, it remains to estimate the time spent in nodes that are not traversed by "report all points" calls. To this end we investigate the number of associated boxes that are intersected by a hyperplane orthogonal to the first coordinate. This gives an upper bound on the boxes intersected by the two side faces of the query box that are orthogonal to the first coordinate. The number of regions intersected by the other faces can be bounded in the same way. Let $Q(N)$ denote the number of intersected regions in a *kd*-tree with N points, whose root has splitting dimension 1. Let H be the hyperplane orthogonal to the first coordinate in question. Since the root has splitting dimension 1, H intersects one of its two children but not both. However, it intersects both children of this child because both have splitting dimension 2. Continuing this argument we arrive after d splitting steps at nodes that again have splitting dimension 1. Hence, the recursion for $Q(N)$ is given by

$$Q(N) = 2^{d-1} + 2^{d-1} Q(N/2^d),$$

because in each splitting step the number of points in the new boxes is reduced by a factor 2. As $Q(1) = \mathcal{O}(1)$ this recursion solves to give $Q(N) = \mathcal{O}(N^{1-1/d})$. □

In most numerical examples the range query behaves much better than is predicted by the last proposition. Nonetheless, an improvement can be achieved, at least theoretically, by using the following *bd*-trees.

If working with cyclic splitting then it is sometimes favorable to subsume one complete cycle into one node. This means that in each step the associated box is divided into 2^d subboxes and that each node has 2^d children. Special cases of the resulting trees are *quadtrees* in a two-dimensional setting and *octrees* in a three-dimensional setting.

Finally, kd-trees can also be used to construct an overlapping domain decomposition, which is useful for a containment query. The only thing that has to be changed is that instead of one cutting hyperplane there now have to be two. All points on the left of the right cutting plane are assigned to the left child and all points on the right of the left cutting plane to the right child.

14.3 *bd*-Trees

While kd-trees work efficiently in moderate space dimensions and if the data points do not deviate too much from quasi-uniformity, something more elaborate has to be done in other cases. In particular, when the points are highly clustered, either it may take many splits to partition them or the splits may result in arbitrary long boxes, which might cause problems when searching is performed.

One possible way of avoiding such problems is to use bd-trees. The name stands for *box decomposition trees*.

A bd-tree is a generalization of a kd-tree. It is still a binary tree but in addition to leaves and splitting nodes a third type of node is introduced. The new type of node represents an additional operation called *shrinking*. Consider a node whose associated box B contains more points than the bucket size. Then, a shrinking rule is invoked. This rule may either perform a split according to the chosen splitting rule or select an axis-parallel box B_1 contained inside B. The points lying in B_1 are assigned to one child and the points lying in $B_2 = B \setminus B_1$ are assigned to the other child. Points on the boundary may be assigned to one or both children. We will describe possible shrinking rules by employing the notation already used in the kd-trees setting.

- **No shrink** Do not perform any shrinking at all, only splitting.
- **Simple shrink** This rule depends on two constants: tn (for example this could equal 2) and *thresh* (for example this could equal 0.5). It first computes the $2d$ distances between each side of the bounding box $B(Y)$ of the current data set Y and the corresponding side of the box B associated with the current node. If at least tn of these distances are larger than the length of the longest side of $B(Y)$ times *thresh* then it shrinks all affected sides. After the shrink, Y is completely contained in the inner box while the outer box contains no point at all. If the criterion is not met then a classical split is performed.
- **Centroid shrink** This rule also depends on two constants: ms (for example 4) and fr (for example 0.5). Theoretically speaking, it applies the splitting rule, without actually generating a splitting node, to find the half space that contains the larger number of points, and it goes on splitting this part until the actual number of points is less than fr times the initial number. If more than ms splits are necessary then a shrinking node corresponding to the final inner box is created; all other points are placed in the outer box. Otherwise, splitting is performed.

While the "no shrink" choice leads to a kd-tree, the other shrinking rules may lead to a different decomposition of space. Figure 14.5 shows the data points of Figure 4.4 but now for a bd-tree with simple shrink and sliding midpoint split. The centroid shrink would lead here to the same decomposition as the splitting rule alone.

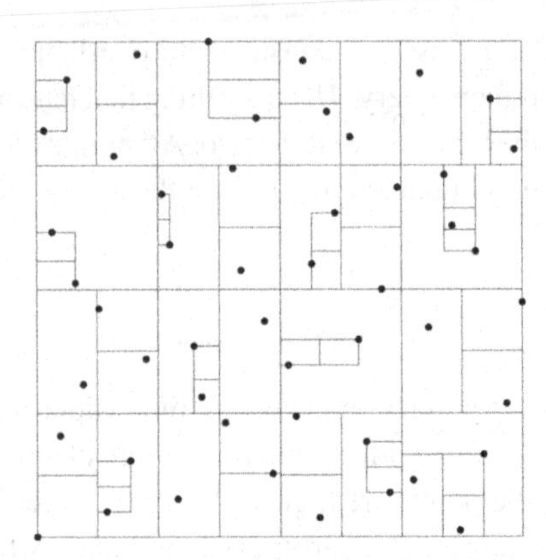

Fig. 14.5 A bd-tree decomposition with simple shrinking rule.

Building a bd-tree structure is very similar to building a kd-tree structure. The necessary modifications of Algorithm 4 should cause no problem. Hence we leave the details to the reader.

We will still call the outer box a *box*, even if it is not a box in the literal sense any more. The concerned reader might replace the word box by "cell" and define a cell to be either a box or the set-theoretical difference of two boxes.

Note that every splitting operation can be seen as a special case of a shrinking operation, with the inner box as (for example) the left-hand box and the outer box as the original box. But this is not optimal because it needs more comparisons to decide whether a point is in a box than to decide on which side of a hyperplane it is situated.

Range queries and a nearest neighbor search can be performed in just the same way as in the case of kd-trees, but modifying the algorithms for the additional nodes. Instead of spelling out these changes, however, we want to point out another possible modification.

For an efficient implementation of the moving least squares method, we need for a given x all points from X that are in the ball $B(x, \delta)$ around x with radius δ. This is clearly a range query problem. But since the weight function has compact support, additional points from X not contained in the ball $B(x, \delta)$ would do no harm. Thus we could also relax our request in the following sense, hoping for faster answers.

- **Approximate range search** Given a region R and $\epsilon > 0$, define the region $R_\epsilon = \cup_{x \in R} B(x, \epsilon)$. A legal answer is an answer that reports all points $X \cap R$ together with some (not necessarily all) points from $X \cap (R_\epsilon \setminus R)$.
- **Approximate nearest neighbor search** Given $x \in \Omega$ and $\epsilon > 0$, report a point $x_j \in X$ whose distance from x is at most $1 + \epsilon$ times that of a true nearest neighbor. Obviously, an approximate ℓ nearest neighbors query can be introduced in a similar fashion.

The algorithms corresponding to these modified problems in the setting of bd-trees are presented in Algorithms 7 and 8, where the latter is based upon a priority queue.

The analysis of these algorithms is beyond the scope of this book. We will collect the necessary results in the next theorem, where midpoint splitting and the centroid shrinking

Algorithm 7 Range query

Input: Query region $R \subseteq \mathbb{R}^d$, an outer range R_O and the root ν of the *bd*-tree.
Output: All points that lie in R, no point outside R_O.

if $B(\nu) \subseteq R_O$ **then** Report all points in $B(\nu)$.
if $B(\nu) \cap R = \emptyset$ **then** Return $\emptyset$.
if *ν is a leaf* **then**
  Return all points assigned to ν that lie in R.
else
  Apply this algorithm to both children of ν.

Algorithm 8 Approximate nearest neighbor search

Input: Query point $x \in \mathbb{R}^d$, $\epsilon > 0$, and the root ν of the *bd*-tree.
Output: A point $q \in X$ with distance from x that is at most $1 + \epsilon$ times the distance of a nearest neighbor from x.

Initialize the distance *dist* as infinity. Initialize the priority queue by the root node ν.
while *the priority queue is not empty* **do**
  Take the first node from the priority queue.
  if *the distance of its associated cell from x exceeds $dist/(1 + \epsilon)$*
  then
    Stop and report the current q.
  if *ν is a leaf* **then**
    Compute the distances from the points stored in this node to x and update *dist* and q.
  else
    Compute the distance from both children of ν and enqueue first the further and then the closer one.

rule are employed to build the data structure. For a proof, we refer the interested reader to the articles of Arya and Mount [3–4] and Arya *et al.* [5].

Theorem 14.10 *For N given points in $\mathbb{R}^d$ the bd-tree can be built in $\mathcal{O}(dN \log N)$ time and $\mathcal{O}(dN)$ space. The tree has depth $\mathcal{O}(\log N)$.*

(1) For $\epsilon > 0$ there is a constant $c_{d,\epsilon} \leq d\lceil 1 + 6d/\epsilon \rceil^d$ such that for every $x \in \mathbb{R}^d$ a $(1 + \epsilon)$-nearest neighbor of x can be reported in $\mathcal{O}(c_{d,\epsilon} \log N)$ time.
(2) Given $\epsilon > 0$ and a range R, the m points in R_ϵ can be reported in $\mathcal{O}(m + 2^d \log N + (3\sqrt{d}/\epsilon)^d)$ time.

It remains to remark that the data structure itself is independent of ϵ. Hence different accuracies can be tested without rebuilding the data structure. Moreover, the constants are worst-case constants. We know already that in the case of quasi-uniform data sets the behavior is much better. This is also true if an average time analysis based on a uniform distribution is performed.

14.4 Range trees

We come finally to a data structure that has been designed for answering rectangular range queries in particular. It is based on ideas from the univariate case, which we will hence investigate first. If $X = \{x_1, \dots, x_N\} \subseteq \mathbb{R}$ consists of N pairwise distinct real numbers and $[x, x']$ is a given interval then we want to report all points from X inside $[x, x']$. This can be done by employing either a sorted array or a balanced binary search tree. Let us use the latter. We will store the points of X in the leaves of our tree $\mathcal{T}$. The internal nodes of $\mathcal{T}$ store certain splitting values to guide the search; let us denote the splitting value stored at node ν by $v(\nu)$. Now, that $\mathcal{T}$ is a search tree means that for every internal node ν its left subtree contains all the points smaller than or equal to $v(\nu)$ and its right subtree contains all the points greater than $v(\nu)$. It is well known that a balanced search tree can be built with $\mathcal{O}(N)$ space and $\mathcal{O}(N \log N)$ time and having $\mathcal{O}(\log N)$ depth.

To report all points in $[x, x']$ we can do the following. First, we search in $\mathcal{T}$ for x and x'. This gives us two leaves μ and μ' at which the searches end. Then, all points in $[x, x']$ are those points stored in leaves between μ and μ' plus, possibly, the points stored at μ and μ' themselves. The leaves between μ and μ' are the leaves of certain subtrees, rooted at nodes ν between the search paths for x and x', whose parents are on one of the search paths. To find these nodes we first find the node ν_{split} at which the two search paths for x and x' split. This is achieved using Algorithm 9.

Algorithm 9 Find split node

Input: A binary search tree $\mathcal{T}$ and a query region $[x, x']$.

Output: The node ν where the search paths to x and x' split, or the leaf where both end.

Set $\nu = root(\mathcal{T})$.

while ν *is not a leaf and* $(x' \leq v(\nu)$ *or* $x > v(\nu))$ **do**

  if $x' \leq v(\nu)$ **then**

    Apply this algorithm to the left child of ν.

  else Apply it to the right child.

Starting from this splitting node we follow the search path for x. At each node where the search path goes left, we report all leaves in the right subtree. Then we follow the path to x' and report the leaves in the left subtree of nodes where the path goes right. Finally, we have to check the points stored at the leaves where the paths end. Details may be found in Algorithm 10.

Algorithm 10 One-dimensional range query

Input: A binary search tree $\mathcal{T}$ and a query region $[x, x']$.
Output: All points from X in $[x, x']$.

```
Use Algorithm 9 to determine the splitting node ν.
if ν is a leaf then
    Report the point stored there if necessary.
else
    Set ν to be its left child.
    while ν is not a leaf do
        if x ≤ v(ν) then
            Report all points in the subtree of the right child of ν.
            Set ν to be its left child.
        else
            Set ν to be its right child.
    Report the point stored at the leaf ν if necessary.
    Similarly, follow the path to x' and report the points in the subtrees left of the path.
    Check whether the point stored at the leaf where the path ends has to be reported.
```

Lemma 14.11 *Algorithm 10 reports exactly those points x_j that lie in $[x, x']$.*

Proof First we show that each point reported by the algorithm is contained in $[x, x']$. Suppose that x_j is one of the reported points. If x_j is stored at one of the leaves where the search paths end then we check explicitly whether $x_j \in [x, x']$. Otherwise it must have been reported inside the 'while' loop. Assume that it was reported on the way to x. Let ν denote the node on the path such that, on the one hand, x_j is in the subtree of ν's right child. Since ν and hence its right child and hence x_j are in the left subtree of the splitting node, we have $x_j \le v(\nu_{\text{split}}) \le x'$. On the other hand, the search path to x goes left at ν, giving $x \le v(\nu) \le x_j$. If x_j is reported while we are on our way to x' then a similar argument holds.

Finally, we have to check whether all points $x_j \in [x, x']$ have been reported. Let $x_j \in X$ be any point in $[x, x']$ and assume that x_j has not been reported. Let μ be the leaf where x_j is stored. Denote by ν the lowest ancestor to μ that has been visited by the query algorithm. Such a ν must exist because the query algorithm starts at the root of the tree. Note that ν cannot be the root of a subtree whose points have been completely reported, because otherwise x_j would also have been reported. Hence, ν must be on the path to x or the path to x' or the path to both. All three cases can be dealt with in a similar manner. Thus, we will investigate only the first. If ν is on the way to x then the search path has to go to the right at ν, because otherwise $x \le v(\nu)$ and all points in the right subtree of ν have been reported, including x_j. But if the search path turns to the right then this means that $x > v(\nu)$. Moreover, since ν is the lowest ancestor of μ on the search path, μ must lie in the left subtree of ν in this situation, which shows that $x_j \le v(\nu) < x$. This contradicts $x_j \in [x, x']$. □

Let us analyze the computational complexity. Because our data structure is a balanced binary search tree it needs $\mathcal{O}(N)$ space and $\mathcal{O}(N \log N)$ time to be built. Assume that k points are contained in $[x, x']$. The number of non-leaf nodes in a balanced binary tree is bounded by the number of leaves, so each report-call in the while-loop can be done in $\mathcal{O}(\#(\text{reported points}))$ time. Since altogether we report k points, all report calls can be done in $\mathcal{O}(k)$ time.

The remaining time is spent in nodes on the two search paths. Since we have a balanced binary tree, each search path has length $\mathcal{O}(\log N)$. Hence, Algorithm 10 needs $\mathcal{O}(k + \log N)$ time to answer a one-dimensional range query. Note also that the number of subtrees that are reported is bounded by the number of nodes on the search paths and hence by $\mathcal{O}(\log N)$.

Proposition 14.12 *Given N points $x_1, \ldots, x_N \in \mathbb{R}$, a balanced binary search tree can be built in $\mathcal{O}(N)$ space and $\mathcal{O}(N \log N)$ time, so that every range query yielding k points can be answered in $\mathcal{O}(\log N + k)$ time.*

How can we use these ideas in the multivariate setting? We have already encountered a possible generalization, in form of the *kd*-tree, that did not lead to the desired result (cf. Proposition 14.9).

Now we want to trade space for time. If we allow the data structure to use more space than $\mathcal{O}(N)$ it is possible to achieve better results even in the worst case. The nature of a range tree needs the additional assumption that all points are pairwise different in every coordinate. One possibility for matching this assumption is *cyclic concatenation*. In other words, we introduce for every coordinate j a new order that allows us to distinguish pairwise-different points $x \ne y \in \mathbb{R}^d$ having the same j-coordinate.

Definition 14.13 *A d-dimensional range tree is defined recursively. For $d = 1$, we have a one-dimensional range tree, which is simply a balanced binary search tree. For $d \ge 2$, a d-dimensional range tree is a balanced binary search tree with respect to the first coordinate. The leaves of this tree contain the d-variate points. The points stored in the leaves of the subtree rooted in node ν are called the canonical subset of ν. We denote this set by $P(\nu)$. Each node contains an additional pointer to a $(d-1)$-dimensional range tree, the associated tree $\mathcal{T}_{\text{assoc}}(\nu)$. This associated range tree $\mathcal{T}_{\text{assoc}}(\nu)$ is built on the points in $P(\nu)$ that are restricted to the last $d-1$ coordinates.*

Algorithm 11 describes how a d-dimensional range tree can be built. However, the algorithm has to be modified to get the optimal construction time, because the construction of a balanced binary search tree from unsorted data costs $\mathcal{O}(N \log N)$ time. In the one-dimensional setting this is optimal, but using it in the multivariate setting leads to a complexity that is too high. Instead, we sort the data points with respect to each coordinate, which costs additional $\mathcal{O}(dN)$ space and $\mathcal{O}(dN \log N)$ time. With presorted points a balanced binary search tree can be constructed bottom-up in linear time. We will use this in the following analysis.

Algorithm 11 Build range tree

Input: Data points $X = \{x_1, \ldots, x_N\} \subseteq \mathbb{R}^d$ and current working dimension j.
Output: The root of a d-dimensional range tree.

if $j = d$ **then**
- Create a balanced binary search tree at the last coordinate.
- Store pointers to the points at the leaves.
- Return the root.

else
- Apply this algorithm to the current set of centers and dimension $j+1$. This yields a pointer ν_{assoc} to the associated data structure.
- **if** *X contains only one point* **then**
  - Create a leaf ν, store a pointer to this point at ν, and make ν_{assoc} the pointer to the associated data structure.
- **else**
  - Split X into two subsets. P_{l} shall contain the points having jth coordinate less than or equal to the median of X with respect to this coordinate; P_{r} shall contain the rest.
  - Apply this algorithm to P_{l} for dimension j. This gives a pointer ν_{l}.
  - Apply this algorithm to P_{r} for dimension j. This gives a pointer ν_{r}.
  - Create a node ν storing the splitting value and having ν_{l} as its left child and ν_{r} as its right child. Let ν_{assoc} be the pointer to the associated data structure of ν.
- Return ν.

Let us discuss the time and space necessary for building a d-dimensional range tree. As mentioned before, the preprocessing step of sorting takes $\mathcal{O}(dN)$ space and $\mathcal{O}(dN \log N)$ time. Let us denote the time and space necessary after the preprocessing step by $T_d(N)$ and $S_d(N)$, respectively. The construction consists of constructing a binary search tree, which takes $\mathcal{O}(N)$ time and space. For each node of this first-level tree we have to build the associated tree. At any depth of the first-level tree every point $x \in X$ belongs to exactly one associated tree. Hence, the time and space needed to build the associated trees for all nodes at a given depth of the first-level tree is at most $\mathcal{O}(T_{d-1}(N))$ and $\mathcal{O}(S_{d-1}(N))$, respectively. Since the depth of the first-level tree is $\mathcal{O}(\log N)$, we derive the recurrence relations

$$T_d(N) = \mathcal{O}(N) + \mathcal{O}(T_{d-1}(N) \log N),$$
$$S_d(N) = \mathcal{O}(N) + \mathcal{O}(S_{d-1}(N) \log N).$$

With the initial values $T_1(N) = S_1(N) = \mathcal{O}(N)$, this reduces to $T_d(N) = S_d(N) = \mathcal{O}(N(\log N)^{d-1})$, which overrules in magnitude the time and space necessary for presorting the data.

Proposition 14.14 *A d-dimensional range tree for N data points in $\mathbb{R}^d$ can be built in $\mathcal{O}(N(\log N)^{d-1})$ space and time.*

This is significantly higher than for building *kd*-trees. Hence, the price we have to pay for a better worst-case range query algorithm, which we will discuss now, is a higher requirement on space and time in building the data structure.

Algorithm 12 Range query

Input: A d-dimensional range tree $\mathcal{T}$, a query range $B = [x, x']$, and the current depth j.

Output: All points from $\mathcal{T}$ that lie inside B.

Apply Algorithm 9 to the jth coordinate to determine the splitting node v.
if *v is a leaf* **then**
  Report the point stored there if necessary.
else
  Set v to be its left child.
  while *v is not a leaf* **do**
    if $x_j \leq v(v)$ **then**
      if $j < d$ **then**
        Apply this algorithm to B, the associated data structure of the right child of v, and dimension $j + 1$.
      else
        Report all points in the subtree of the right child of v.
      Set v to be its left child.
    else
      Set v to be its right child.
  Report the point stored at the leaf v if necessary.
  Similarly, follow the path to x', apply this algorithm to B, the associated structures of subtrees to the left of the path, and dimension $j + 1$. Check whether the point stored at the leaf at which the path ends has to be reported.

The idea of the range query algorithm is quite simple. It first selects the $\mathcal{O}(\log N)$ canonical subsets that together contain those points of X having a first coordinate within the interval of the first coordinate of the query box. This can be done with the one-dimensional range query algorithm, if it is modified in such a way that it can distinguish different points in

$\mathbb{R}^d$ with the same first coordinate. These subsets are queried further by performing a range query on their second-level data structure, which is in each case a range tree on the last $d-1$ coordinates built on the canonical subset. Hence the algorithm uses again the current depth j, which ensures termination in dimension. It is initially called with $j = 1$. Furthermore, it uses Algorithm 9 in an appropriate way. It obviously resembles Algorithm 10, except that where all points in a subtree are reported, we now call the algorithm again for the associated data structure as long as we have not reached the final depth. Algorithm 12 gives the details.

Finally, let us analyze the computational complexity of this algorithm. To this end let $T_d(N)$ denote the query time, not including the time to report the points. This time is determined by searching a first-level tree, which takes $\mathcal{O}(\log N)$ time, and querying a logarithmic number of $(d-1)$-dimensional range trees. Hence, it satisfies

$$T_d(N) = \mathcal{O}(\log N) + \mathcal{O}(T_{d-1}(N) \log N).$$

Since $T_1(N) = \mathcal{O}(\log N)$ this solves to give $\mathcal{O}(\log^d N)$. Adding the time for reporting the points contained yields the following result.

Theorem 14.15 *The range query algorithm for range trees needs, for rectangular query ranges, $\mathcal{O}((\log N)^d + k)$ time, where k denotes the number of points reported.*

This query time can be improved further to $\mathcal{O}((\log N)^{d-1} + k)$ if a technique called *fractional cascading* is used. Details may be found in the book [42] by de Berg *et al.*

14.5 Notes and comments

This chapter differs from all the others since it is related more to computer science than mathematics. But the results are crucial for the development of efficient algorithms for radial basis functions, which are the subject of Chapter 15. Moreover, it becomes in general more and more obvious that the success of any numerical method depends strongly on an efficient implementation and hence on the underlying data structure and the associated algorithms. This chapter gives further strong arguments for scattered data methods that are based merely upon point sets.

In computer science, a run-time analysis of an algorithm is in general done in the worst-case or average case-setting. Here, however, the distribution of the points was taken more into account and most of the analysis was done for quasi-uniform data sets.

The kd-trees that were the subject of the second section were created by Bentley (see for example [24,67]) in 1975 for nearest-neighbor searching. They are an improvement over the classical quadtrees and octrees since they allow a dimensional-independent implementation.

A good source for all kinds of multidimensional tree is the book by Samet [161]. The presentation here borrows from the book [42] of de Berg *et al.* and especially from the work of Mount, Arya and others [3–5, 139]. The examples in the cases of the kd- and bd-trees have been produced by using their *approximate nearest neighbors (ANN)* library. The description of the *bd*-trees in the third section is based on these papers just mentioned of Arya and Mount, who also created the *bd*-trees in the context of approximate range queries and nearest-neighbor searching.

Finally, range trees were independently discovered by various authors around 1980. For more details on them we refer the reader to the book by de Berg *et al.* [42] that has already been mentioned and the comments and references therein.

15
Numerical methods

Now it is time to discuss efficient numerical methods for scattered data interpolation and approximation. So far, we have encountered the moving least squares approximation scheme as one of them. Combined with the data structures of the last chapter, it needs $\mathcal{O}(dN)$ space and time to build a data structure of N points in $\mathbb{R}^d$ provided that the points are uniformly distributed. Moreover, it takes a constant time to evaluate the approximant at a single point.

But what about radial basis functions? The naive approach leads to the problem of solving an $N \times N$ system that costs $\mathcal{O}(N^3)$ time and $\mathcal{O}(N^2)$ space. Furthermore, every evaluation needs another $\mathcal{O}(N)$ time. For a very large N this is definitely too expensive. Thus more efficient methods are necessary. We will review some of the most promising approaches in this chapter.

15.1 Fast multipole methods

In the case of a globally supported basis function and a large number of centers it is impossible to use direct methods for solving the interpolation equations. Instead, iterative methods have to be employed. A popular choice would be the *conjugate gradient* method, or more generally a generalized minimum residual (GMRES) method. We will not describe these standard algorithms here explicitly; It is only necessary to note that the main operation in these methods is a matrix-by-vector multiplication that takes in general $\mathcal{O}(N^2)$ operations. In our situation, the matrix–vector multiplication boils down to the evaluation of N sums of the form

$$s(x) = \sum_{j=1}^{N} \alpha_j \Phi(x, x_j) \tag{15.1}$$

and we have to find efficient algorithms that are able to perform this evaluation in less than $\mathcal{O}(N)$ time, preferably in $\mathcal{O}(\log N)$ or even constant time. This is the subject of this section. Obviously we cannot achieve this goal if we want to reproduce the exact value of s at x. But since s is already an approximation to an unknown function, an additional error might be acceptable. Hence, we try only to approximate s up to a certain accuracy, say $\epsilon > 0$.

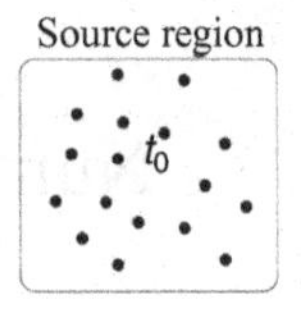

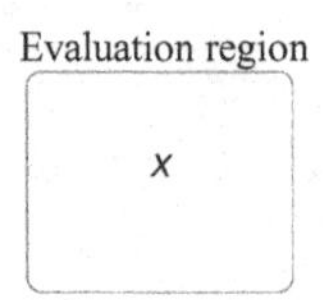

Fig. 15.1 Evaluation and source regions.

In the following, we will call t in $\Phi(x, t)$ a *source point* and x an *evaluation point*. The idea of multipole expansions is based on a far-field expansion of Φ. Suppose that all sources are situated in a certain region, called a *panel*, which is centered at a point t_0. Suppose further that we want to evaluate the function (15.1) at a point x that is sufficiently far from the source panel. Figure 15.1 illustrates the situation.

If we can expand Φ in the form

$$\Phi(x, t) = \sum_{k=1}^{p} \phi_k(x)\psi_k(t) + R(x, t) \tag{15.2}$$

with a remainder R that tends to zero for $\|x - t_0\|_2 \to \infty$, or for $p \to \infty$ if $\|x - t_0\|_2$ is sufficiently large, then we call (15.2) a far-field expansion for Φ around the source t_0. If we use (15.2) to evaluate (15.1) then we can compute

$$\begin{aligned} s(x) &= \sum_{j=1}^{N} \alpha_j \Phi(x, x_j) \\ &= \sum_{j=1}^{N} \alpha_j \sum_{k=1}^{p} \phi_k(x)\psi_k(x_j) + \sum_{j=1}^{N} \alpha_j R(x, x_j) \\ &= \sum_{k=1}^{p} \phi_k(x) \sum_{j=1}^{N} \alpha_j \psi_k(x_j) + \sum_{j=1}^{N} \alpha_j R(x, x_j) \\ &=: \sum_{k=1}^{p} \beta_k \phi_k(x) + \sum_{j=1}^{N} \alpha_j R(x, x_j). \end{aligned}$$

Hence, if we use the approximation $\widetilde{s}(x) = \sum_{k=1}^{p} \beta_k \phi_k(x)$ we have an error bound

$$|s(x) - \widetilde{s}(x)| \leq \|\alpha\|_1 \max_{1 \leq j \leq N} |R(x, x_j)|,$$

which is small if x is far enough away from the sources x_j. Moreover, each coefficient β_k can be computed in advance in linear time. Hence if p is much smaller than N, we can

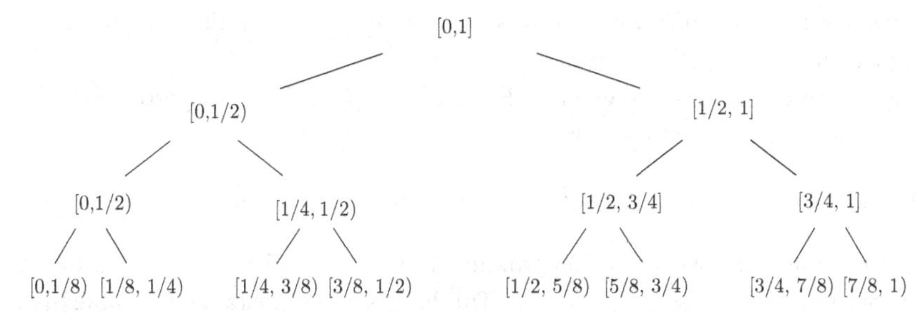

Fig. 15.2 Hierarchical decomposition of the panel [0, 1] into smaller panels. Four levels of the hierarchy are shown.

consider it as constant and we need therefore $\mathcal{O}(N)$ time to compute the coefficients $\{\beta_k\}$ and constant time for each evaluation of $\widetilde{s}$.

One could say that we have averaged the information given at the sources x_j to one single information given at the center t_0 of the panel. In this sense, (15.2) is a unipole expansion of Φ. Note that in the case of translation-invariant kernels it is easy to obtain the far-field expansion of Φ around any t_0 if the far-field expansion around zero is known. To see this, suppose that (15.2) is the far field expansion of $\Phi(x, t) = \Phi(x - t)$ around zero. Then we can write

$$\Phi(x-t) = \Phi((x-t_0)-(t-t_0)) = \sum_{k=1}^{p} \phi_k(x-t_0)\psi_k(t-t_0) + R(x-t_0, t-t_0)$$

to derive the far-field expansion around t_0.

In general, the evaluation points are close to at least a few of the centers. In the next step we will refine our approach to fit to this situation. The solution is an hierarchical subdivision of the region Ω of interest into panels or cells of sources. This is not new to us, since we encountered this concept in the last chapter. The reader should be reminded in particular of kd-trees and bd-trees. Now, how can we use this concept here? Let us explain the idea by an example. Suppose that all data points are contained in the interval [0, 1] and that we divide this interval hierarchically, into "panels" as shown in Figure 15.2. With every source panel T we associate the part of the function s that corresponds to the sources in that panel, by setting

$$s_T = \sum_{x_j \in T} \alpha_j \Phi(\cdot, x_j).$$

Moreover, we also assign the far-field expansion $\widetilde{s}_T$ of s_T to the panel T.

Then the approximation $\widetilde{s}$ to the function s at an evaluation point x is computed by adding the associated functions s_T for the panel that contains x itself and all neighboring panels. Those panels that are well separated from x contribute only by their far-field expansion $\widetilde{s}_T$ to $\widetilde{s}$. Here, we say that a point x is well separated from a panel T if it has distance at least $\operatorname{diam}(T)$ from T. A panel U is called well separated from a panel T if all points in U are well separated from T. Since we want to avoid floating-point operations, we always use the largest possible source panel for the far-field series.

Let us return to the example given in Figure 15.2. If we want to approximate $s(x)$ at a point x in the panel $[0, 1/8)$ we form

$$\widetilde{s}(x) = s_{[0,1/8)}(x) + s_{[1/8,1/4)}(x) + \widetilde{s}_{[1/4,3/8)}(x) + \widetilde{s}_{[3/8,1/2)}(x) + \widetilde{s}_{[1/2,3/4)}(x) + \widetilde{s}_{[3/4,1]}(x).$$

Note, that we use the two level-2 approximants $\widetilde{s}_{[1/2,3/4)}$ and $\widetilde{s}_{[3/4,1]}$ instead of the four level-3 approximants $\widetilde{s}_{[1/2,5/8)}, \ldots, \widetilde{s}_{[7/8,1]}$. This halves the computational complexity in this case. We can do this because the panels $[1/2, 3/4)$ and $[3/4, 1]$ are well separated from $[0, 1/4)$, the panel containing x on the same level as them. However, we could not use the approximant $\widetilde{s}_{[1/2,1]}$ because its panel $[1/2, 1]$ is not well separated from the panel $[0, 1/2]$, which is the panel containing x on the same level.

Similarly, to evaluate $s(x)$ approximately in the panel $[3/8, 1/2)$ we would use

$$\begin{aligned}\widetilde{s}(x) = {} & s_{[3/8,1/2)}(x) + s_{[1/4,3/8)}(x) + s_{[1/2,5/8)}(x) \\ & + \widetilde{s}_{[0,1/8)}(x) + \widetilde{s}_{[1/8,1/4)}(x) + \widetilde{s}_{[5/8,3/4)}(x) + \widetilde{s}_{[3/4,1]}(x).\end{aligned}$$

Having considered this example, we now return to the general situation and describe how to set up the data structure and how to evaluate the approximant.

Since we discussed several kinds of tree in the previous chapter, we do not need to go into the details of setting up the data structure. If we are using kd-trees or bd-trees to decompose space then we have to use a bucket size greater than 1, which is nonetheless small when compared to N. The same is true if quadtrees and their higher-dimensional relatives are used. Another approach would be to fix the maximum depth of the resulting tree. Obviously, for computational reasons this should be $\mathcal{O}(\log N)$ if N is the number of source points, which is always satisfied for certain splitting rules in the case of kd-trees.

Now, to set up the data structure we first choose the accuracy ϵ and from this the number p of necessary terms in the far-field expansion. Then we build the tree as usual, assigning the points to the panels. For the leaf panels we also store the coefficients of the interpolant. After this, we compute the far-field expansions bottom-up. The reason for this is that we can use the results from lower levels to compute the expansions for higher levels. For each panel we have to compute the coefficients β_k, $1 \le k \le p$. If we do this bottom-up then every source x_j is considered only once so that all coefficients can be computed in $\mathcal{O}(N)$ time. But even if we do it top-down then the complexity is bounded by $\mathcal{O}(N \log N)$ and this is the time we need in any case to build the tree. Hence, building the structure takes $\mathcal{O}(N \log N)$ time.

The evaluation of the approximation $\widetilde{s}$ to the radial basis sum s can be done recursively, as will be described in Algorithm 13, and we are going to estimate the computational cost of a single evaluation of $\widetilde{s}$. The far-field part of the work involves the evaluation of only a constant number of far-field expansions on every level of the tree, each consisting of p coefficients. Hence the time needed to sum up the far-field expansion is $\mathcal{O}(p \log N)$. The direct part consists of a finite number (this number, as well as the finite number mentioned in the case of the far-field expansion, depends only upon the space dimension) of panels; thus, this takes only $\mathcal{O}(1)$ time, so that the time needed for one evaluation is $\mathcal{O}(\log N)$.

Algorithm 13 Recursive evaluation of a multipole method

Input: Panel T and evaluation point x.
Output: Approximate value $\widetilde{s}(x)$.

if *T is well separated from x* **then**
 Return $\widetilde{s}_T(x)$.
else
 if *T is a leaf* **then**
  Return $s_T(x)$.
 else
  Apply this algorithm to x and the children of T.
  Return the sum of their results.

Lemma 15.1 *The data structure for a fast multipole algorithm can be built in $\mathcal{O}(N \log N)$ time. One evaluation takes $\mathcal{O}(\log N)$ time.*

In several cases the evaluation time can be further reduced to $\mathcal{O}(1)$, but this will not bother us any further.

The rest of this section is devoted to giving far-field expansions for some of the most interesting kernels. Unfortunately, for optimal results this has to be done explicitly for each kernel and even for each space dimension. In many cases these far-field expansions are truncated Taylor or Laurent expansions. We will give an example for the case of thin-plate splines in two dimensions. After this we will show how the fast Gauss transform gives not only a multipole expansion for the Gaussian but also a general framework for deriving far-field expansions for other conditionally positive definite functions.

The idea of expanding thin-plate splines in two dimensions is based upon identifying $\mathbb{R}^2$ with $\mathbb{C}$. Hence, the Euclidean norm in $\mathbb{R}^2$ is the usual norm in $\mathbb{C}$ and for the moment we will denote it by $|\cdot|$. Let us use the notation $\phi_t(z) = |z-t|^2 \log|z-t|$ for $z, t \in \mathbb{C}$. The idea of the far-field expansion is based upon the simple equation $\phi_t(z) = |z-t|^2 \Re[\log(z-t)]$ and the Taylor expansion for $\log(1-z)$. This gives the Laurent-type expansion

$$\log(z-t) = \log\left[z\left(1-\frac{t}{z}\right)\right] = \log z + \log\left(1-\frac{t}{z}\right) = \log z - \sum_{k=1}^{\infty} \frac{1}{k}\frac{t^k}{z^k},$$

which holds whenever $|t| < |z|$. This proves the first part of

Proposition 15.2 *Let $z, t \in \mathbb{C}$ and $\phi_t(z) = |z-t|^2 \log|z-t|$. Then for $|z| > |t|$ we have*

$$\begin{aligned}\phi_t(z) &= \Re\left[|z-t|^2\left(\log z - \sum_{k=1}^{\infty}\frac{1}{k}\frac{t^k}{z^k}\right)\right] \\ &= \left[|z|^2 - 2\Re(\bar{t}z) + |t|^2\right]\log|z| + \Re\left(\sum_{k=0}^{\infty}(a_k\bar{z} + b_k)z^{-k}\right),\end{aligned}$$

where $b_k = -\bar{t}a_k$, $k \geq 0$, and $a_0 = -t$, $a_k = t^{k+1}/[k(k+1)]$, $k \geq 1$. Moreover, if the approximant $\widetilde{\phi}_t$ is defined for $p \geq 1$ by

$$\widetilde{\phi}_t(z) := \left[|z|^2 - 2\Re(\bar{t}z) + |t|^2\right]\log|z| + \Re\left(\sum_{k=0}^{p}(a_k\bar{z} + b_k)z^{-k}\right)$$

then the approximation error can be bounded for $t \neq 0$ by

$$|\phi_t(z) - \widetilde{\phi}_t(z)| \leq \frac{|t|^2}{(p+1)(p+2)}\frac{c+1}{c-1}c^{-p},$$

where $c = |z/t| > 1$.

Proof In the paragraph before this proposition we discussed the first representation for ϕ_t. To derive the second representation, we write

$$\begin{aligned}\phi_t(z) &= \Re\left[|z-t|^2\left(\log(z) - \sum_{k=1}^{\infty}\frac{1}{k}\frac{t^k}{z^k}\right)\right] \\ &= |z-t|^2\log|z| - \Re\left((\bar{z}-\bar{t})\sum_{k=1}^{\infty}\frac{1}{k}(z-t)\frac{t^k}{z^k}\right) \\ &= |z-t|^2\log|z| - \Re\left((\bar{z}-\bar{t})\sum_{k=1}^{\infty}\frac{1}{k}\left(t^k z^{-(k-1)} - t^{k+1}z^{-k}\right)\right) \\ &= |z-t|^2\log|z| + \Re\left(\sum_{k=0}^{\infty}(a_k\bar{z} + b_k)z^{-k}\right).\end{aligned}$$

To obtain the stated error bound, we note that

$$|a_k\bar{z}z^{-k}| \leq \frac{|t|^{k+1}}{k(k+1)|z|^{k-1}} = \frac{|t|^2}{k(k+1)}\left(\frac{1}{c}\right)^{k-1}$$

and

$$|b_k z^{-k}| \leq \frac{|t|^{k+2}}{k(k+1)|z|^k} = \frac{|t|^2}{k(k+1)}\left(\frac{1}{c}\right)^k.$$

This enables us to bound the error as follows:

$$
\begin{aligned}
|\phi_t(z) - \widetilde{\phi}_t(z)| &\le \sum_{k=p+1}^{\infty} \left(|a_k \overline{z} z^{-k}| + |b_k z^{-k}| \right) \\
&\le \frac{|t|^2}{(p+1)(p+2)} \sum_{k=p+1}^{\infty} \left(\frac{1}{c^{k-1}} + \frac{1}{c^k} \right) \\
&= \frac{|t|^2}{(p+1)(p+2)} \frac{c+1}{c-1} c^{-p}.
\end{aligned}
$$

□

For $t = 0$ the approximation $\widetilde{\phi}_t$ coincides with ϕ_t. Moreover, the error bound is increasing in $|t|$ and decreasing in $c > 1$.

Having the far-field expansion for ϕ_t we can proceed as in the model case to approximate the radial basis sum

$$
s(z) = \sum_{j=1}^{N} \alpha_j \phi_{z_j}(z) = \sum_{j=1}^{N} \alpha_j |z - z_j|^2 \log |z - z_j| \tag{15.3}
$$

with real coefficients $\{\alpha_j\}$, provided that all z_j are close to zero. We only have to replace ϕ_t by $\widetilde{\phi}_t$ and to exchange the order of summation to obtain the following result.

Corollary 15.3 *If ϕ_t in (15.3) is replaced by $\widetilde{\phi}_t$ then the resulting function $\widetilde{s}$ takes the form*

$$
\widetilde{s}(z) = \left[A|z|^2 - 2\Re(\overline{B} z) + C \right] \log |z| + \Re \left(\sum_{k=0}^{p} (A_k \overline{z} + B_k) z^{-k} \right),
$$

where the coefficients are defined by

$$
A := \sum_{j=1}^{N} \alpha_j, \qquad B := \sum_{j=1}^{N} \alpha_j z_j, \qquad C := \sum_{j=1}^{N} \alpha_j |z_j|^2
$$

and

$$
A_k := \begin{cases} -B & \text{for } k = 0, \\ \displaystyle\sum_{j=1}^{N} \frac{\alpha_j z_j^{k+1}}{k(k+1)} & \text{for } k \ge 1, \end{cases}
$$

$$
B_k := \begin{cases} C & \text{for } k = 0, \\ \displaystyle -\sum_{j=1}^{N} \frac{\alpha_j \overline{z}_j z_j^{k+1}}{k(k+1)} & \text{for } k \ge 1. \end{cases}
$$

Finally, if all the centers satisfy $|z_j| \le r$, we have the error bound

$$
|s(z) - \widetilde{s}(z)| \le \frac{\|\alpha\|_1 r^2}{(p+1)(p+2)} \frac{c+1}{c-1} c^{-p}
$$

with $c = |z|/r$.

This shows how a far-field expansion around zero can be derived for thin-plate splines in two dimensions. Unfortunately, in higher dimensions we can no longer use the trick of identifying $\mathbb{R}^2$ with $\mathbb{C}$. Nonetheless, it is possible to develop expansions for higher-dimensional spaces and other basis functions by employing Taylor and Laurent expansions. Here, however, we choose a different approach. We will give a general framework that works for all functions that are (conditionally) positive definite on every $\mathbb{R}^d$. The starting point is the fast Gauss transform, which is based on a Hermite expansion of the Gaussian kernel.

Definition 15.4 *The Hermite polynomials H_n are defined by*

$$H_n(t) := (-1)^n e^{t^2} \frac{d^n}{dt^n} e^{-t^2}.$$

The Hermite functions h_n are defined by $h_n(t) := e^{-t^2} H_n(t)$.

We need two results on Hermite polynomials. The first is the generating function for Hermite polynomials and the second is Cramer's inequality.

Lemma 15.5 *(1) The generating function for Hermite polynomials is given by*

$$e^{2ts-s^2} = \sum_{n=0}^{\infty} \frac{s^n}{n!} H_n(t), \qquad s, t \in \mathbb{R}.$$

(2) For every $n \in \mathbb{N}_0$ and every $t \in \mathbb{R}$ the Hermite polynomials can be bounded by

$$|H_n(t)| \le 2^{n/2} \sqrt{n!}\, e^{t^2}.$$

Proof The first property is easily shown if we regard $w(s,t) := e^{2st-s^2}$ as a function of the first variable. For complex s this is an entire function having Taylor expansion

$$w(s,t) = e^{2st-s^2} = \sum_{n=0}^{\infty} \frac{1}{n!} \left(\frac{\partial^n w}{\partial s^n} \right)_{s=0} s^n,$$

which immediately implies the stated representation since

$$\left(\frac{\partial^n w(srt)}{\partial s^n} \right)_{s=0} = e^{t^2} \left(\frac{\partial^n}{\partial s^n} e^{-(t-s)^2} \right)_{s=0} = (-1)^n e^{t^2} \left(\frac{\partial^n}{\partial u^n} e^{-u^2} \right)_{u=t} = H_n(t).$$

For the second property we start with the representation

$$e^{-t^2} = \frac{1}{2\sqrt{\pi}} \int_{-\infty}^{\infty} e^{-r^2/4} e^{irt}\, dr;$$

see Theorem 6.10. Differentiating under the integral sign leads to a first bound

$$\begin{aligned} |H_n(t)| &\le e^{t^2} \frac{1}{2\sqrt{\pi}} \int_{-\infty}^{\infty} e^{-r^2/4} |r|^n\, dr \\ &= e^{t^2} \frac{1}{\sqrt{\pi}} \int_0^{\infty} e^{-r^2/4} r^n\, dr \end{aligned}$$

$$= \frac{2^n}{\sqrt{\pi}} e^{t^2} \int_0^\infty e^{-s} s^{(n-1)/2} ds$$

$$= \frac{2^n}{\sqrt{\pi}} \Gamma\left(\frac{n+1}{2}\right) e^{t^2}.$$

To derive the stated result it remains to prove that

$$\Gamma\left(\frac{n+1}{2}\right) \le \sqrt{\pi} 2^{-n/2} \sqrt{n!}.$$

This is done as follows. Legendre's duplication formula from Proposition 5.2 gives

$$n! = \frac{2^n}{\sqrt{\pi}} \Gamma\left(\frac{n+1}{2}\right) \Gamma\left(\frac{n}{2}+1\right).$$

By induction we can easily establish that

$$\Gamma\left(\frac{n}{2}+1\right) \ge \frac{1}{\sqrt{\pi}} \Gamma\left(\frac{n+1}{2}\right),$$

so this finishes the proof. □

These two ingredients would allow us to develop a far-field expansion in the one-dimensional setting straight away. But since the multivariate case is no more difficult we immediately proceed to it by introducing the multivariate Hermite polynomials and functions. This is straightforward using tensor products. Hence for $\alpha \in \mathbb{N}_0^d$ and $x = (\mathrm{x}_1, \dots, \mathrm{x}_d)^T$ we set

$$H_\alpha(x) := \prod_{j=1}^{d} H_{\alpha_j}(\mathrm{x}_j), \qquad x \in \mathbb{R}^d,$$

and

$$h_\alpha(x) := \prod_{j=1}^{d} h_{\alpha_j}(\mathrm{x}_j) = e^{-\|x\|_2^2} H_\alpha(x), \qquad x \in \mathbb{R}^d.$$

With these definitions we can find the far-field expansion of the Gaussian around zero easily. Remember that $\alpha \le p$ means $\|\alpha\|_\infty \le p$ for all $\alpha \in \mathbb{N}_0^d$.

Theorem 15.6 *Let $\Phi(x) = e^{-\|x\|_2^2}$. Suppose that the sources x_j, $1 \le j \le N$, satisfy $\|x_j\|_\infty \le r/\sqrt{2}$, $r < 1$. Then the sum*

$$s(x) = \sum_{j=1}^{N} c_j \Phi(x - x_j)$$

can be expressed as

$$s(x) = \sum_{\alpha \in \mathbb{N}_0^d} A_\alpha h_\alpha(x)$$

with

$$A_\alpha = \frac{1}{\alpha!}\sum_{j=1}^{N} c_j x_j^\alpha.$$

Moreover, if $\widetilde{s}$ *denotes the approximation* $\widetilde{s}(x) = \sum_{\alpha \le p} A_\alpha h_\alpha(x)$ *for a* $p \in \mathbb{N}$ *then the approximation error can be bounded as follows:*

$$|s(x) - \widetilde{s}(x)| \le \frac{\|c\|_1}{(1-r)^d} \sum_{k=0}^{d-1} \binom{d}{k} \left(1 - r^{p+1}\right)^k \left(\frac{r^{p+1}}{\sqrt{(p+1)!}}\right)^{d-k}.$$

Proof The alternative representation for $s(x)$ is an immediate consequence of (1) in Lemma 15.5. By definition we have

$$e^{-(t-s)^2} = \sum_{n=0}^{\infty} \frac{s^n}{n!} h_n(t)$$

and the multivariate version gives

$$e^{-\|x-y\|_2^2} = \sum_{\alpha \in \mathbb{N}_0^d} \frac{1}{\alpha!} y^\alpha h_\alpha(x),$$

so that the stated identity follows by exchanging summations.

To derive the approximation error we first split the Hermite representation of $e^{-(t-s)^2}$ into two terms:

$$e^{-(t-s)^2} = \sum_{n=0}^{p} \frac{s!}{n!} h_n(t) + \sum_{n=p+1}^{\infty} \frac{s^n}{n!} h_n(t) =: A_p(t,s) + R_p(t,s).$$

Cramer's inequality gives us a bound on each term for $t \in \mathbb{R}$ and $|s|\sqrt{2} \le r < 1$:

$$|A_p(t,s)| \le \sum_{n=0}^{p} \frac{|s\sqrt{2}|^n}{\sqrt{n!}} \le \sum_{n=0}^{p} r^n = \frac{1 - r^{p+1}}{1-r},$$

$$|R_p(t,s)| \le \frac{1}{\sqrt{(p+1)!}} \sum_{n=p+1}^{\infty} |s\sqrt{2}|^n \le \frac{1}{\sqrt{(p+1)!}} r^{p+1} \sum_{n=0}^{\infty} r^n$$

$$= \frac{1}{\sqrt{(p+1)!}} \frac{r^{p+1}}{1-r}.$$

Since we have

$$e^{-\|x-y\|_2^2} = \prod_{j=1}^{d} \left[A_p(x_j, y_j) + R_p(x_j, y_j)\right]$$

for arbitrary $x, y \in \mathbb{R}^d$ we can conclude for $x \in \mathbb{R}^d$ and $\sqrt{2}\|y\|_\infty \le r < 1$ that

$$\left| e^{-\|x-y\|_2^2} - \sum_{\alpha \le p} \frac{1}{\alpha!} y^\alpha h_\alpha(x) \right| = \left| e^{-\|x-y\|_2^2} - \prod_{j=1}^{d} A_p(x_j, y_j) \right|$$
$$\le \frac{1}{(1-r)^d} \sum_{k=0}^{d-1} \binom{d}{k} (1 - r^{p+1})^k \left(\frac{r^{p+1}}{\sqrt{(p+1)!}} \right)^{d-k},$$

which in turn establishes the stated estimate on $s - \widetilde{s}$. □

This finishes our discussion of the far-field expansion of the Gaussian around zero. Shifting and scaling give us far-field expansions around other centers and with different stretch parameters.

Moreover, as mentioned before, we can use the far-field expansion of the Gaussian and the fast Gauss transform to derive far-field expansions for most of the basis functions in use. The key idea here is to exploit the fact that a radial function $\phi : [0, \infty) \to \mathbb{R}$ is often (conditionally) positive definite on every $\mathbb{R}^d$. First, if ϕ is positive definite on every $\mathbb{R}^d$ then we know from the theory of completely monotone functions that it has a representation

$$\phi(r) = \int_0^\infty e^{-r^2 t} d\alpha(t)$$

with a nonnegative Borel measure α, which has a Lebesgue density in all relevant cases. Hence, the radial sum becomes

$$\sum_{j=1}^{N} c_j \phi(\|x - x_j\|_2) = \int_0^\infty \sum_{j=1}^{N} c_j e^{-\|x-x_j\|_2^2 t} d\alpha(t)$$

and we can use a quadrature rule to boil this down to a finite sum of far-field expansions of the Gaussians.

Second, if ϕ is conditionally positive definite of order $m > 0$ then we have such a representation for the mth-order derivative of ϕ. Integrating this representation gives a representation of ϕ itself. Let us explain this in more detail using two examples.

Proposition 15.7 *The inverse multiquadric $\phi_{\mathrm{inv}}(r) = 1/\sqrt{1+r^2}$ has the representation*

$$\phi_{\mathrm{inv}}(r) = \frac{1}{\sqrt{\pi}} \int_0^\infty t^{-1/2} e^{-r^2 t} e^{-t} dt.$$

Hence, the associated radial sum can be written as

$$\sum_{j=1}^{N} c_j \phi_{\mathrm{inv}}(\|x - x_j\|_2) = \frac{1}{\sqrt{\pi}} \int_0^\infty \sum_{j=1}^{N} c_j e^{-\|x-x_j\|_2^2 t} t^{-1/2} e^{-t} dt.$$

The multiquadric $\phi_{\mathrm{mul}}(r) = \sqrt{1+r^2}$ has the representation

$$\phi_{\mathrm{mul}}(r) = 1 + \frac{1}{2\sqrt{\pi}} \int_0^\infty (1 - e^{-r^2 t}) t^{-3/2} e^{-t} dt.$$

The associated radial sum can be written as

$$\sum_{j=1}^{N} c_j \phi_{\mathrm{mul}}(\|x - x_j\|_2) = 1 + \int_0^\infty \left(1 - \sum_{j=1}^{N} c_j e^{-\|x-x_j\|_2^2 t}\right) t^{-3/2} e^{-t} dt,$$

provided that the coefficients satisfy $\sum c_j = 1$.

Proof The representation for the inverse multquadric has already been proven in a more general form in Theorem 7.15. The representation for the radial sum follows immediately. For the multiquadric note that

$$\frac{d}{dr}\phi_{\mathrm{mul}}(r) = r\phi_{\mathrm{inv}}(r).$$

Hence, using the integral form of ϕ_{inv} yields

$$\begin{aligned}\phi_{\mathrm{mul}}(r) &= 1 + \int_0^r s\phi_{\mathrm{inv}}(s)ds \\ &= 1 + \frac{1}{\sqrt{\pi}} \int_0^\infty \left(\int_0^r s e^{-s^2 t} ds \right) t^{-1/2} e^{-t} dt \\ &= 1 + \frac{1}{2\sqrt{\pi}} \int_0^\infty \frac{1 - e^{-r^2 t}}{t} t^{-1/2} e^{-t} dt,\end{aligned}$$

where the exchange of integration order can be justified by Fubini's theorem. Again, it is easy to see that the radial sum has the stated form, taking into account this time the side condition on the coefficients. □

Similar results can be derived for more general (inverse) multiquadrics and thin-plate splines and related functions.

Both representations of the radial sum in Proposition 15.7 make it necessary to discretize an integral of the form

$$\int_0^\infty f(t) t^a e^{-t} dt$$

with $a = -1/2$ and $a = -3/2$, respectively. Hence, a generalized Gauss–Laguerre quadrature rule is the preferred choice. From classical numerical analysis it is well known that the generalized Laguerre formula gives

$$\int_0^\infty t^a e^{-t} f(t) dt = \sum_{k=1}^{n} w_k f(t_k) + \frac{n!\,\Gamma(n+a+1)}{(2n)!} f^{(2n)}(\xi)$$

with $\xi \in (0, \infty)$. Here, the t_k are the zeros of the Laguerre polynomials $L_n^{(a)}$. These polynomials are the orthonormal polynomials with respect to the weight function $t \mapsto t^a e^{-t}$ on $[0, \infty)$. The weights are given by

$$w_k = \frac{\Gamma(n+a+1) t_k}{n! \left[L_{n+1}^{(a)}(t_k)\right]^2}.$$

This completes the general framework for deriving a far-field expansion for a radial function that is (conditionally) positive definite on every $\mathbb{R}^d$. For special functions there might be better expansions. In any case, the accuracy of the approximate radial sum depends on the ℓ_1-norm of the coefficient vector c. Unfortunately, we know that this can become arbitrarily high.

15.2 Approximation of Lagrange functions

The idea of this iterative method is that every interpolant $s_{f,X}$ to a function f based on a (conditionally) positive definite kernel $\Phi(\cdot,\cdot)$ and a set of centers $X = \{x_1, \dots, x_N\}$ can be written as

$$s_{f,X}(x) = \sum_{k=1}^{N} f_k u_k^*(x)$$

with the cardinal functions $\{u_k^*\}$ from Theorem 11.1 and $f_k = f(x_k)$, $1 \le k \le N$. The cardinal or Lagrange functions come from the space

$$\operatorname{span}\{\Phi(\cdot, x_k) \; : \; 1 \le k \le N\} + \mathcal{P}$$

and satisfy $u_k^*(x_j) = \delta_{jk}$, $1 \le j, k \le N$. They can be found by solving system (11.1). Now, the idea is to replace the cardinal functions by approximations $\widetilde{u}_k$ that are more easily computed and to form the approximation

$$\widetilde{s}_{f,X}(x) = \sum_{k=1}^{N} f_k \widetilde{u}_k(x).$$

A possible way to do this is to fix a nonnegative integer q that is substantially smaller than N. Then for every k we choose a set of exactly q indices $\mathcal{L}_k$ and force $\widetilde{u}_k$ to satisfy the Lagrange conditions $\widetilde{u}_k(x_j) = \delta_{jk}$ for $j \in \mathcal{L}_k$. Each of these new cardinal functions can be computed in constant time (if q is considered constant). But even if in many cases $\widetilde{u}_k$ is a good approximation to u_k^* this will not be true in general and is highly dependent on the choice of $\mathcal{L}_k$. Therefore an iteration on the residuals might be useful. A first possible iterative scheme is given by

$$\begin{aligned} s^{(1)} &= \sum_{k=1}^{N} f_k \widetilde{u}_k, \\ s^{(j+1)} &= s^{(j)} + \sum_{k=1}^{N} \left[f_k - s^{(j)}(x_k) \right] \widetilde{u}_k, \qquad j \ge 1. \end{aligned} \tag{15.4}$$

The iteration stops when all the residuals $f_k - s^{(j)}(x_k)$, $1 \le k \le N$, are sufficiently close to zero.

The iterative definition gives that the residuals of level $j+1$ are connected to the residuals of level j by

$$\begin{aligned} r_i^{(j+1)} &:= f_i - s^{(j+1)}(x_i) \\ &= f_i - s^{(j)}(x_i) - \sum_{k=1}^{N} \left[f_k - s^{(j)}(x_k) \right] \widetilde{u}_k(x_i) \\ &= r_i^{(j)} - \sum_{k=1}^{N} \widetilde{u}_k(x_i) r_k^{(j)}. \end{aligned}$$

Using the $N \times N$ identity matrix I and the matrix R defined by $R_{ik} = \widetilde{u}_k(x_i)$ this can be expressed as

$$r^{(j+1)} = (I - R) r^{(j)}.$$

Hence, the method converges if $I - R$ has norm less than one. To ensure this condition, special care has to be taken in choosing the sets $\mathcal{L}_k$. Before we discuss this in more detail, we will change the method slightly, in a way that allows us to prove convergence almost independently of the particular choice of the sets of centers.

For $k = 1, \ldots, N - q$ let us choose the set of indices such that $\mathcal{L}_k \subseteq \{k, k+1, \ldots, N\}$. Let us further assume that $k \in \mathcal{L}_k$ and that $\{x_j : j \in \mathcal{L}_k\}$ is $\mathcal{P}$-unisolvent. Finally, we assume that the remaining points, i.e. the points corresponding to the indices in $\mathcal{L}^* = \{N - q + 1, \ldots, N\}$, are also $\mathcal{P}$-unisolvent. This can be achieved by rearranging the points. The difference from the iteration (15.4) is that now we do not add all the local cardinal functions in one step to obtain the next iteration of our approximation. Instead, each iteration is divided into three steps. Let us suppose that $s^{(j)} =: s_0^{(j)}$ is the current approximation to $s_{f,X}$.

The first step consists of $N - q$ stages. For $k = 1, \ldots, N - q$, we want to add $\widetilde{u}_k$ times a certain factor to the current approximant, i.e. we want to form $s_k^{(j)} = s_{k-1}^{(j)} + \theta_k^{(j)} \widetilde{u}_k$. The value of $\theta_k^{(j)}$ will be chosen such that the error decreases in a certain way.

In the second step we interpolate in the remaining points. Hence we compute an interpolant $\sigma^{(j)}$ such that $\sigma^{(j)}(x_i) = f_i - s_{N-q}^{(j)}(x_i)$ for $i \in \mathcal{L}^*$. Then the new iteration is given by $s^{(j+1)} = s_{N-q}^{(j)} + \sigma^{(j)}$.

In the final step we update the residuals, i.e. we compute $r^{(j+1)} = [f_i - s^{(j+1)}(x_i)]_{1 \le i \le N}$ and stop when $\|r^{(j+1)}\|_\infty < \epsilon$ for a prescribed accuracy $\epsilon > 0$.

Our choice of $\sigma^{(j)}$ in the second step forces each iteration $s^{(j)}$ to interpolate the data in the remaining points, i.e. $s^{(j)}(x_i) = f_i$ for $N - q + 1 \le i \le N$ and all j.

As mentioned before, the coefficient $\theta_k^{(j)}$ should cause a decrease in the residual. It is easy to compute the optimal value for this coefficient if the residual is measured in the native space (semi-)norm.

Lemma 15.8 *Suppose that s and u are both functions from the native space $\mathcal{N}_\Phi(\Omega)$. Suppose further that u is not in the null space of the semi-norm. Then $\theta \mapsto |s - \theta u|^2_{\mathcal{N}_\Phi(\Omega)}$ attains its* *minimum for*

$$\theta = \frac{(s, u)_{\mathcal{N}_\Phi(\Omega)}}{|u|^2_{\mathcal{N}_\Phi(\Omega)}}.$$

Proof We have to minimize the parabola $\Theta(\theta) := |s - \theta u|^2_{\mathcal{N}_\Phi(\Omega)} = |s|^2_{\mathcal{N}_\Phi(\Omega)} - 2\theta(s, u)_{\mathcal{N}_\Phi(\Omega)} + \theta^2 |u|^2_{\mathcal{N}_\Phi(\Omega)}$. The minimum is found by solving $\Theta'(\theta) = 0$. □

We will apply this lemma to our situation by setting $s = s_{f,X} - s_{k-1}^{(j)}$ and $u = \widetilde{u}_k$. The numbers involved can be computed explicitly. Let us assume that $\widetilde{u}_k$ has the representation

$$\widetilde{u}_k = \sum_{j \in \mathcal{L}_k} \lambda_{kj} \Phi(\cdot, x_j) + p_k.$$

Then, elementary properties of the semi-inner product yield

$$(\widetilde{u}_k, \widetilde{u}_k)_{\mathcal{N}_\Phi(\Omega)} = \sum_{i \in \mathcal{L}_k} \lambda_{ki} \widetilde{u}_k(x_i) = \lambda_{kk}$$

and this is positive because $\{x_i : i \in \mathcal{L}_k\}$ was assumed to be $\mathcal{P}$-unisolvent. Moreover, we have by the same arguments

$$(s_{f,X} - s_{k-1}^{(j)}, \widetilde{u}_k)_{\mathcal{N}_\Phi(\Omega)} = \sum_{i \in \mathcal{L}_k} [f_i - s_{k-1}^{(j)}(x_i)] \lambda_{ki}.$$

Hence, if we choose

$$\theta_k^{(j)} = \frac{1}{\lambda_{kk}} \sum_{i \in \mathcal{L}_k} [f_i - s_{k-1}^{(j)}(x_i)] \lambda_{ki}$$

then Lemma 15.8 tells us that

$$\left| s_{f,X} - s_k^{(j)} \right|_{\mathcal{N}_\Phi(\Omega)} \le \left| s_{f,X} - s_{k-1}^{(j)} \right|_{\mathcal{N}_\Phi(\Omega)}$$

for $1 \le k \le N - q$.

After discussing the choice of $\theta_k^{(j)}$ we formulate the technique completely in Algorithm 14. We assume that the sets $\mathcal{L}_k$ are already chosen and that the coefficients λ_{kj} are known.

It is astonishing that even in this weak setting the method converges. But obviously the convergence rate depends on the choice of the sets $\mathcal{L}_k$. In general, one should not expect too much.

Theorem 15.9 *The sequence $s^{(j)}$ generated by Algorithm 14 converges in the native space (semi-)norm to the radial basis function interpolant $s_{f,X}$.*

Proof We already know that $|s_{f,X} - s_k^{(j)}|_{\mathcal{N}_\Phi(\Omega)} \le |s_{f,X} - s_{k-1}^{(j)}|_{\mathcal{N}_\Phi(\Omega)}$ for all $1 \le k \le N - q$. Moreover, $\sigma^{(j)}$ interpolates $s_{f,X} - s_{N-q}^{(j)}$ for the remaining points. Hence, by Theorem 13.1, we find also that

$$\left| s_{f,X} - s^{(j+1)} \right|_{\mathcal{N}_\Phi(\Omega)} = \left| s_{f,X} - s_{N-q}^{(j)} - \sigma^{(j)} \right|_{\mathcal{N}_\Phi(\Omega)} \le \left| s_{f,X} - s_{N-q}^{(j)} \right|_{\mathcal{N}_\Phi(\Omega)},$$

Algorithm 14 Iterative method based on approximate Lagrange functions

Set $j = 0$ and $s^{(j)} \equiv s_0^{(j)} = 0$.
Compute the residuals $r_i^{(j)} = f_i - s^{(j)}(x_i)$, $1 \le i \le N$, and the maximum error $e = \max_i |r_i^{(j)}|$.
while $e > \epsilon$ **do**
  for $1 \le k \le N - q$ **do**
    Replace $s_{k-1}^{(j)}$ by $s_k^{(j)}$ according to

$$s_k^{(j)} = s_{k-1}^{(j)} + \frac{1}{\lambda_{kk}} \sum_{i \in \mathcal{L}_k} \lambda_{ki} \big[f_i - s_{k-1}^{(j)}(x_i) \big] \widetilde{u}_k.$$

  Generate $s^{(j+1)} =: s_0^{(j+1)}$ by adding to $s_{N-q}^{(j)}$ the solution $\sigma^{(j)}$ of the interpolation problem

$$\sigma^{(j)}(x_i) = f_i - s_{N-q}^{(j)}(x_i), \qquad N - q + 1 \le i \le N.$$

  Compute the new residuals $r^{(j+1)}$ and the new error e.
  Set $j = j + 1$.

showing that the sequence $\{|s_{f,X} - s_k^{(j)}|_{\mathcal{N}_\Phi(\Omega)}\}_{j,k}$ is decreasing. Since obviously this sequence is also bounded from below, it has to converge and we want to show that $s^{(j)} = s_0^{(j)}$ converges to $s_{f,X}$.

All the functions $s_k^{(j)}$ and also $s_{f,X}$ come from the finite-dimensional subspace

$$V_X = \left\{ \sum_{k=1}^{N} \alpha_k \Phi(\cdot, x_k) : \alpha_k \in \mathbb{R} \text{ with } \sum_{k=1}^{N} \alpha_k p(x_k) = 0 \text{ for all } p \in \mathcal{P} \right\} + \mathcal{P}.$$

Hence, all norms on V_X are equivalent to the native space norm restricted to V_X and it suffices to show convergence to zero in any of these norms. For this purpose we choose

$$\|s\|_{L_\infty(X)} := \max\{|s(x_i)| : 1 \le i \le N\}, \qquad s \in V_X.$$

This means that we have to show that $s^{(j)}(x_i)$ tends to $s_{f,X}(x_i)$ for $1 \le i \le N$.

The choice of $\theta_k^{(j)}$ gives us

$$\begin{aligned} \left|s_{f,X} - s_k^{(j)}\right|^2_{\mathcal{N}_\Phi(\Omega)} &= \left|s_{f,X} - s_{k-1}^{(j)} - \theta_k^{(j)} \widetilde{u}_k\right|^2_{\mathcal{N}_\Phi(\Omega)} \\ &= \left|s_{f,X} - s_{k-1}^{(j)}\right|^2 - \frac{\left(s_{f,X} - s_{k-1}^{(j)}, \widetilde{u}_k\right)^2_{\mathcal{N}_\Phi(\Omega)}}{|\widetilde{u}_k|^2_{\mathcal{N}_\Phi(\Omega)}}. \end{aligned}$$

This means for every $1 \le k \le N-q$ that $\left(s_{f,X} - s_{k-1}^{(j)}, \widetilde{u}_k\right)_{\mathcal{N}_\Phi(\Omega)}$ tends to zero as $j \to \infty$. Hence for $k = 1$ we have $\left(s_{f,X} - s^{(j)}, \widetilde{u}_1\right) \to 0$ as $j \to \infty$. Next, for $k = 2$, we find

$$\begin{aligned} 0 &= \lim_{j\to\infty} \left(s_{f,X} - s_1^{(j)}, \widetilde{u}_2\right)_{\mathcal{N}_\Phi(\Omega)} \\ &= \lim_{j\to\infty} \left(s_{f,X} - s^{(j)} - \frac{\left(s_{f,X} - s^{(j)}, \widetilde{u}_1\right)_{\mathcal{N}_\Phi(\Omega)}}{|\widetilde{u}_1|^2_{\mathcal{N}_\Phi(\Omega)}} \widetilde{u}_1, \, \widetilde{u}_2 \right)_{\mathcal{N}_\Phi(\Omega)}. \end{aligned}$$

Since we already know that $\left(s_{f,X} - s^{(j)}, \widetilde{u}_1\right)_{\mathcal{N}_\Phi(\Omega)}$ tends to zero, this shows that $\left(s_{f,X} - s^{(j)}, \widetilde{u}_2\right)_{\mathcal{N}_\Phi(\Omega)}$ also tends to zero for $j \to \infty$. Proceeding in the same way, we can deduce that $\left(s_{f,X} - s^{(j)}, \widetilde{u}_k\right)$ tends to zero as j tends to infinity for every $1 \le k \le N-q$. However, these inner products can be written as

$$\left(s_{f,X} - s^{(j)}, \widetilde{u}_k\right)_{\mathcal{N}_\Phi(\Omega)} = \sum_{i \in \mathcal{L}_k} \lambda_{ki} \left[s_{f,X}(x_i) - s^{(j)}(x_i) \right],$$

so that we now have

$$\lim_{j\to\infty} \sum_{i\in\mathcal{L}_k} \lambda_{ki} \left[f_i - s^{(j)}(x_i) \right] = 0, \qquad 1 \le k \le N-q. \tag{15.5}$$

Finally we consider the equations (15.5) in reverse order, remembering that $s^{(j)}(x_i) = f_i$, $N-q+1 \le i \le N$ and that $\lambda_{kk} > 0$. By induction on k, we conclude that $s_{f,X}(x_k) - s^{(j)}(x_k)$ tends to zero as j tends to infinity for $k = N-q, N-q-1, \ldots, 1$. □

We will end this section by analyzing the computational complexity of this method and giving some hints about choosing the index sets $\mathcal{L}_k$. To start with the latter, we first sort our points in such a way that the last $Q = \dim \mathcal{P}$ points are $\mathcal{P}$-unisolvent. For example, in the case of thin-plate splines in $\mathbb{R}^2$, we would need three noncollinear points. Here, we could choose the two points having the greatest and least first coordinate. The third point would then be the point whose perpendicular distance from the infinite straight line through the other two points is greatest.

Every set $\mathcal{L}_k$ should contain these last Q points. Apart from these Q points, it will contain the data point x_k and the $q - Q - 1$ data points in $\{x_j : j > k\}$ nearest to x_k. With one of the data structures from the last chapter, we can expect that this will be done in $\mathcal{O}(N(q + \log N))$ time.

To analyze the complexity of the entire algorithm, we first have to say a few words about its realization. The main idea is to store and update only the coefficients of the current approximant

$$s(x) = \sum_{k=1}^{N} \lambda_k \Phi(x, x_k) + p(x) \tag{15.6}$$

and the current residual vector r. Moreover, we solve the local problems for the local cardinal functions in advance. This can be done in constant time for each cardinal function and hence in linear time for all of them.

Since we have an iterative algorithm we restrict ourselves here to estimating the complexity for one iteration. Each iteration consists of $\mathcal{O}(N)$ steps. In each step we have to update the coefficients λ_k, which can be done in $\mathcal{O}(1)$ time since at most Q coefficients are affected. More complicated is the update of the residual vector. An update of each entry would cost $\mathcal{O}(N)$ time, so that one iteration sums up to $\mathcal{O}(N^2)$ time. This is comparable to the time needed for one iteration in the CG-algorithm. However, a combination with the multipole expansions of the last section improves the results. If we present the approximant not in the form (15.6) but using a multipole data structure based upon a tree of depth $\mathcal{O}(\log N)$ then updating the coefficients will cost us $\mathcal{O}(\log N)$ since the constant number of points is contained in a constant number of leaf panels and thus only a constant number of coefficients in a constant number of panels on each level of the tree have to be updated. This gives us $\mathcal{O}(\log N)$ complexity, which is slightly worse than the complexity in the direct representation. But the efficiency in updating the residual vector is much greater. The argument just given shows that the new iteration is concentrated on $\mathcal{O}(\log N)$ panels. Hence, its value is zero apart from these panels. This shows that only $\mathcal{O}(\log N)$ residuals have to be updated. Hence, the time necessary for one complete iteration reduces to $\mathcal{O}(N \log N)$. Finally, we have to mention that an additional $\mathcal{O}(N \log N)$ time is necessary for initializing the data structure, but since this is a one-off problem, it is more than compensated by what we gain in the updating phases.

15.3 Alternating projections

In the last section we investigated one possible way of using the far-field expansion theory to derive an efficient method for solving the interpolation equations iteratively. Now we want to discuss another approach, which again exploits the fact that a radial basis function interpolant is also an orthogonal projection.

In the following, we will use only positive definite kernels. This is no restriction, since every conditionally positive definite kernel has an associated positive definite kernel by Theorem 10.20.

Remember that we have assigned to a set $X \subseteq \Omega$ of centers the subspace

$$V_X := \text{span}\{\Phi(\cdot, x) : x \in X\}$$

of the native space $\mathcal{N}_\Phi(\Omega)$. By Theorem 13.1 we know that the orthogonal projection $P : \mathcal{N}_\Phi(\Omega) \to V_X$ is given by the interpolation process, i.e. $Pf = s_{f,X}$. In addition, we need the following simple result on the orthogonal complement of V_X.

Lemma 15.10 *The orthogonal complement of V_X in $\mathcal{N}_\Phi(\Omega)$ is given by*

$$V_X^\perp = \{s \in \mathcal{N}_\Phi(\Omega) : s(x) = 0 \textit{ for all } x \in X\}.$$

Proof An arbitrary element of V_X has the form $g = \sum_{j=1}^{N} \alpha_j \Phi(\cdot, x_j)$. Using the reproducing-kernel property, the inner product can be computed via

$$(s, g)_{\mathcal{N}_\Phi(\Omega)} = \sum_{j=1}^{N} \alpha_j s(x_j).$$

Hence, on the one hand, if s vanishes on X then it is orthogonal to V_X. On the other hand, if s is orthogonal to V_X then we can choose $g = \Phi(\cdot, x_j)$, to see that $s(x_j) = 0$. □

The idea of the algorithm starts with a decomposition of the set X of data sites into subsets $X_1, \ldots, X_k$. These subsets need not be disjoint but their union must be X. The algorithm starts to interpolate on the first set X_1, forms the residual, interpolates this on X_2, and so on. After k steps, one cycle of the algorithm is complete and it starts over again. A more formal description is given in Algorithm 15. The algorithm terminates if a prescribed error bound $\epsilon > 0$ is achieved. Again, we use the notation $P_j f = s_{f,X_j}$ to emphasize the projectional character of the interpolant. We denote the sequence of residuals by f_j and the sequence of interpolants by s_j.

Algorithm 15 Alternating projection

Input: Set of centers X, accuracy $\epsilon > 0$, right-hand side $f|X$.
Output: Approximate interpolant.

Set $f_0 = f$, $s_0 = 0$.
for $n = 0, 1, 2, \ldots$ **do**
  for $r = 1, \ldots, k$ **do**
    $f_{nk+r} = f_{nk+r-1} - P_r f_{nk+r-1}$
    $s_{nk+r} = s_{nk+r-1} + P_r f_{nk+r-1}$
  if $\|f_{(n+1)k}\|_{L_\infty(X)} < \epsilon$ **then**
    stop.

The rest of this section is devoted to showing the convergence of this algorithm. The idea of alternating projections is quite old and convergence results can be established in a general form. To this end we have to introduce the angle between two subspaces of a Hilbert space.

Definition 15.11 *Let U_1 and U_2 be two closed subspaces of a Hilbert space H and denote their intersection by U. The angle α between U_1 and U_2 is defined by*

$$\cos \alpha = \sup\{(u, v) : u \in U_1 \cap U^\perp, v \in U_2 \cap U^\perp, \text{and } \|u\|, \|v\| \le 1\}.$$

With this definition we can prove the main result on alternating projections. This is the major part of proving the convergence of Algorithm 15.

Theorem 15.12 *Suppose that $Q_1, \dots, Q_k$ are orthogonal projections onto closed subspaces $U_1, \dots U_k$ of a Hilbert space H. Let $U = \cap_{j=1}^k U_j$ and let $Q : H \to U$ be the orthogonal projection onto U. Finally, let α_j be the angle between U_j and $A_j := \cap_{i=j+1}^k U_i$. Then, for any $f \in H$,*

$$\left\|(Q_k \cdots Q_1)^\ell f - Qf\right\|^2 \le c^{2\ell} \|f - Qf\|^2,$$

where

$$c^2 \le 1 - \prod_{j=1}^{k-1} \sin^2 \alpha_j.$$

Proof Let us set $R := Q_k \cdots Q_1$. $Qf \in U$ implies that $RQf = Qf$. Hence we have

$$(Q_k \cdots Q_1)^\ell f - Qf = R^\ell f - Qf = R^\ell(f - Qf).$$

Furthermore,

$$(Rv, u) = (Q_k Q_{k-1} \cdots Q_1 v, u) = (Q_{k-1} \cdots Q_1 v, u) = \cdots = (v, u)$$

for all $v \in H$ and $u \in U$ implies that $Rv \in U^\perp$ whenever $v \in U^\perp$. Since $f - Qf \in U^\perp$ it suffices to show that

$$\|Rv\|^2 \le \left(1 - \prod_{j=1}^{k-1} \sin^2 \alpha_j\right) \|v\|^2 \qquad \text{for all } v \in U^\perp.$$

This will be done by induction on k. If $k = 1$ then we have only one subspace and R is the projection onto $U = U_1$. Hence $v \in U^\perp$ implies that $Rv = 0$ and $\|Rv\| \le \|v\|$. Now, for the induction step we set $\widetilde{R} := Q_k \cdots Q_2$. We decompose an arbitrary $v \in U^\perp$ into $v = w + v_1$ with $w \in U_1$ and $v_1 \in U_1^\perp$. This gives in particular $Rv = Rw + Rv_1 = \widetilde{R}w$. Next we set $w = w_1 + w_2$ with $w_1 \in A_1 = U_2 \cap \cdots \cap U_k$ and $w_2 \in A_1^\perp$ and conclude that $\widetilde{R}w = \widetilde{R}w_1 + \widetilde{R}w_2 = w_1 + \widetilde{R}w_2$. Since the last two elements are orthogonal we obtain

$$\|\widetilde{R}w\|^2 = \|w_1\|^2 + \|\widetilde{R}w_2\|^2.$$

Moreover, induction gives

$$\|\widetilde{R}w_2\|^2 \le \left(1 - \prod_{j=2}^{k-1} \sin^2 \alpha_j\right) \|w_2\|^2.$$

From the last two equations and from $\|w\|^2 = \|w_1\|^2 + \|w_2\|^2$ we can conclude that

$$\|\widetilde{R}w\|^2 \le \left(1 - \prod_{j=2}^{k-1} \sin^2 \alpha_j\right) \|w\|^2 + \|w_1\|^2 \prod_{j=2}^{k-1} \sin^2 \alpha_j.$$

Now, w lies in U_1 and is orthogonal to $U = U_1 \cap A_1$ and w_1 lies in A_1 and is also orthogonal to U. Since the angle between U_1 and A_1 is α_1, we have

$$\|w_1\|^2 = (w_1, w) \le \cos \alpha_1 \|w_1\| \|w\|,$$

giving $\|w_1\| \le \cos\alpha_1 \|w\|$. Finally, $\|w\| \le \|v\|$ allows us to derive

$$\begin{aligned}\|Rv\|^2 &= \|\widetilde{R}w\|^2 \\ &\le \left(1 - \prod_{j=2}^{k-1} \sin^2\alpha_j\right)\|v\|^2 + \|v\|^2\cos^2\alpha_1 \prod_{j=2}^{k-1}\sin^2\alpha_j \\ &= \left(1 - \prod_{j=1}^{k-1}\sin^2\alpha_j\right)\|v\|^2.\end{aligned}$$

□

It is time to see how this general result applies to our specific situation. We need to choose $Q_j := I - P_j$, the projection from $\mathcal{N}_\Phi(\Omega)$ onto $V_{X_j}^\perp$. Then we can express complete cycles of Algorithm 15 for the residuals as

$$f^{(n)} := f_{nk} = (Q_k \dots Q_1)^n f,$$

which has the desired form. Moreover, since the residuals and approximants are connected by $f - s_j = f_j$, $j = nk + r$, convergence for the approximants follows by convergence of the residuals. It is not our goal to recover the function $f \in \mathcal{N}_\Phi(\Omega)$, however; this would be doomed to failure. Instead we hope that the sequence $\{s_j\}$ converges to the unique interpolant $s_{f,X}$.

Finally, we have to make sure that the constant c in Theorem 15.12 is strictly less than one, meaning that all angles α_j have to be positive. This is in general obviously wrong and we have to make an additional assumption on the choice of the subsets X_k of X. Fortunately this additional assumption turns out to be very modest.

Definition 15.13 *Let $X_1, \dots, X_k$ be nonempty subsets of $\mathbb{R}^d$ and let $Y_j = \cup_{i=j}^k X_i$, $1 \le j \le k$. The sets $X_1, \dots, X_k$ will be called weakly distinct if $X_j \ne Y_j$ and $Y_{j+1} \ne Y_j$ for each $1 \le j \le k-1$.*

If each set X_j contains a point that is not contained in any of the other sets, the collection of sets is obviously weakly distinct.

Lemma 15.14 *Let $X_1, \dots, X_k$ be finite subsets of $\mathbb{R}^d$ and let $Y_j = \cup_{i=j}^k X_i$. Then we have the following.*

(1) $V_{Y_j}^\perp = \cap_{i=j}^k V_{X_i}^\perp$ for $1 \le j \le k$.
(2) If the sets $X_1, \dots, X_k$ are weakly distinct then the subspaces $V_{X_j}^\perp \cap V_{Y_j}$ and $V_{Y_{j+1}}^\perp \cap V_{Y_j}$ contain nonzero functions for $1 \le j \le k-1$.

Proof First, $V_{X_j}^\perp$ contains exactly those functions that vanish on X_j. Hence, $\cap_{i=j}^k V_{X_i}^\perp$ consists exactly of those functions that vanish on $Y_j = \cup_{i=j}^k X_i$. But this subspace is $V_{Y_j}^\perp$. Now for the second claim. Since $X_1, \dots, X_k$ are weakly distinct, there has to be a point $z \in Y_j$

that is not contained in X_j. Let us consider an element of V_{Y_j} of the form

$$s = \sum_{x_j \in X_j} c_j \Phi(\cdot, x_j) + \Phi(\cdot, z).$$

This element is in $V_{X_j}^{\perp}$ if

$$\sum_{x_j \in X_j} c_j \Phi(x, x_j) = -\Phi(x, z) \qquad \text{for all } x \in X_j,$$

and this system has a unique solution because Φ is positive definite. Hence s is a nontrivial element in $V_{Y_j} \cap V_{X_j}^{\perp}$. The result for $V_{Y_{j+1}}^{\perp} \cap V_{Y_j}$ can be established in a similar way. □

With these preparatory steps we are able to show that the projection algorithm described earlier converges at least linearly.

Theorem 15.15 *Let $f \in \mathcal{N}_\Phi(\Omega)$ be given. Suppose that $X_1, \ldots, X_k$ are weakly distinct subsets of $\Omega \subseteq \mathbb{R}^d$. Set $Y_j = \cup_{i=j}^k X_i$, $1 \le j \le k$. Denote by $s^{(j)}$ the approximant after j completed cycles. Then there exists a constant $c \in (0, 1)$ such that*

$$\|s_{f,Y_1} - s^{(n)}\|_{\mathcal{N}_\Phi(\Omega)} \le c^n \|f\|_{\mathcal{N}_\Phi(\Omega)}.$$

Proof We start our iteration with $f_0 = s_{f,Y_1}$ instead of $f_0 = f$ and set $f^{(n)} = f_{nk}$. In the notation of Theorem 15.12 we then find that $U_j = V_{X_j}^{\perp}$ and

$$U = \bigcap_{j=1}^{k} V_{X_j}^{\perp} = V_{Y_1}^{\perp}$$

or, more generally, $A_j = V_{Y_{j+1}}^{\perp}$. This means in particular that the orthogonal projection of $s_{f,Y_1} \in V_{Y_1}$ to $U = V_{Y_1}^{\perp}$ is given by $Q s_{f,Y_1} = 0$. Hence we have

$$\begin{aligned} \|s_{f,Y_1} - s^{(n)}\|_{\mathcal{N}_\Phi(\Omega)} &= \|f^{(n)} - Q s_{f,Y_1}\|_{\mathcal{N}_\Phi(\Omega)} \\ &\le c^{2n} \|s_{f,Y_1} - Q s_{f,Y_1}\|_{\mathcal{N}_\Phi(\Omega)} \\ &\le c^{2n} \|f\|_{\mathcal{N}_\Phi(\Omega)} \end{aligned}$$

by Theorem 15.12, taking $\|s_{f,Y_1}\|_{\mathcal{N}_\Phi(\Omega)} \le \|f\|_{\mathcal{N}_\Phi(\Omega)}$ (cf. Theorem 10.26) into account also. It remains to show that c is strictly less than one or in other words that all angles α_j are different from zero. α_j is the angle between $U_j = V_{X_j}^{\perp}$ and $A_j = V_{Y_{j+1}}^{\perp}$ and from its definition we have to consider the intersection space

$$B_j := (U_j \cap A_j)^{\perp} = (V_{X_j}^{\perp} \cap V_{Y_{j+1}}^{\perp})^{\perp} = V_{Y_j}.$$

A zero angle means that $\cos \alpha_j = 1$ or

$$\begin{aligned} 1 = \sup \big\{ &(u, v)_{\mathcal{N}_\Phi(\Omega)} : u \in V_{X_j}^{\perp} \cap V_{Y_j}, v \in V_{Y_{j+1}}^{\perp} \cap V_{Y_j}, \\ &\|u\|_{\mathcal{N}_\Phi(\Omega)}, \|v\|_{\mathcal{N}_\Phi(\Omega)} \le 1 \big\}. \end{aligned}$$

Since we are working entirely in the space V_{Y_1} we can use compactness to find two elements $u^* \in V_{X_j}^{\perp} \cap V_{Y_j}$ and $v^* \in V_{Y_{j+1}}^{\perp} \cap V_{Y_j}$ both with norm one and $(u^*, v^*)_{\mathcal{N}_\Phi(\Omega)} = 1$. Equality in

Fig. 15.3 Alternating-projection reconstruction of a relief of Beethoven.

the Cauchy–Schwarz inequality is only achieved if the two elements are linearly dependent, in fact meaning here that $u^* = v^*$. Thus $u^* \in V_{X_j}^{\perp} \cap V_{Y_{j+1}}^{\perp} = V_{Y_j}^{\perp}$. However, u^* is also in V_{Y_j}. Hence it has to be zero, which contradicts the fact that it has norm unity. This means that all the angles have to be different from zero and therefore that $c < 1$. □

The analysis of the complexity of this algorithm is very similar to the analysis for the method of approximated Lagrange functions. We will leave the details to the reader.

Figure 15.3 shows some results of the alternating-projection method when applied to a data set representing a Beethoven relief. This moderate data set consists of about 30 000 regularly distributed points and will serve us as a model in the rest of this chapter. Here we have subdivided the data set into equal-sized slightly overlapping boxes, so that each box contains at most 200 centers. The first picture in Figure 15.3 shows the reconstruction after a half cycle. The next shows the reconstruction after one complete cycle. Here, the sum of residuals has already dropped from 57 708 to 1654.63 and the reconstruction already looks authentic. However, a side view, which is presented in the middle picture, reveals that it is still far from a satisfactory reconstruction. The last two pictures show the final result after 20 cycles. The error has dropped to below 10^{-6}.

15.4 Partition of unity

The idea of the partition-of-unity method is the following. We start with a mildly overlapping covering $\{\Omega_j\}_{j=1}^M$ of the region Ω; we will make the term "mildly overlapping" more precise in just a moment. For this covering we choose a partition of unity, i.e. a family of compactly supported, nonnegative, continuous functions $\{w_j\}$ with $\sum_{j=1}^M w_j = 1$ on Ω and $\operatorname{supp}(w_j) \subseteq \Omega_j$. Moreover, we choose for every cell Ω_j an approximation space V_j. Then a function f is approximated on each cell by a local approximant $s_j \in V_j$, and the local approximants are joined by forming

$$s_f = \sum_{j=1}^{M} s_j w_j. \tag{15.7}$$

To be more precise, we make the following definition.

Definition 15.16 *Let $\Omega \subseteq \mathbb{R}^d$ be a bounded set. Let $\{\Omega_j\}_{j=1}^M$ be an open and bounded covering of Ω. This means that all Ω_j are open and bounded and that Ω is contained in their union. Set $\delta_j = \operatorname{diam}(\Omega_j) = \sup_{x,y\in\Omega_j} \|x-y\|_2$. We call a family of nonnegative functions $\{w_j\}_{j=1}^M$ with $w_j \in C^k(\mathbb{R}^d)$ a k-stable partition of unity with respect to the covering $\{\Omega_j\}$ if*

(1) $\operatorname{supp}(w_j) \subseteq \Omega_j$,

(2) $\sum_{j=1}^M w_j \equiv 1$ *on* Ω,

(3) for every $\alpha \in \mathbb{N}_0^d$ with $|\alpha| \le k$ there exists a constant $C_\alpha > 0$ such that

$$\|D^\alpha w_j\|_{L_\infty(\Omega_j)} \le C_\alpha / \delta_j^{|\alpha|}$$

for all $1 \le j \le M$.

So far, we have not made further assumptions about the covering Ω, but for efficiency it is necessary that

$$I(x) := \#\{j : x \in \Omega_j\} \tag{15.8}$$

is uniformly bounded on Ω. Nonetheless, even without this assumption we can give a first convergence result.

Theorem 15.17 *Let $\Omega \subseteq \mathbb{R}^d$ be bounded. Suppose that $\{\Omega_j\}_{j=1}^M$ is an open and bounded covering of Ω and $\{w_j\}_{j=1}^M$ is a k-stable partition of unity. Let $f \in C^k(\Omega)$ be the function to be approximated. Let $V_j \subseteq C^k(\Omega_j)$ be given. Assume that the local approximation spaces V_j have the following approximation property. In each region $\Omega_j \cap \Omega$, f can be approximated by a function $s_j \in V_j$ such that*

$$\|D^\alpha f - D^\alpha s_j\|_{L_\infty(\Omega\cap\Omega_j)} \le \varepsilon_j(\alpha).$$

Then the function s_f from (15.7) is in $C^k(\Omega)$ and satisfies

$$|(D^\alpha f - D^\alpha s_f)(x)| \le \sum_{j\in I(x)} \sum_{\beta\le\alpha} \binom{\alpha}{\beta} C_{\alpha-\beta} \delta_j^{|\beta|-|\alpha|} \varepsilon_j(\beta), \qquad x \in \Omega, \tag{15.9}$$

for all $|\alpha| \le k$.

Proof The proof is straightforward. We simply use Leibniz' rule and the fact that the $\{w_j\}$ form a partition of unity to derive

$$\begin{aligned}(D^\alpha f - D^\alpha s_f)(x) &= D^\alpha \left(\sum_{j=1}^M w_j(x)[f(x) - s_j(x)] \right) \\ &= \sum_{j\in I(x)} \sum_{\beta\le\alpha} \binom{\alpha}{\beta} D^{\alpha-\beta} w_j(x) D^\beta (f - s_j)(x).\end{aligned}$$

The assumed bounds on the derivatives of w_j and on the derivatives of $D^\beta(f - s_j)$ now yield the stated result. □

It is our goal to use this general result in the context of radial basis functions. Hence, we start with a set of points $X = \{x_1, \dots, x_N\}$ and set $X_j = X \cap \Omega_j$. The local approximation spaces are given by

$$V_j := \text{span}\{\Phi(\cdot, x) : x \in X_j)\} + \mathcal{P},$$

where Φ is a conditionally positive definite kernel with respect to $\mathcal{P}$ on $\widetilde{\Omega} = \cup \Omega_j$. The local approximant s_j is then given by the interpolant s_{f,X_j}. It is interesting to see that the global approximant inherits the interpolation property of the local interpolants, i.e. $s_f(x_k) = f(x_k)$, $1 \le k \le N$; this is a consequence of the partition of unity:

$$s_f(x_k) = \sum_{j=1}^{N} s_{f,X_j}(x_k) w_j(x_k) = \sum_{j \in I(x_k)} f(x_k) w_j(x_k) = f(x_k).$$

In order to use the convergence result from Chapter 11 we have to make some more assumptions on the covering $\{\Omega_j\}$.

Definition 15.18 *Suppose that $\Omega \subseteq \mathbb{R}^d$ is bounded and $X = \{x_1, \dots, x_N\} \subseteq \Omega$ are given. An open and bounded covering $\{\Omega_j\}_{j=1}^N$ is called regular for (Ω, X) if the following properties are satisfied.*

(1) For every $x \in \Omega$ the number of cells Ω_j with $x \in \Omega_j$ is bounded by a global constant K, i.e. $\sum \chi_{\Omega_j}(x) \le K$ for all $x \in \Omega$.
(2) There exists a constant $C_r > C_2$ and an angle $\theta \in (0, \pi/2)$ such that every patch $\Omega_j \cap \Omega$ satisfies an interior cone condition with angle θ and radius $r = C_r h_{X,\Omega}$. Here C_2 is the constant from Theorem 3.14.

This looks technical at first sight. But a closer look at each property shows that these requirements are more or less natural. For example, the first property is necessary to make sure that the outer sum in (15.9) is actually a sum over at most K summands. Since K is independent of N, in contrast with M, which should be proportional to N, this is essential to avoid losing convergence orders. Moreover, it is crucial for an efficient evaluation of the global approximant that only a constant number of local approximants have to be evaluated. To this end, it also has to be possible to locate those K indices in constant time. The second property is important for employing our estimates on radial basis function interpolants.

We will state our convergence results only for (conditionally) positive definite functions with a finite number of continuous derivatives. But it should be clear from the proof that everything works for kernels also. Moreover, if Gaussians or multiquadrics are used then the convergence orders are again spectral.

Theorem 15.19 *Let $\Omega \subseteq \mathbb{R}^d$ be open and bounded and suppose that $X = \{x_1, \dots, x_N\} \subseteq \Omega$. Let $\Phi \in C_\nu^k(\mathbb{R}^d)$ be conditionally positive definite of order m. Let $\{\Omega_j\}$ be a regular covering for (Ω, X) and let $\{w_j\}$ be k-stable for $\{\Omega_j\}$. The error between $f \in \mathcal{N}_\Phi(\Omega)$ and* *its partition-of-unity interpolant* $s_f = \sum_j s_{f,X_j} w_j$ *is bounded as follows*:

$$|D^\alpha f(x) - D^\alpha s_f(x)| \le C h_{X,\Omega}^{(k+\nu)/2-|\alpha|} |f|_{\mathcal{N}_\Phi(\Omega)}$$

for all $x \in \Omega$ *and all* $|\alpha| \le k/2$.

Proof On the one hand, by Theorem 10.46 the function f has a norm-preserving extension $Ef \in \mathcal{N}_\Phi(\mathbb{R}^d)$. On the other hand, the restriction $f_j = Ef|(\Omega_j \cap \Omega)$ satisfies $|f_j|_{\mathcal{N}_\Phi(\Omega_j\cap\Omega)} \le |Ef|_{\mathcal{N}_\Phi(\mathbb{R}^d)} = |f|_{\mathcal{N}_\Phi(\Omega)}$ by Theorem 10.47. Hence, if we denote all these functions by f then we have again $|f|_{\mathcal{N}_\Phi(\Omega_j\cap\Omega)} \le |f|_{\mathcal{N}_\Phi(\Omega)}$.

Referring to the constants from Theorem 3.14 we can see that $C_r \ge C_2$ implies $h = h_{X,\Omega} \le C_r h/C_2$. Hence, by Remark 11.12 we can apply Theorem 11.11 to the local setting $\Omega_j \cap \Omega$, $X_j = \Omega_j \cap X$, and $h = h_{X,\Omega}$ to get

$$|D^\alpha(f - s_{f,X_j})(x)| \le C h^{(k+\nu)/2-|\alpha|} |f|_{\mathcal{N}_\Phi(\Omega)}, \qquad x \in \Omega_j \cap \Omega.$$

Furthermore, the constant C depends on $\Omega_j \cap \Omega$ only via θ, which is the same for all j.

To apply (15.9) we need two more ingredients. Since every patch Ω_j satisfies an interior cone condition with radius $C_r h_{X\Omega}$ we have $\delta_j = \operatorname{diam}(\Omega_j) \ge C_r h_{X,\Omega}$. Moreover, every $x \in \Omega$ is contained in at most K patches Ω_j. Hence, the error bound (15.9) leads to

$$\begin{aligned}
|D^\alpha(f - s_f)(x)| &\le \sum_{j\in I(x)} \sum_{\beta\le\alpha} \binom{\alpha}{\beta} C_{\alpha-\beta} \delta_j^{|\beta|-|\alpha|} \varepsilon_j(\beta) \\
&\le K \sum_{\beta\le\alpha} \binom{\alpha}{\beta} C_{\alpha-\beta} C_r^{|\beta|-|\alpha|} h_{X,\Omega}^{|\beta|-|\alpha|} C h_{X,\Omega}^{(k+\nu)/2-|\beta|} |f|_{\mathcal{N}_\Phi(\Omega)} \\
&= C h_{X,\Omega}^{(k+\nu)/2-|\alpha|} |f|_{\mathcal{N}_\Phi(\Omega)}
\end{aligned}$$

for all $x \in \Omega$ and all $|\alpha| \le k/2$. □

Next let us have a closer look at the computational complexity. Here, it is easily seen that it is not enough to require $\{\Omega_j\}$ to be regular for (Ω, X) to reduce the complexity. For example, if Ω itself satisfied a cone condition we could simply choose $\Omega_j = \Omega$, yielding a regular covering but also the global interpolation problem. Hence, we have to make sure that the cells are small and the covering is local.

Definition 15.20 *Suppose that* $\Omega \subseteq \mathbb{R}^d$ *is bounded and* $X = \{x_1, \dots, x_N\} \subseteq \Omega$ *is given. An open and bounded covering* $\{\Omega_j\}_{j=1}^N$ *is called local for* (Ω, X) *if there exists a constant* $C_{\text{loc}} > 0$ *such that* $\operatorname{diam}(\Omega_j) \le C_{\text{loc}} h_{X,\Omega}$.

If $\{\Omega_j\}$ is local and regular for (Ω, X) and X is quasi-uniform then we know by Corollary 14.2 that the number of centers from X in each Ω_j is bounded by a constant that is independent of the total number of centers in X. Moreover, for every Ω_j we can choose a ball B_j of radius $h_{X,\Omega}$ that is completely contained in $\Omega \cap \Omega_j$. This ball B_j can be contained in at most $K - 1$ other cells $\Omega \cap \Omega_k$. Hence, we can use our standard volume argument to conclude from $\cup_{j=1}^M B_j \subseteq \Omega$ that $M \le C h_{X,\Omega}^{-d} \le CN$.

Theorem 15.21 *Let $X = \{x_1, \ldots, x_N\} \subseteq \Omega$ be quasi-uniform and $\{\Omega_j\}$ a regular and local covering for (Ω, X). Suppose that $\{\Omega_j\}$ can be built in $\mathcal{O}(N)$ time and $\mathcal{O}(N)$ space and that for every $x \in \Omega$ at the most K patches Ω_j with $x \in \Omega_j$ can be reported in constant time. Then the partition-of-unity method based on radial basis functions can be implemented in $\mathcal{O}(N)$ space with $\mathcal{O}(N)$ time needed for the preprocessing step. Furthermore, each evaluation of the global interpolant needs $\mathcal{O}(1)$ time.*

Proof Using the fixed-grid strategy to build our data structure for X, we know that this can be done in $\mathcal{O}(N)$ time, and space. Since the number of centers in each patch is bounded by a constant, we need constant space and constant time for each patch to solve the local interpolation problems. Furthermore, the points in each patch can be reported in constant time. Since the number M of patches Ω_j is bounded by $\mathcal{O}(N)$ this adds up to $\mathcal{O}(N)$ space and time for solving all of them. By assumption we can determine $I(x) = \{j : x \in \Omega_j\}$ in constant time, and the cardinality of $I(x)$ is also bounded by a constant. Thus we have to add up a constant number of local interpolants to get the value of the global interpolant. This can be done in constant time. □

Unfortunately, some of the constants depend exponentially on the space dimension. Thus, in higher dimensions and also for non-quasi-uniform data sets, kd-trees and their extension, bd-trees, seem to have better properties than the fixed-grid strategy.

What about the stability of the process? Our results from Chapter 12 show that the condition number does not depend on the number of centers but on the separation distance. The separation distance of the local center sets X_j is in general of the same size as the separation distance of the whole set of centers X. So we run into the same problem. Fortunately, for basis functions with a finite number of continuous derivatives the problems seem to give much less difficulty than predicted. Moreover, if we are working with thin-plate splines or any other basis function that satisfies the conditions of Theorem 12.13 we know that in fact we could instead work with the kernel $\kappa(\cdot, \cdot)$. Then, by Corollary 12.14 the condition number of the local problems turns out to be constant, i.e. independent of the separation distance.

Figure 15.4 shows the reconstruction of the Beethoven relief using two different partition techniques in the partition-of-unity approach. The reconstruction on the left uses a regularly sized local covering. Owing to this local nature, the interpolant becomes zero when the

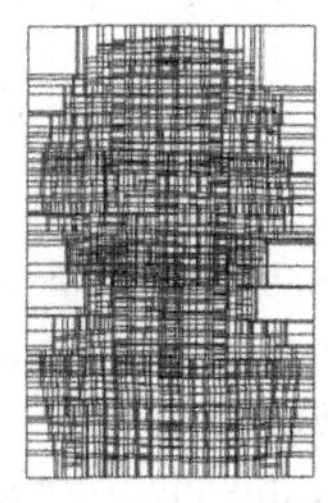

Fig. 15.4 Partition-of-unity reconstruction of the Beethoven relief.

evaluation points are far from the initial data. The second reconstruction (middle picture) is based on an overlapping *kd*-tree decomposition of the bounding box (see the right-hand picture of Figure 15.4). This reconstruction is better adapted to nonuniform data. For example, it was used for the reconstructions of the glacier and the dragon in the first chapter. However, the background in the middle picture of Figure 15.4 indicates that problems may come up when different parts of the surface are merged together.

15.5 Multilevel methods

All the numerical methods discussed so far are for globally supported radial basis functions. In particular all methods except the partition-of-unity technique need a far-field expansion. Even though compactly supported basis functions can be used in the same way as globally supported ones, this does not take the local character of these functions into account. Hence, it is time to discuss ideas for the efficient use of compactly supported functions.

As mentioned earlier, one possibility is to adjust the support radius as a function of the data density. Instead of using the function Φ, it is better to employ the function $\Phi_\delta = \Phi(\cdot/\delta)$ and to choose δ as a function of the fill distance $h_{X,\Omega}$. The choice $\delta = ch_{X,\Omega}$ would always lead to a sparse matrix and we know that its condition number would be independent of the number N of centers and of $h_{X,\Omega}$. Unfortunately, we also know, by the trade-off principle, that we cannot then expect convergence for $h_{X,\Omega} \to 0$. Hence, the right choice of the support radius is a delicate problem and we now want to describe a possible solution.

The idea of a multilevel method is again based an a decomposition of the set of centers X, but this time in a nested sequence of subsets,

$$X_1 \subseteq X_2 \subseteq \cdots \subseteq X_k = X. \tag{15.10}$$

If X is quasi-uniform then the subsets X_j should also be quasi-uniform. Moreover, they should satisfy $q_{X_{j+1}} \approx c_a q_{X_j}$ and $h_{X_{j+1},\Omega} \approx c_a h_{X_j,\Omega}$ with a fixed constant c_a. A good choice for c_a would be $1/2$.

Now, the multilevel method is simply one cycle of the alternating-projection method discussed earlier, but this time we use compactly supported basis functions with a different support radius at each level. We could even use different basis functions at the different levels. Hence, we need to formulate the algorithm more generally. For every $1 \le j \le k$ we choose a basis function Φ_j. As in Section 15.3 we denote the interpolation operator by

$$P_j f = \sum_{x_j \in X_j} c_{x_j}(f)\Phi_j(\cdot - x_j),$$

but using now the basis function Φ_j at level j. We will take Φ_j as $\Phi(\cdot/\delta_j)$ with a compactly supported basis function Φ and scaling parameter δ_j proportional to $h_{X_j,\Omega}$. A more thorough discussion will follow.

The idea behind this algorithm is that one starts with a very thin, widely spread, set of points and uses a smooth basis function to recover the global behavior of the function f. In

Fig. 15.5 Multilevel reconstruction of the Beethoven relief.

the next level a finer set of points is used and a less smooth function possibly with a smaller support is employed to resolve more details, and so on.

As was said before, the algorithm performs one cycle of the alternating-projection algorithm. This means that we set $f_0 = f$ and $s_0 = 0$ and compute

$$s_j = s_{j-1} + P_j f_{j-1},$$
$$f_j = f_{j-1} - P_j f_{j-1}$$

for $1 \le j \le k$.

Even though the multilevel algorithm resembles the alternating-projection algorithm, the idea behind it is completely different. The most obvious differences are that we use different basis functions at each level and that we perform only one cycle. The latter is reasonable since any further cycle would not change our approximant because the data sets are nested.

Figure 15.5 demonstrates the multilevel algorithm in the case of the Beethoven-relief data set. We set up five levels and the support radius on each level was chosen so that on each level the interpolation matrices had a bandwidth of roughly 70–80 points. The interpolation equations were solved using an unpreconditioned conjugate gradient method. The first row of Figure 15.5 shows the accumulated interpolants s_j while the second row shows the level interpolants $P_j f_{j-1}$.

Lemma 15.22 *The function s_k interpolates f on X.*

Proof We first remark that $f_j|X_j \equiv 0$ for $1 \le j \le k$ by definition. Next, we have again the relation $f - s_j = f_j$ for $1 \le j \le k$ between the interpolants and the residuals. Hence, $s_k|X = s_k|X_k = f|X_k - f_k|X_k = f|X_k = f|X$. □

Since f_k is zero on X_k it is also zero on $X_1 \subseteq X_k$, and thus the interpolant to f_k on X_1 is zero and would add nothing to the new approximant. A consequence of this is that we cannot use the technique of Section 15.3 to prove convergence. Another reason why we cannot use those ideas is that we are now using different basis functions, meaning that the interpolation operators are orthogonal projections in different Hilbert spaces with different norms. When scaling a fixed basis function the latter problem can be avoided if the scaling factor is absorbed into the function that is interpolated. But the first problem still remains.

Let us have a closer look at the idea of scaling a compactly supported function. We assume that the decomposition (15.10) of X is such that

$$h_{X_{j+1},\Omega} \approx c_a h_{X_j,\Omega} \qquad \text{and} \qquad q_{X_j} \approx h_{X_j,\Omega} \tag{15.11}$$

for all $1 \le j \le k$. This gives $h_{X_j,\Omega} \approx c_a^{j-1} h_{X_1,\Omega}$. Hence, if we choose as the support radius $\delta_j = c_a^{j-1}$ then the classical volume trick shows that the number of centers in the support of Φ_{δ_j} is bounded by a constant. Thus the interpolation matrix associated with the interpolation operator on level j has only a constant number of nonzero entries in each row, the constant being independent of the level j and of the current number of centers. This means that matrix–vector multiplication can be done in linear time. Moreover, we can interpret interpolation on X with Φ_δ as interpolation on $X/\delta = \{x_1/\delta, \dots, x_N/\delta\}$ with Φ. The separation distance of X/δ is obviously given by $q_{X/\delta} = q_X/\delta$, showing that the condition number of the interpolation matrices is uniformly bounded and independent of the level. Hence, on each level a conjugate gradient method would need only a constant number of iterations to converge.

Proposition 15.23 *Suppose that the decomposition (15.10) of the set $X = \{x_1, \dots, x_N\}$ satisfies (15.11) for $1 \le j \le k$. If we choose a compactly supported radial basis function and set the support on level j as $\delta_j = c_a^{j-1}$ then the multilevel algorithm needs $\mathcal{O}(kN)$ time to determine the interpolant s_k.*

Practical tests show that the first step is crucial for a successful interpolant. They also show that the number of levels k can in general be chosen much smaller than the number of data sites N, making the multilevel algorithm very efficient. Nonetheless, for proving convergence theoretically, k has to tend to infinity. By setting $Q_j := I - P_j$ again, we see that the error has the representation

$$f - s_k = f_k = Q_k Q_{k-1} \cdots Q_1 f.$$

Hence, the method converges if for example all the Q_j have norms less than a constant $c < 1$. Unfortunately, convergence has not yet been proven in the general case and we finish this section by discussing some ideas on how to prove it. But before we do this we want to point out that an improvement can be achieved if a preconditioning technique is used, meaning that Q_j is replaced by $Q_j := R_j(I - P_j)$ with a smoothing operator R_j. But so far it is not really clear which smoothing operator is a good choice.

As pointed out earlier, even though the multilevel algorithm resembles the alternating-projection algorithm we cannot use the ideas developed in Section 15.3 to prove convergence. But we can use the fact that $P_j f$ denotes the best approximation to f from $V_{X_j} = \text{span}\{\Phi_{\delta_j}(\cdot - x) : x \in X_j\}$ with respect to the native space norm $\|\cdot\|_{\mathcal{N}_{\Phi_{\delta_j}}(\Omega)}$. This gives for level j

$$\|f_{j-1} - P_j f_{j-1}\|^2_{\mathcal{N}_{\Phi_{\delta_j}}(\Omega)} + \|P_j f_{j-1}\|^2_{\mathcal{N}_{\Phi_{\delta_j}}(\Omega)} = \|f_{j-1}\|^2_{\mathcal{N}_{\Phi_{\delta_j}}(\Omega)}$$

or

$$\|f_j\|_{\mathcal{N}_{\Phi_{\delta_j}}(\Omega)} \le \|f_{j-1}\|_{\mathcal{N}_{\Phi_{\delta_j}}(\Omega)},$$

and we have to bound this expression by a constant, smaller than one, times $\|f_{j-1}\|_{\mathcal{N}_{\Phi_{\delta_{j-1}}}(\Omega)}$. Hence everything reduces to the problem of finding a constant $c < 1$ such that

$$\|f\|_{\mathcal{N}_{\Phi_\delta}(\Omega)} \le c\|f\|_{\mathcal{N}_\Phi(\Omega)} \tag{15.12}$$

for all $f \in \mathcal{N}_\Phi(\Omega)$ with a given $0 < \delta < 1$. Note that in most cases scaling does not change the native space, only the norm. Things are different if different basis functions are used.

The constant c in equation (15.12) is easily determined if Φ is a thin-plate spline $\Phi(x) = \|x\|_2^{2k} \log \|x\|_2$ or if $\Phi(x) = \|x\|_2^\beta$. For example, in the first case we know from Theorem 8.17 that Φ has a generalized Fourier transform $\widehat{\Phi}(\omega) = c_{d,k}\|\omega\|_2^{-d-2k}$, with a constant $c_{d,k}$ that does not play a role in what we intend to do now. Next, Theorem 10.21 tells us that the native space of Φ on $\mathbb{R}^d$ has the semi-norm

$$|f|^2_{\mathcal{N}_\Phi(\Omega)} = \frac{1}{(2\pi)^{d/2} c_{d,k}} \int_{\mathbb{R}^d} |\widehat{f}(\omega)|^2 \|\omega\|^{d+2k} d\omega$$

so that the semi-norm for the scaled basis function becomes

$$|f|^2_{\mathcal{N}_{\Phi_\delta}(\Omega)} = \frac{\delta^{-d}}{(2\pi)^{d/2} c_{d,k}} \int_{\mathbb{R}^d} |\widehat{f}(\omega)|^2 \|\delta\omega\|^{d+2k} d\omega = \delta^{2k} |f|^2_{\mathcal{N}_\Phi(\Omega)}.$$

Since $\delta < 1$, the constant c can be chosen to be δ^k. Using thin-plate splines in the multilevel algorithm is of course possible and we have just seen that it results in a convergent method. But the multilevel algorithm was tailored for compactly supported functions and hence it is not very surprising that in fact the numerical benefit of using thin-plate splines here is limited. Nonetheless, the success of proving convergence for thin-plate splines might encourage the reader to establish convergence results for other basis functions.

15.6 A greedy algorithm

All the numerical methods we have investigated up to now try to use all the data points. This is of course reasonable since the data often result from expensive experiments and it is hard to explain to the user why some of these measured values should not be used.

Nonetheless, building the approximant on only a subset of the centers might lead to a more efficient way of representing the unknown function. In that situation one has to decide which centers are to be used and to think about strategies to employ all the information given.

In this context one often faces so-called adaptive greedy algorithms. We do not want to discuss the general idea of an adaptive greedy method but we want to explain how it can work in the context of compactly supported radial basis functions. As for the alternating-projection and multilevel methods, the idea is to interpolate on subsets X_j of the original data sets X and to form residuals. But this time the data sets X_j are not chosen in advance. Instead, the choice of data set X_j depends on the residuum of the previous level. As well as the data set X_j, it is also possible to choose the basis function or at least its support radius differently.

In what follows we will use the notation of the previous sections. In particular, we will denote the interpolation operator based on X_j and on a general positive definite kernel Φ_j by P_j. The residuals are denoted by f_j and the accumulated approximants by s_j.

The following proposition is a rephrasing of the thoughts expressed at the end of the last section, but it is the crucial point in what we want to do in this section.

Proposition 15.24 *Suppose that the positive definite kernel Φ is used for all levels j, i.e. $\Phi_j = \Phi$. If there exists a constant $\gamma \in (0, 1)$ such that*

$$\|P_j f_{j-1}\|_{\mathcal{N}_\Phi(\Omega)} \geq \gamma \|f_{j-1}\|_{\mathcal{N}_\Phi(\Omega)}$$

for all $j \geq 1$ then s_j converges linearly to f in the native space.

Proof The proof simply follows from $f - s_j = f_j$ and

$$\begin{aligned}\|f_j\|^2_{\mathcal{N}_\Phi(\Omega)} &= \|f_{j-1} - P_j f_{j-1}\|^2_{\mathcal{N}_\Phi(\Omega)} \\ &= \|f_{j-1}\|^2_{\mathcal{N}_\Phi(\Omega)} - \|P_j f_{j-1}\|^2_{\mathcal{N}_\Phi(\Omega)} \\ &\leq (1 - \gamma^2)\|f_{j-1}\|^2_{\mathcal{N}_\Phi(\Omega)}.\end{aligned}$$

□

Since we want to choose the sets X_j at each step we have to do this in such a way that the condition $\|P_j f_{j-1}\|_{\mathcal{N}_\Phi(\Omega)} \geq \gamma \|f_{j-1}\|_{\mathcal{N}_\Phi(\Omega)}$ is satisfied. This might seem to be problematic at first sight, but things become easier if we concentrate on approximating $s_{f,X}$ instead of f itself. This means that we want to compute an approximation to an interpolant based on a large data set, and this is actually what we have already done several times before.

Thus, we will work in a finite-dimensional space, and we can replace the native space norm by a suitable norm.

Corollary 15.25 *Suppose that the subset X_j of X is chosen such that*

$$\|f_{j-1}\|_{L_p(X_j)} \geq \gamma \|f_{j-1}\|_{L_p(X)} \tag{15.13}$$

for all $j \geq 1$ and for a fixed $\gamma \in (0, 1)$. Then the sequence s_j converges to $s_{f,X}$ linearly in the native space and in the $L_p(X)$-norm.

Proof Since we are working in the finite-dimensional space V_X, all norms are equivalent. In particular, there exist two constants $0 < c_1 \le C_1$ depending on Φ, X, and p such that

$$c_1 \|g\|_{L_p(X)} \le \|g\|_{\mathcal{N}_\Phi(\Omega)} \le C_1 \|g\|_{L_p(X)}$$

holds for all $g \in V_X$. This relation gives us the bound

$$\begin{aligned} \|P_j f_{j-1}\|_{\mathcal{N}_\Phi(\Omega)} &\ge c_1 \|P_j f_{j-1}\|_{L_p(X)} \ge c_1 \|P_j f_{j-1}\|_{L_p(X_j)} \\ &= c_1 \|f_{j-1}\|_{L_p(X_j)} \ge c_1 \gamma \|f_{j-1}\|_{L_p(X)} \\ &\ge \frac{c_1 \gamma}{C_1} \|f_{j-1}\|_{\mathcal{N}_\Phi(\Omega)}. \end{aligned}$$

Since $c_1\gamma/C_1 \in (0, 1)$ we can apply Proposition 15.24 to get linear convergence in the native space norm and, because of norm equivalence, in the $L_p(X)$-norm also. □

Hence, if we fix a $\gamma \in (0, 1)$ and use the Chebychev norm on X we then have to choose X_j in such a way that the maximum of f_{j-1} on X_j is at least γ times the maximum of f_{j-1} on the whole set X. Unfortunately, convergence is not linear in this γ but rather in $\widetilde{\gamma} = \sqrt{1 - (c_1\gamma/C_1)^2}$, and the values of c_1 and C_1 can cause this constant to be rather close to one, making convergence quite slow, at least theoretically.

But if we use orthogonality again, we find

$$\begin{aligned} \|f\|^2_{\mathcal{N}_\Phi(\Omega)} - \|f_k\|^2_{\mathcal{N}_\Phi(\Omega)} &= \sum_{j=1}^{k} \left(\|f_{j-1}\|^2_{\mathcal{N}_\Phi(\Omega)} - \|f_j\|^2_{\mathcal{N}_\Phi(\Omega)} \right) \\ &= \sum_{j=1}^{k} \|P_j f_{j-1}\|^2_{\mathcal{N}_\Phi(\Omega)}, \end{aligned}$$

showing that $\|P_j f_{j-1}\|^2_{\mathcal{N}_\Phi(\Omega)}$ converges to zero at least as fast as $1/j$, and thus also $\|f_{j-1}\|^2_{\mathcal{N}_\Phi(\Omega)}$ and $\|f_{j-1}\|^2_{L_p(X)}$. This does not amount to much, but it does at least allow us to give more accurate estimates on the number of steps necessary to bring the approximation error below a certain level.

Since we have to solve a linear system at each step we want to keep the number of points in X_j small. It is interesting to see that Proposition 15.24 and its corollary give linear convergence, even if we choose $X_j = \{x_{k_j}\}$ to consist of only one point, if we simply make sure that

$$|f_{j-1}(x_{k_j})| \ge \gamma \|f_{j-1}\|_{L_\infty(X)}.$$

For example, we can choose x_{k_j} to be the point where f_{j-1} attains its maximum. Moreover, in this situation we do not have even a linear system to solve; the new interpolant is known explicitly from

$$P_j f_{j-1} = \Phi(\cdot, x_{k_j}) \frac{f_{j-1}(x_{k_j})}{\Phi(x_{k_j}, x_{k_j})}.$$

We summarize this simple approach in Algorithm 16. Our analysis made so far ensures linear convergence and we are going to investigate the complexity of Algorithm 16.

Algorithm 16 Greedy one-point algorithm

Input: Set X of centers, data values $f|X$, accuracy $\epsilon > 0$.
Output: Approximation $\widetilde{s}$.

Set $f_0 = f|X$, $s_0 = 0$, max $= \infty$, $j = 1$.
Choose $x_{k_1} \in X$.
while max $> \epsilon$ **do**
  Set $\beta = f_{j-1}(x_{k_j})/\Phi(x_{k_j}, x_{k_j})$.
  for $1 \le i \le N$ **do**
    Replace $f_{j-1}(x_i)$ by $f_j(x_i) = f_{j-1}(x_i) - \beta\Phi(x_i, x_{k_j})$.
    Replace s_{j-1} by $s_j = s_{j-1} + \beta\Phi(\cdot, x_{k_j})$.
    Keep track of the maximum max of the new residual f_j and the data site $x_{k_{j+1}}$ where it is attained.
  Replace j by $j + 1$.

It is obvious that one iteration of the "while" loop takes $\mathcal{O}(N)$ time if no additional information such as compact support or a far-field expansion is used. With this information, however, we can reduce the time needed to $\mathcal{O}(1)$ or $\mathcal{O}(\log N)$, making the algorithm very fast. Examples show that it produces reasonable approximations even when the number of points is much less than the initial number N. Hence, the algorithm can also be used for compressing data.

We end this section with a short discussion on how additional benefit can be drawn from using a compactly supported basis function with different support radii.

To this end we fix a radial function $\Phi = \phi(\|\cdot\|_2)$ having support in the unit ball and scale it via $\Phi_\delta = \Phi(\cdot/\delta)$. Next we fix certain constants $\alpha, \epsilon > 0$ for recording the accuracy, further constants $0 < \gamma < \beta < 1$ for influencing the support radius, and a constant $\sigma > 0$ for counting points. Finally, we fix a discrete norm on X.

In what follows, a successful try in K steps is defined as a run of K steps of Algorithm 16 at a fixed scale δ such that the discrete norm is reduced by a factor of at least α.

The new algorithm can be described as follows. There is an outer loop that runs over successful tries until the norm of the residuals falls under ϵ. The middle loop uses larger and larger numbers of iterations $K = \sigma, \sigma^2, \ldots$ and the inner loop uses support radii $c, c\beta, c\beta^2, \ldots, c\gamma$ until a successful try is found.

Since we know that Algorithm 16 converges, a sufficiently large number K will lead to a successful try. Hence the new algorithm has to converge. Note that it aims to use as few points as possible in each successful try. Moreover, it uses the largest possible support radius.

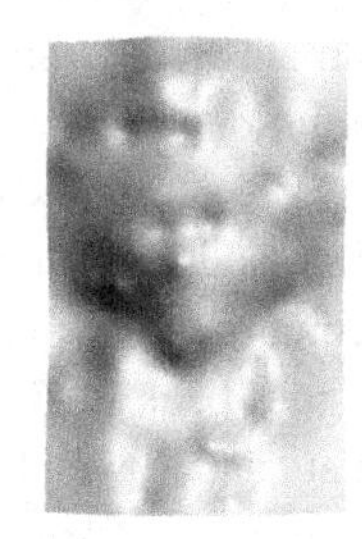

Fig. 15.6 Greedy reconstruction of the Beethoven relief.

Using a large initial radius c and small values for $1-\beta$ and $1-\sigma$ leads to a time-consuming optimization, which tries to reconstruct the data with as few centers as possible. Hence, to speed up the procedure certain improvements have to be implemented. Numerical tests show that often the sequence of support radii is decreasing while at the same time the number of necessary iterations is increasing. Hence, instead of starting with the initial values for each middle and inner loop one could start the new outer loop with the successful values for c and K from the previous step.

Figure 15.6 shows some results for the Beethoven data set. The initial support radius was the diameter of the data set. From left to right we show reconstructions using 10, 100, 1000, and 10 000 points. The final picture shows the distribution of the 10 000 points employed.

15.7 Concluding remarks

The right choice of basis function depends mainly on additional information on the target function, such as its smoothness. The right choice of reconstruction method depends also upon the application. For example, if on the one hand exact interpolation is necessary then any method based on a far-field expansion is of only limited use, since such a method has only an approximate character in the far-field. On the other hand, if the data set has holes which have to be filled (e.g. for mesh repair) then any purely local method would be the wrong choice; even the partition-of-unity approach would have to be handled with care here. A global method such as the approximate-Lagrangian method, the alternating-projection method, or the multilevel method would be more suitable in this case. In all other cases, it seems that, particularly for very large data sets, the partition-of-unity method is fastest.

15.8 Notes and comments

Fast multipole methods have been promulgated mainly by Beatson and various collaborators; see for example [11–14, 16, 17]. They are crucial for almost every efficient algorithm employing globally supported basis functions. The representation given here is based particularly upon [13, 16]. The idea of using the fast Gauss transform for other basis functions comes from Roussos [158], whereas the fast Gauss transform itself was devised by

Greengard and Strain [74] in 1991. Unfortunately the estimate on the truncation error given in that paper is erroneous, as was recently pointed out by Roussos [158] and Baxter and Roussos [10]. In Theorem 15.6 we give the corrected version. The actual inequality of Cramer can be stated in a stronger version; see for example Hille [83] and, for a thorough discussion, Roussos [158]. However, the stronger version has not been necessary in our analysis.

The method of *approximation of Lagrange functions* goes back to Beatson and Powell [18]. The theoretical background was provided later, by Faul and Powell [59] and Faul [58].

The *alternating-projection method* given in the third section is based on a general result by Smith *et al.* [178], given here in Theorem 15.12. The adaptation to the radial basis function setting comes from Beatson *et al.* [15]. They also call their method a *domain decomposition method*, even if actually not a domain but the set of centers is decomposed.

The *partition-of-unity* idea is also quite old. Possibly Maude [118] was the first to use this idea, in 1973, in the context of interpolation, at least in the univariate case. Franke [63] extended the method in 1977 to the multivariate case. He also tested different weight functions and different local approximation processes. One of these local approximation schemes was thin-plate spline interpolation (see [64]). Convergence investigation for the general class of radial basis functions was done by the present author in 2002 [198]. Partition-of-unity methods have recently gained attention again, in the context of meshless methods for partial differential equations; see Babuska and Melenk [7].

One can find the idea of the *multilevel method* in Schaback's paper [164] but it became publicly known through the work of Floater and Iske [60].

Finally, the *greedy algorithm* of the last section was initiated by Schaback and Wendland in [169, 170].

16

Generalized interpolation

Up to now we have dealt only with the problem of recovering an unknown function from certain known function values. But sometimes it might be desirable to recover the function also from other types of data. For example, the function's derivatives might be known at certain points, but not the function itself. This becomes interesting if partial differential equations are considered.

In this chapter we deal with a more general problem than those we have discussed so far. Our approach includes in particular collocation and Galerkin methods for solving partial differential equations. But the methods we will derive are at the present time only able to compete with classical methods to a certain extent. In any case, whenever large data sets are considered one has to combine the methods introduced below with the fast-evaluation ideas of Chapter 15.

In this sense, this chapter should be seen as a unified introduction to a general class of recovery problems.

16.1 Optimal recovery in Hilbert spaces

We start by generalizing results from Chapters 11 and 13. In this section we restrict ourselves for simplicity to the Hilbert space setting, even though everything works in the case of semi-Hilbert function spaces also.

Let H be a Hilbert space and denote its dual by H^*. Suppose that $\Lambda = \{\lambda_1, \dots, \lambda_N\} \subseteq H^*$ is a set of linearly independent functionals on H and that $f_1, \dots, f_N \in \mathbb{R}$ are certain given values. Then a generalized recovery problem would seek to find a function $s \in H$ such that $\lambda_j(s) = f_j$, $1 \le j \le N$. We will call s a *generalized interpolant*.

The optimal-recovery problem searches for the norm-minimal generalized interpolant, i.e. we have to find $s^* \in H$ such that

$$\|s^*\| = \min\{\|s\| : s \in H, \lambda_j(s) = f_j, 1 \le j \le N\}. \tag{16.1}$$

Obviously, this problem is a straightforward generalization of the recovery problem dealt with in Theorem 13.2. There, the solution was given by the (radial) basis function interpolant. Here, the solution can also be given explicitly. To this end let us recall Riesz' representation theorem, which allows us to represent each functional λ_j by a unique element $v_j \in H$ via $\lambda_j = (\cdot, v_j)$.

Theorem 16.1 *Let H be a Hilbert space, $\lambda_1, \dots, \lambda_N \in H^*$ linearly independent functionals with Riesz representers v_j, $1, \le j \le N$, and let $f_1, \dots, f_N \in \mathbb{R}$ be given. The unique solution of (16.1) is given by*

$$s^* = \sum_{j=1}^{N} \alpha_j v_j, \tag{16.2}$$

where the coefficients $\{\alpha_j\}$ are determined by the interpolation conditions

$$\lambda_j(s^*) = f_j, \qquad 1 \le j \le N. \tag{16.3}$$

Proof The proof is very similar to the proof of Theorem 13.2. First of all note that s^* from (16.2) is well defined, by the conditions (16.3). Since the λ_j are supposed to be linearly independent so also are the v_j, showing that the interpolation matrix with entries $(\lambda_i(v_j)) = ((v_i, v_j)) = ((\lambda_i, \lambda_j)_{H^*})$ is positive definite. Next, s^* thus defined is indeed a solution of the optimal-recovery problem (16.1). To see this, assume that $v \in H$ satisfies $\lambda_j(v) = 0$, $1 \le j \le N$, which implies

$$(s^*, v) = \sum_{j=1}^{N} \alpha_j (v, v_j) = \sum_{j=1}^{N} \alpha_j \lambda_j(v) = 0.$$

This shows that

$$\|s^*\|^2 = (s^*, s^* - s + s) = (s^*, s) \le \|s^*\| \, \|s\|$$

for every $s \in H$ with $\lambda_j(s) = f_j$, $1 \le j \le N$.

The solution of our optimal-recovery problem is unique by a classical argument from approximation theory. If $\widetilde{s}$ is any solution of (16.1) then necessarily $(\widetilde{s}, s) = 0$ for all $s \in H$ with $\lambda_j(s) = 0$ for $1 \le j \le N$; otherwise we could form $t = \widetilde{s} + \alpha s$ and would see that $\|t\|^2 = \|\widetilde{s}\|^2 + \alpha^2\|s\|^2 + 2\alpha(\widetilde{s}, s) < \|\widetilde{s}\|^2$ for an appropriate choice of α. Thus, if s_1 and s_2 are both solutions to the recovery problem then we have $(s_1, s_1 - s_2) = 0$ and $(s_2, s_1 - s_2) = 0$, showing that $\|s_1 - s_2\| = 0$ or $s_1 = s_2$. □

Now let us assume that the data values $\{f_j\}$ come from an unknown function $f \in H$, i.e. $f_j = \lambda_j(f)$, $1 \le j \le N$ (it might also be possible to assume that other functionals define the values, i.e. $f_j = \mu_j(f)$, but we do not want to pursue this any further).

Corollary 16.2 *If in addition $f_j = \lambda_j(f)$, $1 \le j \le N$, for an $f \in H$, then s^* is the best approximation from $V = \operatorname{span}\{v_1, \dots, v_N\}$ to f, i.e.*

$$\|f - s^*\| = \min\{\|f - s\| : s \in V\}.$$

Proof This follows directly from $0 = \lambda_j(f) - \lambda_j(s^*) = (f - s^*, v_j)$. □

So far, we have restated two of the three optimal properties of the (radial) basis function interpolant in this more general setting. The third was a reformulation of error estimates with power functions or, to be more precise, of the optimality of the power function and we will generalize this reformulation now. It answers the question how well the optimal recovery s^* matches the initial function f.

Theorem 16.3 *Set* $\Lambda := \operatorname{span}\{\lambda_1, \dots, \lambda_N\}$ *and* $V := \operatorname{span}\{v_1, \dots, v_N\}$. *Then for every linear and continuous functional* $\lambda : H \to \mathbb{R}$ *the estimate*

$$|\lambda(f - s^*)| \le \inf_{\mu \in \Lambda} \|\lambda - \mu\|_{H^*} \inf_{s \in V} \|f - s\| \tag{16.4}$$

is satisfied.

Proof Since every λ_j obviously satisfies $\lambda_j(f - s^*) = 0$ so does every $\mu \in \Lambda$. Thus

$$|\lambda(f - s^*)| = |(\lambda - \mu)(f - s^*)| \le \|\lambda - \mu\|_{H^*} \|f - s^*\|.$$

The rest follows from Corollary 16.2. □

It is interesting to analyze the term on the right-hand side of (16.4). The first factor,

$$P_\Lambda(\lambda) = \inf_{\mu \in \Lambda} \|\lambda - \mu\|_{H^*},$$

is a generalized power function that describes how well the evaluation functional λ can be approximated by the given functionals from Λ. Furthermore, the second factor in (16.4) describes how well the unknown function f can be approximated by the functions from V, and it is obviously bounded by $\|f\|$.

We end this general section by showing that in the case of interpolation by positive definite kernels we have done nothing new.

Suppose that $\Phi \in C(\Omega \times \Omega)$ is a positive definite kernel. For the set of distinct points $X = \{x_1, \dots, x_N\} \subseteq \Omega$ let $\Lambda = \{\delta_{x_j} : x_j \in X\}$ and for $x \in \Omega$ let $\lambda = \delta_x$. Finally let H be the native Hilbert space $\mathcal{N}_\Phi(\Omega)$. Since this space is a reproducing-kernel Hilbert space, the functionals δ_{x_j} have the Riesz representer $\Phi(\cdot, x_j)$. Thus the optimal-recovery problems and their solutions from Theorems 13.2 and 16.1, respectively, coincide. Moreover, since any $\mu \in \Lambda$ has a representation $\mu = \sum u_j \delta_{x_j}$, the new power function

$$P_\Lambda(\lambda) = \min_{u \in \mathbb{R}^N} \left\| \delta_x - \sum_{j=1}^N u_j \delta_{x_j} \right\|_{H^*} = \min_{u \in \mathbb{R}^N} \left\| \Phi(\cdot, x) - \sum_{j=1}^N u_j \Phi(\cdot, x_j) \right\|_{\mathcal{N}_\Phi(\Omega)}$$

coincides with the old one, $P_{\Phi,X}(x)$, by Section 11.1.

Let us point out once more what we have to do to recover a function f that is known only from $\lambda_j(f)$, $1 \le j \le N$, in the case where we are given a positive definite kernel $\Phi \in C(\Omega \times \Omega)$ and where $\lambda_j \in \mathcal{N}_\Phi(\Omega)^*$. We start by assuming an interpolant of the form

$$s_\Lambda = \sum_{j=1}^N \alpha_j \lambda_j^y \Phi(\cdot, y).$$

Then, the interpolation conditions $\lambda_k(s_\Lambda) = f_k$, $1 \le k \le N$, lead to the interpolation matrix

$$A_{\Phi,\Lambda} = (\lambda_k^x \lambda_j^y \Phi(x, y))_{1\le j,k\le N},$$

which is symmetric and positive definite whenever the functionals $\{\lambda_j\}$ are linearly independent over $\mathcal{N}_\Phi(\Omega)$, because

$$\lambda_k^x \lambda_j^y \Phi(x, y) = (\lambda_k^x \Phi(\cdot, x), \lambda_j^y \Phi(\cdot, y))_{\mathcal{N}_\Phi(\Omega)}$$

holds (see Theorem 16.7).

16.2 Hermite–Birkhoff interpolation

In this section we apply the general results of the last section to the specific situation where we are reconstructing a function from Hermite–Birkhoff data.

To be more precise, suppose that $\alpha^{(1)}, \dots, \alpha^{(N)} \in \mathbb{N}_0^d$ are (not necessarily different) multi-indices with $|\alpha^{(j)}| \le k$. Suppose further that we are given certain points $x_1, \dots, x_N \in \Omega$ from an open set $\Omega \subseteq \mathbb{R}^d$. Then we form the functionals $\lambda_j = \delta_{x_j} \circ D^{\alpha^{(j)}}$, i.e. $\lambda_j(f) = D^{\alpha^{(j)}} f(x_j)$, $1 \le j \le N$. To make these functionals pairwise different we assume that for two different indices $j \ne \ell$ we have either $x_j \ne x_\ell$ or $\alpha^{(j)} \ne \alpha^{(\ell)}$.

If $\Phi \in C^{2k}(\Omega \times \Omega)$ is a positive definite kernel on Ω then we know from Theorem 10.45 that $\lambda_j \in \mathcal{N}_\Phi(\Omega)^*$ and that λ_j has the Riesz representer

$$v_j = D_2^{\alpha^{(j)}} \Phi(\cdot, x_j), \qquad 1 \le j \le N,$$

where the additional index 2 on the D-operator again denotes differentiation with respect to the second argument. Hence, our interpolant takes the form

$$s = \sum_{j=1}^N c_j D_2^{\alpha^{(j)}} \Phi(\cdot, x_j) \tag{16.5}$$

and the interpolation matrix has entries of the form

$$D_1^{\alpha^{(\ell)}} D_2^{\alpha^{(j)}} \Phi(x_\ell, x_j). \tag{16.6}$$

This matrix is invertible whenever the functionals are linearly independent, and this depends on how "rich" the native space of the underlying kernel is. Fortunately, all relevant basis functions have a sufficiently rich native space. In accordance to the philosophy of this chapter we give details only for positive definite functions but it should become apparent that the next theorem can also be adapted to the situation of conditionally positive definite functions.

Theorem 16.4 *Suppose that $\Phi \in L_1(\mathbb{R}^d) \cap C^{2k}(\mathbb{R}^d)$ is positive definite. If the functionals $\lambda_j := \delta_{x_j} \circ D^{\alpha^{(j)}}$, $1 \le j \le N$, with $|\alpha^{(j)}| \le k$ are pairwise distinct, meaning that $\alpha^{(j)} \ne \alpha^{(\ell)}$ if $x_j = x_\ell$ for two different $j \ne \ell$, then they are also linearly independent over $\mathcal{N}_\Phi(\mathbb{R}^d)$.*

Proof Suppose we have real numbers c_j such that $\sum_{j=1}^N c_j\lambda_j = 0$ on $\mathcal{N}_\Phi(\mathbb{R}^d)$. This means in particular that

$$0 = \left\| \sum_{j=1}^N c_j\lambda_j \right\|_{\mathcal{N}_\Phi(\mathbb{R}^d)^*} = \left\| \sum_{j=1}^N c_j\lambda_j^y \Phi(\cdot - y) \right\|_{\mathcal{N}_\Phi(\mathbb{R}^d)}.$$

By Theorem 10.12 we know that we can evaluate the last norm via Fourier transforms. To this end we compute

$$\lambda_j^y \Phi(\cdot - y) = (-1)^{|\alpha^{(j)}|}\big(D^{\alpha^{(j)}}\Phi\big)(\cdot - x_j)$$

and

$$(\lambda_j^y \Phi(\cdot - y))^\wedge(\omega) = (-i\omega)^{\alpha^{(j)}} e^{-ix_j^T\omega}\widehat{\Phi}(\omega) = \lambda_j^y\big(e^{-iy^T\omega}\big)\widehat{\Phi}(\omega).$$

This yields

$$\left\| \sum_{j=1}^N c_j\lambda_j^y \Phi(\cdot - y) \right\|^2_{\mathcal{N}_\Phi(\mathbb{R}^d)} = (2\pi)^{-d/2} \int_{\mathbb{R}^d} \left| \sum_{j=1}^N c_j\lambda_j^y\big(e^{-iy^T\omega}\big) \right|^2 \widehat{\Phi}(\omega)d\omega.$$

Since Φ is positive definite there exists an open set $U \subseteq \mathbb{R}^d$ where $\widehat{\Phi}(\omega) > 0$. Hence, we must necessarily have

$$\sum_{j=1}^N c_j(i\omega)^{\alpha^{(j)}} e^{ix_j^T\omega} = 0 \tag{16.7}$$

for all ω with $-\omega \in U$.

Now we can proceed as in Lemma 6.7. Thus by analytic continuation we can see that (16.7) is true for all $\omega \in \mathbb{R}^d$, not only for the ω with $-\omega \in U$. Then we can choose a test function $f \in \mathcal{S}$ and get

$$0 = \sum_{j=1}^N c_j(i\omega)^{\alpha^{(j)}} e^{ix_j^T\omega}\widehat{f}(\omega) = \left(\sum_{j=1}^N c_j\big(D^{\alpha^{(j)}} f\big)(\cdot + x_j) \right)^\wedge (\omega)$$

for all $\omega \in \mathbb{R}^d$, which implies

$$\sum_{j=1}^N c_j D^{\alpha^{(j)}} f(x + x_j) = 0, \qquad x \in \mathbb{R}^d,$$

and in particular, setting $x = 0$,

$$\sum_{j=1}^N c_j\lambda_j(f) = 0.$$

Finally, we have to choose the test function f appropriately. We let f_0 be a compactly supported test function having support contained in the ball around zero with radius $0 < \epsilon < \min_{j\neq k} \|x_j - x_k\|_2$ and $f_0(x) = 1$ if $\|x\|_2 < \epsilon/2$. The latter means in particular that $D^\alpha f_0(0) = 0$ for all $\alpha \neq 0$. For $1 \le \ell \le N$ we then define the function $f = f_\ell$ to be

$$f_\ell(x) = \frac{(x - x_\ell)^{\alpha^{(\ell)}}}{\alpha^{(\ell)}!} f_0(x - x_\ell), \qquad x \in \mathbb{R}^d.$$

Leibniz' rule for multivariate functions now gives $\lambda_j(f_\ell) = \delta_{j,\ell}$, showing that all coefficients c_j are zero. □

Our approach differs crucially from the following naive, unsymmetric, approach. Since at first sight it might seem undesirable that the functionals $\{\lambda_j\}$ are applied twice to the kernel to form the generalized interpolation matrix, one could be tempted to start instead with a function of the form

$$s = \sum_{j=1}^{N} c_j \Phi(\cdot, x_j).$$

The interpolation condition would then lead to an interpolation matrix with entries of the form

$$D_1^{\alpha^{(\ell)}} \Phi(x_\ell, x_j).$$

This matrix is obviously not symmetric; moreover, it may not be invertible. It is possible to construct a counter-example that leads to a singular matrix; interestingly, though such counter examples are found rather seldom and with difficulty. Nonetheless, this easier approach is doomed to fail in the general setting. To save it, one could start to discuss conditions for the functionals to ensure invertibility of the generalized interpolation matrix, but we do not want to pursue this further here.

Instead, we want to discuss the use of conditionally positive definite kernels. As mentioned earlier we could use the kernel $K(\cdot,\cdot)$ from Theorem 10.20 instead of the initial kernel $\Phi(\cdot,\cdot)$. Since K is positive definite we can form an interpolant using (16.5) with interpolation matrix (16.6), simply by replacing Φ by K.

But in the case of these specific functionals we want to introduce another possibility, which might be more natural. We start by forming an interpolant of the form

$$s = \sum_{j=1}^{N} c_j D_2^{\alpha^{(j)}} \Phi(\cdot, x_j) + \sum_{\ell=1}^{Q} d_\ell p_\ell, \tag{16.8}$$

where $\{p_1, \dots, p_Q\}$ denotes the usual basis of $\mathcal{P}$. The interpolation conditions

$$D^{\alpha^{(i)}} s(x_i) = f_i, \qquad 1 \le i \le N, \tag{16.9}$$

together with the additional conditions

$$\sum_{j=1}^{N} c_j D^{\alpha^{(j)}} p_\ell(x_j) = 0, \qquad 1 \le \ell \le Q, \tag{16.10}$$

lead to a system

$$\begin{pmatrix} A_{\Phi,\Lambda} & \Lambda(P) \\ \Lambda(P)^T & 0 \end{pmatrix} \begin{pmatrix} c \\ d \end{pmatrix} = \begin{pmatrix} f \\ 0 \end{pmatrix}, \tag{16.11}$$

with matrices $A_{\Phi,\Lambda} = (\lambda_j^x \lambda_\ell^y \Phi(x, y)) \in \mathbb{R}^{N\times N}$ and $\Lambda(P) = (\lambda_i(p_\ell)) \in \mathbb{R}^{N\times Q}$. This system is very similar to the classical interpolation system (8.10) and we have to check whether or when it is uniquely solvable.

Theorem 16.5 *Suppose that $\Phi \in C^{2k}(\Omega \times \Omega)$ is a conditionally positive definite kernel with respect to $\mathcal{P} \subseteq C^k(\Omega)$. Suppose further that the functionals $\lambda_j = \delta_{x_j} \circ D^{\alpha^{(j)}}$ are linearly independent and that $\lambda_j(p) = 0$ for all $1 \le j \le N$ and $p \in \mathcal{P}$ implies that $p = 0$. Then there exists exactly one interpolant of the form (16.8) that satisfies the conditions (16.9) and (16.10).*

Proof Since we are working in a finite-dimensional setting it suffices to prove that the matrix in (16.11) is injective. So let us assume that

$$A_{\Phi,\Lambda}c + \Lambda(P)d = 0,$$
$$\Lambda(P)^T c = 0.$$

Multiplying the first equation by c^T and using the second yields $c^T A_{\Phi,\Lambda}c + c^T \Lambda(P) = c^T A_{\Phi,\Lambda}c = 0$. Now it is easy to see that

$$\begin{aligned} c^T A_{\Phi,\Lambda}c &= \sum_{j,\ell=1}^{N} c_j c_\ell D_1^{\alpha^{(j)}} D_2^{\alpha^{(\ell)}} \Phi(x_j, x_\ell) \\ &= \sum_{j,\ell=1}^{N} c_j c_\ell D_1^{\alpha^{(j)}} D_2^{\alpha^{(\ell)}} K(x_j, x_\ell) \end{aligned}$$

for all c with $c^T \Lambda(P) = 0$. But we already know that the last quadratic form is positive definite and hence equals zero if and only if $c = 0$. The additional conditions imposed on the λ_j lead now to $d = 0$ also. □

In the case of pure interpolation (i.e. all $\alpha^{(j)} = 0$) we know that the classical interpolant and the interpolant formed using the kernel K are the same if $\Xi \subseteq X$ (see Corollary 12.10). This is no longer true if derivatives come into play. Since

$$\begin{aligned} K(x, y) = \Phi(x, y) &- \sum_{j=1}^{Q} p_j(x)\Phi(\xi_j, y) - \sum_{\ell=1}^{Q} p_\ell(y)\Phi(x, \xi_\ell) \\ &+ \sum_{j,\ell=1}^{Q} p_j(x)p_\ell(y)\Phi(\xi_j, \xi_\ell) + \sum_{j=1}^{Q} p_j(x)p_j(y), \end{aligned}$$

the interpolant built from the basis functions $D_2^{\alpha} K(\cdot, y)$ is contained in

$$\operatorname{span}\left\{\{D_2^{\alpha^{(j)}}\Phi(\cdot, x_j) : 1 \le j \le N\} \cup \{\Phi(\cdot, \xi_\ell) : 1 \le \ell \le Q\}\right\} + \mathcal{P},$$

whereas the interpolant (16.8) obviously comes from the space

$$\operatorname{span}\left\{D_2^{\alpha^{(j)}}\Phi(\cdot, x_j) : 1 \le j \le N\right\} + \mathcal{P}.$$

16.3 Solving PDEs by collocation

The previous section allows us to consider a general class of boundary-value problems. Suppose that $\Omega \subseteq \mathbb{R}^d$ is an open and bounded region. We want to solve a problem of the type

$$\begin{aligned} Lu &= f \quad \text{in} \quad \Omega, \\ Bu &= g \quad \text{on} \quad \partial\Omega. \end{aligned}$$

Here, the right-hand sides f, g, the partial differential operator L, and the boundary operator B are given and we are looking for the unknown solution u.

Definition 16.6 *An operator $L : C^k(\Omega) \to C(\Omega)$ is called a linear differential operator of order k if it has the the form*

$$L = \sum_{|\alpha| \le k} c_\alpha D^\alpha$$

with $c_\alpha \in C(\Omega)$. L is said to be a linear differential operator with constant coefficients if additionally all the functions c_α are constant.

Note that a linear differential operator is indeed linear by definition.

The ideas of the last section suggest the following approach. Suppose that L is a linear differential operator of order k; then we choose a positive definite function $\Phi \in C^{2k}(\Omega \times \Omega)$ such that $L : \mathcal{N}_\Phi(\Omega) \to C(\Omega)$. We choose points $X = X_1 \cup X_2$ with $X_1 = \{x_1, \ldots, x_n\} \subseteq \Omega$ and $X_2 = \{x_{n+1}, \ldots, x_N\} \subseteq \partial\Omega$ and use the functionals

$$\lambda_j = \begin{cases} \delta_{x_j} \circ L, & 1 \le j \le n, \\ \delta_{x_j} \circ B, & n+1 \le j \le N. \end{cases}$$

Even if this has not been specified so far, we are assuming that the boundary operator B behaves nicely in the sense that it is well defined on $\mathcal{N}_\Phi(\Omega)$. The reader might think of Dirichlet boundary conditions, i.e. $Bu = u$, as a prototype. But more complicated boundary operators given by Neumann or mixed boundary conditions are possible.

Under the assumption that the functionals $\Lambda = \{\lambda_1, \ldots, \lambda_N\}$ are linearly independent, we know that an interpolant of the form

$$\begin{aligned} s_{u,\Lambda} &= \sum_{j=1}^{N} \alpha_j \lambda_j^y \Phi(\cdot, y) \\ &= \sum_{j=1}^{n} \alpha_j L_2 \Phi(\cdot, x_j) + \sum_{j=n+1}^{N} \alpha_j B_2 \Phi(\cdot, x_j) \end{aligned} \tag{16.12}$$

uniquely exists that satisfies

$$\begin{aligned} L s_{u,\Lambda}(x_j) &= f(x_j), \quad & 1 \le j \le n, \\ B s_{u,\Lambda}(x_j) &= g(x_j), \quad & n+1 \le j \le N. \end{aligned}$$

The subscripts 2 on B and L denote again the fact that these operators act with respect to the second argument.

In the rest of this section we want to analyze the error for our approximation by giving bounds on both $L(u - s_{u,\Lambda})(x)$, $x \in \Omega$, and $B(u - s_{u,\Lambda})(x)$, $x \in \partial\Omega$. In the case of an elliptic problem with Dirichlet boundary conditions these two bounds lead to a bound for $u - s_{u,\Lambda}$ on $\overline{\Omega}$. We will end the section with such an example.

To derive our estimates we have to prove a couple of results that are interesting in themselves. Even if we have a differential operator L in mind, all results hold in the more general case of a linear operator L with $\delta_x \circ L \in \mathcal{N}_\Phi(\Omega)^*$, $x \in \Omega$. Thus L could be for example an integral operator of Volterra type. We will concentrate mainly on the analysis for L. Similar steps have to be undertaken for B.

Theorem 16.7 *Suppose that H is a real Hilbert space with reproducing kernel Φ. Let $\lambda, \mu \in H^*$. Then $\lambda^y \Phi(\cdot, y) \in H$ and*

$$\lambda(f) = (f, \lambda^y \Phi(\cdot, y))$$

for all $f \in H$. Moreover

$$(\lambda, \mu)_{H^*} = \lambda^x \mu^y \Phi(x, y).$$

Proof Riesz' representation theorem guarantees the existence of an h_λ such that $(f, h_\lambda) = \lambda(f)$ for all $f \in H$. Since $f_x := \Phi(\cdot, x)$ is an element of H we find that

$$\lambda(f_x) = (f_x, h_\lambda) = (h_\lambda, f_x) = (h_\lambda, \Phi(\cdot, x)) = h_\lambda(x)$$

by the reproducing property of the kernel. This establishes $\lambda^y \Phi(\cdot, y) = h_\lambda \in H$ and the first property. For the second, note that

$$(\lambda, \mu)_{H^*} = (h_\lambda, h_\mu) = (h_\mu, h_\lambda) = \lambda(h_\mu) = \lambda^x \mu^y \Phi(x, y).$$

□

Under our assumptions on L (or B) and Φ the functionals $\delta_x \circ L$ are in $\mathcal{N}_\Phi(\Omega)^*$. Thus we can define

$$\Phi_L(x, y) := (\delta_x \circ L)^u (\delta_y \circ L)^v \Phi(u, v), \qquad x, y \in \Omega. \tag{16.13}$$

By Theorem 16.7 we know that Φ_L is symmetric. It is also positive definite under the following mild condition: from now on we will assume that

$$\{\delta_x \circ L : x \in \Omega\} \text{ is linearly independent over } \mathcal{N}_\Phi(\Omega). \tag{16.14}$$

Theorem 16.8 *Suppose that $\Phi \in C(\Omega \times \Omega)$ is positive definite. Suppose further that the linear operator L satisfies $\delta_x \circ L \in \mathcal{N}_\Phi(\Omega)^*$ for $x \in \Omega$. Finally, assume that L satisfies (16.14). Then Φ_L is positive definite on Ω.*

Proof Let $\lambda = \sum_j \alpha_j \delta_{y_j}$ with pairwise distinct $y_j \in \Omega$ be given. Then

$$\begin{aligned}\sum_{j,k} \alpha_j \alpha_k \Phi_L(y_j, y_k) &= \sum_{j,k} \alpha_j \alpha_k (\delta_{y_j} \circ L)^u (\delta_{y_k} \circ L)^v \Phi(u, v)\\ &= (\lambda \circ L)^u (\lambda \circ L)^v \Phi(u, v)\\ &= \|\lambda \circ L\|_{\mathcal{N}_\Phi(\Omega)^*},\end{aligned}$$

which is clearly nonnegative and vanishes only if $\alpha = 0$, by the conditions imposed on L. □

Our next result enlightens the connection between the kernels Φ and Φ_L. Now that we know that Φ_L is also a positive definite function we have the whole machinery of native spaces, power functions, etc. at hand.

Theorem 16.9 *Suppose that $\Phi \in C(\Omega \times \Omega)$ is a positive definite kernel and that $L : \mathcal{N}_\Phi(\Omega) \to C(\Omega)$ satisfies (16.14). Then $L(\mathcal{N}_\Phi(\Omega)) = \mathcal{N}_{\Phi_L}(\Omega)$, and the following mappings,*

$$\begin{aligned} L:\quad & \mathcal{N}_\Phi(\Omega) \to \mathcal{N}_{\Phi_L}(\Omega), && f \mapsto Lf,\\ & \mathcal{N}_{\Phi_L}(\Omega)^* \to \mathcal{N}_\Phi(\Omega)^*, && \lambda \mapsto \lambda \circ L,\end{aligned}$$

are isometric isomorphisms. In particular, if $f \in \mathcal{N}_\Phi(\Omega)$ and $\lambda \in \mathcal{N}_{\Phi_L}(\Omega)^*$ then $Lf \in \mathcal{N}_{\Phi_L}(\Omega)$ and $\lambda \circ L \in \mathcal{N}_\Phi(\Omega)^*$ with

$$\|Lf\|_{\mathcal{N}_{\Phi_L}(\Omega)} = \|f\|_{\mathcal{N}_\Phi(\Omega)} \quad \textit{and} \quad \|\lambda\|_{\mathcal{N}_{\Phi_L}(\Omega)^*} = \|\lambda \circ L\|_{\mathcal{N}_\Phi(\Omega)^*}.$$

Proof The proof will be given in several steps. First of all let us introduce the following spaces:

$$\begin{aligned}\mathcal{L}_0 &:= \operatorname{span}\{\delta_x : x \in \Omega\},\\ \mathcal{L}_T^0 &:= \{\lambda \circ L : \lambda \in \mathcal{L}_0\} \subseteq \mathcal{N}_\Phi(\Omega)^*,\\ F_\Phi(\Omega) &= \{\lambda^y \Phi(\cdot, y) : \lambda \in \mathcal{L}_0\} \subseteq \mathcal{N}_\Phi(\Omega),\\ \mathcal{F}_T^0 &:= \{(\lambda \circ L)^y \Phi(\cdot, y) : \lambda \in \mathcal{L}_0\} \subseteq \mathcal{N}_\Phi(\Omega).\end{aligned}$$

When equipped with the inner product of the space $\mathcal{N}_\Phi(\Omega)^*$, the linear space $\mathcal{L}_0$ becomes a subspace of $\mathcal{N}_\Phi(\Omega)^*$; we will use the notation $\mathcal{L}_{0,\Phi}$ to make this clear. Each of the other spaces is assumed to carry the inner product of the space of which it is a subspace. The space $F_\Phi(\Omega)$ is not new to us at all. We have already used this space to construct the native space $\mathcal{N}_\Phi(\Omega)$. We now proceed as follows.

(1) Since $F_\Phi(\Omega)$ is dense in $\mathcal{N}_\Phi(\Omega)$ by construction, Riesz' representation theorem ensures that $\mathcal{L}_{0,\Phi}$ is dense in $\mathcal{N}_\Phi(\Omega)^*$. One of our goals here is to show that $\mathcal{F}_T^0$ is also dense in $\mathcal{N}_\Phi(\Omega)$.

(2) Let us introduce the isometric isomorphism

$$\begin{aligned} T &: F_{\Phi_L}(\Omega) \to \mathcal{F}_T^0, && \lambda^y \Phi_L(\cdot, y) \mapsto (\lambda \circ L)^y \Phi(\cdot, y),\\ T^* &: \mathcal{L}_{0,\Phi_L} \to \mathcal{L}_T^0, && \lambda \mapsto \lambda \circ L,\end{aligned}$$

induced by the operator L. Both mappings are indeed isometric, because

$$(\lambda \circ L)^x(\lambda \circ L)^y\Phi(x,y) = \lambda^u\lambda^v(\delta_u \circ L)^x(\delta_v \circ L)^y\Phi(x,y) = \lambda^u\lambda^v\Phi_L(u,v)$$

for all $\lambda \in \mathcal{L}_0$. Theorem 16.7 and the norm-preserving property of the Riesz representer are also helpful. Moreover, both mappings are surjective by construction.

(3) The restriction of the Riesz mapping

$$R_\Phi|\mathcal{L}_T^0 : \mathcal{L}_T^0 \to \mathcal{F}_T^0, \qquad \lambda \circ L \mapsto (\lambda \circ L)^y\Phi(\cdot, y)$$

is also an isomorphic isomorphism. Thus we can describe the situation so far by the following commutative diagram:

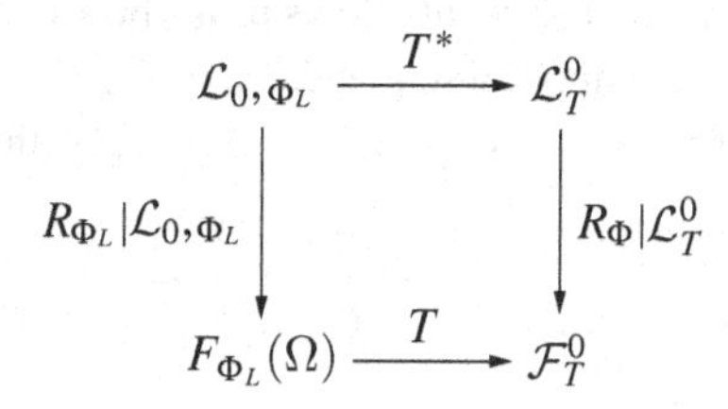

(4) $T^{-1} : \mathcal{F}_T^0 \to F_{\Phi_L}(\Omega)$ coincides with L. To see this, choose an arbitrary $f = (\lambda \circ L)^y\Phi(\cdot, y) \in \mathcal{F}_T^0$, with $\lambda \in \mathcal{L}_0$. Then we have $T^{-1}f = \lambda^y\Phi_L(\cdot, y)$, leading to

$$\begin{aligned} T^{-1}f(x) &= \lambda^y\Phi(x,y) \\ &= \lambda^y(\delta_x \circ L)^u(\delta_y \circ L)^v\Phi(u,v) \\ &= (\delta_x \circ L)^u(\lambda \circ L)^v\Phi(u,v) \\ &= (\delta_x \circ L)^u f(u) \\ &= Lf(x). \end{aligned}$$

(5) Since both T and T^* map dense subspaces of Hilbert spaces, into Hilbert spaces, they possess unique isometric extensions

$$\begin{aligned} \widetilde{T^*} : \mathcal{N}_{\Phi_L}(\Omega)^* &\to \mathcal{L}_T := \widetilde{T^*}(\mathcal{N}_{\Phi_L}(\Omega)^*) \subseteq \mathcal{N}_\Phi(\Omega)^*, \\ \widetilde{T} : \mathcal{N}_{\Phi_L}(\Omega) &\to \mathcal{F}_T := \widetilde{T}(\mathcal{N}_{\Phi_L}(\Omega)) \subseteq \mathcal{N}_\Phi(\Omega). \end{aligned}$$

Since both $\widetilde{T} \circ R_{\Phi_L} \circ \widetilde{T^*}^{-1}$ and $R_\Phi|\mathcal{L}_T$, are isometric extensions of $R_\Phi|\mathcal{L}_T^0$, they have to coincide. Thus, the following diagram has to be commutative also:

$$\begin{array}{ccc} \mathcal{N}_{\Phi_L}(\Omega)^* & \xrightarrow{\ \widetilde{T^*}\ } & \mathcal{L}_T \\ \downarrow {\scriptstyle R_{\Phi_L}} & & \downarrow {\scriptstyle R_\Phi|\mathcal{L}_T} \\ \mathcal{N}_{\Phi_L}(\Omega) & \xrightarrow{\ \widetilde{T}\ } & \mathcal{F}_T \end{array}$$

The next step is to show that both the extensions $\widetilde{T}$ and $\widetilde{T^*}$ are surjective, i.e. that $\mathcal{L}_T = \mathcal{N}_\Phi(\Omega)^*$ and $\mathcal{F}_T = \mathcal{N}_\Phi(\Omega)$, respectively.

(6) Since $\mathcal{N}_{\Phi_L}(\Omega)$ is complete and since $\widetilde{T} : \mathcal{N}_{\Phi_L}(\Omega) \to \mathcal{F}_T \subseteq \mathcal{N}_\Phi(\Omega)$ is an isometric isomorphism, $\mathcal{F}_T$ has also to be complete. But this means that $\mathcal{F}_T$ is a reproducing-kernel Hilbert function space with reproducing kernel Φ. Thus, by Theorem 10.11 we have $\mathcal{F}_T = \mathcal{N}_\Phi(\Omega)$. Moreover, if we have an arbitrary $\lambda \in \mathcal{N}_\Phi(\Omega)^*$ then the Riesz mapping gives $R_\Phi(\lambda) = \lambda^y \Phi(\cdot, y) \in \mathcal{N}_\Phi(\Omega) = \mathcal{F}_T$ and therefore $(R_\Phi|\mathcal{L}_T)^{-1} R_\Phi(\lambda) \in \mathcal{L}_T$, showing that $R_\Phi|\mathcal{L}_T = R_\Phi$ or, in other words, $\mathcal{L}_T = \mathcal{N}_\Phi(\Omega)^*$.

The equality $\mathcal{F}_T = \mathcal{N}_\Phi(\Omega)$ has an important side effect. The space $\mathcal{F}_T$ is the image of $\mathcal{N}_{\Phi_L}(\Omega)$ under the mapping $\widetilde{T}$, which is the extension of the mapping T from the dense subspace $F_{\Phi_L}(\Omega)$ of $\mathcal{N}_{\Phi_L}(\Omega)$. Hence, $T(F_{\Phi_L}(\Omega)) = \mathcal{F}_T^0$ is dense in $\mathcal{N}_\Phi(\Omega)$.

(7) The density mentioned in the last point allows us to show that $\widetilde{T}^{-1} = L$, as follows. If $f \in \mathcal{N}_\Phi(\Omega)$ is given then we can choose a sequence $f_n = (\lambda_n \circ L)^y \Phi(\cdot, y) \in \mathcal{F}_T^0$, $\lambda_n \in \mathcal{L}_0$, with $\|f - f_n\|_{\mathcal{N}_\Phi(\Omega)} \to 0$ for $n \to \infty$. Since $\delta_x \circ L \in \mathcal{N}_\Phi(\Omega)^*$ this means that on the one hand

$$|Lf(x) - Lf_n(x)| \le \|\delta_x \circ L\|_{\mathcal{N}_\Phi(\Omega)^*} \|f - f_n\|_{\mathcal{N}_\Phi(\Omega)} \to 0, \qquad n \to \infty.$$

On the other hand, we know by step (4) that $T^{-1} f_n(x) = Lf_n(x)$, giving

$$|\widetilde{T}^{-1} f(x) - Lf_n(x)| \le \|\delta_x \circ \widetilde{T}^{-1}\|_{\mathcal{N}_\Phi(\Omega)^*} \|f - f_n\|_{\mathcal{N}_\Phi(\Omega)^*} \to 0, \qquad n \to \infty.$$

This establishes the identity $L = \widetilde{T}^{-1}$. Thus for every $f \in \mathcal{N}_\Phi(\Omega)$ we have $Lf = \widetilde{T}^{-1} f \in \mathcal{N}_{\Phi_L}(\Omega)$ and $\|Lf\|_{\mathcal{N}_{\Phi_L}(\Omega)} = \|f\|_{\mathcal{N}_\Phi(\Omega)}$.

(8) Similarly, we can establish the identity $\widetilde{T}^*(\lambda) = \lambda \circ L$ for all $\lambda \in \mathcal{N}_{\Phi_L}(\Omega)^*$. Namely, for any $\lambda \in \mathcal{N}_{\Phi_L}(\Omega)^*$ there exists owing to step (1) a sequence $\{\lambda_n\} \subseteq \mathcal{L}_{0,\Phi_L}$ approximating λ for $n \to \infty$. Thus if $f \in \mathcal{N}_\Phi(\Omega)$ is given then we know by step (7) that $Lf \in \mathcal{N}_{\Phi_L}(\Omega)$, showing that

$$|\lambda \circ L(f) - \lambda_n \circ L(f)| \le \|\lambda - \lambda_n\|_{\mathcal{N}_{\Phi_L}(\Omega)^*} \|Lf\|_{\mathcal{N}_{\Phi_L}(\Omega)} \to 0, \qquad n \to \infty.$$

By definition we have $T^*(\lambda) = \lambda \circ L$ so that

$$\begin{aligned} |\widetilde{T}^*(\lambda)(f) - \lambda_n \circ L(f)| &\le \|\widetilde{T}^*(\lambda - \lambda_n)\|_{\mathcal{N}_\Phi(\Omega)^*} \|f\|_{\mathcal{N}_\Phi(\Omega)} \\ &= \|\lambda - \lambda_n\|_{\mathcal{N}_\Phi(\Omega)^*} \|f\|_{\mathcal{N}_\Phi(\Omega)} \to 0, \quad n \to \infty. \end{aligned}$$

This allows us to conclude the proof. For $\lambda \in \mathcal{N}_{\Phi_L}(\Omega)^*$ we now know that $\lambda \circ L = \widetilde{T}^*(\lambda) \in \mathcal{N}_\Phi(\Omega)^*$ and $\|\lambda \circ L\|_{\mathcal{N}_\Phi(\Omega)^*} = \|\lambda\|_{\mathcal{N}_{\Phi_L}(\Omega)^*}$. □

After discussing the relations between the native spaces of Φ and Φ_L we now investigate the connection between their power functions.

Theorem 16.10 *Suppose that $\Lambda = \{\lambda_1, \dots, \lambda_N\} \subseteq \mathcal{N}_{\Phi_L}(\Omega)^*$ and $\lambda \in \mathcal{N}_{\Phi_L}(\Omega)^*$ are given. Then*

$$P_{\Phi_L,\Lambda}(\lambda) = P_{\Phi,\Lambda\circ L}(\lambda \circ L).$$

Proof By definition and by Theorem 16.9 we have that

$$\begin{aligned}
P_{\Phi,\Lambda\circ L}(\lambda\circ L) &= \inf_{\mu\in\text{span}\{\Lambda\circ L\}} \|\lambda\circ L-\mu\|_{\mathcal{N}_\Phi(\Omega)^*}\\
&= \inf_{\mu\in\text{span}\{\Lambda\}} \|\lambda\circ L-\mu\circ L\|_{\mathcal{N}_\Phi(\Omega)^*}\\
&= \inf_{\mu\in\text{span}\{\Lambda\}} \|\lambda-\mu\|_{\mathcal{N}_{\Phi_L}(\Omega)^*}\\
&= P_{\Phi_L,\Lambda}(\lambda).
\end{aligned}$$

□

Our next result also deals with the power function. This time we want to investigate what happens if we drop some of the functionals. The result should no longer astonish us.

Theorem 16.11 *Suppose that $\Lambda=\{\lambda_1,\dots,\lambda_N\}\subseteq\mathcal{N}_\Phi(\Omega)^*$ and $\Lambda'\subseteq\Lambda$ are given. Then*

$$P_{\Phi,\Lambda}(\lambda)\le P_{\Phi,\Lambda'}(\lambda)$$

for all $\lambda\in\mathcal{N}_\Phi(\Omega)^*$.

Proof This simply follows from the definition of the power function,

$$\begin{aligned}
P_{\Phi,\Lambda}(\lambda) &= \inf_{\mu\in\text{span}\{\Lambda\}}\|\lambda-\mu\|_{\mathcal{N}_\Phi(\Omega)^*}\le \inf_{\mu\in\text{span}\{\Lambda'\}}\|\lambda-\mu\|_{\mathcal{N}_\Phi(\Omega)^*}\\
&= P_{\Phi,\Lambda'}(\lambda).
\end{aligned}$$

□

We now come back to our initial boundary-value problem, described at the beginning of this section. Suppose that on the one hand condition (16.14) is satisfied. Then we know that Φ_L is positive definite, and the theory derived so far allows us to bound the error in the interior simply by dropping all boundary functionals in the power function. On the other hand, to derive estimates on the boundary we drop all interior functionals. As a consequence the power function is reduced to the well-investigated power function for point evaluations. Introducing the following additional notation,

$$\begin{aligned}
\Delta_1 &= \{\delta_{x_1},\dots,\delta_{x_n}\},\\
\Delta_2 &= \{\delta_{x_{n+1}},\dots,\delta_{x_N}\},\\
\Lambda_1 &= \{\delta_{x_1}\circ L,\dots,\delta_{x_n}\circ L\}=\Delta_1\circ L,\\
\Lambda_2 &= \{\delta_{x_{n+1}}\circ B,\dots,\delta_{x_N}\circ B\}=\Delta_2\circ B,
\end{aligned}$$

we can give an estimate of the error bound in the interior:

$$\begin{aligned}
|Lu(x)-Ls_{u,\Lambda}(x)| &\le P_{\Phi,\Lambda}(\delta_x\circ L)\|u\|_{\mathcal{N}_\Phi(\Omega)}\\
&\le P_{\Phi,\Lambda_1}(\delta_x\circ L)\|u\|_{\mathcal{N}_\Phi(\Omega)}\\
&= P_{\Phi,\Delta_1\circ L}(\delta_x\circ L)\|u\|_{\mathcal{N}_\Phi(\Omega)}\\
&= P_{\Phi_L,\Delta_1}(\delta_x)\|u\|_{\mathcal{N}_\Phi(\Omega)}\\
&= P_{\Phi_L,X_1}(x)\|u\|_{\mathcal{N}_\Phi(\Omega)}.
\end{aligned}$$

The same is possible on the boundary, leading to

$$|Bu(x) - Bs_{u,\Lambda}(x)| \le P_{\Phi_B,X_2}(x)\|u\|_{\mathcal{N}_\Phi(\Omega)}.$$

Let us demonstrate in more detail how this works in the case of positive definite functions in $L_1(\mathbb{R}^d)$. Since L is a differential operator and also since B contains in general at most also certain derivatives, the first step is to answer the question when are the functionals linearly independent. But this has already been done in Theorem 16.4. A consequence for a linear differential operator is

Corollary 16.12 *Let $\Phi \in L_1(\mathbb{R}^d) \cap C^{2k}(\mathbb{R}^d)$ be a positive definite function. Suppose that $L : C^k(\mathbb{R}^d) \to C(\mathbb{R}^d)$ is a linear differential operator of order k, i.e. $L = \sum_{|\alpha|\le k} c_\alpha D^\alpha$ with $c_\alpha \in C(\Omega)$. Suppose further that either $c_\alpha \equiv 0$ or c_α is nonzero everywhere and that not all the c_α vanish. Then Φ_L is also a positive definite kernel.*

Proof Let $\alpha^{(1)}, \dots, \alpha^{(M)}$ be a numeration of all $\alpha \in \mathbb{N}_0^d$ with $|\alpha| \le k$ and c_α nonzero, so that the operator L takes the form $L = \sum_{\ell=1}^M \tilde{c}_\ell D^{\alpha^{(\ell)}}$ and the coefficient functions satisfy $\tilde{c}_\ell(x) = c_{\alpha^{(\ell)}}(x) \ne 0$ for all $x \in \Omega$ and all $1 \le \ell \le M$. We want to apply Theorem 16.8. Thus we have to show that for arbitrary but distinct $x_1, \dots, x_N \in \mathbb{R}^d$ and arbitrary $b_1, \dots, b_N \in \mathbb{R}$ the assumption

$$0 = \sum_{j=1}^N b_j Lf(x_j) = \sum_{j=1}^N \sum_{\ell=1}^M b_j \tilde{c}_\ell(x_j) D^{\alpha^{(\ell)}} f(x_j) \tag{16.15}$$

for all $f \in \mathcal{N}_\Phi(\mathbb{R}^d)$ leads to $b_1 = \cdots = b_N = 0$. To this end, let us set $y_{(j-1)M+\ell} = x_j$, $\beta^{((j-1)M+\ell)} = \alpha^{(\ell)}$, and $d_{(j-1)M+\ell} = b_j\tilde{c}_\ell(x_j)$, each time for $1 \le j \le N$ and $1 \le \ell \le M$. Then assumption (16.15) becomes

$$\sum_{k=1}^{NM} d_k D^{\beta^{(k)}} f(y_k) = 0$$

for all $f \in \mathcal{N}_\Phi(\mathbb{R}^d)$. But the functionals $\lambda_k = \delta_{y_k} \circ D^{\beta^{(k)}}$, $1 \le k \le NM$, are pairwise distinct in the sense of Theorem 16.4. Hence $d_k = 0$, $1 \le k \le NM$, by that theorem. Since $\tilde{c}_\ell(x_j) \ne 0$ this means that all the b_j are zero. □

Thus the kernel Φ_L is also positive definite. But since it has the form

$$\Phi_L(x, y) = L^x L^y \Phi(x - y) = \sum_{|\alpha|,|\beta|\le k} c_\alpha(x) c_\beta(y)(-1)^{|\beta|}(D^{\alpha+\beta}\Phi)(x - y)$$

it is no longer a translation-invariant kernel; the property of translation invariance is only guaranteed if L has constant coefficients.

For the rest of this section we will assume Dirichlet boundary values, i.e. $B = I$, to make life easier. Moreover, we restrict ourselves to polygonal regions, which is standard in finite-element theory. More general regions need interpolation by positive definite functions on $(d - 1)$-dimensional manifolds, which is the subject of the next chapter.

Definition 16.13 *A bounded region $\Omega \subseteq \mathbb{R}^d$ is said to be a simple polygonal region if it is the intersection of a finite number of half spaces. A half space in $\mathbb{R}^d$ is a set $H_{a,b} = \{x \in \mathbb{R}^d : a^T x \le b\}$ with $a \in \mathbb{R}^d \setminus \{0\}$ and $b \in \mathbb{R}$. A region $\Omega \subseteq \mathbb{R}^d$ is said to be a polygonal region if it is the union of a finite number of simple polygonal regions.*

A polygonal region is therefore bounded and obviously satisfies an interior cone condition. Moreover, its boundary is the union of a finite number of $(d-2)$-variate simple polygonal regions.

To bound the error on the boundary, we need the following lemma.

Lemma 16.14 *Suppose that $\Phi \in C(\mathbb{R}^d)$ is a positive definite function and that $X = \{x_1, \dots, x_N\} \subseteq \mathbb{R}^d$ is a set of pairwise distinct points. Suppose further that $T : \mathbb{R}^d \to \mathbb{R}^d$ is a bijective affine mapping, i.e. $Tx = Sx + c$, $x \in \mathbb{R}^d$, with an invertible matrix $S \in \mathbb{R}^{d\times d}$ and a constant $c \in \mathbb{R}^d$. Then the the following relation for the power function holds:*

$$P_{\Phi,X}(x) = P_{\Phi\circ S^{-1},T(X)}(Tx), \qquad x \in \mathbb{R}^d.$$

Here $T(X)$ denotes the set $\{Tx_1, \dots, Tx_N\}$.

Proof Let $u_j^*(x;\Phi,X)$, $1 \le j \le N$, denote the cardinal functions with respect to Φ and X, i.e. $u^*(x;\Phi,X) = A_{\Phi,X}^{-1} R_{\Phi,X}(x)$ with $A_{\Phi,X} = (\Phi(x_j - x_k)) \in \mathbb{R}^{N\times N}$ and $R_{\Phi,X}(x) = (\Phi(x - x_j)) \in \mathbb{R}^N$. Set $Z = T(X)$, $z = T(x)$. Because $z_j - z_k = Tx_j - Tx_k = S(x_j - x_k)$, we have obviously $A_{\Phi,X} = A_{\Phi\circ S^{-1},Z}$ and $R_{\Phi,X}(x) = R_{\Phi\circ S^{-1},Z}(z)$ or, in other words,

$$\begin{aligned} u^*(x;\Phi,X) &= A_{\Phi\circ S^{-1},Z}^{-1} R_{\Phi\circ S^{-1},Z}(z) = u^*(z;\Phi\circ S^{-1}, Z) \\ &= u^*(Tx;\Phi\circ S^{-1}, T(X)). \end{aligned}$$

This means that the power function is found from

$$\begin{aligned} P_{\Phi,X}^2(x) &= \Phi(0) - 2\sum_{j=1}^N u_j^*(x;\Phi,X)\Phi(x-x_j) \\ &\quad + \sum_{j,k=1}^N u_j^*(x;\Phi,X)u_k^*(x;\Phi,X)\Phi(x_j - x_k) \\ &= \Phi\circ S^{-1}(0) - 2\sum_{j=1}^N u_j^*(z;\Phi\circ S^{-1},Z)\Phi\circ S^{-1}(z - z_j) \\ &\quad + \sum_{j,k=1}^N u_j^*(z;\Phi\circ S^{-1},Z)u_k^*(z;\Phi\circ S^{-1},Z)\Phi\circ S^{-1}(z_j - z_k) \\ &= \mathcal{P}_{\Phi\circ S^{-1},T(X)}(Tx), \end{aligned}$$

finishing the proof. □

After this preparatory result, we can state and prove our convergence estimates for the solution of a partial differential equation by collocation with positive definite functions.

Theorem 16.15 *Let $\Omega \subseteq \mathbb{R}^d$ be a polygonal and open region. Let $L \neq 0$ be a linear differential operator of order $\ell \leq k$, with coefficients $c_\alpha \in C^{2(k-\ell)}(\overline{\Omega})$ that either vanish on $\overline{\Omega}$ or have no zero there. Suppose that $\Phi \in C^{2k}(\mathbb{R}^d)$ is a positive definite function. Suppose further that the boundary-value problem*

$$\begin{aligned} Lu &= f \quad \text{in} \quad \Omega, \\ u &= g \quad \text{on} \quad \partial\Omega \end{aligned}$$

has a unique solution $u \in \mathcal{N}_\Phi(\Omega)$ for given $f \in C(\Omega)$ and $g \in C(\partial\Omega)$. Let $s_{u,\Lambda}$ be the interpolant (16.12) based on Φ. Then the following error estimates,

$$|Lu(x) - Ls_{u,\Lambda}(x)| \leq Ch_{X_1,\Omega}^{k-\ell} \|u\|_{\mathcal{N}_\Phi(\Omega)}, \qquad x \in \Omega, \tag{16.16}$$

$$|u(x) - s_{u,\Lambda}(x)| \leq Ch_{X_2,\partial\Omega}^{k} \|u\|_{\mathcal{N}_\Phi(\Omega)}, \qquad x \in \partial\Omega, \tag{16.17}$$

are satisfied for all sufficiently dense sets of data sites. Here C denotes a generic constant.

Proof We use the notation from the beginning of this section and that introduced in the paragraph following the proof of Theorem 16.11. By Theorem 10.46 the function u has a natural extension to the whole of $\mathbb{R}^d$, and the extended function has the same norm as the original one. Thus we can assume that $u \in \mathcal{N}_\Phi(\mathbb{R}^d)$.

The function Φ_L corresponding to the linear operator $L = \sum_{|\beta| \leq \ell} c_\beta D^\beta$ is a positive definite kernel in $C^{2k-2\ell}(\Omega \times \Omega)$. Moreover, the assumptions imposed on the coefficient functions of L and on Φ itself show that the number $C_{\Phi_L}(x)$ from Theorem 11.13 is uniformly bounded on $\overline{\Omega}$, if we replace k by $k - \ell$ in that theorem. Hence, our analysis made so far together with the theorem just cited yields

$$|Lu(x) - Ls_{u,\Lambda}(x)| \leq P_{\Phi_L, X_1}(x) \|u\|_{\mathcal{N}_\Phi(\Omega)} \leq Ch_{X_1,\Omega}^{k-\ell} \|u\|_{\mathcal{N}_\Phi(\Omega)}, \quad x \in \Omega,$$

which is (16.16).

Let us now turn to the estimate on the boundary. First of all, our general theory leads us to

$$|u(x) - s_{u,\Lambda}(x)| \leq P_{\Phi,X_2}(x) \|u\|_{\mathcal{N}_\Phi(\Omega)}, \qquad x \in \partial\Omega.$$

Unfortunately, we cannot apply either Theorem 11.13 or Theorem 11.11 directly. Why is this so? The estimates on the power function require that the region where x comes from has to satisfy an interior cone condition in $\mathbb{R}^d$. But our region of interest is $\partial\Omega$, which definitely does not satisfy an interior cone condition. It does not even contain an interior point. The remedy to this problem is to use the fact that $\partial\Omega$ is locally a hyperplane, which can be mapped affinely to $\mathbb{R}^{d-1}$. The image of this mapping satisfies an interior cone condition in $\mathbb{R}^{d-1}$; thus we can use this to work in $\mathbb{R}^{d-1}$. Let us make this more precise. The boundary $\partial\Omega$ is the union of a finite number of surfaces $H \subseteq \tilde{H} = \{y \in \mathbb{R}^d : a^T y = b\}$. Each surface H is a simple polygonal region in $\mathbb{R}^{d-1}$, in the sense that there exists an affine bijective mapping $Ty = Sy + c$ such that $T(\tilde{H}) = \{z \in \mathbb{R}^d : z_d = 0\} = \mathbb{R}^{d-1}$, where $T(H)$ is a simple polygonal region in $\mathbb{R}^{d-1}$ satisfying therefore an interior cone condition in $\mathbb{R}^{d-1}$.

Since we have only a finite number of these regions, we can assume that the angles and radii of all the cone conditions are the same.

We will bound the error now for one of these surfaces, say H. Let $Y = \{y_1, \dots, y_M\} = X_2 \cap H$. Let $Z = T(Y) = \{z_1, \dots, z_M\}$ and $z = Tx$ for $x \in H$. If $h_{X_2,\partial\Omega}$ is sufficiently small then we can find for $x \in H$ an $y_{j_0} \in X_2 \cap H$ with $\|x - y_{j_0}\|_2 \le 2h_{X_2,\partial\Omega}$. Hence we have

$$\|z - z_{j_0}\|_2 \le \|S\| \|x - y_{j_0}\|_2 \le Ch_{X_2,\partial\Omega},$$

which means that $h_{Z,T(H)} \le Ch_{X_2,\partial\Omega}$. Since $\Phi \circ S^{-1}$ is a positive definite function even when restricted to $\mathbb{R}^{d-1}$, which has the same smoothness properties as Φ, we can now apply Theorem 11.11 to $T(H) \subseteq \mathbb{R}^{d-1}$ and $Z = H(Y)$ to get

$$P_{\Phi\circ S^{-1},Z}(z) \le Ch^k_{Z,T(H)} \le Ch^k_{X_2,\partial\Omega}, \qquad z \in T(H),$$

provided that $h_{X_2,\partial\Omega}$ is sufficiently small. By Lemma 16.14 and Theorem 16.11 we can estimate, for $x \in H$,

$$\begin{aligned}|u(x) - s_{u,\Lambda}(x)| &\le P_{\Phi,X_2}(x)\|u\|_{\mathcal{N}_\Phi(\Omega)} \le P_{\Phi,Y}(x)\|u\|_{\mathcal{N}_\Phi(\Omega)}\\ &\le P_{\Phi\circ S^{-1},Z}(z)\|u\|_{\mathcal{N}_\Phi(\Omega)} \le Ch^k_{X_2,\partial\Omega}\|u\|_{\mathcal{N}_\Phi(\Omega)}.\end{aligned}$$

Since this can be done for every surface and since the number of surfaces is finite, this completes our proof. □

Theorem 16.15 shows that in order to get good approximation results the interior Ω should be discretized more finely than the boundary. A good choice is obviously

$$h^{k-\ell}_{X_1,\Omega} \approx h^k_{X_2,\partial\Omega}.$$

Moreover, more information on L and Φ might lead to a better estimate on $C_{\Phi_L}(x)$, yielding additional approximation orders.

As stated before, in the case of an elliptic second-order operator estimates (16.16) and (16.17) together lead to an estimate of $u - s_{u,\Lambda}$ on $\overline{\Omega}$.

Definition 16.16 *A linear operator $L : C^2(\Omega) \to C(\Omega)$ of the form*

$$Lu(x) = -\sum_{j,k=1}^{d} \frac{\partial}{\partial x_k}\left(a_{j,k}(x)\frac{\partial u}{\partial x_j}(x)\right) + \sum_{j=1}^{d} b_j(x)\frac{\partial u}{\partial x_j}(x) + b_0(x)u(x)$$

with $a_{j,k} \in C^1(\Omega)$ and $b_j \in C(\Omega)$ is called an elliptic differential operator of the second order if the matrix $A(x) = (a_{j,k}(x)) \in \mathbb{R}^{d\times d}$ is uniformly positive definite on Ω. This means that there exists an $\alpha > 0$ such that

$$c^T A(x)c \ge \alpha c^T c$$

for all $c \in \mathbb{R}^d$ and all $x \in \Omega$.

If all the functions $a_{j,k}$ are bounded on Ω by a constant $C > 0$, the maximum principle for elliptic operators gives

$$\|u - s_{u,\Lambda}\|_{L_\infty(\Omega)} \le \|u - s_{u,\Lambda}\|_{L_\infty(\partial\Omega)} + \frac{C}{\alpha}\|Lu - Ls_{u,\Lambda}\|_{L_\infty(\Omega)}.$$

Hence we have the following corollary to Theorem 16.15:

Corollary 16.17 *If in addition to the assumptions of Theorem 16.15 the operator L is of second order and elliptic and if*

$$h = \max\{h_{X_1,\Omega}, h_{X_2,\partial\Omega}\}$$

then

$$\|u - s_{u,\Lambda}\|_{L_\infty(\Omega)} \le Ch^{k-2}\|u\|_{\mathcal{N}_\Phi(\Omega)}$$

for all sufficiently small h.

16.4 Notes and comments

In this chapter we have aimed to demonstrate the potential of (generalized) scattered data approximation. The ultimate goal is the solution of time-dependent partial differential equations with moving boundaries, where classical methods such as finite elements encounter severe problems because of the necessary remeshing. As already mentioned in Chapter 1, first promising steps can be found in Lorentz *et al.* [109] and Behrens and Iske [21]. Other possible applications come from the financial sciences, where differential equations in high-dimensional spaces have to be solved.

The crucial point in the first section was showing that the use of positive definite kernels is so flexible that it essentially makes no difference whether pure interpolation or more general functionals are investigated. The Hermite–Birkhoff interpolation served as an example here. It was initially investigated by Wu [202] and Narcowich and Ward [146].

Besides the collocation method introduced here, which obviously produces symmetric coefficient matrices and which was introduced by Fasshauer [56] and investigated by Franke and Schaback [61,62], there is another method on the market that is often used. This method was introduced by Kansa [96,97] in 1990 and simply uses the Ansatz from pure interpolation,

$$s = \sum_{j=1}^{N} \alpha_j \Phi(\cdot, x_j),$$

but determines the coefficients via general functionals $\lambda_j(s) = \lambda_j(f)$, $1 \le j \le N$. Unless the functionals λ_j are point evaluation functionals at the sites x_j, this method produces a nonsymmetric coefficient matrix, which might even become singular; see Hon and Schaback [85]. But these cases seem to be rare, and in all other cases Kansa's method often behaves better than the symmetric one.

Collocation is not the only method that has been investigated for solving partial differential equations, however. The present author studied in [193, 194] Galerkin methods. In applications, the so-called dual reciprocity method combined with a boundary-element method has often been used. The dual reciprocity method was introduced by Nardini and Brebbia [150] in 1982 and brought into the context of radial basis functions by Golberg [71]. The idea behind this method is to homogenize the differential equation, for example by radial basis functions, and then to use specific boundary-element methods to solve the remaining problem.

17

Interpolation on spheres and other manifolds

So far w have been concerned with interpolation on an arbitrary domain $\Omega \subseteq \mathbb{R}^d$. However, we have not used any topological information about Ω. Instead, we have employed only the fact that it is a subset of $\mathbb{R}^d$. As a matter of fact, without having more information on Ω this is the only way. But many applications provide us with additional information on the underlying domain. For example, problems coming from geology often relate to the entire earth, so that the unit sphere would be an appropriate model and the additional information should lead to a better approximant.

Hence, in this chapter, we want to give an introduction to the theory of scattered data interpolation on spheres and other compact manifolds by radial or zonal functions.

17.1 Spherical harmonics

Generally, functions on the sphere are expressed as Fourier series with respect to an orthonormal family called *spherical harmonics*. In this section we will review the results on these functions. Since this material is only necessary for the present chapter we did not incorporate it into Chapter 5. Moreover, we have to skip the proofs once again. The interested reader is referred to Müller's book [140].

The domain of interest is the d-variate unit sphere $S^{d-1} := \{x \in \mathbb{R}^d : \|x\|_2 = 1\} \subseteq \mathbb{R}^d$. It has surface area

$$\omega_{d-1} = \frac{2\pi^{d/2}}{\Gamma(d/2)}.$$

On S^{d-1}, we will employ the usual inner product

$$(f, g)_{L_2(S^{d-1})} := \int_{S^{d-1}} f(x)g(x)dS(x), \tag{17.1}$$

where $dS(x)$ is given by the standard measure on the sphere.

The distance between two points $x, y \in S^{d-1}$ is the geodesic distance, which is the length of the shorter part of the great circle joining x and y or, in other words, $\operatorname{dist}(x, y) = \arccos(x^T y)$.

There are different ways of introducing spherical harmonics. We start by defining the set of spherical polynomials of degree ℓ, $\pi_\ell(S^{d-1})$, as the restriction of the classical d-variate polynomials of degree ℓ to the sphere, i.e. $\pi_\ell(S^{d-1}) := \pi_\ell(\mathbb{R}^d)|S^{d-1}$. Note that a basis of $\pi_\ell(\mathbb{R}^d)$ is no longer a basis for $\pi_\ell(S^{d-1})$ because of the additional requirement $\|x\|_2^2 = x_1^2 + \cdots + x_d^2 = 1$. A possible basis for $\pi_\ell(S^{d-1})$ is given by the set of spherical harmonics.

Definition 17.1 *The orthogonal complement of $\pi_{\ell-1}(S^{d-1})$ in $\pi_\ell(S^{d-1})$ with respect to the inner product (17.1) is denoted by $\pi_{\ell-1}^\perp(S^{d-1})$. The spherical harmonics $\{Y_{\ell,k} : 1 \le k \le N(d,\ell)\}$ are an orthonormal basis for $\pi_{\ell-1}^\perp(S^{d-1})$.*

Here, we use $\pi_{-1}(S^{d-1}) = \{0\}$, so that $\pi_{-1}^\perp(S^{d-1})$ is the one-dimensional space of constants. It is known that

$$N(d,\ell) = \dim \pi_{\ell-1}^\perp(S^{d-1}) = \begin{cases} 1 & \text{if } \ell = 0, \\ \dfrac{2\ell+d-2}{\ell}\dbinom{\ell+d-3}{\ell-1} & \text{if } \ell \ge 1, \end{cases}$$

and that $N(d,\ell) = \mathcal{O}(\ell^{d-2})$ for $\ell \to \infty$. Since $\pi_\ell(S^{d-1})$ is the disjoint union of the $\pi_{j-1}^\perp(S^{d-1})$ for $0 \le j \le \ell$, the dimension of the space $\pi_\ell(S^{d-1})$ is given by

$$\dim \pi_\ell(S^{d-1}) = \sum_{j=0}^{\ell} N(d,j) = N(d+1,\ell),$$

which is easily established by induction on ℓ.

Another way of introducing spherical harmonics explains the name better. Remember that a harmonic function f satisfies $\Delta f = 0$. A spherical harmonic of order ℓ is in this definition the restriction of a homogeneous harmonic polynomial of degree ℓ to the sphere.

Since the spherical harmonics form an orthonormal basis for $L_2(S^{d-1})$, every function $f \in L_2(S^{d-1})$ has a Fourier representation of the form

$$f = \sum_{\ell=0}^{\infty} \sum_{k=1}^{N(d,\ell)} \widehat{f}_{\ell,k} Y_{\ell,k} \qquad \text{with} \qquad \widehat{f}_{\ell,k} = (f, Y_{\ell,k})_{L_2(S^{d-1})},$$

and we will discuss in particular expansions of this form for the basis functions. But before we do this we have to introduce another class of polynomials, which are important for introducing the analogues of radial functions.

Definition 17.2 *The (generalized) Legendre polynomial of degree ℓ in $d \ge 2$ dimensions is denoted by $P_\ell = P_\ell(d;\cdot)$. It is normalized by $P_\ell(d;1) = 1$ and satisfies*

$$\int_{-1}^{1} P_\ell(t) P_k(t)(1-t^2)^{(d-3)/2} dt = \frac{\omega_{d-1}}{\omega_{d-2} N(d,\ell)} \delta_{\ell,k}.$$

There exists an intrinsic relation between generalized Legendre polynomials and spherical harmonics.

Lemma 17.3 (Addition theorem) *Between the spherical harmonics of order ℓ and the Legendre polynomial of degree ℓ there exists the relation*

$$\sum_{k=1}^{N(d,\ell)} Y_{\ell,k}(x)Y_{\ell,k}(y) = \frac{N(d,\ell)}{\omega_{d-1}} P_\ell(d; x^T y), \qquad x, y \in S^{d-1}.$$

We need another result concerning the asymptotic behavior of both the Legendre functions and the spherical harmonics.

Lemma 17.4 *Let $Y \in \pi_\ell^\perp(S^{d-1}) = \text{span}\{Y_{\ell,k} : 1 \le k \le N(d,\ell)\}$; then*

$$|Y(x)| \le \sqrt{\frac{N(d,\ell)}{\omega_{d-1}}} \|Y\|_{L_2(S^{d-1})}, \qquad x \in S^{d-1}.$$

Moreover, the Legendre polynomials satisfy $|P_\ell(d;t)| \le 1$ for $t \in [-1, 1]$.

17.2 Positive definite functions on the sphere

In this section, we discuss and characterize positive definite functions on the sphere. Of course, the restriction of a positive definite function on $\mathbb{R}^d$ to S^{d-1} forms a positive definite function on the sphere, and we will use this as a standard example. However, this does not take the special situation of points on the sphere into account. Hence, we will investigate positive definite kernels of the form

$$\Phi(x, y) = \sum_{\ell=0}^{\infty} \sum_{k=1}^{N(d,\ell)} a_{\ell,k} Y_{\ell,k}(x) Y_{\ell,k}(y), \qquad x, y \in S^{d-1}. \tag{17.2}$$

As in the $\mathbb{R}^d$ case, radial kernels will play an important role. Note that radial now means with respect to the geodesic distance.

Definition 17.5 *A kernel $\Phi : S^{d-1} \times S^{d-1}$ is called radial or zonal if $\Phi(x, y) = \phi(\text{dist}(x, y)) = \psi(x^T y)$ with univariate functions ϕ, ψ. The function ψ is called the shape function of the kernel Φ.*

A first example of zonal functions comes from the $\mathbb{R}^d$ case. Suppose that $\Phi = \phi(\|\cdot\|_2) : \mathbb{R}^d \to \mathbb{R}$ is a positive definite and radial function on $\mathbb{R}^d$. Since we have, for $x, y \in S^{d-1}$, that $\|x - y\|_2^2 = 2 - 2x^T y$, we can see that the restriction of Φ to S^{d-1} has the representation $\Phi(x - y) = \phi(\|x - y\|_2) = \phi(\sqrt{2 - 2x^T y})$. Thus it is indeed a zonal function with shape function $\psi = \phi(\sqrt{2 - 2\,\cdot})$. Note that the function ϕ here does not coincide with the function ϕ in Definition 17.5. That function is given by $\psi \circ \cos = \phi(\sqrt{2 - 2\cos\cdot})$.

Zonal functions have the remarkable property that all the Fourier coefficients at a given ℓ-level are the same.

Proposition 17.6 *A kernel Φ of the form (17.2) is radial if and only if $a_{\ell,k} = a_\ell$, $1 \le k \le N(d,\ell)$.*

Proof Suppose that $a_{\ell,k} = a_\ell$, $1 \le k \le N(d,\ell)$. Then by the addition theorem we have

$$\Phi(x, y) = \sum_{\ell=0}^{\infty} \frac{a_\ell N(d,\ell)}{\omega_{d-1}} P_\ell(d; x^T y),$$

which shows that Φ is radial. Conversely, if Φ is radial then we can expand the shape function ψ using the orthogonal basis $P_\ell(d; \cdot)$ for $L_2[-1, 1]$ to get

$$\Phi(x, y) = \sum_{\ell=0}^{\infty} b_\ell P_\ell(d; x^T y).$$

The addition theorem and the uniqueness of the Fourier series give the rest. □

Given an expansion of the form (17.2), it is natural to characterize a positive definite function by its Fourier coefficients. To allow point evaluations – so far all expansions have been in the L_2-sense – we have to assume that the coefficients decay fast enough. In the case of a zonal kernel it obviously suffices to require that

$$\sum_{\ell=0}^{\infty} |a_\ell| N(d,\ell) < \infty,$$

since then

$$|\Phi(x, y)| \le \sum_{\ell=0}^{\infty} \frac{|a_\ell| N(d,\ell)}{\omega_{d-1}} |P_\ell(d; x^T y)| < \infty,$$

because of the bound on the Legendre polynomials from Lemma 17.4. The Weierstrass M-test proves continuity. To state the corresponding assumption in the case of nonzonal kernels we first define

$$\widetilde{a}_\ell := \max_{1 \le k \le N(d,\ell)} |a_{\ell,k}| \tag{17.3}$$

and assume that

$$\sum_{\ell=0}^{\infty} \widetilde{a}_\ell N(d,\ell) < \infty. \tag{17.4}$$

Since $N(d,\ell)$ grows as $\mathcal{O}(\ell^{d-2})$ we see that (17.4) is satisfied if, for example,

$$\widetilde{a}_\ell = \mathcal{O}(\ell^{-(d-1)-\epsilon}), \qquad \ell \to \infty, \tag{17.5}$$

with an $\epsilon > 0$.

Lemma 17.7 *Suppose that the coefficients of the kernel (17.2) satisfy (17.4); then Φ is a continuous function in both arguments.*

Proof Let us fix $y \in S^{d-1}$ and define $S_\ell := \sum_{k=1}^{N(d,\ell)} a_{\ell,k} Y_{\ell,k}(y) Y_{\ell,k}$. The norm of this function satisfies

$$\|S_\ell\|_{L_2(S^{d-1})}^2 = \sum_{k=1}^{N(d,\ell)} a_{\ell,k}^2 |Y_{\ell,k}(y)|^2 \le \widetilde{a}_\ell^2 \sum_{k=1}^{N(d,\ell)} |Y_{\ell,k}(y)|^2 = \widetilde{a}_\ell^2 \frac{N(d,\ell)}{\omega_{d-1}},$$

by Lemma 17.3. Hence, from Lemma 17.4 we can conclude that

$$|S_\ell(x)| \le \sqrt{\frac{N(d,\ell)}{\omega_{d-1}}} \|S_\ell\|_{L_2(S^{d-1})} \le \widetilde{a}_\ell \frac{N(d,\ell)}{\omega_{d-1}}, \qquad x \in S^{d-1},$$

so that

$$|\Phi(x,y)| \le \sum_{\ell=0}^{\infty} |S_\ell(x)| \le \sum_{\ell=0}^{\infty} \widetilde{a}_\ell \frac{N(d,\ell)}{\omega_{d-1}} < \infty$$

by assumption (17.4). Continuity now follows by the Weierstrass M-test again. □

Next, we come to the problem of finding positive definite kernels. Remember that a positive definite kernel is by definition continuous and symmetric. Because of the results in the $\mathbb{R}^d$ case the following theorem should not be a surprise. Since the kernel is symmetric and real-valued we can restrict ourselves to real coefficient vectors in the quadratic form.

Theorem 17.8 *Suppose that the kernel (17.2) is continuous. Then it is positive semi-definite if and only if all coefficients are nonnegative. Moreover, if all coefficients are positive then Φ is positive definite.*

Proof For given pairwise distinct points $X = \{x_1, \dots, x_N\} \subseteq S^{d-1}$ and a vector $\alpha \in \mathbb{R}^N$, we can express the quadratic form as follows:

$$\sum_{i,j=1}^{N} \alpha_i \alpha_j \Phi(x_i, x_j) = \sum_{\ell=0}^{\infty} \sum_{k=1}^{N(d,\ell)} a_{\ell,k} \left| \sum_{j=1}^{N} \alpha_j Y_{\ell,k}(x_j) \right|^2.$$

Hence, if all coefficients are nonnegative then clearly the quadratic form is nonnegative. Moreover, if the quadratic form vanishes and if all coefficients are positive we must have

$$\sum_{j=1}^{N} \alpha_j Y(x_j) = 0,$$

for every spherical polynomial Y. Since N is finite, we can find for each $1 \le j \le N$ a spherical polynomial Y_j with $Y_j(x_i) = \delta_{ij}$, which shows that $\alpha_j = 0$.

It remains to demonstrate that a positive semi-definite function has nonnegative Fourier coefficients. The easiest way to do this is to use an equivalent characterization of positive semi-definite functions, namely integrally positive semi-definite functions. Even though we cannot apply Proposition 6.4 directly, its proof implies that

$$\int_{S^{d-1}} \int_{S^{d-1}} \gamma(x)\gamma(y)\Phi(x,y) dS(x) dS(y) \ge 0$$

for all $\gamma \in C(S^{d-1})$. Inserting the expansion (17.2) for Φ and setting $\gamma = Y_{\lambda,\kappa}$ shows that $a_{\lambda,\kappa} \geq 0$. □

In the case of radial functions this reduces to the following result by Schoenberg [174].

Corollary 17.9 (Schoenberg) *A radial function* $\Phi(x, y) = \phi(\text{dist}(x, y))$, $x, y \in S^{d-1}$, *is positive semi-definite if and only if*

$$\phi(r) = \sum_{\ell=0}^{\infty} b_\ell P_\ell(d; \cos r)$$

with

$$b_\ell \geq 0 \text{ for all } \ell \in \mathbb{N}_0 \qquad \text{and} \qquad \sum_{\ell=0}^{\infty} b_\ell < \infty. \tag{17.6}$$

Moreover, if all coefficients b_ℓ *are positive then* Φ *is positive definite.*

Proof Since Φ is radial, its shape function ψ has the representation $\psi(r) = \sum_{\ell=0}^{\infty} b_\ell P_\ell(d; r)$. The coefficients are given by $b_\ell = a_\ell N(d, \ell)/\omega_{d-1}$. Hence, from Theorem 17.8 we have immediately that on the one hand (17.6) implies that Φ is positive semi-definite and that Φ is positive definite if all coefficients are positive.

On the other hand, in the case of a positive semi-definite kernel, Theorem 17.8 shows that all coefficients b_ℓ have to be nonnegative. But then, since Φ is continuous, we can conclude that $\phi(0) = \sum_{\ell=0}^{\infty} b_\ell < \infty$. □

Considering radial or zonal basis functions, one can start with a univariate function ϕ and ask the question whether it is positive definite on every sphere S^{d-1}. Such functions must exist because every radial function that is positive definite on every $\mathbb{R}^d$ is one of them. Moreover, the function $\phi(r) = \cos r$ is positive semi-definite on every sphere S^{d-1} since the quadratic form becomes simply

$$\sum_{j,k=1}^{N} \alpha_j \alpha_k \phi(\text{dist}(x_j, x_k)) = \sum_{j,k=1}^{N} \alpha_j \alpha_k x_j^T x_k = \left\| \sum_{j=1}^{N} \alpha_j x_j \right\|_2^2 \geq 0.$$

Furthermore, the product of two positive semi-definite functions is in turn positive semi-definite and the same is true if we form linear combinations with nonnegative coefficients. This establishes the sufficient part of the next result. For the necessary part, we refer the interested reader again to Schoenberg's paper [174].

Theorem 17.10 (Schoenberg) *A function* ϕ *is positive definite on every sphere* S^{d-1}, $d \geq 2$, *if it has the representation*

$$\phi(r) = \sum_{\ell=0}^{\infty} b_\ell \cos^\ell r,$$

with nonnegative coefficients b_ℓ *that satisfy* $\sum b_\ell < \infty$.

17.3 Error estimates

The investigation of the interpolation error follows the lines of the $\mathbb{R}^d$ case. This means that we employ the native Hilbert space $\mathcal{N}_\Phi(S^{d-1})$ for a positive definite kernel $\Phi \in C(S^{d-1} \times S^{d-1})$ and bound the interpolation error $f - s_{f,X}$ for a function $f \in \mathcal{N}_\Phi(S^{d-1})$ in terms of the fill distance.

Throughout this section we will assume that the Fourier coefficients $a_{\ell,k}$ of the kernel Φ are positive and that they satisfy the decay condition (17.4). From our general theory on native spaces it should be clear what they are in this case. In particular, Theorem 10.29 gives the characterization

$$\mathcal{N}_\Phi(S^{d-1}) = \left\{ f = \sum_{\ell=0}^{\infty} \sum_{k=1}^{N(d,\ell)} \widehat{f}_{\ell,k} Y_{\ell,k} : \sum_{\ell=0}^{\infty} \sum_{k=1}^{N(d,\ell)} \frac{|\widehat{f}_{\ell,k}|^2}{a_{\ell,k}} < \infty \right\},$$

and the inner product takes the form

$$(f, g)_{\mathcal{N}_\Phi(S^{d-1})} = \sum_{\ell=0}^{\infty} \sum_{k=1}^{N(d,\ell)} \frac{\widehat{f}_{\ell,k} \widehat{g}_{\ell,k}}{a_{\ell,k}}.$$

To derive error estimates, we split the error into a product of two terms, the power function and the native space norm of f, as we did in Theorem 11.4. The condition that Ω is open was only necessary for deriving estimates on the derivatives. Here, we want to restrict ourselves to the case of a pure interpolation error, so we do not need to incorporate this condition.

The next step is to bound the power function in terms of the fill distance, which now takes the form

$$h_X := \sup_{x \in S^{d-1}} \inf_{x_j \in X} \operatorname{dist}(x, x_j),$$

paying tribute to the special topology of the sphere. For h_X we have dropped the additional index that indicated the domain, since we are now just working on the sphere.

The original idea behind bounding the power function was to use its minimization property with respect to its coefficients and to construct a local polynomial reproduction. The same is possible here, but the compactness of the sphere makes the locality unnecessary. Hence, we only have to construct a family of functions that reproduces polynomials and has a uniformly bounded Lebesgue function. This is done by employing norming sets once again.

Lemma 17.11 *Suppose that the knot set $X = \{x_1, \dots, x_N\} \subseteq S^{d-1}$ has fill distance $h_X \leq 1/(2m)$. Then, $Z = \operatorname{span}\{\delta_x : x \in X\}$ is a norming set for $\pi_m(S^{d-1})$ with norming constant $c = 1/2$.*

Proof The proof uses the same ideas as the proof of the corresponding result in $\mathbb{R}^d$. Since the restriction of any spherical polynomial $Y \in \pi_m(S^{d-1})$ to a great circle is a univariate trigonometric polynomial of degree less than or equal to m, we can apply the classical Bernstein inequality to get

$$|Y(x) - Y(y)| \leq m \operatorname{dist}(x, y) \, \|Y\|_{L_\infty(S^{d-1})}, \qquad x, y \in S^{d-1}.$$

Hence, if $Y \in \pi_m(S^{d-1})$ satisfies $\|Y\|_{L_\infty(S^{d-1})} = 1$ then there exists a point $x \in S^{d-1}$ such that $|Y(x)| = 1$. By the condition imposed on X we can find a data site x_j such that $\operatorname{dist}(x, x_j) \le h_X \le 1/(2m)$. This, together with Bernstein's inequality, shows that $|Y(x) - Y(x_j)| \le 1/2$ or $|Y(x_j)| \ge 1/2$. □

From the general theory on norming sets, in particular Theorem 3.4, we can immediately conclude

Corollary 17.12 *Suppose the knot set $X = \{x_1, \ldots, x_N\} \subseteq S^{d-1}$ has fill distance $h_X \le 1/(2m)$. Then there exist functions $u_j : S^{d-1} \to \mathbb{R}$ such that*

(1) $\sum_{j=1}^N u_j(x)Y(x_j) = Y(x)$ for all $Y \in \pi_m(S^{d-1})$ and $x \in S^{d-1}$,
(2) $\sum_{j=1}^N |u_j(x)| \le 2$ for all $x \in S^{d-1}$.

We will use this "global" (if compared to the $\mathbb{R}^d$ case) result to derive our first bound on the interpolation error. We express the error in terms of the Fourier coefficients of the basis kernel.

Theorem 17.13 *Suppose that the kernel Φ has only positive Fourier coefficients $a_{\ell,k}$, which satisfy the decay condition (17.4). Suppose further that $X = \{x_1, \ldots, x_N\} \subseteq S^{d-1}$ has fill distance $1/(2m+2) < h_X \le 1/(2m)$. Then the error between $f \in \mathcal{N}_\Phi(S^{d-1})$ and its interpolant $s_{f,X}$ can be bounded by*

$$|f(x) - s_{f,X}(x)|^2 \le \frac{9}{\omega_{d-1}} \sum_{\ell=m+1}^{\infty} \widetilde{a}_\ell N(d,\ell) \|f\|^2_{\mathcal{N}_\Phi(S^{d-1})}, \qquad x \in S^{d-1}.$$

Proof As usual, at the start we bound the interpolation error by $|f(x) - s_{f,X}(x)| \le P_{\Phi,X}(x)\|f\|_{\mathcal{N}_\Phi(\Omega)}$. Next, also as usual, we use the cardinal functions $\{u_j^*\}$ from Theorem 11.1 and the kernel expansion (17.2) to express the power function as

$$P^2_{\Phi,X}(x) = \sum_{\ell=0}^{\infty} \sum_{k=1}^{N(d,\ell)} a_{\ell,k} \left| Y_{\ell,k}(x) - \sum_{j=1}^{N} u_j^*(x) Y_{\ell,k}(x_j) \right|^2.$$

In the final step we employ the minimal property of the power function from Theorem 11.5 and replace the functions $\{u_j^*\}$ by the functions $\{u_j\}$ from Corollary 17.12. Furthermore, we set $u_0(x) := -1$ and $x_0 = x$, to derive

$$\begin{aligned} P^2_{\Phi,X}(x) &\le \sum_{\ell=m+1}^{\infty} \sum_{k=1}^{N(d,\ell)} a_{\ell,k} \left| Y_{\ell,k}(x) - \sum_{j=1}^{N} u_j(x) Y_{\ell,k}(x_j) \right|^2 \\ &\le \sum_{\ell=m+1}^{\infty} \widetilde{a}_\ell \sum_{k=1}^{N(d,\ell)} \sum_{i,j=0}^{N} u_i(x)u_j(x) Y_{\ell,k}(x_i) Y_{\ell,k}(x_j) \\ &= \sum_{\ell=m+1}^{\infty} \widetilde{a}_\ell \sum_{i,j=0}^{N} u_i(x)u_j(x) \frac{N(d,\ell)}{\omega_{d-1}} P_\ell(d; x_i^T x_j). \end{aligned}$$

The last equality follows from Lemma 17.3. If finally we take into account that the generalized Legendre polynomial P_ℓ is bounded by one and that

$$\sum_{i,j=0}^{N} |u_i(x)u_j(x)| \le \left(1+\sum_{j=1}^{N}|u_j(x)|\right)^2 \le 9,$$

then by Corollary 17.12 we have completed the proof. □

It is now easy to express the error estimates in terms of the fill distance if additional assumptions on the decay of the Fourier coefficients are made.

Corollary 17.14 *Suppose that the assumptions of Theorem 17.13 hold.*
(1) If $\widetilde{a}_\ell N(d,\ell) \le c(1+\ell)^{-\alpha}$ *with* $\alpha > 1$ *then*

$$\|f - s_{f,X}\|_{L_\infty(S^{d-1})} \le C h_X^{(\alpha-1)/2} \|f\|_{\mathcal{N}_\Phi(S^{d-1})}.$$

(2) If $\widetilde{a}_\ell N(d,\ell) \le c e^{-\alpha(1+\ell)}$ *with* $\alpha > 0$ *then*

$$\|f - s_{f,X}\|_{L_\infty(S^{d-1})} \le C e^{-\alpha/(4h_X)} \|f\|_{\mathcal{N}_\Phi(S^{d-1})}.$$

Proof In the first case, the assumption imposed on $\widetilde{a}_\ell$ gives

$$\sum_{\ell=m+1}^{\infty} \widetilde{a}_\ell N(d,\ell) \le c \int_m^\infty (1+\ell)^{-\alpha} d\ell = \frac{c}{1-\alpha}(1+m)^{-\alpha+1} = \frac{c\, 2^{\alpha-1}}{\alpha-1} h_X^{\alpha-1}.$$

In the second case, the same argument yields

$$\sum_{\ell=m+1}^{\infty} \widetilde{a}_\ell N(d,\ell) \le c \int_m^\infty e^{-\alpha(1+\ell)} d\ell = \frac{c}{\alpha} e^{-\alpha(1+m)} = \frac{c}{\alpha} e^{-\alpha/(2h_X)}.$$

□

By now, it should be clear that other results, for example those on doubling the approximation order, can be carried over to the sphere in the same way. We leave the details to the reader.

17.4 Interpolation on compact manifolds

The sphere is one possible example of a compact smooth manifold. In this short section we will point out some ideas on how the results that we have reached so far can be extended to other manifolds. As in the last two sections we will concentrate on positive definiteness and error estimates.

For the convenience of the reader we review the necessary material on manifolds. A good source for this is the book [30] by Boothby.

Definition 17.15 *A set* $M \subseteq \mathbb{R}^d$ *is called a topological manifold of dimension* n *if it is a Hausdorff space with a countable basis of open sets such that for every* $x \in M$ *there exist an open set* $U \subseteq M$ *with* $x \in U$ *and a mapping* $\varphi : U \to \mathbb{R}^n$ *that maps* U *homeomorphically* *to the open set* $V := \varphi(U) \subseteq \mathbb{R}^n$. *The pair* (U, φ) *is called a coordinate neighborhood of* x *or a chart. A chart is of class* C^k *if* $\varphi^{-1} \in C^k(\varphi(U))$. *A collection* $\mathcal{A} = \{(U_\alpha, \varphi_\alpha)\}$ *of* C^k*-charts is called a* C^k*-atlas of* M *if*

(1) the sets U_α *cover* M,
(2) for any U_α, U_β *with* $U_\alpha \cap U_\beta \neq \emptyset$ *the functions* $\varphi_\beta \circ \varphi_\alpha^{-1}$ *and* $\varphi_\alpha \circ \varphi_\beta^{-1}$ *are in* C^k *on* $\varphi_\alpha(U_\alpha \cap U_\beta)$ *and* $\varphi_\beta(U_\alpha \cap U_\beta)$, *respectively.*

Finally, a manifold M *is called a* C^k*-manifold if it possesses a* C^k*-atlas.*

The smoothness of a function $f : M \to \mathbb{R}$ is defined by the smoothness of $f \circ \varphi^{-1}$ with a chart (U, φ). To be more precise, we will say that f is k times differentiable on M, or $f \in C^k(M)$, if $f \circ \varphi^{-1} \in C^k(\varphi(U))$ for every chart (U, φ) of M. In what follows we will assume that the underlying manifold is sufficiently smooth.

Most relevant examples such as the sphere and the torus are submanifolds of $\mathbb{R}^d$. This means in particular that they inherit the standard metric of $\mathbb{R}^d$, which is induced by the Euclidean norm. However, everything we have in mind works in the more general setting of Riemannian manifolds. To introduce them, we have to recall concepts regarding curves on manifolds and tangent spaces.

For $x \in M$, the tangent space $T_x(M)$ consists of all tangent vectors v to M in x. Here, a vector v is a tangent vector if there exists a differentiable curve $\gamma : [-\epsilon, \epsilon] \to M$ with $\gamma(0) = x$ and $\gamma'(0) = v$. It turns out that $T_x(M)$ is an n-dimensional subspace of $\mathbb{R}^d$ and that a basis is given by

$$\frac{\partial \varphi^{-1}}{\partial v_1}(\varphi(x)), \quad \ldots, \quad \frac{\partial \varphi^{-1}}{\partial v_n}(\varphi(x)).$$

Definition 17.16 *A* C^k*-manifold is called a* C^k *Riemannian manifold if for every* $x \in M$ *there exists an inner product* $g_x : T_x(M) \times T_x(M) \to \mathbb{R}$ *such that for every coordinate neighborhood* (U, φ) *the* n^2 *functions*

$$g_{ij}^{\varphi}(v) := g_{\varphi^{-1}(v)}\left(\frac{\partial \varphi^{-1}}{\partial v_i}(v), \frac{\partial \varphi^{-1}}{\partial v_j}(v)\right), \qquad v \in V = \varphi(U),$$

are in $C^k(V)$.

Finally, we use the Riemannian metric to define the length of a curve on M.

Definition 17.17 *Suppose that* M *is a* C^k *Riemannian manifold. Let* $x, y \in M$ *be two distinct points and let* $\gamma : [a, b] \to M$ *be a piecewise* C^1 *curve that connects these points, i.e.* $\gamma(a) = x$, $\gamma(b) = y$. *Then the length of* γ *is*

$$L(\gamma) = L(\gamma; a, b) := \int_a^b \left[g_{\gamma(t)}\left(\frac{d\gamma}{dt}(t), \frac{d\gamma}{dt}(t)\right)\right]^{1/2} dt.$$

Finally, we set $\operatorname{dist}(x, y)$ *to be the infimum over the length of all such curves connecting* x *and* y. *The shortest such curve is called the shortest path for* x *and* y, *and* $\operatorname{dist}(x, y)$ *is their geodesic or Riemannian distance.*

Several remarks are necessary. First of all, if $M = \mathbb{R}^n$ and if g_x is the canonical inner product on $\mathbb{R}^n$ then our definition of the length of a curve coincides with the classical definition. In this case $\mathrm{dist}(x, y) = \|x - y\|_2$, i.e the shortest curve between two points in $\mathbb{R}^n$ is the straight line between them. In general, dist defines a metric on M, if M is connected. The latter assumption is obviously necessary to make dist always well defined. Even more, the topology induced by dist is equivalent to the initial topology on M. We will come back to this later. On the sphere, our new definition of dist coincides with the old one, since both denote the length of the shorter portion of the great circle connecting the two points.

The inner products g_x, $x \in M$, are necessary for introducing the concept of integration on M. Suppose M is a compact C^k Riemannian manifold with C^k-atlas $\mathcal{A} = \{(U_j, \varphi_j)\}_{j=1}^L$. For this atlas one can choose a partition of unity, i.e. a family of functions $\{\chi_j\}$ such that $\sum \chi_j = 1$ on M and $\mathrm{supp}(\chi_j) \subseteq U_j$. Moreover, a reasonable choice makes $\chi_j \circ \varphi_j^{-1}$ integrable over $\varphi_j(U)$. Then the integral for a measurable function f on M is defined by

$$\int_M f(x)dS(x) = \sum_{j=1}^{L} \int_{\varphi_j(U_j)} (\chi_j f) \circ \varphi_j^{-1}(v) g_j^{1/2}(v) dv,$$

with $g_j(v) := \det(g_{ik}^{\varphi_j}(v))$. Of course, one has to show that the integral is independent of the chosen atlas and the chosen partition of unity. In this sense, spaces of integrable functions can be introduced.

After reviewing the basic concepts of Riemannian manifolds, we return to the study of positive definite kernels on them. A first obvious but also intrinsic observation is the following.

Proposition 17.18 *Suppose that M is a differentiable manifold and (U, φ) is a chart. If $\Phi : M \times M \to \mathbb{R}$ is a positive definite kernel on M then*

$$\Psi(u, v) := \Phi(\varphi^{-1}(u), \varphi^{-1}(v)), \qquad u, v \in \varphi(U),$$

is a positive definite kernel on $\varphi(U)$.

Reversing the argument, one could use this result to prove the positive definiteness of a kernel defined on M. But since everything depends on the charts that are chosen, the use of Proposition 17.18 in this context is restricted. However, in providing error estimates it will be very helpful.

On the sphere, we used the expansion (17.2) of Φ in terms of spherical harmonics to characterize positive definite kernels. Spherical harmonics can also be interpreted as the system of eigenfunctions of the Laplace–Beltrami operator on S^{d-1}. Hence, a possible way of generalizing (17.2) would be to use the eigenfunctions of the Laplace–Beltrami operator on M; these form an orthonormal basis of $L_2(M)$. But any other orthonormal basis of $L_2(M)$, which is also dense in $C(M)$, will do.

Theorem 17.19 *Let M be a compact C^k Riemannian manifold and let $\{Y_\ell\}_{\ell=1}^\infty$ be an orthonormal basis for $L_2(M)$, which is also dense in $C(M)$. Suppose that*

$$\Phi(x, y) = \sum_{\ell=1}^{\infty} a_\ell Y_\ell(x) Y_\ell(y)$$

is in $C(M \times M)$. Then Φ is positive semi-definite on M if and only if all the coefficients a_ℓ are nonnegative. Moreover, if they are all positive then Φ is positive definite.

Proof The proof is more or less the same as the proof for Theorem 17.8. Hence, we leave the details for the reader. □

After having characterized positive definite kernels on manifolds, we come to error estimates. Of course, we could split the interpolation error again, using the power function; then the next step would be to use the optimality of the power function with respect to its coefficients. Hence, if we wanted to follow this path we would have to construct a better-suited family of functions, which reproduce polynomials and have uniformly bounded Lebesgue functions. In fact, though, we want to use a different approach based on Proposition 17.18. To this end we need to relate the distance measure on the manifold to that in the range of the charts. As pointed out earlier, the main implication of this is that the dist topology is equivalent to the initial topology.

Lemma 17.20 *Let M be a C^k Riemannian manifold with $k \geq 1$. For every $x \in M$ there exists a chart (U_x, φ_x) with $x \in U_x$ and constants $0 < m_x \leq M_x$ such that*

$$m_x \|\varphi_x(y) - \varphi_x(z)\|_2 \leq \text{dist}(y, z) \leq M_x \|\varphi_x(y) - \varphi_x(z)\|_2, \qquad y, z \in U_x.$$

Moreover, $\varphi(U_x)$ can be chosen as $\{v \in \mathbb{R}^n : \|v\|_2 < r\}$ with $r = r_x > 0$. Finally, φ_x^{-1} is in C^k up to the boundary of $\varphi_x(U_x)$.

Proof Let (U, φ) be a chart with $x \in U$. Without restriction we can assume that $\varphi(x) = 0$. Moreover, there exists an $r > 0$ such that the closed ball $B(0, 3r) = \{v \in \mathbb{R}^n : \|v\|_2 \leq 3r\}$ is contained in $V := \varphi(U)$. We define $\widetilde{V} := \{v \in \mathbb{R}^n : \|v\|_2 < r\}$ and $\widetilde{U} := \varphi^{-1}(\widetilde{V})$.

Next, we note that the function

$$(y, \alpha) \mapsto \sum_{j,k=1}^{n} \alpha_j \alpha_k g_y \left(\frac{\partial \varphi^{-1}}{\partial v_j}(\varphi(y)), \frac{\partial \varphi^{-1}}{\partial v_k}(\varphi(y)) \right)$$

is continuous on the compact set $\varphi^{-1}(B(0, 3r)) \times S^{n-1}$ and hence attains a minimum and a maximum. Moreover, since g_y is positive definite, the minimum is positive. In other words, there exist constants $0 < m_x \leq M_x < \infty$ such that

$$m_x \leq \sum_{j,k=1}^{n} \alpha_j \alpha_k g_y \left(\frac{\partial \varphi^{-1}}{\partial v_j}(\varphi(y)), \frac{\partial \varphi^{-1}}{\partial v_k}(\varphi(y)) \right) \leq M_x, \tag{17.7}$$

holds for all $y \in \varphi^{-1}(B(0, 3r))$, $\alpha \in S^{n-1}$.

Now suppose that $y, z \in \widetilde{U}$ are given and that $\gamma : [a, b] \to M$ is a connecting piecewise differentiable curve. We are going to bound its length from below.

For the moment, we will suppose that γ stays completely within the set $\varphi^{-1}(B(0, 3r))$. Then $v := \varphi \circ \gamma$ denotes a piecewise differentiable curve that connects $\varphi(y)$ with $\varphi(z)$ and is completely contained in $B(0, 3r)$. Moreover, $\gamma = \varphi^{-1} \circ v$ gives

$$\frac{d\gamma}{dt} = \sum_{j=1}^{n} \frac{dv_j}{dt} \frac{\partial \varphi^{-1}}{\partial v_j} \circ v,$$

so that

$$\begin{aligned} g_{\gamma(t)}\left(\frac{d\gamma}{dt}, \frac{d\gamma}{dt}\right) &= \sum_{j,k=1}^{n} \frac{dv_j}{dt} \frac{dv_k}{dt} g_{\gamma(t)}\left(\frac{\partial \varphi^{-1}}{\partial v_j}(v(t)), \frac{\partial \varphi^{-1}}{\partial v_k}(v(t))\right) \\ &\geq m_x \left\| \frac{dv}{dt} \right\|_2 . \end{aligned}$$

But this means in particular that

$$L(\gamma) \geq m_x \int_a^b \left\| \frac{dv}{dt} \right\|_2 \geq m_x \|b - a\|_2,$$

where the last inequality follows from the fact that the shortest curve between two points in $\mathbb{R}^n$ is the connecting line segment.

If γ leaves $\varphi^{-1}(B(0, 3r))$ then there exists a point $\widetilde{z} = \varphi^{-1}(c)$, $\|c\|_2 = 3r$, on the curve. But since $\varphi(y)$ and $\varphi(z)$ are contained in the ball around zero with radius r, the computations just made show that

$$L(\gamma) \geq L(\gamma; a, c) \geq m_x \|c - a\|_2 \geq 2r m_x \geq m_x \|b - a\|_2.$$

Since γ is an arbitrary curve, we have proven that

$$\operatorname{dist}(y, z) \geq m_x \|\varphi(y) - \varphi(z)\|_2.$$

For the upper bound, we can choose an arbitrary curve that connects y and z, for example, $\gamma(t) = \varphi^{-1}(t\varphi(y) + (1 - t)\varphi(z))$, $t \in [0, 1]$. Since this particular curve is everywhere contained within $\varphi^{-1}(B(0, 3r))$, we can conclude from (17.7) that

$$\operatorname{dist}(y, z) \leq L(\gamma) \leq M_x \|\varphi(y) - \varphi(z)\|_2.$$

□

The preceding result makes it easy to reduce the error estimates to those of the $\mathbb{R}^d$ case. As usual, we give the error estimates for functions from the native space $\mathcal{N}_\Phi(M)$ and express them in terms of the fill distance $h_{X,M}$, which is now defined using the geodesic distance $\operatorname{dist}(\cdot, \cdot)$. In the proof, we will also use the fill distance for sets in $\mathbb{R}^n$, which will be defined using the Euclidean norm. The reader should notice where each definition is employed.

Theorem 17.21 *Let M be a compact C^ℓ Riemannian manifold. Suppose that $\Phi \in C^{2k}(M \times M)$ is positive definite and we have $\ell \geq 2k$. Then there exist $h_0 > 0$ and $C > 0$ such that for all $f \in \mathcal{N}_\Phi(M)$ and all $X \subseteq M$ with $h_X \leq h_0$ the error between f and its interpolant $s_{f,X}$ can be bounded by*

$$|f(x) - s_{f,X}(x)| \leq C h_{X,M}^k \|f\|_{\mathcal{N}_\Phi(M)}, \qquad x \in M.$$

Proof For every $x \in M$ we choose a chart (U_x, φ_x) according to Lemma 17.20. Since M is compact we need only a finite number of them to cover M, say $(U_{y_j}, \varphi_{y_j})_{1 \leq j \leq L}$. Let (U, φ) be one of these. By Proposition 17.18, the kernel $\Psi(u, v) := \Phi(\varphi^{-1}(u), \varphi^{-1}(v))$, $u, v \in V := \varphi(U)$, is positive definite on V. Moreover, because of the smoothness assumptions, we have $\Psi \in C^{2k}(V \times V)$. As in Lemma 16.14, one sees that the power functions are related by

$$P_{\Phi, X \cap U}(x) = P_{\Psi, \varphi(X \cap U)}(\varphi(x)), \qquad x \in U,$$

and we can study the power function on the right-hand side, which now exists on an open ball in $\mathbb{R}^n$. Hence, Theorem 11.13 and Lemma 17.20 give

$$\begin{aligned} P_{\Psi, \varphi(X \cap U)}(\varphi(x)) &\leq C C_\Psi(\varphi(x))^{1/2} h^k_{\varphi(X \cap U), \varphi(U)} \\ &\leq C C_\Psi(\varphi(x))^{1/2} h^k_{X \cap U, U}. \end{aligned}$$

The number $C_\Psi(\varphi(x))$ can be uniformly bounded since φ^{-1} is in C^ℓ up to the boundary of $\varphi(U)$. Moreover, $h_{X \cap U, U}$ can be bounded by a constant times $h_{X,M}$ for sufficiently small $h_{X,M}$. Since we have only a finite number of such charts, this completes the proof. □

17.5 Notes and comments

Schoenberg [174] was the first to study positive semi-definite, and in particular zonal, functions on the sphere. Almost every other paper concerned with this subject is based on his work. Since Schoenberg was interested in positive semi-definite rather than positive definite functions, this has left plenty of room for other authors. Interestingly, besides the confusion already mentioned about the term *positive definiteness*, in the context of the sphere there are two different definitions of positive definite functions on the market. While some authors define this as we have done, others, for example Narcowich [141], use an integral characterization. The latter approach is more restrictive than ours, as pointed out by Ron and Sun [157].

The first papers concerned with conditions for positive definiteness were by Light and Cheney [106] and by Xu and Cheney [205]. In the first paper, the authors restricted themselves to the unit sphere in $\mathbb{R}^2$ or, in other words, to periodic basis functions. In these two papers, positive definiteness for a finite number N of centers was first considered: rather than formulating conditions on Φ that give rise to positive definite interpolation matrices for all N, the authors were looking for conditions only for a fixed N. Most of the subsequent papers followed this approach. Besides the authors already mentioned, Menegatto has done most of the investigations on positive definiteness; see for example [127, 128].

There are in the main three different approaches for providing error estimates for interpolation by positive definite functions on the sphere. The first, which we presented in Section 17.3, is due to Jetter *et al.* [89] with recent improvements by Morton and Neamtu [138]. The second mimics the $\mathbb{R}^d$ ideas of a local polynomial reproduction; details can be found in the paper [72] by Golitschek and Light. Finally, the third approach is the one that works for arbitrary smooth Riemannian manifolds. The idea of using local coordinates was employed for radial basis functions by Levesley and Ragozin [103], even if their arguments differ slightly from ours. The present author [197] has used the local coordinate argument in the case of moving least squares approximation on the sphere.

A thorough general discussion of positive definite functions on arbitrary manifolds started with the paper [141] by Narcowich, that has been mentioned already.

Recent overviews of other approximation methods on the sphere were given by Freden *et al.* in [65, 66] and by Fasshauer and Schumaker in [57].

References

[1] E. J. Akutowicz. On extrapolating a positive definite function from a finite interval. *Math. Scand.*, **7**: 157–169, 1959.

[2] N. Aronszajn. Theory of reproducing kernels. *Trans. Am. Math. Soc.*, **68**: 337–404, 1950.

[3] S. Arya and D. M. Mount. Approximate nearest neighbor searching. In *Proc. 4th Ann. ACM-SIAM Symposium on Discrete Algorithms*, pp. 271–280, New York, ACM Press, 1993.

[4] S. Arya and D. M. Mount. Approximate range searching. In *Proc. 11th Annual ACM Symp. on Computational Geometry*, pp. 172–181, New York, ACM Press, 1995.

[5] S. Arya, D. M. Mount, N. S. Netanyahu, R. Silverman, and A. Y. Wu. An optimal algorithm for approximate nearest neighbor searching. *J. ACM*, **45**: 891–923, 1998.

[6] R. Askey. Radial characteristic functions. Technical Report 1262, University of Wisconsin, Mathematics Research Center, 1973.

[7] I. Babuska and J. M. Melenk. The partition of unity method. *Int. J. Numer. Methods Eng.*, **40**: 727–758, 1997.

[8] K. Ball. Eigenvalues of Euclidean distance matrices. *J. Approx. Theory*, **8** pp. 74–82, 1992.

[9] H. Bauer. *Measure and Integration Theory*. Berlin, de Gruyter, 2001.

[10] B. J. C. Baxter and G. Roussos. A new error estimate of the fast Gauss transform. *SIAM J. Sci. Comput.*, **24**: 257–259, 2002.

[11] R. K. Beatson and E. Chacko. Fast evaluation of radial basis functions: a multivariate momentary evaluation scheme. In A. Cohen, C. Rabut, and L. L. Schumaker, eds., *Curve and Surface Fitting: Saint-Malo 1999*, pp. 37–46, Nashville, Vanderbilt University Press, 2000.

[12] R. K. Beatson, J. B. Cherrie, and D. L. Ragozin. Polyharmonic splines in $\mathbb{R}^d$: tools for fast evaluation. In A. Cohen, C. Rabut, and L. L. Schumaker, eds., *Curve and Surface Fitting: Saint-Malo 1999*, pp. 47–56, Nashville, Vanderbilt University Press, 2000.

[13] R. K. Beatson and L. Greengard. A short course on fast multipole methods. In M. Ainsworth et al., eds., *Wavelets, Multilevel Methods and Elliptic PDEs. 7th EPSRC Numerical analysis Summer School, University of Leicester, Leicester, GB, July 8–19, 1996*, pp. 1–37. Oxford, Clarendon Press, 1997.

[14] R. K. Beatson and W. A. Light. Fast evaluation of radial basis functions: methods for two-dimensional polyharmonic splines. *IMA J. Numer. Anal.*, **17**: 343–372, 1997.

[15] R. K. Beatson, W. A. Light, and S. Billings. Fast solution of the radial basis function interpolation equations: domain decomposition methods. *SIAM J. Sci. Comput.*, **22**: 1717–1740, 2000.

[16] R. K. Beatson and G. N. Newsam. Fast evaluation of radial basis functions: I. *Comput. Math. Appl.*, **24**: 7–19, 1992.

[17] R. K. Beatson and G. N. Newsam. Fast evaluation of radial basis functions: Moment-based methods. *SIAM J. Sci. Comput.*, **19**: 1428–1449, 1998.

[18] R. K. Beatson and M. J. D. Powell. An iterative method for thin plate spline interpolation that employs approximations to Lagrange functions. In D. F. Griffiths *et al.*, eds., *Numerical analysis 1993. Proc. 15th Dundee Biennial Conf. on Numerical Analysis at University of Dundee (United Kingdom), June 29–July 2, 1993*, pp. 17–39, Harlow, Longman Scientific & Technical, 1994.

[19] A. Beckert. Coupling fluid (CFD) and structural (FE) models using finite interpolation elements. *Aerosp. Sci. Technol.*, **1**: 13–22, 2000.

[20] A. Beckert and H. Wendland. Multivariate interpolation for fluid-structure-interaction problems using radial basis functions. *Aerosp. Sci. Technol.*, **5**: 125–134, 2001.

[21] J. Behrens and A. Iske. Grid-free adaptive semi-Lagrangian advection using radial basis functions. *Comput. Math. Appl.*, **43**: 319–327, 2002.

[22] A. Bejancu. Local accuracy for radial basis function interpolation on finite uniform grids. *J. Approx. Theory*, **99**: 242–257, 1999.

[23] T. Belytschko, Y. Krongauz, D. Organ, M. Fleming, and P. Krysl. Meshless methods: an overview and recent developments. *Comput. Methods in Appl. Mechanics and Engineering*, **139**: 3–47, 1996.

[24] J. L. Bentley. Multidimensional binary search trees used for associative searching. *Comm. ACM*, **18**: 509–517, 1975.

[25] S. Bernstein. Sur la définition et les propriétés des fonctions analytiques d'une variable réelle. *Math. Ann.*, **75**: 449–468, 1914.

[26] S. Bernstein. Sur les fonctions absolument monotones. *Acta Math.*, **51**: 1–66, 1928.

[27] J. F. Blinn. A generalization of algebraic surface drawing. *ACM Transaction on Graphics*, **1**: 235–256, 1982.

[28] S. Bochner. *Vorlesungen über Fouriersche Integrale*. Leipzig, Akademischer Verlagsgesellschaft, 1932.

[29] S. Bochner. Monotone Funktionen, Stieltjes Integrale und harmonische Analyse. *Math. Ann.*, **108**: 378–410, 1933.

[30] W. M. Boothby. *An Introduction to Differentiable Manifolds and Riemannian Geometry*. New York, Academic Press, 1975.

[31] S. Brenner and L. Scott. *The Mathematical Theory of Finite Element Methods*. New York, Springer, 1994.

[32] R. Brownlee and W. Light. Approximation orders for interpolation by surface splines to rough functions. Preprint, University of Leicester, *IMA J. Numer. Anal*, **24**: 179–192, 2004.

[33] M. D. Buhmann. Multivariable interpolation using radial basis functions. Ph.D. thesis, University of Cambridge, 1989.

[34] M. D. Buhmann. Radial functions on compact support. *Proc. Edinb. Math. Soc.*, **41**: 33–46, 1998.

[35] M. D. Buhmann. A new class of radial basis functions with compact support. *Math. Comput.*, **70**: 307–318, 2001.

[36] S. Cambanis, R. Keener, and G. Simons. On α-symmetric multivariate distributions. *J. Multivariate Anal.*, **13**: 213–233, 1983.

[37] J. C. Carr, R. K. Beatson, J. B. Cherrie, T. J. Mitchell, W. R. Fright, B. C. McCallum, and T. R. Evans. Reconstruction and representation of 3D objects with radial basis functions. In *Computer Graphics Proceedings, Annual Conference Series*, pp. 67–76. Addison Wesley, 2001.

[38] J. R. Cebral and R. Löhner. Conservative load projection and tracking for fluid-structure problems. *AIAA Journal*, **35**: 687–692, 1997.

[39] K. F. Chang. Strictly positive definite functions. *J. Approx. Theory*, **87**: 148–158, 1996.

[40] A. Y. Chanysheva. Positive definite functions of a special form. *Moscow University Math. Bull.*, **45**: 57–59, 1990.

[41] E. W. Cheney. *Introduction to Approximation Theory*. New York, McGraw-Hill, 1966.

[42] M. de Berg, M. van Kreveld, M. Overmars, and O. Schwarzkopf. *Computational Geometry*. Berlin, Springer, 1997.

[43] C. de Boor. *A Practical Guide to Splines*, revised edition. New York, Springer, 2001.

[44] C. de Boor, K. Höllig, and S. Riemenschneider. *Box Splines*. New York, Springer, 1993.

[45] J. Deny and J.-L. Lions. Les espaces du type de Beppo Levi. *Ann. Inst. Fourier*, **5**: 305–370, 1953/54.

[46] W. F. Donoghue. *Distributions and Fourier transforms*. New York, Academic Press, 1969.

[47] J. Duchon. Interpolation des fonctions de deux variables suivant le principe de la flexion des plaques minces. *Rev. Française Automat. Informat. Rech. Opér. Anal. Numer.*, **10**: 5–12, 1976.

[48] J. Duchon. Splines minimizing rotation-invariant semi-norms in Sobolev spaces. In W. Schempp and K. Zeller, eds., *Constructive Theory of Functions of Several Variables*, pp. 85–100. Springer, Berlin, 1977.

[49] J. Duchon. Sur l'erreur d'interpolation des fonctions de plusieurs variables par les D^m-splines. *Rev. Française Automat. Informat. Rech. Opér. Anal. Numer.*, **12**: 325–334, 1978.

[50] M. Falcone and R. Ferretti. Convergence analysis for a class of high-order semi-Lagrangian advection schemes. *SIAM J. Numer. Anal.*, **35**: 909–940, 1998.

[51] C. Farhat and M. Chandesris. Time-decomposed parallel time-integrators: theory and feasibility studies for fluid, structure, and fluid-structure applications. *Int. J. Numer. Methods Eng.*, **58**: 1397–1434, 2003.

[52] C. Farhat and M. Lesoinne. Higher-order staggered and subiteration free algorithms for coupled dynamic aeroelasticity problems. In *36th Aerospace Sciences Meeting and Exhibit., AIAA 98-0516, Reno, New Virginia*, 1998.

[53] R. Farwig. Multivariate interpolation of arbitrarily spaced data by moving least squares methods. *J. Comput. Appl. Math.*, **16**: 79–83, 1986.

[54] R. Farwig. Multivariate interpolation of scattered data by moving least squares methods. In J. C. Mason and M. G. Cox, eds., *Algorithms for Approximation*, pp. 193–211. Clarendon Press, Oxford, 1987.

[55] R. Farwig. Rate of convergence of moving least squares interpolation methods: the univariate case. In P. Nevai and A. Pinkus, eds., *Progress in Approximation Theory*, pp. 313–327. Boston, Academic Press, 1991.

[56] G. E. Fasshauer. Solving partial differential equations by collocation with radial basis functions. In A. Le Méhauté, C. Rabut, and L. L. Schumaker, eds., *Surface Fitting and Multiresolution Methods*, pp. 131–138. Vanderbilt University Press, Nashville, 1997.

[57] G. E. Fasshauer and L. L. Schumaker. Scattered data fitting on the sphere. In M. Dæhlen, T. Lyche, and L. L. Schumaker, eds., *Mathematical Methods for Curves and Surfaces II*, pp. 117–166. Vanderbilt University Press, Nashville, 1998.

[58] A. Faul. *Iterative techniques for radial basis function interpolation*. Ph.D. thesis, Churchill College Cambridge, 2000.

[59] A. Faul and M. J. D. Powell. Proof of convergence of an iterative technique for thin plate spline interpolation in two dimensions. *Adv. Comput. Math.*, **11**: 183–192, 1999.

[60] M. S. Floater and A. Iske. Multistep scattered data interpolation using compactly supported radial basis functions. *J. Comput. Appl. Math.*, **73**: 65–78, 1996.

[61] C. Franke and R. Schaback. Convergence order estimates of meshless collocation methods using radial basis functions. *Adv. Comput. Math.*, **8**: 381–399, 1998.

[62] C. Franke and R. Schaback. Solving partial differential equations by collocation using radial basis functions. *Appl. Math. Comput.*, **93**: 73–82, 1998.

[63] R. Franke. Locally determined smooth interpolation at irregularly spaced points in several variables. *J. Inst. Math. Applic.*, pp. 471–482, 1977.

[64] R. Franke. Smooth interpolation of scattered data by local thin plate splines. *Comput. Math. Appl.*, **8**: 273–281, 1982.

[65] W. Freeden, T. Gervens, and M. Schreiner. *Constructive Approximation on the Sphere*. Oxford, Clarendon Press, 1998.

[66] W. Freeden, M. Schreiner, and R. Franke. A survey on spherical spline approximation. *Surv. Math. Ind.*, **7**: 29–85, 1997.

[67] J. H. Friedman, J. L. Bentley, and R. A. Finkel. An algorithm for finding best matches in logarithmic expected time. *ACM Trans. Math. Software*, **3**: 209–226, 1977.

[68] G. Gasper. Positive integrals of Bessel functions. *SIAM J. Math. Anal.*, **6**: 868–881, 1975.

[69] I. M. Gel'fand and N. Y. Vilenkin. *Generalized Functions,* Volume 4, *Applications of Harmonic Analysis*. New York, Academic Press, 1964.

[70] T. Gneiting. On the derivatives of radial positive definite functions. *J. Math. Anal. Appl.*, **236**: 86–93, 1999.

[71] M. A. Golberg. The theory of radial basis functions applied to the BEM for inhomogeneous partial differential equations. *Boundary Elements Comm.*, **5**: 57–61, 1994.

[72] M. von Golitschek and W. A. Light. Interpolation by polynomials and radial basis functions on spheres. *Constr. Approx.*, **17**: 1–18, 2001.

[73] M. Golomb and H. F. Weinberger. Optimal approximation and error bounds. In R. F. Langer, ed., *On Numerical Approximation*, pp. 117–190. Madison, University of Wisconsin Press, 1959.

[74] L. Greengard and J. Strain. The fast Gauss transform. *SIAM J. Sci. Statist. Comput.*, **12**: 79–94, 1991.

[75] T. N. E. Greville. Introduction to spline functions. In T. N. E. Greville, ed., *Theory and Applications of Spline Functions*, pp. 1–35, New York, Academic Press, 1969.

[76] M. Griebel and M. A. Schweitzer, eds. *Adaptive Meshfree Method of Backward Characteristics for Nonlinear Transport Equations*. Heidelberg, Springer, 2002.

[77] K. Guo, S. Hu, and X. Sun. Conditionally positive definite functions and Laplace–Stieltjes integrals. *J. Approx. Theory*, **74**: 249–265, 1993.
[78] P. R. Halmos. *Measure Theory*. New York, Van Nostrad Reinhold, 1950.
[79] R. L. Hardy. Multiquadric equations of topography and other irregular surfaces. *J. Geophys. Res.*, **76**: 1905–1915, 1971.
[80] R. L. Hardy. Theory and applications of the multiquadric–biharmonic method. 20 years of discovery 1968–1988. *Comput. Math. Appl.*, **19**: 163–208, 1990.
[81] F. Hausdorff. Summationsmethoden und Momentfolgen I. *Math. Z.*, **9**: 74–109, 1921.
[82] F. Hausdorff. Summationsmethoden und Momentfolgen II. *Math. Z.*, **9**: 280–299, 1921.
[83] E. Hille. A class of reciprocal functions. *Ann. Math.*, **27**: 427–464, 1926.
[84] H. Hishimura, M. Hirau, T. Kawai, T. Kawata, I. Shirkawa, and K. Omura. Object modeling by distribution function and a method of image generation. *Transactions of the Institute of Electronics and Communcation Engineers of Japan*, **J68-D**(4): 718–725, 1985.
[85] Y. C. Hon and R. Schaback. On unsymmetric collocation by radial basis functions. *J. Appl. Math. Comp.*, **119**: 177–186, 2001.
[86] H. Hoppe. Surface Reconstruction from Unorganized Points. Ph.D. thesis, University of Washington, 1994.
[87] H. Hoppe, T. DeRose, T. Duchamp, J. McDonald, and W. Stuetzle. Surface reconstruction from unorganized points. *In Computer Graphics, Proc SIGGRAPH'92*, **26**: 71–78, 1992.
[88] M. H. L. Hounjet and J. J. Meijer. Evaluation of elastomechanical and aerodynamic data transfer methods for non-planar configurations in computational aeroelastic analysis. Technical report, ICAS publication, 1994.
[89] K. Jetter, J. Stöckler, and J. Ward. Error estimates for scattered data interpolation on spheres. *Math. Comput.*, **68**: 733–747, 1999.
[90] M. J. Johnson. A bound on the approximation order of surface splines. *Constr. Approx.*, **14**: 429–438, 1998.
[91] M. J. Johnson. An improved order of approximation for thin-plate spline interpolation in the unit disc. *Numer. Math.*, **84**: 451–474, 2000.
[92] M. J. Johnson. Overcoming the boundary effects in surface spline interpolation. *IMA J. Numer. Anal.*, **20**: 405–422, 2000.
[93] M. J. Johnson. The L_2-approximation order of surface spline interpolation. *Math. Comput.*, **70**: 719–737, 2001.
[94] M. J. Johnson. The L_p-approximation order of surface spline interpolation for $1 \leq p \leq 2$. *Constr. Approx.* **20**(2), 2004.
[95] D. S. Jones. *The Theory of Generalized Functions*. Cambridge, Cambridge University Press, 1982.
[96] E. J. Kansa. Multiquadrics – a scattered data approximation scheme with applications to computational fluid-dynamics. i. surface approximations and partial derivative estimates. *Comput. Math. Appl.*, **19**: 127–145, 1990.
[97] E. J. Kansa. Multiquadrics – a scattered data approximation scheme with applications to computational fluid-dynamics – ii: solutions to parabolic, hyperbolic and elliptic partial differential equations. *Comput. Math. Appl.*, **19**: 147–161, 1990.
[98] M. Klein. Spezielle Probleme der Rekonstruktion multivariater Funktionen. Diplomarbeit, University Göttingen, 1998.

[99] J. Kuelbs. Positive definite symmetric functions on linear spaces. *J. Math. Anal. Appl.*, **42**: 413–426, 1973.
[100] P. Kutler. Multidisciplinary computational aerosciences. In *Proc. 5th Int. Symp. on Comp. Fluid Dynamics*, pp. 109–119, 1993.
[101] P. Lancaster and K. Salkauskas. Surfaces generated by moving least squares methods. *Math. Comput.*, **37**: 141–158, 1981.
[102] N. N. Lebedev. *Special Functions and their Applications*. Englewood Cliffs, Prentice-Hall, 1965.
[103] J. Levesley and D. L. Ragozin. Radial basis interpolation on homogenous manifolds: convergence rates. Technical report No. 2001/04, University of Leicester, 2001.
[104] D. Levin. The approximation power of moving least-squares. *Math. Comput.*, **67**: 1517–1531, 1998.
[105] D. Levin. Stable integration rules with scattered integration points. *J. Comput. Appl. Math.*, **112**: 181–187, 1999.
[106] W. A. Light and E. W. Cheney. Interpolation by periodic radial basis functions. *J. Math. Anal. Appl.*, **168**: 111–130, 1992.
[107] W. A. Light and M. Vail. Extension theorems for spaces arising from approximation by translates of a basic function. *J. Approx. Theory*, **114**: 164–200, 2002.
[108] W. A. Light and H. Wayne. Spaces of distributions, interpolation by translates of a basis function and error estimates. *Numer. Math.*, **81**: 415–450, 1999.
[109] R. Lorentz, F. J. Narcowich, and J. D. Ward. Collocation discretization of the transport equation with radial basis functions. *Appl. Math. Comput.*, **145**: 97–116, 2003.
[110] E. Lukacs. *Characteristic Functions*, second edition, London, Griffin, 1970.
[111] W. R. Madych and S. A. Nelson. Multivariate interpolation: a variational theory. Unpublished manuscript, 1983.
[112] W. R. Madych and S. A. Nelson. Multivariate interpolation and conditionally positive definite functions. *Approximation Theory Appl.*, **4**: 77–89, 1988.
[113] W. R. Madych and S. A. Nelson. Multivariate interpolation and conditionally positive definite functions II. *Math. Comput.*, **54**: 211–230, 1990.
[114] W. R. Madych and S. A. Nelson. Bounds on multivariate polynomials and exponential error estimates for multiquadric interpolation. *J. Approx. Theory*, **70**: 94–114, 1992.
[115] J. C. Mairhuber. On Haar's theorem concerning Chebychev approximation problems having unique solutions. *Proc. Am. Math. Soc.*, **7**: 609–615, 1956.
[116] G. Matheron. *Les variables régionalisées et leur estimation*. Paris, Masson, 1965.
[117] M. Mathias. Über positive Fourier-Integrale. *Math. Z.*, **16**: 103–125, 1923.
[118] A. D. Maude. Interpolation – mainly for graph plotters. *Comput. J.*, **16**: 64–65, 1973.
[119] V. Maz'ya and G. Schmidt. On approximate approximations using Gaussian kernels. *IMA J. Numer. Anal.*, **16**: 13–29, 1996.
[120] D. H. McLain. Drawing contours from arbitrary data points. *Comput. J.*, **17**: 318–324, 1974.
[121] D. H. McLain. Two dimensional interpolation from random data. *Comput. J.*, **19** pp. 178–181, 1976.
[122] J. Meinguet. Basic mathematical aspects of surface spline interpolation. In G. Hämmerlin, ed., *Numerische Integration, Oberwolfach 1978*, volume 45 of *Int. Ser. Numer. Math.*, pp. 211–220. Basel, Birkhäuser, 1979.
[123] J. Meinguet. A convolution approach to multivariate representation formulas. In W. Schempp and K. Zeller, eds., *Multivariate Approximation Theory, Oberwohlfach*

1979, volume 51 of *Int. Ser. Numer. Math.*, pp. 198–210. Basel, Birkhäuser, 1979.

[124] J. Meinguet. An intrinsic approach to multivariate spline interpolation at arbitrary points. In B. N. Sahney, ed., *Polynomial and Spline Approximation*, pp. 163–190, Dordrecht, Reidel, 1979.

[125] J. Meinguet. Multivariate interpolation at arbitrary points made simple. *J. Appl. Math. Phys.*, **30**: 292–304, 1979.

[126] J. Meinguet. Surface spline interpolation: basic theory and computational aspects. In S. P. Singh, J. W. H. Burry and B. Watson, eds., *Approximation Theory and Spline Functions*, pp. 127–142, Dordrecht, Reidel, 1984.

[127] V. A. Menegatto. Strictly positive definite kernels on the Hilbert sphere. *Appl. Anal.*, **55**: 91–101, 1994.

[128] V. A. Menegatto. Strict positive definiteness on spheres. *Analysis*, **19**: 217–233, 1999.

[129] H. Meschkowski. *Hilbertsche Räume mit Kernfunktionen*. Berlin, Springer, 1962.

[130] H. N. Mhaskar, F. J. Narcowich, N. Sivakumar, and J. D. Ward. Approximation with interpolatory constraints. *Proc. Am. Math. Soc.*, **130**: 1355–1364, 2002.

[131] H. N. Mhaskar, F. J. Narcowich, and J. D. Ward. Corrigendum to spherical Marcinkiewicz–Zygmund inequalities and positive quadrature. *Math. Comput.*, **71**: 453–454, 2001.

[132] H. N. Mhaskar, F. J. Narcowich, and J. D. Ward. Spherical Marcinkiewicz–Zygmund inequalities and positive quadrature. *Math. Comput.*, **70**: 1113–1130, 2001.

[133] C. A. Micchelli. Interpolation of scattered data: distance matrices and conditionally positive definite functions. *Constr. Approx.*, **2**: 11–22, 1986.

[134] C. A. Micchelli and T. J. Rivlin. A survey of optimal recovery. In C. A. Micchelli and T. J. Rivlin, eds., *Optimal Estimation in Approximation Theory*, pp. 1–54. New York, Plenum Press, 1977.

[135] Y. Mizuta. Integral representations of Beppo Levi functions of higher order. *Hiroshima Math. J.*, **4**: 375–396, 1974.

[136] Y. Mizuta. Integral representations of Beppo Levi functions and the existence of limits at infinity. *Hiroshima Math. J.*, **19**: 259–279, 1989.

[137] B. Morse, T. S. Yoo, P. Rheingans, D. T. Chen, and K. R. Subramanian. Interpolating implicit surfaces from scattered surface data using compactly supported radial basis functions. In *2001 International Conf. on Shape Modeling and Applications (SMI 2001), Los Alamitos*, CA, pp. 89–98. IEEE Computer Society Press, 2001.

[138] T. M. Morton and M. Neamtu. Error bounds for solving pseudodifferential equatons on spheres by collocation with zonal kernels. *J. Approx. Theory*, **114**: 242–268, 2002.

[139] D. M. Mount. *ANN Programming Manual*. Maryland, College Park, 1998.

[140] C. Müller. *Spherical Harmonics*. Berlin, Springer, 1966.

[141] F. J. Narcowich. Generalized Hermite interpolation and positive definite kernels on a Riemannian manifold. *J. Math. Anal. Appl.*, **190**: 165–193, 1995.

[142] F. J. Narcowich and J. D. Ward. Norms of inverses and condition numbers for matrices associated with scattered data. *J. Approx. Theory*, **64**: 69–94, 1991.

[143] F. J. Narcowich and J. D. Ward. Norms of inverses for matrices associated with scattered data. In P.-J. Laurent, A. Le Méhauté, and L. L. Schumaker, eds., *Curves and Surfaces*, pp. 341–348. Boston, Academic Press, 1991.

[144] F. J. Narcowich and J. D. Ward. Scattered-data interpolation on $\mathbb{R}^n$: error estimates for radial basis and band-limited functions. *SIAM J. Math. Anal.*, **36**: 284–300, 2004.

[145] F. J. Narcowich and J. D. Ward. Norm estimates for the inverse of a general class of scattered-data radial-function interpolation matrices. *J. Approx. Theory*, **69**: 84–109, 1992.

[146] F. J. Narcowich and J. D. Ward. Generalized Hermite interpolation via matrix-valued conditionally positive definite functions. *Math. Comput.*, **63**: 661–687, 1994.

[147] F. J. Narcowich and J. D. Ward. On condition numbers associated with radial-function interpolation. *J. Math. Anal. Appl.*, **186**: 457–485, 1994.

[148] F. J. Narcowich, J. D. Ward, and H. Wendland. Refined error estimates for radial basis function interpolation. *Constr. Approx.*, **19**: 541–564, 2003.

[149] F. J. Narcowich, J. D. Ward, and H. Wendland. Sobolev bounds on functions with scattered zeros, with applications to radial basis function surface fitting. Preprint, 2003. Accepted for publication in *Math. Comput.*

[150] D. Nardini and C. A. Brebbia. A new approach to free vibration analysis using boundary elements. In C. A. Brebbia, ed., *Boundary Element Methods in Engineering*, pp. 312–326. New York, Springer, 1982.

[151] Y. Ohtake, A. Belyaev, M. Alexa, G. Turk, and H.-P. Seidel. Multi-level partition of unity implicits. *ACM Transactions on Graphics*, **22**: 463–470, 2003.

[152] Y. Ohtake, A. Belyaev, and H.-P. Seidel. A multi-scale approach to 3d scattered data interpolation with compactly supported basis functions. In *Shape Modeling International*, pp. 153–164. Washington DC, IEEE Computer Society Press, 2003.

[153] A. A. Pasko and V. V. Savchenko. Blending operations for the functionally based constructive geometry. In *Set-Theoretic Solid Modeling: Techniques and Applications*, pp. 151–161, 1994.

[154] W. Pogorzelski. *Integral Equations and their Applications,* Volume *1*. Oxford, Pergamon Press, 1966.

[155] M. J. D. Powell. The theory of radial basis function approximation in 1990. In W.A. Light, ed., *Advances in Numerical Analysis II: Wavelets, Subdivision Algorithms, and Radial Basis Functions*, pp. 105–210. Oxford, Oxford University Press, 1992.

[156] A. Ron. The L_2-approximation orders of principal shift-invariant spaces generated by a radial basis function. In D. Braess *et al.*, eds., *Numerical Methods in Approximation Theory,* Volume 9, *Proc. conf. Oberwolfach, Germany, November 24–30, 1991*, volume 105 of *Int. Ser. Numer. Math.*, pp. 245–268. Basel, Birkhäuser, 1992.

[157] A. Ron and X. Sun. Strictly positive definite functions on spheres in Euclidean spaces. *Math. Comput.*, **65**: 1513–1530, 1996.

[158] G. Roussos. Computation with radial basis functions. Ph.D. thesis, Imperial College of Science Technology and Medicine, University of London, 1999.

[159] W. Rudin. The extension problem for positive-definite functions. *Ill. J. Math.*, **7**: 532–539, 1963.

[160] W. Rudin. An extension theorem for positive-definite functions. *Duke Math. J.*, **37**: 49–53, 1970.

[161] H. Samet. *The Design and Analysis of Spatial Data Structures*. Reading, Addison-Wesley, 1990.

[162] R. Schaback. Creating surfaces from scattered data using radial basis functions. In M. Dæhlen, T. Lyche, and L. L. Schumaker, eds., *Mathematical Methods for Curves and Surfaces*, pp. 477–496. Nashville, Vanderbilt University Press, 1995.

[163] R. Schaback. Error estimates and condition number for radial basis function interpolation. *Adv. Comput. Math.*, **3**: 251–264, 1995.

[164] R. Schaback. Multivariate interpolation and approximation by translates of a basis function. In C. K. Chui and L. L. Schumaker, eds., *Approximation Theory VIII,* Volume 1, *Approximation and Interpolation*, pp. 491–514. World Scientific Publishing, Singapore, 1995.

[165] R. Schaback. Approximation by radial basis functions with finitely many centers. *Constr. Approx.*, **12**: 331–340, 1996.

[166] R. Schaback. Improved error bounds for scattered data interpolation by radial basis functions. *Math. Comput.*, **68**: 201–216, 1999.

[167] R. Schaback. Native Hilbert spaces for radial basis functions I. In M. W. Müller *et al.*, eds., *New Developments in Approximation Theory. 2nd International Dortmund Meeting (IDoMAT '98), Germany, February 23–27, 1998*, volume 132 of *Int. Ser. Numer. Math.*, pp. 255–282. Basel, Birkhäuser Verlag, 1999.

[168] R. Schaback. A unified theory of radial basis functions: native Hilbert spaces for radial basis functions II. *J. Comput. Appl. Math.*, **121**: 165–177, 2000.

[169] R. Schaback and H. Wendland. Adaptive greedy techniques for approximate solution of large RBF systems. *Numer. Algorithms*, **24**: 239–254, 2000.

[170] R. Schaback and H. Wendland. Numerical techniques based on radial basis functions. In A. Cohen, C. Rabut, and L. L. Schumaker, eds., *Curve and Surface Fitting, Saint-Malo 1999*, pp. 359–374. Nashville, Vanderbilt University Press, 2000.

[171] R. Schaback and H. Wendland. Inverse and saturation theorems for radial basis function interpolation. *Math. Comput.*, **71**: 669–681, 2002.

[172] R. Schaback and Z. Wu. Operators on radial functions. *J. Comput. Appl. Math.*, **73**: 257–270, 1996.

[173] I. J. Schoenberg. Metric spaces and completely monotone functions. *Ann. Math.*, **39**: 811–841, 1938.

[174] I. J. Schoenberg. Positive definite functions on spheres. *Duke Math. J.*, **9**: 96–108, 1942.

[175] L. L. Schumaker. *Spline Functions – Basic Theory*. New York, Wiley-Interscience, 1981.

[176] L. L. Schumaker. Recent progress on multivariate splines. In J. R. Whiteman, ed., *The Mathematics of Finite Elements and Applications VII*, pp. 535–562. London, Academic Press, 1991.

[177] D. Shepard. A two dimensional interpolation function for irregularly spaced data. In *Proc. ACM National conference*, pp. 517–524. New York, ACM Press. 1968.

[178] K. T. Smith, D. C. Solmon, and S. L. Wagner. Practical and mathematical aspects of the problem of reconstructing objects from radiographs. *Bull. Amer. Math. Soc.*, **83**: 1227–1270, 1977.

[179] T. Sonar. Optimal recovery using thin plate splines in finite volume methods for the numerical solution of hyperbolic conservation laws. *IMA J. Numer. Anal.*, **16**: 549–581, 1996.

[180] E. M. Stein and G. Weiss. *Fourier Analysis in Euclidean Spaces*. Princeton, New Jersey, Princeton University Press, 1971.

[181] J. Stewart. Positive definite functions and generalizations, an historical survey. *Rocky Mt. J. Math.*, **6**: 409–434, 1976.

[182] X. Sun. Conditionally positive definite functions and their application to multivariate interpolations. *J. Approx. Theory*, **74**: 159–180, 1993.

[183] X. Sun. Conditional positive definiteness and complete monotonicity. In C. K. Chui and L. L. Schumaker, eds., *Approximation Theory VIII,* volume 1. *Approximation and Interpolation*, pp. 537–540. Singapore, World Scientific, 1995.

[184] G. Turk, H. Q. Dinh, J. O'Brien, and G. Yngve. Implicit surfaces that interpolate. In *2001 International Conf. on Shape Modeling and Applications (SMI 2001)*, pp. 62–71. Los Alamitos, Carolina, IEEE Computer Society Press, 2001.

[185] G. Turk and J. F. O'Brien. Variational implicit surfaces. Technical Report GITGVU 9915, Georgia Institute of Technology, 1999.

[186] G. Turk and J. F. O'Brien. Modelling with implicit surfaces that interpolate. *ACM Transactions on Graphics*, **21**: 855–873, 2002.

[187] G. N. Watson. *A Treatise on the Theory of Bessel Functions*. Cambridge, Cambridge University Press, 1966.

[188] J. Wells and R. Williams. *Embeddings and Extensions in Analysis*. New York, Springer, 1975.

[189] H. Wendland. Ein Beitrag zur Interpolation mit radialen Basisfunktionen. Diplomarbeit, University Göttingen, 1994.

[190] H. Wendland. Piecewise polynomial, positive definite and compactly supported radial functions of minimal degree. *Adv. Comput. Math.*, **4**: 389–396, 1995.

[191] H. Wendland. *Konstruktion und Untersuchung radialer Basisfunktionen mit kompaktem Träger*. Ph.D. thesis, University of Göttingen, 1996.

[192] H. Wendland. Error estimates for interpolation by compactly supported radial basis functions of minimal degree. *J. Approx. Theory*, **93**: 258–272, 1998.

[193] H. Wendland. Numerical solutions of variational problems by radial basis functions. In C. K. Chui and L. L. Schumaker, eds., *Approximation Theory IX,* Volume 2, *Computational Aspects*, pp. 361–368. Nashville, Vanderbilt University Press, 1998.

[194] H. Wendland. Meshless Galerkin methods using radial basis functions. *Math. Comput.*, **68**: 1521–1531, 1999.

[195] H. Wendland. Gaussian interpolation revisited. In K. Kopotun, T. Lyche, and M. Neamtu, eds., *Trends in Approximation Theory*, pp. 427–436. Nashville, Vanderbilt University Press, 2001.

[196] H. Wendland. Local polynomial reproduction and moving least squares approximation. *IMA J. Numer. Anal.*, **21**: 285–300, 2001.

[197] H. Wendland. Moving least squares approximation on the sphere. In T. Lyche and L. L. Schumaker, eds., *Mathematical Methods for Curves and Surfaces: Oslo 2000*, pp. 517–526. Nashville, Vanderbilt University Press, 2001.

[198] H. Wendland. Fast evaluation of radial basis functions: methods based on partition of unity. In C. K. Chui, L. L. Schumaker, and J. Stöckler, eds., *Approximation Theory X: Wavelets, Splines, and Applications*, pp. 473–483. Nashville, Vanderbilt University Press, 2002.

[199] D. V. Widder. Necessary and sufficient conditions for the representation of a function as a Laplace integral. *Trans. Am. Math. Soc.*, **33**: 851–894, 1931.

[200] D. V. Widder. *The Laplace Transform*. Princeton, Princeton University Press, 1946.

[201] R. E. Williamson. Multiply monotone functions and their Laplace transform. *Duke Math. J.*, **23**: 189–207, 1956.

[202] Z. Wu. Hermite–Birkhoff interpolation of scattered data by radial basis functions. *Approximation Theory Appl.*, **8**: 1–10, 1992.
[203] Z. Wu. Compactly supported positive definite radial functions. *Adv. Comput. Math.*, **4**: 283–292, 1995.
[204] Z. Wu and R. Schaback. Local error estimates for radial basis function interpolation of scattered data. *IMA J. Numer. Anal.*, **13**: 13–27, 1993.
[205] Y. Xu and E. W. Cheney. Strictly positive definite functions on spheres. *Proc. Am. Math. Soc.*, **116**: 977–981, 1992.
[206] G. Yngve and G. Turk. Creating smooth implicit surfaces from polygonal meshes. Technical Report GITGVU9942, Georgia Institute of Technology, 1999.
[207] J. Yoon. Approximation in $L_p(\mathbb{R}^d)$ from a space spanned by the scattered shifts of radial basis function. *Constr. Approx.*, **17**: 227–248, 2001.
[208] J. Yoon. Interpolation by radial basis functions on Sobolev space. *J. Approx. Theory*, **112**: 1–15, 2001.
[209] V. P. Zastavnyi. Positive definite functions depending on the norm. *Russian J. Math. Phys.*, **1**: 511–522, 1993.

Index